Bioelectrochemistry

Bioelectrochemistry

Editor:

Valeria Severino, Ph.D.

ArclerPress

www.arclerpress.com

Bioelectrochemistry

Editor: Valeria Severino, Ph.D.

© 2017 Arcler Press LLC
708 3rd Avenue, 6th Floor
New York
NY 10017
United States of America
www.arclerpress.com

ISBN: 978-1-68094-413-6

Library of Congress Control Number: 2016951062

About the Editor

Valeria Severino, Ph.D.

Valeria Severino obtained her PhD in Biological Processes and Biomolecules from the Second University of Naples in 2009. Her scientific interests concern the structural and functional characterization of proteins and peptides by using biochemical approaches. Fields of expertise comprises food chemistry, cellular biology, natural products and related applications. She published more than 40 papers on international peer-reviewed journals.

Contents

List of Contributors

Frauke Kracke
Centre for Microbial Electrochemical Systems, The University of Queensland, Brisbane, QLD, Australia
Advanced Water Management Centre, The University of Queensland, Brisbane, QLD, Australia

Igor Vassilev
Centre for Microbial Electrochemical Systems, The University of Queensland, Brisbane, QLD, Australia
Advanced Water Management Centre, The University of Queensland, Brisbane, QLD, Australia

Jens O. Krömer
Centre for Microbial Electrochemical Systems, The University of Queensland, Brisbane, QLD, Australia
Advanced Water Management Centre, The University of Queensland, Brisbane, QLD, Australia

Pier-Luc Tremblay
Novo Nordisk Foundation Center for Biosustainability, Technical University of Denmark, Hørsholm, Denmark

Tian Zhang
Novo Nordisk Foundation Center for Biosustainability, Technical University of Denmark, Hørsholm, Denmark

Okkyoung Choi
Department of Chemical Engineering, Hanyang University, 222 Wangshimniro, Seongdonggu, Seoul 04763, South Korea

ByoungIn Sang
Department of Chemical Engineering, Hanyang University, 222 Wangshimniro, Seongdonggu, Seoul 04763, South Korea

Wenzong Liu
Key Laboratory of Environmental Biotechnology, Research Center for EcoEnvironmental Sciences, Chinese Academy of Sciences, Beijing 100085, China

Zhangwei He
State Key Laboratory of Urban Water Resource and Environment, Har bin Institute of Technology, Harbin 150090, China

Chunxue Yang
State Key Laboratory of Urban Water Resource and Environment, Har bin Institute of Technology, Harbin 150090, China

Aijuan Zhou
College of Environmental Science and Engineering, Taiyuan University of Technology, Taiyuan 030024, China

Zechong Guo
State Key Laboratory of Urban Water Resource and Environment, Har bin Institute of Technology, Harbin 150090, China

Bin Liang
Key Laboratory of Environmental Biotechnology, Research Center for EcoEnvironmental Sciences, Chinese Academy of Sciences, Beijing 100085, China

Cristiano Varrone
Department of Chemical and Biochemical Engineering, Center for Bio Process Engineering, Technical University of Denmark, Lyngby, Denmark

AiJie Wang
Key Laboratory of Environmental Biotechnology, Research Center for EcoEnvironmental Sciences, Chinese Academy of Sciences, Beijing 100085, China
State Key Laboratory of Urban Water Resource and Environment, Har bin Institute of Technology, Harbin 150090, China

Heyang Yuan
Department of Civil and Environmental Engineering, Virginia Polytechnic Institute and State University, Blacksburg, VA 24061, USA

Yaobin Lu
Department of Civil and Environmental Engineering, Virginia Polytechnic Institute and State University, Blacksburg, VA 24061, USA

Ibrahim M AbuReesh
Department of Chemical Engineering, College of Engineering, Qatar University, Doha, Qatar

Zhen He
Department of Civil and Environmental Engineering, Virginia Polytechnic Institute and State University, Blacksburg, VA 24061, USA

Souichiro Kato
Bioproduction Research Institute, National Institute of Advanced Industrial Science and Technology (AIST), 2-17-2-1 Tsukisamu-Higashi, Toyohira-ku, Sapporo, Hokkaido 062-8517, Japan
Research Center for Advanced Science and Technology, The University of Tokyo, 4-6-1 Komaba, Meguro-ku, Tokyo 153-8904, Japan
Division of Applied Bioscience, Graduate School of Agriculture, Hokkaido University, Kita-9 Nishi-9,Kita-ku, Sapporo, Hokkaido 060-8589, Japan

Qi Zheng
College of Food Science and Engineering, Harbin Institute of Technology, Harbin 150001, China

Lei Xiong
Department of Electrical Engineering, Huaqiao University, Xiamen 361021, China

Bing Mo
Department of Electrical Engineering, Huaqiao University, Xiamen 361021, China

Weihong Lu
College of Food Science and Engineering, Harbin Institute of Technology, Harbin 150001, China

Suki Kim
Department of Electrical Engineering, Korea University, Seoul 361021, Korea

Zhenyu Wang
College of Food Science and Engineering, Harbin Institute of Technology, Harbin 150001, China

Pau Rodenas Motos
Wetsus, European Centre of Excellence for Sustainable Water Technology, Leeuwarden, Netherlands
Sub-department of Environmental Technology, Wageningen University, Wageningen, Netherlands

Annemiek ter Heijne
Sub-department of Environmental Technology, Wageningen University, Wageningen, Netherlands

Renata van der Weijden
Wetsus, European Centre of Excellence for Sustainable Water Technology, Leeuwarden, Netherlands
Sub-department of Environmental Technology, Wageningen University, Wageningen, Netherlands

Michel Saakes
Wetsus, European Centre of Excellence for Sustainable Water Technology, Leeuwarden, Netherlands

Cees J. N. Buisman
Wetsus, European Centre of Excellence for Sustainable Water Technology, Leeuwarden, Netherlands
Sub-department of Environmental Technology, Wageningen University, Wageningen, Netherlands

Tom H. J. A. Sleutels
Wetsus, European Centre of Excellence for Sustainable Water Technology, Leeuwarden, Netherlands

Wen-Cheng Wang,
Key Laboratory of Tropical Medicinal Plant Chemistry of Ministry of Education, College of Chemistry and Chemical Engineering, Hainan Normal University, Haikou 571158, China

Li-Jun Yan,
Key Laboratory of Tropical Medicinal Plant Chemistry of Ministry of Education, College of Chemistry and Chemical Engineering, Hainan Normal University, Haikou 571158, China

Fan Shi,
Key Laboratory of Tropical Medicinal Plant Chemistry of Ministry of Education, College of Chemistry and Chemical Engineering, Hainan Normal University, Haikou 571158, China

Xue-Liang Niu,
Key Laboratory of Tropical Medicinal Plant Chemistry of Ministry of Education, College of Chemistry and Chemical Engineering, Hainan Normal University, Haikou 571158, China

Guo-Lei Huang,
Key Laboratory of Tropical Medicinal Plant Chemistry of Ministry of Education, College of Chemistry and Chemical Engineering, Hainan Normal University, Haikou 571158, China

Cai-Juan Zheng
Key Laboratory of Tropical Medicinal Plant Chemistry of Ministry of Education, College of Chemistry and Chemical Engineering, Hainan Normal University, Haikou 571158, China

Wei Sun
Key Laboratory of Tropical Medicinal Plant Chemistry of Ministry of Education, College of Chemistry and Chemical Engineering, Hainan Normal University, Haikou 571158, China

Lu Lu
Department of Civil, Environmental, and Architectural Engineering, University of Colorado Boulder, Boulder, CO 80309, USA

Cuiping Zeng
Department of Civil, Environmental, and Architectural Engineering, University of Colorado Boulder, Boulder, CO 80309, USA
The Key Laboratory of Orogenic Belts and Crustal Evolution, School of Earth and Space Sciences, Peking University, Beijing 100871, P.R. China

Luda Wang
Department of Mechanical Engineering, University of Colorado Boulder, Boulder, CO 80309, USA

XiaoboYin
Department of Mechanical Engineering, University of Colorado Boulder, Boulder, CO 80309, USA

Song Jin
Department of Civil and Architectural Engineering, University of Wyoming, Laramie, WY 82071, USA

Anhuai Lu
The Key Laboratory of Orogenic Belts and Crustal Evolution, School of Earth and Space Sciences, Peking University, Beijing 100871, P.R. China

Zhiyong Jason Ren
Department of Civil, Environmental, and Architectural Engineering, University of Colorado Boulder, Boulder, CO 80309, USA

Shuai Luo
Department of Civil and Environmental Engineering, Virginia Polytechnic Institute and State University, Blacksburg, VA 24061, USA

Hongyue Sun
Grado Department of Industrial and Systems Engineering, Virginia Polytechnic Institute and State University, Blacksburg, VA 24061, USA

Qingyun Ping
Department of Civil and Environmental Engineering, Virginia Polytechnic Institute and State University, Blacksburg, VA 24061, USA

Ran Jin
Grado Department of Industrial and Systems Engineering, Virginia Polytechnic Institute and State University, Blacksburg, VA 24061, USA

Zhen He
Department of Civil and Environmental Engineering, Virginia Polytechnic Institute and State University, Blacksburg, VA 24061, USA

Elliot S. Friedman
Department of Biological and Environmental Engineering, Cornell University, Ithaca, NY, USA

Lauren E. McPhillips
[1]Department of Biological and Environmental Engineering, Cornell University, Ithaca, NY, USA

Jeffrey J. Werner
Department of Biological and Environmental Engineering, Cornell University, Ithaca, NY, USA
Department of Chemistry, State University of New York College at Cortland, Cortland, NY, USA

Angela C. Poole
Department of Molecular Biology and Genetics, Cornell University, Ithaca, NY, USA

Ruth E. Ley
Department of Molecular Biology and Genetics, Cornell University, Ithaca, NY, USA

M. Todd Walter
Department of Biological and Environmental Engineering, Cornell University, Ithaca, NY, USA

Largus T. Angenent
Department of Biological and Environmental Engineering, Cornell University, Ithaca, NY, USA

Ying Liu
Department of Mechanical Engineering, Virginia Polytechnic Institute and State University, Blacksburg, VA 24061, USA

MohanQin
Department of Civil and Environmental Engineering, Virginia Polytechnic Institute and State University, Blacksburg, VA 24061, USA

Shuai Luo
Department of Civil and Environmental Engineering, Virginia Polytechnic Institute and State University, Blacksburg, VA 24061, USA

Zhen He
Department of Civil and Environmental Engineering, Virginia Polytechnic Institute and State University, Blacksburg, VA 24061, USA

RuiQiao
Department of Mechanical Engineering, Virginia Polytechnic Institute and State University, Blacksburg, VA 24061, USA

Bin Lai
Centre for Microbial Electrochemical Systems (CEMES), The University of Queensland, Office 618, Gehrmann Building (60), St. Lucia, Brisbane, QLD 4072, Australia
Advanced Water Management Centre (AWMC), The University of Queensland, Brisbane, Australia

Shiqin Yu
Centre for Microbial Electrochemical Systems (CEMES), The University of Queensland, Office 618, Gehrmann Building (60), St. Lucia, Brisbane, QLD 4072, Australia
Advanced Water Management Centre (AWMC), The University of Queensland, Brisbane, Australia

Paul V. Bernhardt
School of Chemistry and Molecular Biosciences, The University of Queensland, Brisbane, Australia

Korneel Rabaey
Laboratory of Microbial Ecology and Technology (LabMET), Ghent University, Ghent, Belgium

Bernardino Virdis
Centre for Microbial Electrochemical Systems (CEMES), The University of Queensland, Office 618, Gehrmann Building (60), St. Lucia, Brisbane, QLD 4072, Australia
Advanced Water Management Centre (AWMC), The University of Queensland, Brisbane, Australia

Jens O. Krömer
Centre for Microbial Electrochemical Systems (CEMES), The University of Queensland, Office 618, Gehrmann Building (60), St. Lucia, Brisbane, QLD 4072, Australia
Advanced Water Management Centre (AWMC), The University of Queensland, Brisbane, Australia

Min-HuaCui
State Key Laboratory of Urban Water Resource and Environment, Harbin Institute of Technology, Harbin 150090, PR China

DanCui
Key Laboratory of Environmental Biotechnology, Research Center for Eco-Environmental Sciences, Chinese Academy of Sciences, Beijing 100085, PR China

Hyung-Sool Lee
Department of Civil and Environmental Engineering, University of Waterloo, 200 University Avenue West Waterloo, Ontario N2L 3G1, Canada

Bin Liang
Key Laboratory of Environmental Biotechnology, Research Center for Eco-Environmental Sciences, Chinese Academy of Sciences, Beijing 100085, PR China

Ai-JieWang
State Key Laboratory of Urban Water Resource and Environment, Harbin Institute of Technology, Harbin 150090, PR China
Key Laboratory of Environmental Biotechnology, Research Center for Eco-Environmental Sciences, Chinese Academy of Sciences, Beijing 100085, PR China

Hao-Yi Cheng
Key Laboratory of Environmental Biotechnology, Research Center for Eco-Environmental Sciences, Chinese Academy of Sciences, Beijing 100085, PR China

Wei-Jhe Ma
Instrument System and Chip Group, Department of Electric Engineering, National Cheng Kung University, Tainan 70101, Taiwan

Ching-Hsing Luo
Instrument System and Chip Group, Department of Electric Engineering, National Cheng Kung University, Tainan 70101, Taiwan
Institute of Medical Science and Technology, National Sun Yat-sen University, Kaohsiung 80424, Taiwan

Jiun-Ling Lin
Biotechnology and Biochemical Engineering Group, Department of Chemical Engineering, National Cheng Kung University, Tainan 70101, Taiwan

Sin-Houng Chou
Biotechnology and Biochemical Engineering Group, Department of Chemical Engineering, National Cheng Kung University, Tainan 70101, Taiwan

Ping-Hung Chen
Biotechnology and Biochemical Engineering Group, Department of Chemical Engineering, National Cheng Kung University, Tainan 70101, Taiwan

Mei-Jywan Syu
Biotechnology and Biochemical Engineering Group, Department of Chemical Engineering, National Cheng Kung University, Tainan 70101, Taiwan

Shin-Hung Kuo
Instrument System and Chip Group, Department of Electric Engineering, National Cheng Kung University, Tainan 70101, Taiwan

Shin-Chi Lai
Department of Computer Science and Information Engineering, Nan Hua University, Chiayi 62249, Taiwan

Kouta Takeda
Department of Biotechnology and Life Science, Tokyo University of Agriculture and Technology, Tokyo, Japan

Takayuki Uchihashi
Department of Physics, Kanazawa University, Kakuma-machi, Kanazawa, Japan
Advanced Low Carbon Technology Research and Development Program (ALCA), Japan Science and Technology Agency (JST), Tokyo, Japan

Hiroki Watanabe
Department of Physics, Kanazawa University, Kakuma-machi, Kanazawa, Japan

Takuya Ishida
Department of Biomaterials Sciences, Graduate School of Agriculture and Life Sciences, The University of Tokyo, Bunkyo-ku, Tokyo, Japan
Advanced Low Carbon Technology Research and Development Program (ALCA), Japan Science and Technology Agency (JST), Tokyo, Japan

Kiyohiko Igarashi
Department of Biomaterials Sciences, Graduate School of Agriculture and Life Sciences, The University of Tokyo, Bunkyo-ku, Tokyo, Japan
Advanced Low Carbon Technology Research and Development Program (ALCA), Japan Science and Technology Agency (JST), Tokyo, Japan

Nobuhumi Nakamura
Department of Biotechnology and Life Science, Tokyo University of Agriculture and Technology, Tokyo, Japan
Advanced Low Carbon Technology Research and Development Program (ALCA), Japan Science and Technology Agency (JST), Tokyo, Japan

Hiroyuki Ohno
Department of Biotechnology and Life Science, Tokyo University of Agriculture and Technology, Tokyo, Japan

JohannesGladisch
Biosystems Technology, Institute for Applied Life Sciences, Technical University of Applied Sciences Wildau, Hochschulring 1, D-15745, Wildau, Germany

David Sarauli
Biosystems Technology, Institute for Applied Life Sciences, Technical University of Applied Sciences Wildau, Hochschulring 1, D-15745, Wildau, Germany
Department of Chemistry and Centre for NanoScience (CeNS), University of Munich (LMU), Butenandtstraße 5-13 (E), D-81377, Munich, Germany

Daniel Schäfer
Biosystems Technology, Institute for Applied Life Sciences, Technical University of Applied Sciences Wildau, Hochschulring 1, D-15745, Wildau, Germany

Birgit Dietzel
Institute for Thin Film and Microsensor Technologies, Kantstraße 55, D-14513 Teltow, Germany

Burkhard Schulz
Institute for Thin Film and Microsensor Technologies, Kantstraße 55, D-14513 Teltow, Germany

Fred Lisdat
Biosystems Technology, Institute for Applied Life Sciences, Technical University of Applied Sciences Wildau, Hochschulring 1, D-15745, Wildau, Germany

Preface

Bioelectrochemistry is a fast growing field at the interface between electrochemistry and other sciences such as biochemistry, analytical chemistry and medicinal chemistry. In the recent years, the methods and the understanding of the fundamentals have seen significant progress, which has led to rapid development in the field.

Bioelectrochemistry represents a scientific field in which electrochemical principles and approaches are exploited to study biological processes. This topic is an incredible fast growing field situated at the interface among electrochemistry, biochemistry, analytical and medicinal chemistry, environmental and food science. Recently, significant progress in understanding, development and application of bioelectrochemistry has been reached, thus inducing an always more rapid growth in the field. Several efforts have been made to develop particular bioelectrochemical systems, such as the microbial fuel cells and microbial electrolysis cells, able to generate electricity, hydrogen or other chemicals.

In this book recent developments in this hot interdisciplinary field have been selected. The resulting overview comprises several different examples of innovative applications of bioelectrochemistry. The work covers the fundamental aspects of the chemistry, physics and biology which underline the subject area. It describers some of the different experimental techniques that can be used to study bioelectrochemical problems and it describes various applications of bioelectrochemistry including aerometric biosensors, immunoassays, electrochemistry of DNA, whole cell biosensors, in vivo applications and bioelectrosynthesis. In particular, the first part is focused on the use of microbes as source of useful products as biofuel, electricity or as novel eco-sustainable agents. The second part is aimed to analyze particular technological solutions, as the production of hydrogen starting from the oxidation of graphene or the developmental of a bioelectrochemical sensor to detect urea concentration. Finally, a particularly interesting and stable bioelectrocatalytic platform useful for biosensors and biofuel cells is reported. This book is intended for Electrochemists, Biochemists, Analytical Chemists, and Medicinal Chemists.

Editor
VALERIA SEVERINO, Ph.D.

INTRODUCTION

Bioelectrochemistry is a branch of electrochemistry and biophysical chemistry concerned with electrophysiological topics like cell electron-proton transport, cell membrane potentials and electrode reactions of redox enzymes. The beginnings of bioelectrochemistry, as well as those of electrochemistry, are closely related to physiology through the works of Luigi Galvani and then Alessandro Volta. The first modern work in this field is considered that of the German physiologist Julius Bernstein (1902) concerning the source of biopotentials due to different ion concentration through the cell's membrane. The domain of bioelectrochemistry has grown considerably over the past century, maintaining the close connections to various medical and biological and engineering disciplines like electrophysiology and biomedical engineering.

Biomedical Engineering

Biomedical engineering is a scientific discipline which brings the principles of engineering to biology and medical treatment. Engineers are world-famous for coming up with innovative approaches to problems, and they are fond of saying that no problem is too large for an engineer. Turning this can-do attitude to the field of medicine, biomedical engineers work on a wide variety of things, from artificial hearts to cultured skin grafts, in the hopes of advancing medical treatment.

People who wish to study biomedical engineering must pursue training in both engineering and biology. Many universities have biomedical engineering departments to meet the need for new biomedical engineers and the growing interest in this field, and students typically pursue doctoral degrees in this field so that they can learn as much as possible.

A wide variety of things fall under the purview of biomedical engineering. Most new medical devices, for example, are constructed by people in this field, including things like equipment used in medical treatment, medical imaging systems, artificial organs, prosthetic limbs, and hearing aids. The ability to develop new medical devices is a critical aspect of biomedical engineering, with researchers also working on things like biocompatible materials so that the body will not reject implants, along with advanced prosthetics which can interface with the brain and nervous system.

Biomedical engineers also work in the field of genetics, researching the human genome and looking into the possibility of growing organs and skin for grafts and transplant, and they research pharmaceuticals, life support systems, and even cosmetic surgery. Biomedical engineering has brought about better artificial hip joints, improved breast implants, and new technology for everything from bone grafts to blood drawing equipment.

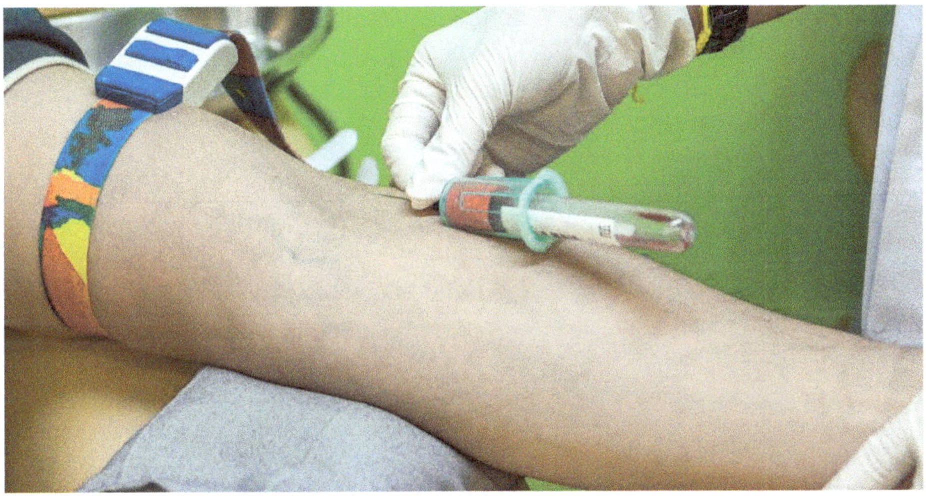

The field of biomedical engineering has contributed immensely to medical science since the 20th century, when biomedical engineers began to formally organize themselves and commit to working on problems which faced the medical community. Today, biomedical engineers can be found working for private companies, cooperating in labs with other medical researchers, performing research for the government, teaching new generations in their field, and working with non-profits to develop low-cost medical solutions for the developing world.

This field is constantly expanding, and generally, someone with a degree in biomedical engineering will have no trouble getting a job. A number of trade and professional organizations for biomedical engineers hold annual conferences and continuing education seminars for people to hone their skills in the field and connect with each other.

What does a Biomedical Engineering Tech do?

A biomedical engineering technician installs, maintains, and troubleshoots complicated medical equipment at a hospital or clinical laboratory. He or she is responsible for ensuring medical personnel know how to safely operate different machines and what to do if equipment does not function properly. Some techs specialize by working with particular machinery such as x-ray scanners, but most professionals are skilled at programming and fixing a wide variety of hospital electronics.

While a biomedical engineering tech does not work directly with patients, he or she still plays an important role in their health care. Doctors and nurses depend on modern electronic equipment to help them make accurate diagnoses and provide essential inpatient treatments. A biomedical engineering tech regularly inspects devices to make sure they are in proper working order. Routine maintenance tasks may include testing batteries, checking for damaged or burnt wires, and replacing electronic chips.

When devices malfunction or stop working altogether, a biomedical engineering tech is normally consulted to troubleshoot them. A tech usually follows instruction manuals, schematics, and detailed technical guides when checking parts and diagnosing problems. Most professionals have access to work stations that are stocked with hand and power tools so they can be prepared for any repair job. If a machine cannot be fixed, a tech might submit a purchase request to his or her supervisors.

Techs work with many kinds of medical technology. They set up diagnostic imaging equipment such as computerized tomography scanners and ultrasounds. Techs also perform repairs on such emergency medical equipment as defibrillators and dialysis machines. In addition, many biomedical engineering techs are skilled at working on pathology lab equipment such as microscopes and cell counters.

A person who wants to become a biomedical engineering tech can develop the practical skills needed for the job in a technical or vocational school program. In school, students take courses in electrical theory, math, computers, and electronics design. They also participate in hands-on activities working with various types of medical equipment. Most degree programs take between one and two years to complete, and they culminate in certificates or associate's degrees.

With the proper educational training, an individual can apply at local hospitals, doctor's offices, and medical labs. A new biomedical engineering tech can expect to spend several weeks in on-the-job training to learn about the facility's specific devices and how to obtain work orders, tools, and equipment. Techs usually start working independently within their first six months on the job, and they may have the opportunity to advance to supervisory roles in large hospitals with ongoing experience.

What are the Different Biomedical Engineering Jobs?

Biomedical engineers are an important part of the medical community. The knowledge, inventions, and people that are behind many biomedical engineering jobs are responsible for improving lives across the globe by creating new theories on life systems or designing medical instruments. The contributions made by those employed in biomedical engineering jobs are countless:

minuscule devices to inhibit cell growth; artificial bones, tendons, and discs; highly sensitive monitors and medical imaging systems; artificial hearts; synthetic blood; medical robotics; and tissue engineering – to name just a few.

There are three areas where biomedical engineering jobs seem to originate: research, design, and instruction. Each has specific duties and specialized fields of work. It is up to the individual to decide which area interests her most. Biomedical research is a broad term. Biomedical engineers who fall in the research category must use utilize a wide range of knowledge in their everyday jobs, pulling from mechanical, chemical and electrical engineering backgrounds. In addition, it is important to understand all aspects of living systems, including anatomy and physiology.

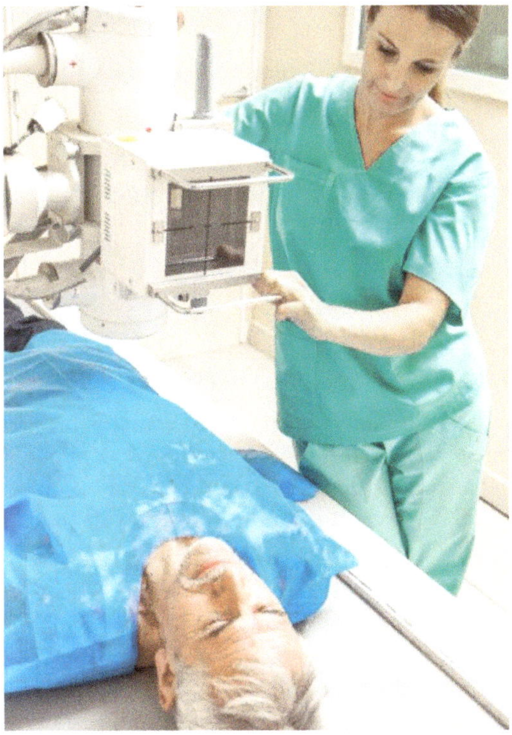

For those who are interested in biomedical engineering jobs with an emphasis in design, medical devices will become their primary focus. They will work to design artificial limbs, organ, special surgical lasers, and high tech machines. They will also build systems to keep hospitals, labs, and clinic current with the various procedures. In addition, those involved in the design aspect of biomedical engineering may teach staff how to use the new devices that they created.

The third facet of biomedical engineering is instruction. There are jobs available on the university level. Those who accept biomedical engineering jobs and an instructor will teach classes, advice students, join committees related to biomedical engineering, and become involved in university research projects.

Luckily, there are additional biomedical engineering jobs that are specific to a given area, yet falling amongst one of the three categories mentioned above. For example, a bioinstrumentation engineer will create devices that are used to analyze and treat various diseases. Computers – from microprocessors to microcomputers involved in medical imaging - are a valuable part to those involved in bioinstrumentation.

Other biomedical engineering jobs include biomaterials engineers, biomechanics engineers, clinical engineers, and systems physiology engineers. Biomaterials engineers research and develop artificial materials that are used inside the human body to make sure they are compatible and not toxic. Biomechanics engineers use the laws of mechanics to study the way fluids work inside the body and study the way mechanics is applied to medical problems. Clinical engineers purchase medical devices and instruments for hospitals and work with doctors to ensure that the equipment meets their needs. Systems physiology engineers use engineering strategies and techniques to understand a wide range of living organisms – from viruses to people.

Biomedical Engineer

A biomedical engineer researches and develops new medical equipment, using his or her knowledge of biology, medicine, technology, and mechanical engineering. He or she might create new prosthetic devices, artificial organs, or metal joint replacements for hips, shoulders, and knees. A biomedical engineer also improves the safety and efficiency of hospital equipment, such as medical resonance imaging (MRI) and x-ray machines.

Many biomedical engineers work for medical research and development institutions, studying the existing medical technology and thinking of ways to improve upon it. An engineer may spend months researching a particular instrument, such as an MRI machine, to learn about its complicated parts and processes. He or she then devises a plan on how to make the machine more safe, effective, and easier to use. The biomedical engineer might draw up plans for a new machine and work with a team of technicians to design a prototype device. Once the prototype proves effective in several laboratory and clinical trials, it can be marketed and implemented in medical facilities.

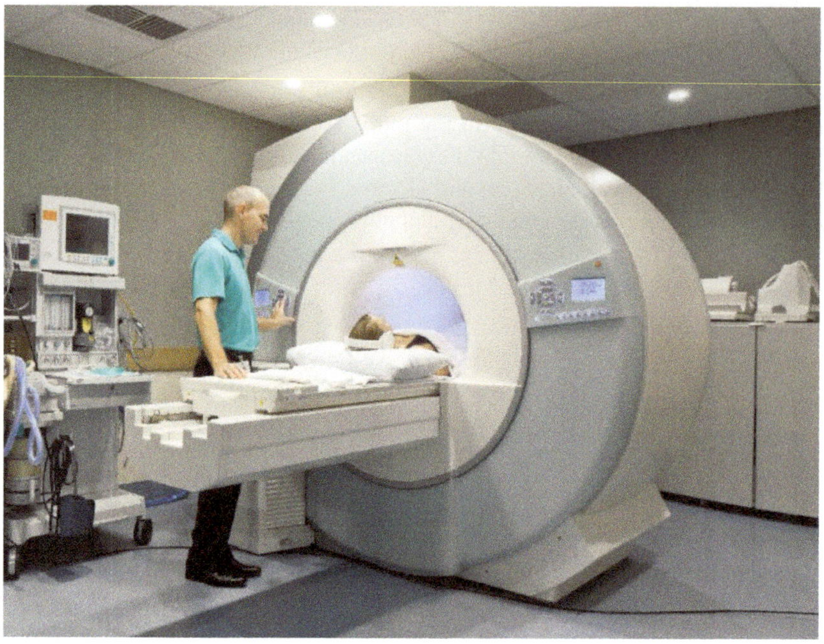

Advancements in medical technology lead to the development of complex prosthetic organs and limbs, as well as braces, wheelchairs, and support systems. A biomedical engineer considers the most common problems that patients face, such as heart issues, and construct instruments that can be used to prevent or overcome their problems. The artificial heart, for example, was the product of years of intensive biomedical research. It is now used worldwide to help people overcome life-threatening heart conditions.

To become a biomedical engineer, a person is usually required to obtain at least a bachelor's degree in engineering, although many research and development institutions give preference to scientists with master's or doctoral degrees. Undergraduates typically major in mechanical engineering, though they often take multiple classes in biology, chemistry, and medicine to supplement their engineering courses. Many graduates take internship positions at research facilities, where they learn firsthand about the creation and implementation of new devices. Some states and countries require that new engineers pass a licensing examination in order to conduct independent research and develop new products.

Experienced biological engineers are generally in high demand in order to apply the newest technology and principles to the creation of safe, effective medical devices. The increasing population of elderly people and injured war veterans is creating an immediate need for new equipment, including prosthetic devices and wheelchairs. Skilled biomedical engineers ensure that people can live as comfortably and independently as possible by providing them with the latest medical technology.

Bioelectronics

Bioelectronics is a field of research in the convergence of biology and electronics. At the first C.E.C. Workshop, in Brussels in November 1991, bioelectronics was defined as 'the use of biological materials and biological architectures for information processing systems and new devices'. Bioelectronics, specifically bio-molecular electronics, were described as 'the research and development of bio-inspired (i.e. self-assembly) inorganic and organic materials and of bio-inspired (i.e. massive parallelism) hardware architectures for the implementation of new information processing systems, sensors and actuators, and for molecular manufacturing down to the atomic scale'. The National Institute of Standards and Technology (NIST), an agency of the U.S. Department of Commerce, defined bioelectronics in a 2009 report as "the discipline resulting from the convergence of biology and electronics".

Sources for information about the field include the Institute of Electrical and Electronics Engineers (IEEE) with its Elsevier journal Biosensors and Bioelectronics published since 1990. The journal describes the scope of bioelectronics as seeking to: " exploit biology in conjunction with electronics in a wider context encompassing, for example, biological fuel cells, bionics and biomaterials for information processing, information storage, electronic components and actuators. A key aspect is the interface between biological materials and micro- and nano-electronics."[

Bioelectronics is the application of the principles of electronics to biology and medicine. Hollywood has expanded the horizons with its imagination, and movies like Star Wars, Star Trek, iRobot, Minority Report and many other sci-fi movies have shown us what modern technology is capable of. The potential of bioelectronics is also shown in these movies. Artificial limbs, humanoids, the various sensors that are attached to the body, etc. are the applications of bioelectronics.

Enzymatic biofuel cell

An Enzymatic biofuel cell is a specific type of fuel cell that uses enzymes as a catalyst to oxidize its fuel, rather than precious metals. Enzymatic biofuel cells, while currently confined to research facilities, are widely prized for the promise they hold in terms of their relatively inexpensive components and fuels, as well as a potential power source for bionic implants.

Operation

Enzymatic biofuel cells work on the same general principles as all fuel cells: use a catalyst to separate electrons from a parent molecule and force it to go around an electrolyte barrier through a wire to generate an electric current. What makes the enzymatic biofuel cell distinct from more conventional fuel cells are the catalysts they use and the fuels that they accept. Whereas most fuel cells use metals like platinum and nickel as catalysts, the enzymatic biofuel cell uses enzymes derived from living cells (although not within living cells; fuel cells that use whole cells to catalyze fuel are called microbial fuel cells). This offers a couple of advantages for enzymatic biofuel cells: Enzymes are relatively easy to mass-produce and so benefit from economies of scale, whereas precious metals must be mined and so have an inelastic supply. Enzymes are also specifically designed to process organic compounds such as sugars and alcohols, which are extremely common in nature. Most organic compounds cannot be used as fuel by fuel cells with metal catalysts because the carbon monoxide formed by the interaction of the carbon molecules with oxygen during the fuel cell's functioning will quickly "poison" the precious metals that the cell relies on, rendering it useless. Because sugars and other biofuels can be grown and harvested on a massive scale, the fuel for enzymatic biofuel cells is extremely cheap and can be found in nearly any part of the world, thus making it an extraordinarily attractive option from a logistics standpoint, and even more so for those concerned with the adoption of renewable energy sources.

Enzymatic biofuel cells also have operating requirements not shared by traditional fuel cells. What is most significant is that the enzymes that allow the fuel cell to operate must be "immobilized" near the anode and cathode in order to work properly; if not immobilized, the enzymes will diffuse into the cell's fuel and most of the liberated electrons will not reach the electrodes, compromising its effectiveness. Even with immobilization, a means must also be provided for electrons to be transferred to and from the electrodes. This can be done either directly from the enzyme to the electrode ("direct electron transfer") or with the aid of other chemicals that transfer electrons from the enzyme to the electrode ("mediated electron transfer"). The former technique is possible only with certain types of enzymes whose activation sites are close to the enzyme's surface, but doing so presents fewer toxicity risks for fuel cells intended to be used inside the human body. Finally, completely processing the complex fuels used in enzymatic biofuel cells requires a series of different enzymes for each step of the 'metabolism' process; producing some of the required enzymes and maintaining them at the required levels can pose problems

Feasibility of enzymes as catalysts

With respect to fuel cells, enzymes have several advantages to their incorporation. An important enzymatic property to consider is the driving force or potential necessary for successful reaction catalysis. Many enzymes operate at potentials close to their substrates which is most suitable for fuel cell applications. Furthermore, the protein matrix surrounding the active site provides many vital functions; selectivity for the substrate, internal electron coupling, acidic/basic properties and the ability to bind to other proteins (or the electrode). Enzymes are more stable in the absence of proteases and enzymes from thermophilic organisms and thus offer a wider range of temperatures. Usual operating conditions is generally between 20-50 °C and pH 4.0 to 8.0.

A drawback with the use of enzymes is size; given the large size of enzymes, they yield a low current density per unit electrode area due to the limited space. Since it is not possible to reduce enzyme size, it has been argued that these types of cells will be lower in activity. One solution has been to use three-dimensional electrodes or immobilization on conducting carbon supports which provide high surface area. These electrodes are extended into three-dimensional space which greatly increases the surface area for enzymes to bind thus increasing the current.

Biomagnetism

Biomagnetism is the phenomenon of magnetic fields produced by living organisms; it is a subset of bioelectromagnetism. In contrast, organisms' use of magnetism in navigation is magnetoception and the study of the magnetic fields' effects on organisms is magneto biology. (The word biomagnetism has also been used loosely to include magnetobiology, further encompassing almost any combination of the words magnetism, cosmology, and biology, such as "magneto astrobiology".)

The origin of the word biomagnetism is unclear, but seems to have appeared several hundred years ago, linked to the expression "animal magnetism." The present scientific definition took form in the 1970s, when an increasing number of researchers began to measure the magnetic fields produced by the human body. The first valid measurement was actually made in 1963, but the field of research began to expand only after a low-noise technique was developed in 1970. Today the community of biomagnetic researchers does not have a formal organization, but international conferences are held every two years, with about 600 attendees. Most conference activity centers on the MEG (magneto encephalogram), the measurement of the magnetic field of the brain.

Biomagnetic Therapy

Biomagnetic therapy is a healing practice that uses the application of magnets placed strategically on a person's body in order to alleviate pain or cure an illness. The practice has been around for centuries and tends to be most popular in Asia and South America, though there are practitioners trained in the use of magnets for health in most places. Proponents often claim that the therapy can bring about wildly successful results, including everything from alleviating sleep disorders to curing cancers. Most classically trained medical doctors do not recommend magnetic therapy as a sole means of treatment, however, and the practice is usually considered an "alternative" form of medicine.

Basic Theory

The main idea behind the practice is that the human body is made up of several different energy fields that can become imbalanced. Importantly, human blood is usually about 4% iron, and magnets are believed to attract the iron in the blood. This is said to increase blood flow to certain areas of the body which may bring balance to the body's systems might jump-start its natural ability to heal itself. It is also believed that placing magnets at certain critical places can draw impurities out of the blood and restore health and vital energy.

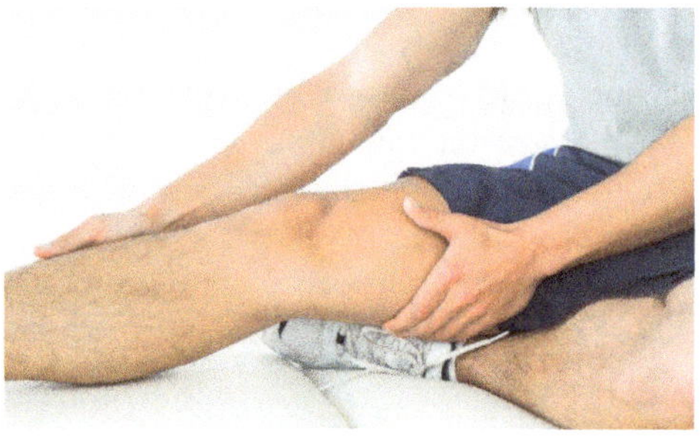

Magnets are generally used in combination with the reduction of stressful electromagnetic fields, but they can also be used on their own. Patients sometimes wear magnets in jewelry, particularly necklaces and bracelets, and may also sit with magnets balanced on parts of their body during therapy sessions.

Purported Benefits

Proponents of biomagnetic therapy claim that the practice has many tangible benefits. Some of these include increased oxygen and nutrients carried to the cells, relaxation of muscles, and the reduction of pain due to swelling of muscles and inflammation of joints. Regular treatments are also said to increase bone density, prevent and reversal of chronic disease, help overcome sleep disorders, and provide an overall increase to the body's natural immune system.

The therapy is routinely used as something of a "cure-all." Practitioners often say that the technique can be tailored for basically any ailment. It's perhaps most commonly used as a general method of good health and to provide relief from ordinary ailments like headaches and muscle stiffness, but there have been a number of cases where magnet therapy has been used to treat much more serious conditions, including various cancers. The success rate is not widely agreed upon. Many medical experts are skeptical, but proponents often say that the method works even if it hasn't been widely proved.

Historical Use

Magnets in the form of lodestones were used for healing by ancient cultures in China, Egypt, India, North America, and elsewhere. Scholars often pinpoint 200 BC as the first recorded use; in this year, magnets are said to have been used by a Greek doctor named Galan to treat pain. Queen Cleopatra, during her reign from 69-30 BC, is also said to have worn a magnet in an amulet in an attempt to maintain her beauty. The Swiss physician Paracelsus, who lived from 1493-1541 and who is considered to be one of the founders of modern medicine, believed that magnets could influence the body's life force.

Regulation and Standardization

In the 1800s, Carl Gauss created his system of magnetic units called the gauss. A gauss measurement indicates the strength of a magnet. Using this system, the World Health Organization (WHO) set standards as to what strengths are safe. In the US, the Food and Drug Administration (FDA) has classified biomagnetic therapy as "not essentially harmful." This is not exactly an endorsement, but it isn't a ban, either.

There are a few formal traning programs for people interested in becoming biomagnetic specialists, though the standards for training usually aren't nearly as rigorous as they are for most other medical arts. Practitioners don't usually need a license, and they aren't usually overseen or regulated by any larger body. This isn't necessarily problematic, but does often mean that treatment may not be consistent from person to person or provider to provider.

Common Precautions

As with almost any medical therapy, there are a number of precautions people need to take before beginning. The simple fact that it is alternative and doesn't depend on medications doesn't automatically mean that it is safe for everyone. In general, people shouldn't try to treat themselves unless they've undertaken some formal study of the effect of magnetic power on the body. Practitioners can often be found through biomagnetic therapy organizations or local alternative medicine associations.

In addition, magnets should not be used by anyone who has a pacemaker, or near a computer or magnetic data storage devices such as bank and credit cards. Magnetic forces can cause these items to become degaussed, or wiped of information. Some of the stronger magnets can also have a negative impact on flight instrumentation on aircraft, which means that they may not be suitable for use during travel.

CHAPTER 1

MICROBIAL ELECTRON TRANSPORT AND ENERGY CONSERVATION – THE FOUNDATION FOR OPTIMIZING BIOELECTROCHEMICAL SYSTEMS

Frauke Kracke, Igor Vassilev and Jens O. Krömer

Keywords: Microbial electron transport, redox potential, microbial electrosynthesis, microbial fuel cell, bioelectrochemical system, bio electrochemistry, acetogenic bacteria, ATP yield

INTRODUCTION

The fact that some bacteria are able to transport electrons beyond their cell wall and therefore electrically interact with their environment is known for over a century (Potter, 1911). This feature can be used to develop advanced electrically enhanced bio-processes. In so called BESs the organisms interact with electrodes via the exchange of electrons, which are either supplied or removed through an external electrical circuit. The application possibilities for electrode–bacteria interactions include the production of electricity, waste and wastewater treatment, bioremediation and the production of valuable products therefore opening a wide research field (Franks, 2012;Lovley, 2012; Yang et al., 2012; Harnisch et al., 2014). Initial research mainly focused on the application of BESs for power production. In so called microbial fuel cells microbes donate electrons to electrodes and therefore generate an electrical current (Rabaey et al., 2004; Franks and Nevin, 2010; Janicek et al., 2014). Another application is presented by bioremediation of aquatic sediments and groundwater where metal-reducing microbes catalyze the transformation of organic contaminants to carbon dioxide (Williams et al., 2009; Zhang et al., 2010). Within these systems the anodic oxidation by bacteria is coupled to production of chemicals on the cathode, usually hydrogen or methane, and they are referred to as microbial electrolysis cells (Wagner et al., 2009).

In recent years another approach gets more and more attention which intends a shift in the current: the technique termed microbial electrosynthesis refers the microbial production of multi-carbon compounds in a BES under current consumption. Novel sophisticated techniques like wind turbines and photovoltaic cells enable sustainable and cheap energy production and therefore allow bringing energy consuming technologies into focus (Lovley and Nevin, 2013). Initially the term microbial electrosynthesis was used exclusively for the microbial reduction of carbon dioxide with the help of electricity (Nevin et al., 2010; Rabaey and Rozendal, 2010). But the research field was quickly widened by multiple studies that follow the same approach of optimizing microbial production by electrical enhancement from other substrates than CO_2, often referred to as electro fermentation (Kim and Kim, 1988; Emde and Schink, 1990; Shin et al., 2002; Steinbusch et al., 2009; Rabaey and Rozendal, 2010; Choi et al., 2012). Common in all studies is the aim of overcoming metabolic redox limitations by electron exchange between microbes and electrodes for increased production. In a recent metabolic modeling study, it was shown that not necessarily the degree of reduction of a product but rather the metabolic pathway that leads from the chosen substrate to the target compound decides if there is an electron surplus or demand inside the metabolic network of a cell. This was highlighted with the production of 1,4-butanediol and 2,3-butanediol from sugar. Both products

Citation: Kracke F, Vassilev I and Krömer JO (2015) Microbial electron transport and energy conservation – the foundation for optimizing bioelectrochemical systems. Front. Microbiol. 6:575. doi: 10.3389/fmicb.2015.00575.

have the same degree of reduction, but the different metabolic pathways employed for production will lead to different by-products and a different overall electron balance, such that in case of 2,3-butanediol a surplus of electrons needs to be balanced, while in the 1,4-butanediol case a shortage of electron limits the theoretical yield (Kracke and Krömer, 2014). Therefore cathodic as well as anodic BES are of interest and we suggest extending the term microbial electrosynthesis to all processes that intend an electrically induced shift of carbon flow toward production of valuable chemicals.

Whether one is considering current consuming or current producing bio-processes, crucial in each application is the performance of the microbial host. The ability and especially the efficiency of the organism to exchange electrons with an electrode and connect this EET to its cmetabolism significantly influences the overall process performance. Many applications in BESs are so far restricted to lab-scale research projects as the electron transfer rates are simply too low to design a viable process scale-up (Logan and Rabaey, 2012). In order to optimize and advance bioelectrochemical techniques a more thorough understanding of possible extracellular electron exchange mechanisms in both directions is needed (Tremblay and Zhang, 2015). Especially cathodic systems and the fundamentals of electron transport towards microbes are poorly understood (Rosenbaum et al., 2011; Sydow et al., 2014). In this article we discuss the natural electron transport chains of different organisms and discuss their potential benefits and limitations if used in a BES. Special focus lies on the connection of redox and energy metabolism in each species.

VARIETIES OF MICROBIAL ELECTRON TRANSPORT CHAINS

In microbial electron transport chains electrons are transferred from a low potential electron donor to an acceptor with more positive redox potential by redox reactions. These reactions are usually catalyzed by membrane-bound compounds that use the energy difference between donor and acceptor to establish an ion-gradient across the membrane, which in turn is used for ATP synthesis and thus converts the difference in electrical potential into chemical energy for the cell (Anraku, 1988).

In order to adapt to different environmental conditions microbes developed an enormous variety of electron transport chains (Hernandez and Newman, 2001). Important systems catalyzing these redox reactions include primary dehydrogenases that supply high energetic electrons from a donor such as NADH and usually couple electron transport to H^+ or Na^+ transport across the membrane (Anraku, 1988). Also involved in transmembrane ion-transport are membrane-localized (multi-) protein complexes such as cytochromes and terminal oxidases (reductases) that transfer electrons to a final acceptor such as oxygen, nitrate or fumarate (Hannemann et al., 2007; Vignais and Billoud, 2007; Richter et al., 2012). Most transmembrane reductases and oxidases function as ion pumps but some do not. Electron carrying co-factors such as quinones, flavines, heme, iron–sulfur clusters or copper ions also play an important role in microbial electron transport. Some of these are soluble lipophilic molecules that shuttle electrons between the relatively large enzymatic complexes inside the membrane (e.g., quinones) while others are catalytic cofactors bound to proteins (e.g., heme groups of cytochromes; Deller et al., 2008; Dolin, 2012). There are also membrane bound complexes that use the exergonic electron bifurcation of soluble cellular redox compounds such as ferredoxin (Fd) and NADH for transmembrane ion transport and therefore establishment of a motive force across the membrane that can drive ATP synthesis (Herrmann et al., 2008; Müller et al., 2008; Schuchmann and Müller, 2012).

The achievable energy gain (Gibbs free energy, ΔG) of each electron transport chain is depending on the redox potential difference (ΔE) of all reactions between electron donor and acceptor. Some bacteria incorporate several electron transport chains, which they can use sometimes even simultaneously in order to respond to different electron acceptors and donors available in the environment (Haddock and Schairer, 1973; Anraku, 1988). Others are restricted to only one respiratory pathway (Das and Ljungdahl, 2003; Müller et al., 2008). This diversity of microbial electron transport mechanisms illustrates the complexity of the approach of bioelectrochemical techniques. In order to interfere efficiently with the redox metabolism of an organism one needs to understand the targeted site of EET and its metabolic impact.

A wide range of microbes has been discovered to be able to exchange electrons with solid surfaces (direct EET) and/or soluble mediators (indirect EET) but only a few have been studied in depth. In fact the mechanisms of electron transport that are found in different species can differ significantly from one another. Dissimilatory metal reducing bacteria are amongst the most studied being, able to "respire" insoluble metals in anaerobic environments. The model organisms *Geobacter sulfurreducens* and *Shewanella oneidensis* were studied by various research groups for decades and evidence for direct and indirect electron transfer between the organism and electrodes could be found (Bond and Lovley, 2003; Thrash and Coates, 2008; Rosenbaum et al., 2011; Ross et al., 2011). For both bacteria outer-membrane cytochromes were identified as essential compounds to enable EET (Mehta et al., 2005; Shi et al., 2007, 2009; Breuer et al., 2015). However, there are several differences in their electron transport chains, for example *Shewanella* excretes soluble electron carriers while similar compounds are missing in *Geobacter* sp. (Holmes et al., 2006;Marsili et al., 2008a). Furthermore it could be shown that the redox chains catalyzing an inward current rely on different mechanisms than current producing reactions (Dumas et al., 2008;Strycharz et al., 2008). Another group of dissimilatory metal reducing bacteria is presented by the obligate anaerobe *Thermincola*, which were also found to be capable of directly transferring electrodes to anodes (Marshall and May, 2009; Wrighton et al., 2011). Their EET mechanisms also seem to rely on cytochromes that in this case are cell-wall associated of the Gram-positive bacteria (Carlson et al., 2012). Interestingly there are also organisms such as *Clostridium ljungdahlii*, which do not have any cytochromes but were tested positive on EET (Köpke et al., 2010; Nevin et al., 2011). The exact mechanisms by which electrons are transferred between the electrode surface and the microbial metabolism still remain unclear (Patil et al., 2012; Sydow et al., 2014). Therefore we have a look at the native electron transport chains of several organisms that were studied in BESs. An overview is given in Table 1 and the following sections discuss the mechanisms of the presented bacteria in detail.

Table 1: Electron transport characteristics and behavior in bioelectrochemical systems of organisms discussed in this study

Organism	Gram staining	Key components in electron transport chain and coupling to energy conservation	Experience in BES		Important reference
			Anode	Cathode	
Shewanella oneidensis 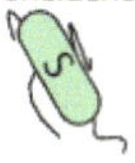	–	Mtr pathway: Proton gradient created by cytochromes (c-type), soluble electron carriers and membrane bound NADH-hydrogenase; ATP via H^+-ATPase	Various studies, model organism; direct and self-mediated electron transfer	Direct use of electrons by thin biofilms for reduction of fumarate to succinate	Coursolle and Gralnick (2010), Ross et al. (2011)
Geobacter sulfurreducens 	–	Branched OMCs system: Proton gradient created by cytochromes (c-, d-types), soluble electron carriers and membrane bound NADH-hydrogenase; ATP via H^+-ATPase	Various studies, model organism; generation of comparatively high current densities by direct electron transfer through biofilms	Direct use of electrons by thin biofilms for reduction of fumarate to succinate	Bond and Lovley (2003), Gregory et al. (2004)
Thermincola ferriacetica	+	Putative electron transport chain based on multiheme c-type cytochromes associated with periplasm and cell surface; ATP via H^+-ATPase	First proof of anodic current production by direct contact of a Gram-positive organism	No report	Marshall and May (2009), Parameswaran et al. (2013)
Sporomusa ovata	–	H^+ gradient via membrane-bound cytochromes (b-, c-types) and quinones; ATP via H^+-ATPase	No report	Direct use of electrons from an electrode for CO_2 reduction to acetate and 2-oxobutyrate	Nevin et al. (2010)
Moorella thermoacetica 	–	H^+ gradient via membrane-bound cytochromes (b, d-type), quinones and/or Ech-complex; ATP via H^+-ATPase	No report	Direct use of electrons from an electrode for CO_2 reduction to acetate at high columbic efficiencies (>80%)	Nevin et al. (2011)
Acetobacterium woodii 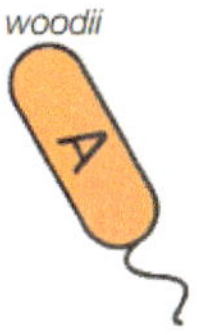	+	Electron bifurcating ferredoxin reduction Na^+ gradient via membrane-bound Rnf complex (Ferredoxin:NAD^+-oxidoreductase), Membrane bound corrinoids (No cytochromes, no quinones); ATP via Na^+-ATPase	No report	*A. woodii* was shown not to be able to directly accept electrons from a cathode; however was also determined as a dominant species in a cathodic mixed culture producing acetate from CO_2 and microbial and/or electrochemically produced hydrogen	Nevin et al. (2011), Marshall et al. (2012)
Clostridium ljungdahlii 	+	Electron bifurcating ferredoxin reduction H^+ gradient via membrane-bound Rnf complex (Ferredoxin:NAD^+-oxidoreductase) (No cytochromes, no quinones); ATP via H^+-ATPase	No report (however, a close relative *C. acetobutylicum* was shown to be able to oxidate acetate under current production)	Direct use of electrons from an electrode for CO_2 reduction to acetate	Logan (2009), Nevin et al. (2011)
Escherichia coli 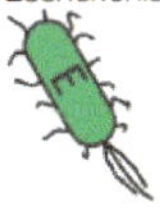	–	H^+ gradient via membrane-bound cytochromes (a-, b-, d-, o-type) dehydrogenases, quinones, flavins (bound); ATP via H^+-ATPase	*E. coli* is able to produce current after long acclimation times without mediator, or on modified electrodes	No report	Schröder et al. (2003), Zhang et al. (2006, 2007)
Pseudomonas aeruginosa 	–	H^+ gradient via membrane-bound cytochromes (a-, b-, c-, o-type), phenazines, flavines (soluble and bound), quinones and dehydrogenases; ATP via H^+-ATPase	Current production mediated by self-secreted phenazines	No report	Hernandez et al. (2004), Rabaey et al. (2005)
Corynebacterium glutamicum 	+	H^+ gradient via membrane-bound cytochromes (a-, b-, c-, d-type), quinones, flavins (bound) and dehydrogenases; ATP via H^+-ATPase	No report	Increased lactic acid production with a mediator in a cathodic system	Sasaki et al. (2014)

Specific EET in Metal Respiring Bacteria

The first microorganisms that we want to introduce and discuss belong to the group of dissimilatory metal reducing bacteria. As their name indicates, the microorganisms have the ability to interact with solid minerals, e.g., Fe(III) or Mn(IV), to obtain energy by using those minerals as electron acceptors and/or donors for their respiration process. Here a transport of electrons from a low redox potential donor to an acceptor with a higher redox potential can result in a proton gradient to drive ATP synthesis. That characteristic of breathing metals plays an important role in biogeochemical cycles and has the potential to be used in bioremediation and BES (Lovley et al., 2004; Richardson et al., 2012).

But how does the microorganisms respire insoluble metals and extract energy? This question has not been fully clarified in detail yet. In contrast to other respiration processes, where freely diffusible gas or readily soluble substances can easily enter the cell and be used as electron acceptors/donors, the major challenge for metal respiring bacteria is the interaction with the extracellular minerals, which cannot pass the cell membrane and do not have access to the periplasm nor cytoplasm. To overcome this barrier the bacteria need redox-active molecules in their outer membrane or have to excrete redox-active shuttle molecules to transfer electrons between the cellular interior and the extracellular metals (Hernandez and Newman, 2001;Hartshorne et al., 2009). There exists a great diversity of mechanisms for such electron-shuttling pathways, which is explained in more detail in the section below by describing the EET of two model organisms.

G. sulfurreducens:Branched OMCs System

The Gram-negative obligate anaerobic δ-proteobacterium *G. sulfurreducens* is used as a model organism for investigating electroactive microorganisms (Levar et al., 2012). Since its genome was sequenced it is easier to analyze the detailed molecular mechanism of EET and to construct molecular models. In its genome more than 110 genes coding for putative c-type cytochromes have been identified, which likely play an important role in the electron transport pathway of this bacterium (Methe et al., 2003). It is assumed that several multiheme c-type cytochromes enable the transport of redox equivalents between the cellular menaquinone (MQ) pool and the extracellular insoluble metals to create a proton gradient for energy conservation (Levar et al., 2012). The interaction between the cytochrome complexes in the electron transport chain is based on the redox potential of the different multiheme molecules of the cytochromes, whereby each heme has its own specific redox potential. In this way wide windows of potential ranges are created that overlap with each other and allow a bio-energetic transfer of electrons (Inoue et al., 2010; Qian et al., 2011).

Figure 1A illustrates a model of the EET mechanism with the participating proteins, which were assigned a central role. Here a diheme cytochrome c peroxidase, designated 'metal-reduction-associated cytochrome' (MacA), functions as a transmitter of electrons from the inner membrane to the triheme periplasmic c-type cytochrome (PpcA) in the periplasm. Following on PpcA passes the electrons to the outer membrane cytochromes, termed OMCs (e.g., OmcB, OmcC, OmcS, OmcZ), which transfer the electrons to the extracellular acceptor. The branched OMCs system is very complex and is still not fully understood. It seems that different OMCs are required to interact with different extracellular metals or electrodes (Levar et al., 2012). For example, the octaheme cytochrome OmcZ is more abundant in biofilms grown on an electrode and a deletion of *omcZ* gene leads to current decrease of more than 90% while there is no impact on the reduction of other electron acceptors as Fe(III) oxide (Nevin et al., 2009; Richter et al., 2009). The dodecaheme cytochrome OmcB and the hexaheme cytochrome OmcS are essential for Fe(III) reduction while knock-out of the *omcB* or *omcS* genes results in no or hardly any effect on current generation through biofilms grown on an anode (Kim et al., 2008; Richter et al., 2012). The proposed model reflects a greatly simplified EET mechanism. In fact the EET pathway is subjected to a complex regulatory mechanism and there exist many more homologous *omc* genes, which can replace lacking cytochromes genes in generated mutants (Kim et al., 2006; Schrott et al., 2011). For instance OmcB is important for the reduction of soluble Fe(III), but an Δ*omcB* mutant can express homologs such as a dodecaheme OmcC, which allows the use of a parallel pathway to respire Fe(III) (Leang and Lovley, 2005).

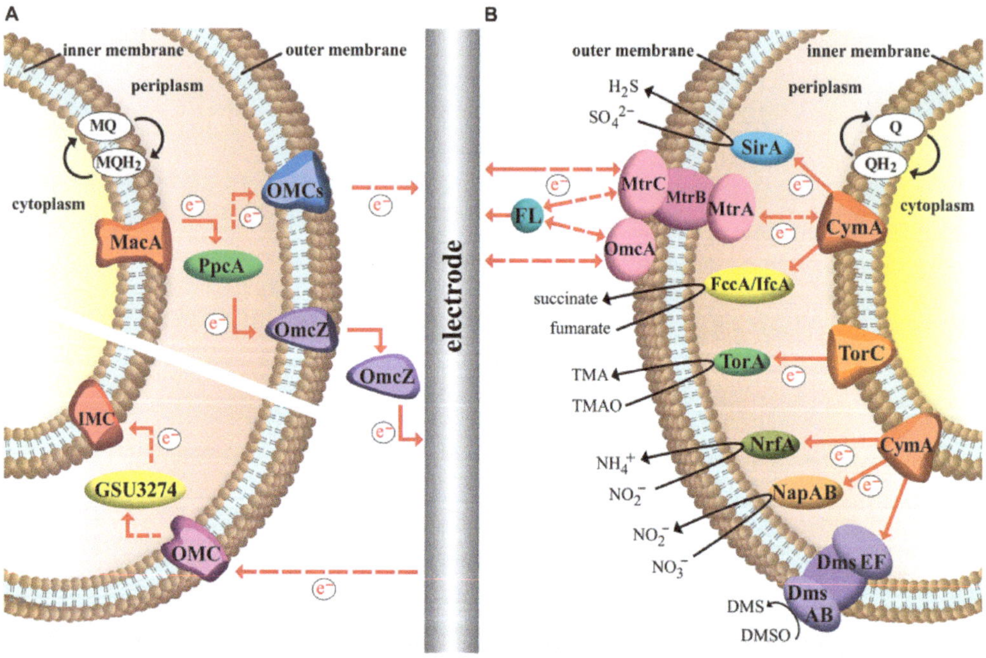

Figure 1: Schematic image of the proposed EET of two metal respiring bacteria and their interactions with an electrode in a bioelectrochemical system. Dashed arrows indicate hypothetical electron flow and solid arrows indicate experimental proved electron flow. (A) Branched outer membrane cytochromes (OMCs) system of Geobacter sulfurreducens. Electrons can be transported between inner membrane, periplasm, outer membrane, and an electrode via a chain of cytochromes and menaquinones (MQ). Terminal OMCs can vary depending on the environmental conditions.(B) Unique Mtr-pathway and terminal reductases of Shewanella oneidensis. Quinones (Q) pass electrons to CymA or TorC, which transfer the electrons to terminal reductases or a MtrCAB complex. MtrCAB complex can interact with the electrode direct or via flavin molecules (FL).

In comparison to the transport of electrons to the anode, much less is known about the mechanism by which the cell takes up electrons from the cathode. Due to the reverse electron chain pathway the practicing shuttle molecules have to operate at different redox potentials. In contrast to current-generating biofilms, in current-consuming biofilms a lower expression of OMCs such as OmcZ were detected. Deletion of those associated genes, which were essential for anodic biofilms, had no impact to cathodic biofilms. In further studies it was observed, that in current-consuming biofilms a gene (GSU3274) encoding a putative monoheme c-type cytochrome was strongly expressed. Mutants lacking this gene lost their ability to take up electrons from a cathode, but did not show differences in EET to an anode. So it seems that GSU3274 plays a significant role in the EET from a cathode (Rosenbaum et al., 2011; Strycharz et al., 2011).

Geobacter sulfurreducens shows best EET performance as a biofilm grown on the electrode based on direct electron transfer. This bacterium possesses excellent biofilm formation properties and the thickness of the biofilm is linked with the amount of generated current in a linear correlation up to a certain thickness limit (Reguera et al., 2006). The transport of electrons in a multilayer biofilm to an anode can be achieved by two combined mechanisms. One transfer way is based on secretion of non-diffusing 'mediators' (e.g., cytochromes such as OmcZ) into the biofilm matrix, which can act as electron shuttles (Richter et al., 2009). The second way depends on 'nano-wires,' which are electrically conductive appendages that enable physical connection between cells and/or cells and surface of the electrode (Reguera et al., 2006). The mechanism of the EET in such nano-wires has not been clarified yet. One model proposes that the nano-wires are pili with a metallic-like conductivity, which is based on aromatic amino acid residues within the appendages, which enable electron delocalization due to π-stacking (Malvankar et al., 2011, 2015). Another theory describes a 'superexchange' model, in which the electrons are 'hopping' along a chain of redox active proteins to the final electron acceptor (Malvankar et al., 2012; Strycharz-Glaven and Tender, 2012).

In comparison to anodic-biofilms, it was reported, that cathodic-biofilms are less-developed and much thinner (Gregory et al., 2004; Strycharz et al., 2008). The reason might be limiting redox surplus under non-autotrophic anaerobic growth conditions, which is enhanced by cathodic electron input as suggested by a recent *in silico* study (Kracke and Krömer, 2014). However, this observation needs to be studied in depth and be validated experimentally for a better understanding.

S. oneidensis: Characteristic Mtr-Pathway

The second electroactive model organism that we want to introduce and discuss its characteristic EET mechanism is the Gram-negative facultative anaerobic bacterium *S. oneidensis*. The interesting characteristic of *S. oneidensis* is its ability to utilize a great variety of inorganic and organic compounds as final electron acceptor in the absence of oxygen. As in the case for *G. sulfurreducens* that property is based on interaction of a large number of multiheme cytochromes. The sequenced genome of *S. oneidensis* shows 42 putative cytochromes, of which 80% are located in the outer membrane (Heidelberg et al., 2002; Lower et al., 2005). The first step of the electron transfer through the cell membrane is the oxidation of small electron carriers (quinols), which enable the transport of electrons between NADH-dehydrogenase and cytochromes in the inner membrane to create a proton gradient for energy conservation in form of ATP. That oxidation reaction can be catalyzed by the tetraheme cytochromes TorC and CymA, whose sequence is very similar and both are attached to the inner membrane by a single α-helix. The next link in the electron transfer chain of TorC is a periplasmic reductase TorA, which can utilize the outer membrane permeable trimethylamine *N*-oxide compound as terminal electron acceptor and reduces it to trimethoxyamphetamine (Dos Santos et al., 1998). In comparison to TorC, CymA can interact with different redox partners in the periplasm and as a consequence outer membrane-crossing molecules like sulfite can be reduced by the octaheme redox partner SirA, (Shirodkar et al., 2011) nitrite by pentaheme cytochrome NrfA, (Gao et al., 2009) nitrate by NapAB reductase (Simpson et al., 2010) and fumarate by FccA and IfcA reductase (Maier et al., 2003). Furthermore an octaheme cytochrome OTR was detected and showed *in vitro* capability of the reduction of a range of N and S oxides and oxyanions, but *in vivo* the function of OTR has not been confirmed yet (Atkinson et al., 2007).

Additionally as a metal respiring bacterium like *G. sulfurreducens*, *S. oneidensis* is able to use extracellular terminal electron acceptors, e.g., electrodes, Fe(III) and Mn(IV) (see Figure 1B). To overcome the outer membrane barrier the bacterium possesses a Mtr-pathway. Homologous genes for that pathway were also found in other dissimilatory metal reducing and oxidizing bacteria (Shi et al., 2012). However, *S. oneidensis* Mtr-pathway is one of the best investigated EET chains. It has been suggested that a decaheme cytochrome MtrA takes up electrons from CymA via an electron transport chain and passes them on to an extracellular decaheme cytochrome MtrC, which transfers the redox equivalents to final exogenous electron acceptors (Myers and Myers, 2000;Coursolle and Gralnick, 2010). Here MtrA, MtrC and a third element, MtrB form a complex, MtrCAB. MtrB is a porin molecule in the outer membrane and serves to organize and stabilize MtrA and MtrC to enable electron transfer. Besides MtrC a second decaheme cytochrome, OmcA was detected anchored as a lipoprotein in the outer membrane that was able to transfer electrons to exogenous electron acceptors as well (White et al., 2013; Breuer et al., 2015). In addition to the*mtrCAB* genes the genome of *S. oneidensis* encloses homologs like *mtrFDE*. The homologous cytochromes can support the EET or can partly take over the activity of the MtrCAB complex depending on the electron acceptor and the growth conditions (Coursolle and Gralnick, 2010). Likewise, the DmsABEF complex is based on homologs. Here the porin cytochrome complex DmsEF transfer the electrons to the DmsAB complex, which is then able to reduce DMSO to DMS (Gralnick et al., 2006). The periplasmic tetraheme cytochrome STC has to be mention as well, which seems to support the transfer of electrons between inner membrane and outer membrane, however, the exact mechanism is unclear (Fonseca et al., 2013). In contrast to EET to an anode, the information about taking up electrons from a cathode is limited. The interacting compounds in the cathodic process were not established yet. It was suggested due to *in vivo* studies, where electrons were provided from a graphite electrode to reduce fumarate, that an electron uptake is possible through a reverse Mtr-pathway (Ross et al., 2011).

In contrast to *G. sulfurreducens*, *S. oneidensis* shows not only the ability of direct electron transfer, but can also perform indirect electron transfer due to excretion of redox active mediators. Direct electron transfer is based on physical connection with the electrode by forming a biofilm on the surface (Marsili et al., 2008a,b) and through extensions in form of nano-wires. While *Geobacter*nano-wires are assumed to be type IV pili, for *S. oneidensis* nano-wires it was shown, that their structure is similar to outer membrane vesicles and those nano-wires can be seen as extensions of the outer membrane and periplasm that include the multiheme cytochromes responsible for EET (Pirbadian et al., 2014). In case of indirect electron transfer, *S. oneidensis* secretes flavin molecules (FL) that act as small diffusible shuttle molecules to transfer electrons between electrode and OMCs (Marsili et al., 2008a) or as bounded cofactors for OMCs (Okamoto et al., 2013). *G. sulfurreducens*is also able to produce FL, but here the flavins are preferentially bound to the OMCs and are not mobile like free shuttle units (Okamoto et al., 2014).

Carbon Respiration of Acetogenic Bacteria

Acetogenic bacteria, short acetogens, are anaerobic organisms that are able to assimilate CO_2 or CO via the Wood–Ljungdahl pathway, also called carbonate-respiration or acetyl–CoA pathway. This autotrophic pathway offers the possibility to develop biotechnological processes that combine the usage of cheap ubiquitous substrates (i.e., syngas) with greenhouse gas reduction and therefore makes acetogens attractive hosts for biotechnology (Köpke et al., 2010; Tracy et al., 2012). This feature put the bacteria into focus of the research community trying to establish an artificial 'photo'synthesis process by using CO_2 and electrons from an electrode (Nevin et al., 2010). Acetogens are also able to utilize a great variety of heterotrophic molecules such as sugars, glycerol, and cellulose, which broadens the spectrum of possible substrates to waste streams from biodiesel industry (e.g., glycerol) or dairy industry (e.g., whey) and many more (Tracy et al., 2012). The main product is usually acetate (hence the name) but acetogens also bear the feature to naturally produce a broad spectrum of other chemicals, which are of industrial use either directly or as precursors such as ethanol, butanol, 2,3-butanediol or butyrate (Tracy et al., 2012). The mayor intermediate from carbon fixation via Wood–Ljungdahl pathway is acetyl–CoA and is linked to other metabolic pathways such as Embden–Meyerhof–Parnas pathway, therefore offering a great metabolic diversity for metabolic engineering of other production pathways.

In the Wood–Ljungdahl pathway two CO_2 molecules are merged to form one molecule Acetyl-Coenzyme-A, which is either converted to acetate or assimilated in biomass. The overall conversion of CO_2 to acetate uses one ATP (in the step of formyltetrahydrofolate synthetase) and creates one ATP by acetate kinase reaction. Therefore no net energy gain can be achieved via substrate-level phosphorylation. Since acetogens are able to grow autotrophically, the pathway must be coupled to a chemiosmotic mechanism that provides additional energy. By calculating the Gibbs free energies of each reaction in the Wood–Ljungdahl pathway a net energy benefit of about –95 kJ/mol can be determined (Diekert and Wohlfarth, 1994). This energy could support the synthesis of 1–2 mol ATP via chemiosmosis as anaerobic bacteria require –60 to –80 kJ of free energy for the synthesis of 1 mol of ATP (Das and Ljungdahl, 2003). In recent years experimental evidence for membrane driven ATP synthesis in acetogens could be found, however, the sites and mechanisms of energy conservation differ between organisms. Over 100 different species of acetogens have been identified and despite their common feature of CO_2 assimilation via the Wood–Ljungdahl pathway they are very diverse in terms of metabolism, phylogenetics, or preferred habitat (Drake et al., 2008). The best-studied organisms belong to the genera *Acetobacterium* and *Clostridium* and include a few species with fully available genome sequences (genome published: *Moorella thermoacetica*, *Clostridium ljungdahlii*, *Clostridium carboxidivorans*, *Eubacterium limosum;*under preparation: *Acetobacterium woodii, Clostridium aceticum*). Genetic and genomic tools are under intense development and promise fast advancing metabolic engineering platforms for acetogens (Lütke-Eversloh and Bahl, 2011; Lee et al., 2012; Schiel-Bengelsdorf and Dürre, 2012;Banerjee et al., 2014).

M. thermoacetica: Cytochromes or Ech-Complex?

From an energetic point of view and in regards to electron-transport properties acetogens are often divided in two groups: Na^+-gradient and H^+-gradient dependent species. *M. thermoacetica*(formerly *Clostridium thermoaceticum*) an example for the latter group was the model organism for HG Wood and LG Ljungdahl for their studies of the acetyl–CoA pathway. They identified several membrane-integrated electron carriers MQ MK-7, a flavoprotein and two b-type cytochromes that are believed to play major parts in creating a proton gradient over the membrane (Das and Ljungdahl, 2003). Later such a proton motive force that could drive ATP synthesis in *M. thermoacetica* was measured, however, which components exactly transfer protons across the membrane remains unknown (Ragsdale and Pierce, 2008). Genome studies of the organism revealed also membrane-bound components such as hydrogenases and NADH-dehydrogenase, which are known to transfer protons across the membrane in other organisms. Therefore a membrane integrated electron transport chain via these complexes with H_2 as electron donor and methylene-THF as electron acceptor is proposed (Müller and Frerichs, 2013). However, so far no experimental data supports this hypothesis. Recent studies also deliver evidence for electron-bifurcating enzymes that play important roles in electron transfer of autotrophic and heterotrophic pathways of *M. thermoacetica* (Huang et al., 2012). The soluble complex HydABC oxidizes hydrogen with simultaneous reduction of Fd and NAD^+ (Wang et al., 2013). Additionally a second soluble transhydrogenase (NfnAB) catalyzes the electron bifurcation from reduced Fd and NADH to $NADP^+$. Furthermore the genome of *M. thermoacetica* codes for a membrane-bound energy converting hydrogenase, called Ech-complex. For the methanogenic archaea *Methanosarcina mazei* the Ech-complex is responsible for establishment of a proton gradient across the membrane, which leads to the theory that could also be the case in acetogens (Welte et al., 2010; Schuchmann and Müller, 2014). This complex uses the excess energy that is freed from electron transfer from reduced Fd to H^+ to transport ions across the membrane. A very recent report states that electron stoichiometry is only balanced with involvement of the Ech-complex in energy conservation while the membrane-bound dehydrogenases and cytochromes play no major part (Schuchmann and Müller, 2014). As this hypothesis also lacks distinct experimental validation we included both theories in our summary of possible electron transport mechanisms of *M. thermoacetica* (Figure2A).

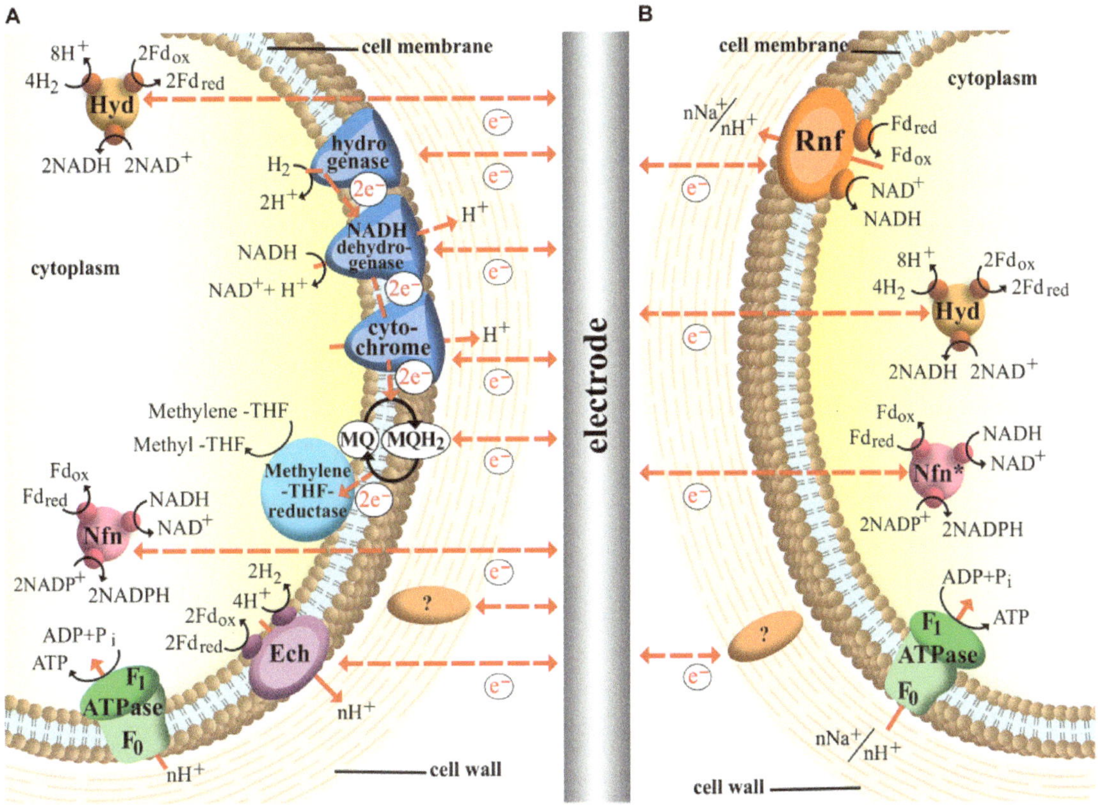

Figure 2: Schematic image of acetogenic electron transport chains and possible interactions with an electrode in a bioelectrochemical system. Dashed arrows indicate hypothetical electron and proton flow. (A) Electron transport mechanisms in *Moorella thermoacetica* via membrane-bound cytochromes and hydrogenases, MQ, soluble electron-bifurcating complexes (Hyd and Nfn) and proton pumping Ech-complex;(B) Electron transport of *Clostridium ljungdahlii* (H^+) and *Acetobacterium woodii* (Na^+) based on membrane-bound Rnf-complex and soluble electron-bifurcating complexes (*Nfn-complex is not found in *A. woodii*). ? represents hypothetical cell-wall associated proteins that could facilitate electron transfer.

A. woodii and *C. ljungdahlii*: Electron Transport without Cytochromes?

Nevin et al. (2010) reported that the electrons needed for CO_2 fixation via Wood–Ljungdahl pathway could be provided directly by an electrode, a breakthrough work in the field of bioelectrochemical techniques (Nevin et al., 2010). The acetogen *Sporomosa ovata*, a close relative to *M. thermoacetica* was able to directly accept electrons from a cathode and convert carbon dioxide to acetate and 2-oxobutyrate. Following studies showed similar abilities of other acetogens of the *Sporomosa* and *Clostridium* genera (Nevin et al., 2011). The acetogen *A. woodii*, however, was found unable to consume current and showed different behavior compared to other acetogens in Nevins experiments. *A. woodii* is an example strain for Na^+-dependent acetogens that typically lack cytochromes. It could be shown that the conversion of CO_2 to acetate via Wood–Ljungdahl pathway is coupled to the generation of a sodium gradient across the cytoplasmic membrane (Müller, 2003; Fast and Papoutsakis, 2012). Since *A. woodii* does not contain cytochromes or quinones energy conservation must be secured by a different electron transport system. In 2008 evidence for a novel membrane-bound Fd:NAD^+ oxidoreductase (Rnf complex) was reported that seems to be responsible for transmembrane Na^+ transport (Müller et al., 2008). In this complex the electrons from reduced Fd are transferred to NAD^+ to form NADH. Since the redox potential of Fd ($E^{0'}_{Fd}$= -500 to -420 mV) is more negative than that of the NAD^+/NADH couple ($E^{0'}_{NADH}$= –320 mV), the energy surplus (equivalent to –20 to –35 kJ/mol, three to four times more than released by the Ech-complex) is available for transmembrane ion transport (Biegel et al., 2011). This sodium gradient is harvested by the Na^+-dependent F_1F_0ATP synthase of *A. woodii*. It was suggested that some steps of the Wood–Ljungdahl pathway also add to the transmembrane ion gradient. The reduction of methylene-H_4F to methyl-H_4F was discussed as a likely site since the reaction is exergonic and coupled to a corrinoid iron sulfur protein (Ragsdale and Pierce, 2008). But recent studies could identify the Rnf complex as the only membrane-bound electron transfer system and rather suggest a Fd reduction by the methylene-THF reductase (Poehlein et al., 2012).

Interestingly a similar Rnf complex was detected in *C. ljungdahlii* even though its membrane gradient is proton based, which would put the organism into the H^+-acetogen-group together with*M. thermoacetica*. However, *C. ljungdahlii* does not contain any cytochromes and therefore it seems more reasonable to categorize anaerobic homoacetogenic organisms into Ech- and Rnf-containing groups with subgroups of Na^+- and H^+- dependent species (Köpke et al., 2010;Schuchmann and Müller, 2014). It could be shown that the Rnf complex is an electron bifurcating Fd:NAD^+ oxidoreductase and is essential for developing a proton gradient over the membrane under autotrophic as well as heterotrophic growth conditions (Tremblay et al., 2013). With fructose as substrate and electron donor Rnf-deficient strains were growth limited with significantly reduced ATP yields as a result of disruptions in the membrane gradient development. Autotrophic growth without Rnf complex was completely inhibited, indicating the Rnf-complex being a major if not the sole electron transport mechanism linked to energy conservation. The soluble electron bifurcating complexes HydABC and NfnAB are also found in *C. ljungdahlii* while *A. woodii* is believed to only use HydABC (Schuchmann and Müller, 2014). The proposed electron transport mechanisms for both organisms are shown in Figure 2B.

Other Respiratory Pathways

In the section above we introduced typical electroactive microorganisms, which were used in BES. But what about other model and/or industrial microorganisms like *Escherichia coli* and*Corynebacterium glutamicum* or pathogenic microorganisms like *Pseudomonas aeruginosa*? Can we influence/manipulate their redox and energy metabolism? In order to answer this question, it is necessary to understand their respiratory pathways.

E. coli: Model Organism with Respiratory Flexibility

The respiratory system of *E. coli* is very well known and in many studies the bacterium is used as a model for investigation of energetics and regulation of respiration. The respiratory chains show a great diversity and variability enclosing 15 primary dehydrogenases to oxidize electron donors and ten terminal reductases (or oxidases) to reduce electron acceptors (including isoenzymes; Berlyn, 1998). Those primary dehydrogenases and terminal reductases are linked by three quinones: ubiquinone ($E^{0'}$ = +110 mV), demethyl MQ ($E^{0'}$ = +40 mV) and MQ ($E^{0'}$ = –80 mV). Depending on the enzymes various quantities of energy can be conserved due to the build-up of a proton gradient through proton pumps, or by arranging catalytic sites in a certain way to release the protons on opposite sides of the membrane to create a charge separation. The H^+/e^- ratios vary from 0 to 4 H^+/e^- (Unden and Bongaerts, 1997). Under aerobic conditions *E. coli* can conserve most energy by using quinol oxidases ($E^{0'}$ = +820 mV) to reduce O_2 to H_2O. Here O_2 represses the terminal reductases of anaerobic respiration. However, in the absence of O_2, energy can be generated by nitrate reductases ($E^{0'}$ = +420 mV), nitrite reductase ($E^{0'}$ = +360 mV), DMSO reductase ($E^{0'}$ = +160 mV), TMAO reductase ($E^{0'}$ = +130 mV) or fumarate reductase ($E^{0'}$ = +30 mV) with nitrate delivering the most energy and fumarate the least energy (Ingledew, 1983; Unden and Bongaerts, 1997; Berlyn, 1998). So depending on which level or in which step the electron chain is targeted or manipulated, the metabolism will gain more or less energy.

Furthermore it was demonstrated that *E. coli*, evolved under electrochemical tension in a microbial fuel cell, can generate current by using the electrode as an electron sink. It was proposed that the evolved cells possess the ability to excrete hydroquinone derivatives through a highly permeable outer membrane, which act as mediators to transport the electrons between cell and electrode (Qiao et al., 2008; Qian et al., 2011). Another approach to obtain *E. coli* cells that show electro activity is via metabolic engineering, which is discussed in Section "Key Compounds in Different Electron Transport Chains."

P. aeruginosa: Secretion of Redox Carriers

Another interesting microorganism is the Gram-negative aerobic bacterium *P. aeruginosa*. It is an opportunistic pathogen, which can live in various environments due its ability to catabolize a large number of substances (Baltch and Smith, 1994). Additionally the bacterium is a good biofilm-builder and has branched respiratory chains to use oxygen as an electron acceptor, involving five oxidases (*bo_3* oxidase, *aa_3* oxidase, *cbb_3* oxidase 1, *cbb_3* oxidase 2, and cyanide-insensitive oxidase), which are adapted to the varying availability of oxygen in the different biofilm stages due to biofilm thickness (Jo et al., 2014). Preferentially *P. aeruginosa* obtains its energy from aerobic respiration; however, under anaerobic conditions the bacterium can also survive in presence of nitrate or nitrite by using reductases to reduce the N-molecules (Hassett et al., 2002).

When *P. aeruginosa* is cultivated in a microbial fuel cell, the bacterium shows an interesting behavior. Instead of oxygen, nitrate or nitrite *P. aeruginosa* can use the anode as an electron sink to generate energy for an active growth (Rabaey et al., 2005). The anode stimulates the production of phenazine derivatives, e.g., phenazine-1-carboxylic acid ($E^{0'}$ = –275 mV), phenazine-1-carboxamide ($E^{0'}$ = -150 mV) and procyanin ($E^{0'}$ = -32 mV). The secreted phenazine derivatives operate as soluble mediators, which significantly enhance the electron transfer between electrode and cells, resulting in an increased current generation. In addition the diffusible phenazine derivatives enable the use of the electrode as an electron sink for cells in thick multilayer biofilms (Rabaey et al., 2005; Yang et al., 2012). Another attractive phenomenon; that in a mixed culture the secreted mediators can be use not only by *P. aeruginosa* itself, but also by other microorganism, which generally are not able to produce redox active mediators (Boon et al., 2008).

C. glutamicum: Oxygen Dependency

The third bacterium, *C. glutamicum*, is an important Gram-positive industrial microorganism, which has been widely used as a microbial cell factory for the production of various amino acids, nucleic acids and other chemicals in food, pharmaceutical, cosmetics and chemical industries (Tatsumi and Inui, 2012). *C. glutamicum* can utilize various carbon sources for energy conversion and oxygen as the preferable electron acceptor by using three oxidases. The *bc_1* oxidase can take up electrons from menaquinol and pass the electrons to the *aa_3* oxidase by forming a supercomplex that has a low oxygen affinity, whereas the third oxidase, a *bd* oxidase, has a high affinity to oxygen (Bott and Niebisch, 2003). The bacterium can also survive under anaerobic conditions in the presence of nitrate. However, the growth is very limited, because the bacterium has a nitrate reductase, but lacks enzymes to degrade the toxic product of the nitrate reductase (nitrite;Nishimura et al., 2007).

Manipulating the redox metabolism by cultivating *C. glutamicum* in a BES can result in higher yields of the target product (Sasaki et al., 2014). Experiments demonstrated that the bacterium grown under a cathodic potential ($E^{0'}$ = –600 mV) using glucose as the carbon source showed a decreased growth rate and an increased lactate yield. Here the anthraquinone 2,6-disulfonate was added to the medium as an artificial mediator to shuttle electrons between cathode and cells. The mechanisms, how the redox metabolism is influenced by the artificial mediator is still unclear (Sasaki et al., 2014).

MAKING THE CONNECTION: MICROBE–ELECTRODE INTERACTION

The previous chapter demonstrates the impressive diversity and complexity of microbial solutions for cellular electron transport. Similar to the ability to interact with many different electron donors and acceptors in the environment one can assume that microbes are also able to exchange electrons with electrodes via different cellular components and mechanisms. In order to achieve successful EET the specific characteristics of the electron transport chains of the target organism should be considered. While it is questionable if there are groups or microbes that are better adapted for EET than others it is for certain that a bio electrochemical approach is challenged by different conditions depending on the catalytic host. Organisms that feature outer membrane redox-components might be able to perform direct electron transfer with electrodes while soluble intracellular complexes such as the electron bi-furcating Nfn and Hyd complexes of acetogens are most likely only targetable via soluble mediators. For each organism the specific redox-window of its electron transport chain(s) might dictate required potentials for use in BES. The following two sections discuss key components in electron transport chains of the presented organisms and the possible metabolic impact if EET at different sites can be achieved.

Key Compounds in Different Electron Transport Chains

With Figure 3 we created an overview of important redox-reactions that are catalyzed by the presented organisms. Actual environmental conditions in living cells differ from the standard biochemical conditions (25°C, 1 atm, pH = 7), which might influence the actual redox potentials of each couple. The potential of the proton hydrogen couple for example is -414 mV under standard conditions. In acetogenic environments, however, this potential lies closer to –350 mV as the partial hydrogen pressure is around ~200 Pa (Müller and Frerichs, 2013). Many intracellular redox-carriers also show different redox potential than the corresponding pure compounds. The standard redox potential of Fds with one or more iron–sulfur clusters for example lies around –400 mV (Buckel and Thauer, 2013). Under physiological conditions in acetogens, however, Fds are usually >90% reduced and therefore reported to be able to catalyze reducing reactions at redox potentials as low as –500 mV (Buckel and Thauer, 2013; Hess et al., 2013). A similar effect shifts the true redox potential of the NAD^+/NADH couple to around –280 mV as the majority of molecules (>90%) are oxidized even though the $E^{0'}$ under standard conditions is –320 mV. Very close to that of $NADP^+$/NADPH (–324 mV), which in turn is shifted to –360 mV due to the intracellular ratio of $NADP^+$/NADPH being 1/40 (Buckel and Thauer, 2013). In these cases we indicated a redox-area rather than one known midpoint potential to illustrate the potential range at which these reactions might occur inside the organisms. We allocated a fixed standard redox potential (dashed lines) or potential range for each reaction depending on availability in the literature. However the reader might want to keep in mind that environmental conditions such as pH, redox potential of the solution and specific concentrations can result in a further redox potential shift. The ion couple Fe^{3+}/Fe^{2+} as an inorganic electron acceptor/donor is a good example for how the redox potential can be influenced by environmental conditions. The midpoint potential is about +770 mV at a low pH and in the absence of precipitation. However, depending on pH, concentration and in which form iron is available the midpoint potential can vary strongly (Straub et al., 2001). For example the midpoint potential of Fe_3O_4/Fe(II) is significantly lower (–314 mV), because magnetite is a less soluble mineral. For the organic chelate complex Fe(III)-citrate/Fe(II)-citrate the solubility is higher and, respectively, the midpoint potential (+372 mV; Thamdrup, 2000; Straub et al., 2001).

Some of the listed cellular redox compounds have been shown to play a major part in electron transfer between organisms and electrodes and therefore others might as well. As discussed above direct EET in *Geobacter* and *Shewanella* can be achieved via a network of cytochromes with different midpoint potentials [e.g., PpcA: –170 mV (Lloyd et al., 2003), OmcZ: –180 mV (Inoue et al., 2010), OmcS: -212 mV (Qian et al., 2011), CymA: -200 mV, MtrA: -100 mV, MtrC: -138 mV (Firer-Sherwood et al., 2008)]. However, for a better characterization and a functional understanding of the cytochromes their wide potential windows should be considered, because the cytochromes are complex molecules not showing narrow midpoint potentials but more wide ranges of potentials suggested to be due to the multiheme molecules in the cytochromes. Each heme has its own specific redox potential and affects the potential of the neighbor hemes creating a specific wide window of potential ranges for the overall molecule (Morgado et al., 2010; Breuer et al., 2015). For example the potential windows of OmcZ, OmcS, CymA, and MtrC are from –420 to –60 mV (Inoue et al., 2010), –360 to –40 mV (Qian et al., 2011), –0.350 to –0.080 mV and –0.280 to –0 mV (Firer-Sherwood et al., 2008), respectively. It is assumed that the wide windows allow an overlapping of the redox potentials of the cytochromes in an electron transport chain and make a thermodynamic downhill process of electron transport along the chain possible. Furthermore these wide potential ranges allow the interaction with a broad spectrum of external electron acceptors and donors (Firer-Sherwood et al., 2008; Inoue et al., 2010; Breuer et al., 2015). However, the potential windows can vary depending on the environmental conditions. For the cytochromes of *S. oneidensis* it was shown that the potential windows are shifted as a function of pH. It was observed that under higher pH conditions in most cases the redox potentials are shifted to more negative values (Firer-Sherwood et al., 2008). In Figure 3 we allocated a broad potential range to c-type cytochromes based on the discussed literature and indicated all organisms that contain c-type cytochromes. However, this does not necessarily mean that all these bacteria are able to catalyze reactions in the indicated redox potential window. *G. sulfurreducens'* genome encodes for around 110 different c-type cytochromes (Methe et al., 2003), *S. oneidensis* for 42 (Heidelberg et al., 2002), and *P. aeruginosa* for 27 (Bertini et al., 2006), while *C. glutamicum* genome only encodes for a single c-type cytochrome (Bertini et al., 2006). Therefore the flexibility of the metal reducing organisms to adapt to different electron acceptors and donors is much higher.

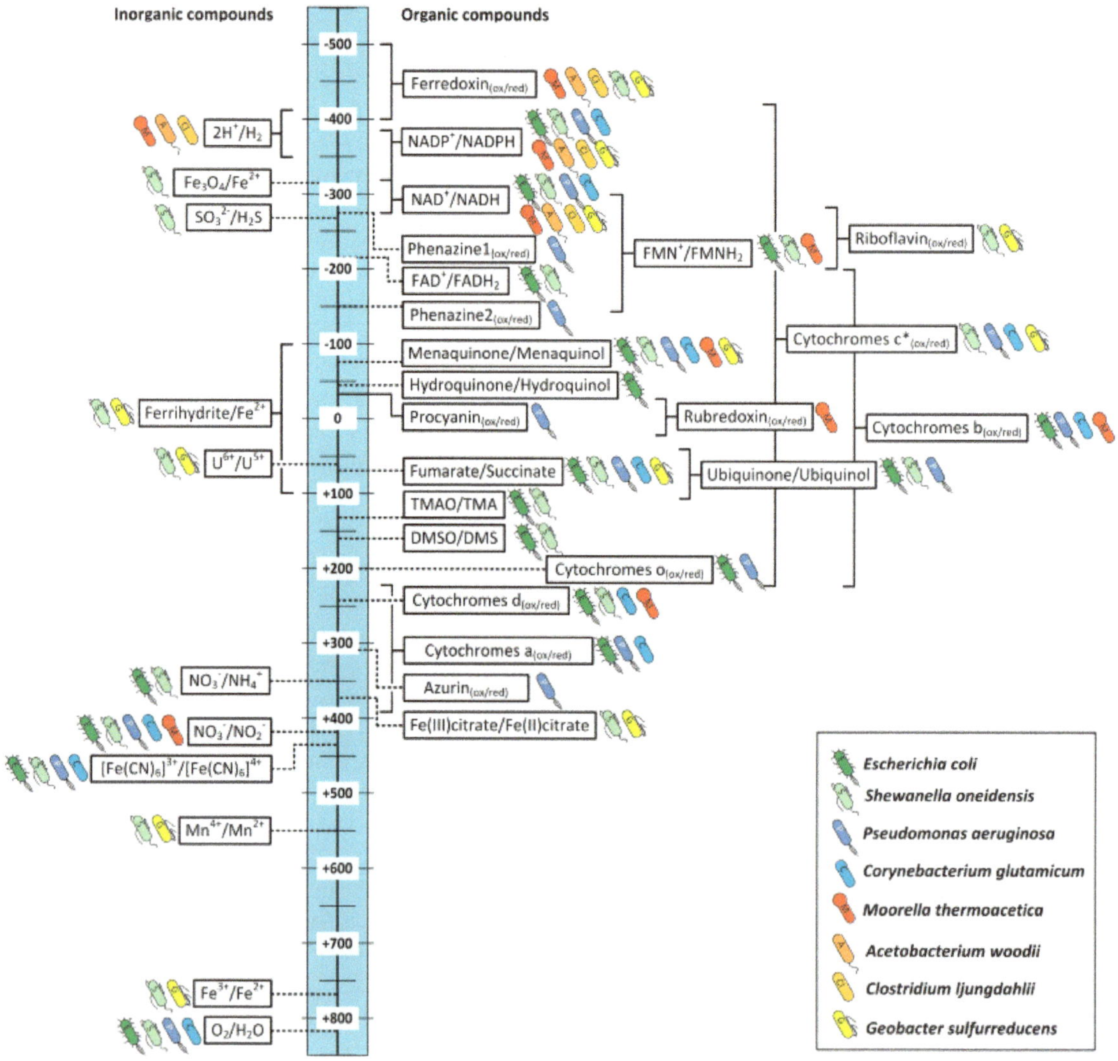

Figure 3: Redox potentials of important redox reactions in electron transport chains catalyzed by the bacteria discussed in this study. Standard redox potential ($E^{0'}$ [mV, 25°C, pH = 7]) are indicated by dashed lines. If physiological or environmental conditions are known to shift the potential from the $E^{0'}$, redox windows are indicated (solid lines). The bacterial symbol behind each reaction shows the organisms that are known to catalyze the reaction naturally. Blue: aerobes; green: facultative anaerobes; red–yellow: obligate anaerobes; Phenazine1 = Phenazine-1-carboxylic acid; Phenazine2 = Phenazine-1-carboxamide. *c-type cytochromes can cover a broad range of redox potentials as indicated. Not all bacteria mentioned will cover the whole range. For detailed discussion refer to main text.

In addition to the described cytochromes of *Geobacter* and *Shewanella* comparable complexes of other organisms might be able to perform EET via similar mechanisms. The membrane of the Gram-positive acetogen *M. thermoacetica* for example contains b-type cytochromes that could be a possible site of interaction between electrode and organism (Pierce et al., 2008). In fact it could be shown that *M. thermoacetica* together with other cytochrome containing acetogens is able to directly use electrons from a cathode for reduction of CO_2 (Nevin et al., 2011). Interestingly the same study reports *C. ljungdahlii* is also capable of direct EET even though it is not known to be able to produce any cytochromes (see Figures 2B and 3; Köpke et al., 2010). Could this electron transport rely on other membrane-bound enzymes, or be mediated by an unknown soluble molecule, or enzyme secreted by the organism? Since there is no evidence of such the actual mechanism of EET in *C. ljungdahlii* remains pure speculation to date. However, for other microbes self-excreted redox mediators are indeed an important way to exchange electrons with electrodes in both ways. *P. aeruginosa* for example produces several compounds of different redox potentials such as azurin ($E^{0'}$ = +310 mV), procyanin ($E^{0'}$ = -31 mV), phenazine-1-carbox amide ($E^{0'}$ = –150 mV) and phenazine-1-carboxylic acid ($E^{0'}$ = –275 mV). These enable the organism to adapt to different electron acceptors in the environment, including solid state electrodes (Vijgenboom et al., 1997; Rabaey et al., 2005). A very recent study of Yang et al. (2015) demonstrates that the concentration of such endogenous redox compounds can mediate electron flow in both directions and that the concentration directly correlates with achievable current densities. By heterologous expression of a synthetic flavin biosynthesis pathway from *Bacillus subtilis* in *S. oneidensis* the secreted flavin concentration could be increased ~26 times, which increased power output as well as inward current of the organism in BES (Yang et al., 2015). Another possibility could be the secretion of whole enzymes that facilitate electron flow from the electrode surface toward organisms. In a very recent study Deutzmann et al. (2015) were the first to report evidence of this mechanism of direct EET by cell-derived free enzymes in a cathodic BES using methanogens. They could show that small surface associated enzymes such as hydrogenases and fumarate reductase were released from *Methanococcus maripaludis* cells and accumulated at the cathode surface where they catalyzed the formation of small mediating molecules such as hydrogen or formate which in turn were immediately taken up by the cells.

The location of the target site of electron transfer decides if an organism might be able to perform direct EET or if a (endogenous or exogenous) mediator is required to transport the charge across the cellular membrane. Additionally each organism can be allocated with a specific redox-potential-window in which its electron transport chains are able to operate, which is also visualized in Figure3. It becomes obvious that compounds of facultative aerobe organisms such as *E. coli* and *S. oneidensis* are represented over a wide range of potentials. This visualizes the high flexibility of both organisms to adapt their electron transport chains to multiple electron donors and acceptors.*G. sulfurreducens* even though a strict anaerobe still covers an impressive range of redox potentials with its many cytochromes and the ability to use soluble metals such as iron and manganese as final electron acceptors. The rnf-complex containing acetogens *C. ljungdahlii* and *A. woodii*, however, are only able to transfer electrons in a limited potential window as they do not contain any cytochromes and are restricted to only one way of respiration at relatively low potentials.

Transferring Electrogenic Capabilities between Bacteria

To transfer the ability of one organism that is known to interact with an electrode in a specific way to another organism, synthetic biology tools can be used. It could be shown that expressing enzymes of the Mtr pathway of *S. oneidensis* inside *E. coli* increased current production of the optimized strain (Pitts et al., 2003; Jensen et al., 2010). In 2008 researchers could show that the expression of one single enzyme from *Shewanella's* Mtr-pathway is enough to transform *E. coli* into a metal respiring bacterium (Gescher et al., 2008). Expression of the cytoplasmic-membrane tetraheme c-type cytochrome CymA enabled the mutant strain to reduce Fe(III) and sustain growth while the native strain lacked this ability. This was independent of the presence of periplasmic or outer-membrane cytochromes for electron transfer. However, it could also be shown that activities of CymA for iron reduction were much lower in complete cells, indicating that diffusion limitations of solid state electron acceptors also play a significant role in this electron transport chain (Gescher et al., 2008). A very important follow up study was able to find evidence for a connection of this transformed electron transport to intracellular carbon flow in *E. coli* (TerAvest et al., 2014). Expression of CymA and Mtr cytochromes from *Shewanella oneidensis* in *E. coli* resulted in a strain that coupled current production to a shift in its metabolic fluxes towards more oxidized products (TerAvest et al., 2014). The possibility to design a target electron flow via artificial complexes across the outer and inner membranes and their connection to cellular redox balance and carbon flow promises many possibilities to design optimized electro active organisms. The above cited example also shows that co-expression of multiple linked enzymes can greatly enhance the electron transfer rates (TerAvest et al., 2014). Therefore the specific electron transport chains of the target organism and their possible connection to the enzymes to be introduced need to be considered when designing electro active bacteria.

Targeting Electron Transport Impacts the Cellular Energy Metabolism

The achievable energy yield of an electron transport chain is dependent on the difference in electrical potential between electron donor and acceptor. Therefore organisms that are able to respire in multiple ways will always choose available acceptors with the biggest potential difference to the donor (e.g., *E. coli* O_2 > NO_3^-> fumarate). To compare energy efficiencies of different electron transport chains it is referred to the ratio of phosphate to oxygen (P/O quotient), which describes the ratio of mol ATP produced per mol oxygen reduced. The P/O ratio depends on the amount of ions that are transported across the membrane during the corresponding electron-transport chain and are available for ATP-synthesis via gradient-driven ATPase. It is also influenced by the efficiency of the specific ATPase as the amount of ions that are needed for synthesis of one mol ATP differ between organisms (Møller et al., 1996). For example the aerobic electron transport chain of *E. coli* transports up to eight protons across the membrane with NADH as electron donor (2 e^-) and oxygen as final acceptor (see Figure 4; Unden and Bongaerts, 1997). With an ATPase that requires three protons for the conversion of one mol ADP to ATP a P/O ratio of ~2.7 is calculated (Noguchi et al., 2004).

The ATP yield is an important factor for bacterial growth and can also significantly influence production if the metabolic pathway towards the target compound requires energy. Therefore we present similar to the P/O ratio a hypothetical P/2e^- ratio that reflects the ATP yield per pair of electrons transferred between the organism and an electrode in a BES. Some examples are presented in Table 2. The electrode can either function as electron donor (cathode) or as final electron acceptor (anode). Since the exact sites of electrode-interaction with the compounds of each electron transport chain remain unknown we assume different scenarios. In column 2 of Table 2the assumed site of electrode–bacteria interaction is given together with the corresponding hypothetical reaction. The P/2e$^-$ ratio is calculated based on the resulting amount of protons carried across the membrane, which are assumed to be available to drive ATP synthesis via ATPase. An example is explained in detail in the following paragraph.

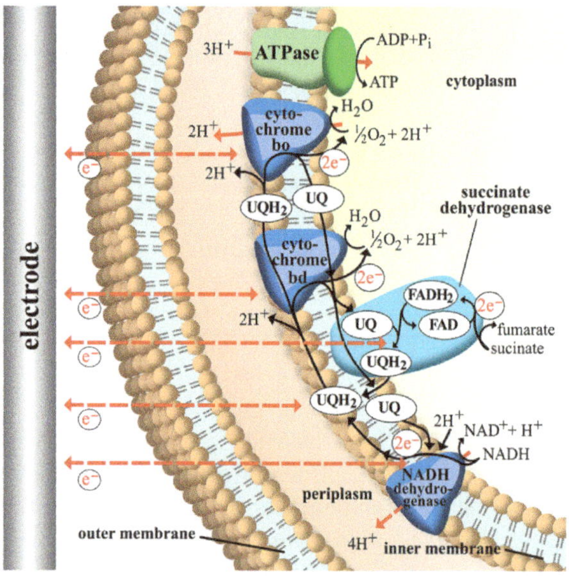

Figure 4: Schematic image of electron transport chains in *Escherichia coli*.NADH as electron donor via NADH dehydrogenase (NuoA-N), ubiquinone pool (UQ), succinate-dehydrogenase, and cytochromes *bd* (CydAB) and *bo* (CyoABCD). ATP is generated via F_1F_0-ATPase (3H$^+$/ATP) from the membrane proton gradient. Possible sites of interaction with an electrode are indicated with dashed arrows. For detailed discussion and references refer to main text.

Table 2: Theoretical P/2e$^-$ **ratios for extracellular electron transfer between electrodes and different sites in the electron transport chains of** *E. coli*, *M. thermoacetica* **and** *C. ljungdahlii*

Organism	Site of electron transfer	P/2e$^-$
Anodic electron transport		
E. coli[a]	Cytochrome *bo*	2.7
	Cytochrome *bd*	2
	Ubiquinone-pool ($UQH_2 \rightarrow 2e^- + UQ + 2H^+_{periplasm}$)	2
	FAD reduction ($FADH_2 \rightarrow 2e^- + FAD^+ + 2H^+_{cytoplasm}$)	0.6
	NADH oxidation ($NADH \rightarrow 2e^- + NAD^+ + 2H^+_{cytoplasm}$)	0
Cathodic electron transfer		
E. coli[a]	NADH reduction ($2e^- + NAD^+ + 2H^+_{cytoplasm} \rightarrow NADH$)	2.7
	FAD reduction ($2e^- + FAD^+ + 2H^+_{cytoplasm} \rightarrow FADH_2$)	1.3
	Ubiquinone reduction ($2e^- + UQ + 2H^+_{cytoplasm} \rightarrow UQH_2$)	1.3
	Ubiquinone reduction ($2e^- + UQ + 2H^+_{periplasm} \rightarrow UQH_2$)	0.6
M. thermoacetica[b,c]	Ferredoxin reduction ($2e^- + Fd_{ox} \rightarrow Fd_{red}$)	0.75
	Hydrogen evolution ($2e^- + 2H^+_{cytoplasm} \rightarrow H_2$)	0.375
	NADH reduction ($2e^- + NAD^+ \rightarrow NADH$)	0
C. ljungdahlii[b]	Ferredoxin reduction ($2e^- + Fd_{ox} \rightarrow Fd_{red}$)	0.5
	Hydrogen evolution ($2e^- + 2H^+_{cytoplasm} \rightarrow H_2$)	0.25
	NADH reduction ($2e^- + NAD^+ + 2H^+_{cytoplasm} \rightarrow NADH$)	0

Figure 4 shows a simplified image of electron transport chains in *E. coli* including possible sites of EET (dashed red arrows). As discussed above it could be shown that cytochromes of *G. sulfurreducens* are able to use electrodes as final electron acceptor. If we assume the cytochromes of *E. coli* are also possible sites of EET we observe that the ATP yield of the electron transport chain depends on the specific cytochrome that is performing the electron transfer to the electrode. In case of cytochrome *bo* transferring electrons to an anode as final electron acceptor 8 H$^+$ are transported across the membrane (4 H$^+$ from NADH dehydrogenase and 4 H$^+$ from UQH_2 and cytochrome *bo*). With an ATP synthase that requires ~3 H$^+$ per ATP (Noguchi et al., 2004) this leads to a P/2e$^-$ ratio of 2.7. Running the electron transport chain with cytochrome *bd* as the final oxidase, however, leads to a P/2e$^-$ ratio of 2 as fewer protons are transferred across the membrane (6 H$^+$ in total: 4H$^+$from NADH dehydrogenase and 2 H$^+$ from cytochrome *bd*). By including soluble redox carriers such as ubiquinones, FAD and NADH to the scenario we can address any site of *E. coli's* electron transport chain. The observation from this theoretical exercise is to internalize that the achievable ATP yield depends on the actual site of EET. If for example electrons are drawn directly from the cellular NADH pool of *E. coli* no energy gain via chemiosmotic coupling is possible as no proton gradient is established (see Figure 4; Table 2).

This becomes especially crucial in regards to cathodic electron transfer where processes often aim on microbial metabolism with electrons from electrodes as sole energy and redox source (Rabaey and Rozendal, 2010; Nevin et al., 2011). As discussed in Section "Carbon Respiration of Acetogenic Bacteria" the metabolism of acetogens for example relies on ATP synthesis via membrane-based electron transport complexes. Now we can pick again an example strain and theoretically discuss metabolic impact of EET at different sites. *C. ljungdahlii* does not transfer electrons via cytochromes but was shown to interact with electrodes so we can assume EET via mediators or unknown membrane bound compounds towards its cellular redox compounds. In case electrons from a cathode would lead to hydrogen formation the organism could produce 2 mol of reduced Fd per 4 mol of H_2 via electron bifurcation in the Hyd-complex (see Figure 2; Buckel and Thauer, 2013). The exergonic reaction of reduced Fd and NAD$^+$ is used for proton translocation in the rnf complex (2 H$^+$/Fd_{red}) which in turn enables energy conservation via the membrane bound ATPase. With recent estimates of four protons per mol of ATP a P/2e$^-$ ratio of 0.25 is calculated (Schuchmann and Müller, 2014). If it is possible to transfer electrons from a cathode directly to Fd the ratio increases to 0.5 mol ATP per pair of electrons transferred. An electron input toward the NADH pool of the bacterium, however, would not deliver enough energy to establish a proton motive force and therefore ATP generation (see Table 2).

Remembering Figure 3 one can draw the connection between required redox potential and target reaction in a specific electron transport chain of an organism to optimize the conditions for a bio electrochemical process. For a cathodic process of *C. ljungdahlii* for example the applied electrode potential needs to be at least low enough to allow hydrogen formation to enable ATP formation and therefore bacterial growth.

CONCLUSION

The remaining question after the presented analysis of microbial electron transport chains is: can we identify an ultimate organism for microbial electrochemical techniques? Requirements for an advantageous organism would be first of all high electro-activity. But depending on the target application different other features can be of crucial importance. Microbial electrosynthesis for example requires a strain that produces industrial relevant products preferably combined with the ability to utilize a variety of cheap substrates. Metabolic engineering tools should be available to optimize production and perform advanced strain designing. For application in bioremediation processes the ability of breaking down a wide range of organic contaminates has priority together of course with efficiency of EET. Other influencing factors are general characteristics that simplify process design such as high oxygen tolerance for anaerobes, no high risk organism, no complicated fermentation conditions such as expensive media supplements, pressurized reactor etc.

The diversity of applications for BES makes it impossible to identify one organism that features all required properties. Still for each application a specific species might outperform others. The highly flexible electron transport chains of metal respiring bacteria such as *Geobacter* and *Shewanella*make them excellent current producers over a wide range of potentials. And acetogens seem to be a very promising group of target organisms for microbial electrosynthesis with their ability to use CO_2 as sole carbon source. But our study also shows that the complexity of microbial electron transport possibilities bears many challenges for bioelectrochemical techniques. For the successful design of an electrically enhanced bio process the specific electron transport properties of the involved species needs to be considered. Environmental conditions such as applied potentials should be adjusted according to specific target sites for EET. Since these are unknown in most cases an analysis as presented in this manuscript will help to identify best- and worst-case scenarios for microbe-electrode interactions and identification of optimized windows for process parameters. Especially when looking at CO_2 as a feedstock, the available energy gain through EET will limit the feasibility of many organisms and constrain the list of feasible products. There is currently a lot of interest in microbial electrochemical technologies; however, we believe that without deeper understanding of the underlying electron transfer process development remains a trial and error exercise.

AUTHOR CONTRIBUTIONS

FK, IV, and JK designed the study. FK and IV collected the data and drafted the manuscript. FK performed the calculations. All authors edited the manuscript and approved the final version.

ACKNOWLEDGMENTS

The authors want to thank the University of Queensland and the Australian Research Council (LP120100517) for financial support. Special thanks go to Helena Reiswich for designing the layout of Figures 1, 2, and 4.

REFERENCES

1. Anraku, Y. (1988). Bacterial electron transport chains. *Annu. Rev. Biochem.* 57, 101–132. doi: 10.1146/annurev.bi.57.070188.000533
2. Atkinson, S. J., Mowat, C. G., Reid, G. A., and Chapman, S. K. (2007). An octaheme c-type cytochrome from *Shewanella oneidensis* can reduce nitrite and hydroxylamine. *FEBS Lett.* 581, 3805–3808. doi: 10.1016/j.febslet.2007.07.005
3. Baltch, A. L., and Smith, R. P. (eds). (1994). *Pseudomonas aeruginosa: Infections and Treatment.* New York, NY: Marcel Dekker, Inc.
4. Banerjee, A., Leang, C., Ueki, T., Nevin, K. P., and Lovley, D. R. (2014). Lactose-inducible system for metabolic engineering of*Clostridium ljungdahlii*. *Appl. Environ. Microbiol.* 80, 2410–2416. doi: 10.1128/AEM.03666-13
5. Berlyn, M. K. (1998). Linkage map of *Escherichia coli* K-10: the traditional map. *Microbiol. Mol. Biol. Rev.* 62, 814–984.
6. Bertini, I., Cavallaro, G., and Rosato, A. (2006). Cytochrome c: occurrence and functions. *Chem. Rev.* 106, 90–115. doi: 10.1021/cr050241v
7. Biegel, E., Schmidt, S., González, J., and Müller, V. (2011). Biochemistry, evolution and physiological function of the Rnf complex, a novel ion-motive electron transport complex in prokaryotes. *Cell. Mol. Life Sci.* 68, 613–634. doi: 10.1007/s00018-010-0555-8
8. Bond, D. R., and Lovley, D. R. (2003). Electricity production by *Geobacter sulfurreducens* attached to electrodes. *Appl. Environ. Microbiol.* 69, 1548–1555. doi: 10.1128/aem.69.3.1548-1555.2003
9. Boon, N., De Maeyer, K., Höfte, M., Rabaey, K., and Verstraete, W. (2008). Use of *Pseudomonas* species producing phenazine-based metabolites in the anodes of microbial fuel cells to improve electricity generation. *Appl. Microbiol. Biotechnol.* 80, 985–993. doi: 10.1007/s00253-008-1619-7
10. Bott, M., and Niebisch, A. (2003). The respiratory chain of *Corynebacterium glutamicum*. *J. Biotechnol.* 104, 129–153. doi: 10.1016/S0168-1656(03)00144-5
11. Breuer, M., Rosso, K. M., Blumberger, J., and Butt, J. N. (2015). Multi-haem cytochromes in *Shewanella oneidensis* MR-1: structures, functions and opportunities. *J. R. Soc. Interface* 12, 20141117. doi: 10.1098/rsif.2014.1117
12. Buckel, W., and Thauer, R. K. (2013). Energy conservation via electron bifurcating ferredoxin reduction and proton/Na+ translocating ferredoxin oxidation. *Biochim. Biophys. Acta* 1827, 94–113. doi: 10.1016/j.bbabio.2012.07.002
13. Carlson, H. K., Iavarone, A. T., Gorur, A., Yeo, B. S., Tran, R., Melnyk, R. A., et al. (2012). Surface multiheme c-type cytochromes from *Thermincola potens* and implications for respiratory metal reduction by Gram-positive bacteria. *Proc. Natl. Acad. Sci. U.S.A.* 109, 1702–1707. doi: 10.1073/pnas.1112905109
14. Choi, O., Um, Y., and Sang, B.-I. (2012).

Butyrate production enhancement by *Clostridium tyrobutyricum* using electron mediators and a cathodic electron donor. *Biotechnol. Bioeng.* 109, 2494–2502. doi: 10.1002/bit.24520

15. Coursolle, D., and Gralnick, J. A. (2010). Modularity of the Mtr respiratory pathway of *Shewanella oneidensis* strain MR-1. *Mol. Microbiol.* 77, 995–1008. doi: 10.1111/j.1365-2958.2010.07266.x
16. Das, A., and Ljungdahl, L. G. (2003). "Electron-transport system in acetogens," in *Biochemistry and Physiology of Anaerobic Bacteria*, eds L. G. Ljungdahl, M. W. Adams, L. Barton, J. G. Ferry, and M. K. Johnson (New York, NY: Springer-Verlag), 191–204. doi: 10.1007/0-387-22731-8_14
17. Deller, S., Macheroux, P., and Sollner, S. (2008). Flavin-dependent quinone reductases. *Cell. Mol. Life Sci.* 65, 141–160. doi: 10.1007/s00018-007-7300-y
18. Deutzmann, J. S., Sahin, M., and Spormann, A. M. (2015). Extracellular enzymes facilitate electron uptake in biocorrosion and bioelectrosynthesis. *MBio* 6:e00496-15. doi: 10.1128/mBio.00496-15
19. Diekert, G., and Wohlfarth, G. (1994). Metabolism of homoacetogens. *Antonie Van Leeuwenhoek* 66, 209–221. doi: 10.1007/BF00871640
20. Dolin, M. (2012). Survey of microbial electron transport mechanisms. *Bacteria* 2:319.
21. Dos Santos, J.-P., Iobbi-Nivol, C., Couillault, C., Giordano, G., and Méjean, V. (1998). Molecular analysis of the trimethylamine N-oxide (TMAO) reductase respiratory system from a *Shewanella* species. *J. Mol. Biol.* 284, 421–433. doi: 10.1006/jmbi.1998.2155
22. Drake, H. L., Gößner, A. S., and Daniel, S. L. (2008). Old acetogens, new light. *Ann. N. Y. Acad. Sci.* 1125, 100–128. doi: 10.1196/annals.1419.016
23. Dumas, C., Basseguy, R., and Bergel, A. (2008). Microbial electrocatalysis with *Geobacter sulfurreducens* biofilm on stainless steel cathodes. *Electrochim. Acta* 53, 2494–2500. doi: 10.1016/j.electacta.2007.10.018
24. Emde, R., and Schink, B. (1990). Enhanced propionate formation by *Propionibacterium freudenreichii* subsp. freudenreichii in a three-electrode amperometric culture system. *Appl. Environ. Microbiol.* 56, 2771–2776.
25. Fast, A. G., and Papoutsakis, E. T. (2012). Stoichiometric and energetic analyses of non-photosynthetic CO2-fixation pathways to support synthetic biology strategies for production of fuels and chemicals. *Curr. Opin. Chem. Eng.* 1, 380–395. doi: 10.1016/j.coche.2012.07.005
26. Firer-Sherwood, M., Pulcu, G. S., and Elliott, S. J. (2008). Electrochemical interrogations of the Mtr cytochromes from*Shewanella*: opening a potential window. *J. Biol. Inorg. Chem.* 13, 849–854. doi: 10.1007/s00775-008-0398-z
27. Fonseca, B. M., Paquete, C. M., Neto, S. E., Pacheco, I., Soares, C. M., and Louro, R. O. (2013). Mind the gap: cytochrome interactions reveal electron pathways across the periplasm of *Shewanella oneidensis* MR-1. *Biochem. J.* 449, 101–108. doi: 10.1042/BJ20121467
28. Franks, A. (2012). What's current with electric microbes. *J. Bacteriol. Parasitol.* 3:e116. doi: 10.4172/2155-9597.1000e116
29. Franks, A. E., and Nevin, K. P. (2010). Microbial fuel cells, a current review. *Energies* 3, 899–919. doi: 10.3390/en3050899
30. Gao, H., Yang, Z. K., Barua, S., Reed, S. B., Romine, M. F., Nealson, K. H., et al. (2009). Reduction of nitrate in *Shewanella oneidensis* depends on atypical NAP and NRF systems with NapB as a preferred electron transport protein from CymA to NapA.*ISME J.* 3, 966–976. doi: 10.1038/ismej.2009.40
31. Gescher, J. S., Cordova, C. D., and Spormann, A. M. (2008). Dissimilatory iron reduction in *Escherichia coli*: identification of CymA of *Shewanella oneidensis* and NapC of *E. coli* as ferric reductases. *Mol. Microbiol.* 68, 706–719. doi: 10.1111/j.1365-2958.2008.06183.x
32. Gralnick, J. A., Vali, H., Lies, D. P., and Newman, D. K. (2006). Extracellular respiration of dimethyl sulfoxide by *Shewanella oneidensis* strain MR-1. *Proc. Natl. Acad. Sci. U.S.A.* 103, 4669–4674. doi: 10.1073/pnas.0505959103
33. Gregory, K. B., Bond, D. R., and Lovley, D. R. (2004). Graphite electrodes as electron donors for anaerobic respiration.*Environ. Microbiol.* 6, 596–604. doi: 10.1111/j.1462-2920.2004.00593.x
34. Haddock, B. A., and Schairer, H. U. (1973). Electron-transport chains of *Escherichia coli*. *Eur. J. Biochem.* 35, 34–45. doi: 10.1111/j.1432-1033.1973.tb02806.x
35. Hannemann, F., Bichet, A., Ewen, K. M., and Bernhardt, R. (2007). Cytochrome P450 systems—biological variations of electron transport chains. *Biochim. Biophys. Acta* 1770, 330–344. doi: 10.1016/j.bbagen.2006.07.017
36. Harnisch, F., Rosa, L. F., Kracke, F., Virdis, B., and Krömer, J. O. (2014). Electrifying white biotechnology: engineering and economic potential of electricity-driven bio-production. *ChemSusChem* 8, 758–766. doi: 10.1002/cssc.201402736
37. Hartshorne, R. S., Reardon, C. L., Ross, D., Nuester, J., Clarke, T. A., Gates, A. J., et al. (2009). Characterization of an electron conduit between bacteria and the extracellular environment. *Proc. Natl. Acad. Sci. U.S.A.* 106, 22169–22174. doi: 10.1073/pnas.0900086106
38. Hassett, D. J., Cuppoletti, J., Trapnell, B., Lymar, S. V., Rowe, J. J., Yoon, S. S., et al. (2002). Anaerobic metabolism and quorum sensing by *Pseudomonas aeruginosa* biofilms in chronically infected cystic fibrosis airways: rethinking antibiotic treatment strategies and drug targets. *Adv. Drug Del. Rev.* 54, 1425–1443. doi: 10.1016/S0169-409X(02)00152-7
39. Heidelberg, J. F., Paulsen, I. T., Nelson, K. E., Gaidos, E. J., Nelson, W. C., Read, T. D., et al. (2002). Genome sequence of the dissimilatory metal ion–reducing bacterium *Shewanella oneidensis*. *Nat. Biotechnol.* 20, 1118–1123. doi: 10.1038/nbt749
40. Hernandez, M. E., Kappler, A., and Newman, D. K. (2004). Phenazines and other redox-active antibiotics promote microbial mineral reduction. *Appl. Environ. Microbiol.* 70, 921–928. doi: 10.1128/AEM.70.2.921-928.2004

41. Hernandez, M., and Newman, D. (2001). Extracellular electron transfer. *Cell. Mol. Life Sci.* 58, 1562–1571. doi: 10.1007/PL00000796

42. Herrmann, G., Jayamani, E., Mai, G., and Buckel, W. (2008). Energy conservation via electron-transferring flavoprotein in anaerobic bacteria. *J. Bacteriol.* 190, 784–791. doi: 10.1128/JB.01422-07

43. Hess, V., Schuchmann, K., and Müller, V. (2013). The ferredoxin: NAD+ oxidoreductase (Rnf) from the acetogen*Acetobacterium woodii* requires Na+ and is reversibly coupled to the membrane potential. *J. Biol. Chem.* 288, 31496–31502. doi: 10.1074/jbc. M113.510255

44. Holmes, D. E., Chaudhuri, S. K., Nevin, K. P., Mehta, T., Methé, B. A., Liu, A., et al. (2006). Microarray and genetic analysis of electron transfer to electrodes in *Geobacter sulfurreducens*. *Environ. Microbiol.* 8, 1805–1815. doi: 10.1111/j.1462-2920.2006.01065.x

45. Huang, H., Wang, S., Moll, J., and Thauer, R. K. (2012). Electron bifurcation involved in the energy metabolism of the acetogenic bacterium *Moorella thermoacetica* growing on glucose or H_2 plus CO_2. *J. Bacteriol.* 194, 3689–3699. doi: 10.1128/JB.00385-12

46. Ingledew, W. J. (1983). The electron transport chain of *Escherichia coli* grown anaerobically with fumarate as terminal electron acceptor: an electron paramagnetic resonance study. *J. Gen. Microbiol.* 129, 1651–1659. doi: 10.1099/00221287-129-6-1651

47. Inoue, K., Qian, X., Morgado, L., Kim, B.-C., Mester, T., Izallalen, M., et al. (2010). Purification and characterization of OmcZ, an outer-surface, octaheme c-type cytochrome essential for optimal current production by *Geobacter sulfurreducens*. *Appl. Environ. Microbiol.* 76, 3999–4007. doi: 10.1128/AEM.00027-10

48. Janicek, A., Fan, Y., and Liu, H. (2014). Design of microbial fuel cells for practical application: a review and analysis of scale-up studies. *Biofuels* 5, 79–92. doi: 10.4155/bfs.13.69

49. Jensen, H. M., Albers, A. E., Malley, K. R., Londer, Y. Y., Cohen, B. E., Helms, B. A., et al. (2010). Engineering of a synthetic electron conduit in living cells. *Proc. Natl. Acad. Sci. U.S.A.* 107, 19213–19218. doi: 10.1073/pnas.1009645107

50. Jo, J., Price-Whelan, A., and Dietrich, L. E. (2014). An aerobic exercise: defining the roles of *Pseudomonas aeruginosa* terminal oxidases. *J. Bacteriol.* 196, 4203–4205. doi: 10.1128/JB.02336-14

51. Kim, B.-C., Postier, B. L., Didonato, R. J., Chaudhuri, S. K., Nevin, K. P., and Lovley, D. R. (2008). Insights into genes involved in electricity generation in *Geobacter sulfurreducens* via whole genome microarray analysis of the OmcF-deficient mutant.*Bioelectrochemistry* 73, 70–75. doi: 10.1016/j.bioelechem.2008.04.023

52. Kim, B.-C., Qian, X., Leang, C., Coppi, M. V., and Lovley, D. R. (2006). Two putative c-type multiheme cytochromes required for the expression of OmcB, an outer membrane protein essential for optimal Fe (III) reduction in *Geobacter sulfurreducens*. *J. Bacteriol.* 188, 3138–3142. doi: 10.1128/JB.188.8.3138-3142.2006

53. Kim, T. S., and Kim, B. H. (1988). Electron flow shift in *Clostridium acetobutylicum* fermentation by electrochemically introduced reducing equivalent. *Biotechnol. Lett.* 10, 123–128. doi: 10.1007/BF01024638

54. Köpke, M., Held, C., Hujer, S., Liesegang, H., Wiezer, A., Wollherr, A., et al. (2010). *Clostridium ljungdahlii* represents a microbial production platform based on syngas. *Proc. Natl. Acad. Sci. U.S.A.* 107, 13087–13092. doi: 10.1073/pnas.1004716107

55. Kracke, F., and Krömer, J. O. (2014). Identifying target processes for microbial electrosynthesis by elementary mode analysis.*BMC Bioinformatics* 15:6590. doi: 10.1186/s12859-014-0410-2

56. Leang, C., and Lovley, D. R. (2005). Regulation of two highly similar genes, omcB and omcC, in a 10 kb chromosomal duplication in *Geobacter sulfurreducens*. *Microbiology* 151, 1761–1767. doi: 10.1099/mic.0.27870-0

57. Lee, J., Jang, Y.-S., Choi, S. J., Im, J. A., Song, H., Cho, J. H., et al. (2012). Metabolic engineering of *Clostridium acetobutylicum*ATCC 824 for isopropanol-butanol-ethanol fermentation. *Appl. Environ. Microbiol.* 78, 1416–1423. doi: 10.1128/AEM.06382-11

58. Levar, C., Rollefson, J., and Bond, D. (2012). "Energetic and molecular constraints on the mechanism of environmental Fe (III) reduction by *Geobacter*," in *Microbial Metal Respiration*, eds J. Gescher and A. Kappler (Berlin: Springer), 29–48.

59. Lloyd, J., Leang, C., Hodges, M. A., Coppi, M., Cuifo, S., Methe, B., et al. (2003). Biochemical and genetic characterization of PpcA, a periplasmic c-type cytochrome in *Geobacter sulfurreducens*. *Biochem. J.* 369, 153–161. doi: 10.1042/BJ20020597

60. Logan, B. E. (2009). Exoelectrogenic bacteria that power microbial fuel cells. *Nat. Rev. Microbiol.* 7, 375–381. doi: 10.1038/nrmicro2113

61. Logan, B. E., and Rabaey, K. (2012). Conversion of wastes into bioelectricity and chemicals by using microbial electrochemical technologies. *Science* 337, 686–690. doi: 10.1126/science.1217412

62. Lovley, D. R. (2012). Electromicrobiology. *Annu. Rev. Microbiol.* 66, 391–409. doi: 10.1146/annurev-micro-092611-150104

63. Lovley, D. R., Holmes, D. E., and Nevin, K. P. (2004). Dissimilatory fe (iii) and mn (iv) reduction. *Adv. Microb. Physiol.* 49, 219–286. doi: 10.1016/s0065-2911(04)49005-5

64. Lovley, D. R., and Nevin, K. P. (2013). Electrobiocommodities: powering microbial production of fuels and commodity chemicals from carbon dioxide with electricity. *Curr. Opin. Biotechnol.* 24, 385–390. doi: 10.1016/j.copbio.2013.02.012

65. Lower, B. H., Hochella, M. F., and Lower, S. K. (2005). Putative mineral-specific proteins synthesized by a metal reducing bacterium. *Am. J. Sci.* 305, 687–710. doi: 10.2475/ajs.305.6-8.687

66. Lütke-Eversloh, T., and Bahl, H. (2011). Metabolic

engineering of *Clostridium acetobutylicum*: recent advances to improve butanol production. *Curr. Opin. Biotechnol.* 22, 634–647. doi: 10.1016/j.copbio.2011.01.011

67. Maier, T. M., Myers, J. M., and Myers, C. R. (2003). Identification of the gene encoding the sole physiological fumarate reductase in *Shewanella oneidensis* MR-1. *J. Basic Microbiol.* 43, 312–327. doi: 10.1002/jobm.200390034

68. Malvankar, N. S., Tuominen, M. T., and Lovley, D. R. (2012). Comment on "*On Electrical Conductivity of Microbial Nanowires and Biofilms*" by S. M. Strycharz-Glaven, R. M. Snider, A. Guiseppi-Elie, and L. M. Tender, *Energy Environ. Sci.* 2011, 4, 4366.*Energy Environ. Sci.* 5, 6247–6249. doi: 10.1039/c2ee02613a

69. Malvankar, N. S., Vargas, M., Nevin, K. P., Franks, A. E., Leang, C., Kim, B.-C., et al. (2011). Tunable metallic-like conductivity in microbial nanowire networks. *Nat. Nanotechnol.* 6, 573–579. doi: 10.1038/nnano.2011.119

70. Malvankar, N. S., Vargas, M., Nevin, K., Tremblay, P.-L., Evans-Lutterodt, K., Nykypanchuk, D., et al. (2015). Structural basis for metallic-like conductivity in microbial nanowires. *MBio* 6:e00084-15. doi: 10.1128/mBio.00084-15

71. Marshall, C. W., and May, H. D. (2009). Electrochemical evidence of direct electrode reduction by a thermophilic gram-positive bacterium, *Thermincola ferriacetica*. *Energy Environ. Sci.* 2, 699–705. doi: 10.1039/b823237g

72. Marshall, C. W., Ross, D. E., Fichot, E. B., Norman, R. S., and May, H. D. (2012). Electrosynthesis of commodity chemicals by an autotrophic microbial community. *Appl. Environ. Microbiol.* 78, 8412–8420. doi: 10.1128/AEM.02401-12

73. Marsili, E., Baron, D. B., Shikhare, I. D., Coursolle, D., Gralnick, J. A., and Bond, D. R. (2008a). *Shewanella* secretes flavins that mediate extracellular electron transfer. *Proc. Natl. Acad. Sci. U.S.A.* 105, 3968–3973. doi: 10.1073/pnas.0710525105

74. Marsili, E., Rollefson, J. B., Baron, D. B., Hozalski, R. M., and Bond, D. R. (2008b). Microbial biofilm voltammetry: direct electrochemical characterization of catalytic electrode-attached biofilms. *Appl. Environ. Microbiol.* 74, 7329–7337. doi: 10.1128/AEM.00177-08

75. Mehta, T., Coppi, M. V., Childers, S. E., and Lovley, D. R. (2005). Outer membrane c-type cytochromes required for Fe (III) and Mn (IV) oxide reduction in *Geobacter sulfurreducens*. *Appl. Environ. Microbiol.* 71, 8634–8641. doi: 10.1128/AEM.71.12.8634-8641.2005

76. Methe, B., Nelson, K. E., Eisen, J., Paulsen, I., Nelson, W., Heidelberg, J., et al. (2003). Genome of *Geobacter sulfurreducens*: metal reduction in subsurface environments. *Science* 302, 1967–1969. doi: 10.1126/science.1088727

77. Møller, J. V., Juul, B., and Le Maire, M. (1996). Structural organization, ion transport, and energy transduction of P-type ATPases. *Biochim. Biophys. Acta Rev. Biomembr.* 1286, 1–51. doi: 10.1016/0304-4157(95)00017-8

78. Morgado, L., Bruix, M., Pessanha, M., Londer, Y. Y., and Salgueiro, C. A. (2010). Thermodynamic characterization of a triheme cytochrome family from *Geobacter sulfurreducens* reveals mechanistic and functional diversity. *Biophys. J.* 99, 293–301. doi: 10.1016/j.bpj.2010.04.017

79. Müller, V. (2003). Energy conservation in acetogenic bacteria. *Appl. Environ. Microbiol.* 69, 6345–6353. doi: 10.1128/AEM.69.11.6345-6353.2003

80. Müller, V., and Frerichs, J. (2013). "Acetogenic bacteria," in *Encyclopedia of Life Sciences* (Chichester: John Wiley & Sons), 1–9. doi: 10.1002/9780470015902.a0020086.pub2

81. Müller, V., Imkamp, F., Biegel, E., Schmidt, S., and Dilling, S. (2008). Discovery of a Ferredoxin: NAD+-Oxidoreductase (Rnf) in*Acetobacterium woodii*. *Ann. N. Y. Acad. Sci.* 1125, 137–146. doi: 10.1196/annals.1419.011

82. Myers, J. M., and Myers, C. R. (2000). Role of the tetraheme cytochrome CymA in anaerobic electron transport in cells of*Shewanella putrefaciens* MR-1 with normal levels of menaquinone. *J. Bacteriol.* 182, 67–75. doi: 10.1128/JB.182.1.67-75.2000

83. Nevin, K. P., Hensley, S. A., Franks, A. E., Summers, Z. M., Ou, J., Woodard, T. L., et al. (2011). Electrosynthesis of organic compounds from carbon dioxide is catalyzed by a diversity of acetogenic microorganisms. *Appl. Environ. Microbiol.* 77, 2882–2886. doi: 10.1128/em.02642-10

84. Nevin, K. P., Kim, B.-C., Glaven, R. H., Johnson, J. P., Woodard, T. L., Methé, B. A., et al. (2009). Anode biofilm transcriptomics reveals outer surface components essential for high density current production in *Geobacter sulfurreducens*fuel cells. *PLoS ONE* 4:e5628. doi: 10.1371/journal.pone.0005628

85. Nevin, K. P., Woodard, T. L., Franks, A. E., Summers, Z. M., and Lovley, D. R. (2010). Microbial electrosynthesis: feeding microbes electricity to convert carbon dioxide and water to multicarbon extracellular organic compounds. *MBio* 1, e00103–e00110. doi: 10.1128/mBio.00103-10

86. Nishimura, T., Vertès, A. A., Shinoda, Y., Inui, M., and Yukawa, H. (2007). Anaerobic growth of*Corynebacterium glutamicum*using nitrate as a terminal electron acceptor. *Appl. Microbiol. Biotechnol.* 75, 889–897. doi: 10.1007/s00253-007-0879-y

87. Noguchi, Y., Nakai, Y., Shimba, N., Toyosaki, H., Kawahara, Y., Sugimoto, S., et al. (2004). The energetic conversion competence of *Escherichia coli* during aerobic respiration studied by 31P NMR using a circulating fermentation system. *J. Biochem.* 136, 509–515. doi: 10.1093/jb/mvh147

88. Okamoto, A., Hashimoto, K., Nealson, K. H., and Nakamura, R. (2013). Rate enhancement of bacterial extracellular electron transport involves bound flavin semiquinones. *Proc. Natl. Acad. Sci. U.S.A.* 110, 7856–7861. doi: 10.1073/pnas.1220823110

89. Okamoto, A., Saito, K., Inoue, K., Nealson, K. H., Hashimoto, K., and Nakamura, R. (2014). Uptake of

self-secreted flavins as bound cofactors for extracellular electron transfer in *Geobacter* species. *Energy Environ. Sci.* 7, 1357–1361. doi: 10.1039/c3ee43674h

90. CrossRef Full Text | Google Scholar

91. Parameswaran, P., Bry, T., Popat, S. C., Lusk, B. G., Rittmann, B. E., and Torres, C. I. (2013). Kinetic, electrochemical, and microscopic characterization of the thermophilic, anode-respiring bacterium *Thermincola ferriacetica*. *Environ. Sci. Technol.*47, 4934–4940. doi: 10.1021/es400321c

92. Patil, S. A., Haegerhaell, C., and Gortin, L. (2012). Electron transfer mechanisms between microorganisms and electrodes in bioelectrochemical systems. *Bioanal. Rev.* 4, 159–192. doi: 10.1007/s12566-012-0033-x

93. Pierce, E., Xie, G., Barabote, R. D., Saunders, E., Han, C. S., Detter, J. C., et al. (2008). The complete genome sequence of*Moorella thermoacetica* (f. *Clostridium thermoaceticum*). *Environ. Microbiol.* 10, 2550–2573. doi: 10.1111/j.1462-2920.2008.01679.x

94. Pirbadian, S., Barchinger, S. E., Leung, K. M., Byun, H. S., Jangir, Y., Bouhenni, R. A., et al. (2014). *Shewanella oneidensis* MR-1 nanowires are outer membrane and periplasmic extensions of the extracellular electron transport components. *Proc. Natl. Acad. Sci. U.S.A.* 111, 12883–12888. doi: 10.1073/pnas.1410551111

95. Pitts, K. E., Dobbin, P. S., Reyes-Ramirez, F., Thomson, A. J., Richardson, D. J., and Seward, H. E. (2003). Characterization of the *Shewanella oneidensis* MR-1 decaheme cytochrome MtrA: expression in *Escherichia coli* confers the ability to reduce soluble Fe(III) chelates. *J. Biol. Chem.* 278, 27758–27765. doi: 10.1074/jbc.M302582200

96. Poehlein, A., Schmidt, S., Kaster, A.-K., Goenrich, M., Vollmers, J., Thürmer, A., et al. (2012). An ancient pathway combining carbon dioxide fixation with the generation and utilization of a sodium ion gradient for ATP synthesis. *PLoS ONE* 7:e33439. doi: 10.1371/journal.pone.0033439

97. Potter, M. C. (1911). Electrical effects accompanying the decomposition of organic compounds. *Proc. R. Soc. Lond. B* 1911, 260–276. doi: 10.1098/rspb.1911.0073

98. Qian, X., Mester, T., Morgado, L., Arakawa, T., Sharma, M. L., Inoue, K., et al. (2011). Biochemical characterization of purified OmcS, a c-type cytochrome required for insoluble Fe (III) reduction in *Geobacter sulfurreducens*. *Biochim. Biophys. Acta* 1807, 404–412. doi: 10.1016/j.bbabio.2011.01.003

99. Qiao, Y., Li, C. M., Bao, S.-J., Lu, Z., and Hong, Y. (2008). Direct electrochemistry and electrocatalytic mechanism of evolved*Escherichia coli* cells in microbial fuel cells. *Chem. Commun.* 11, 1290–1292. doi: 10.1039/b719955d

100. Rabaey, K., Boon, N., Höfte, M., and Verstraete, W. (2005). Microbial phenazine production enhances electron transfer in biofuel cells. *Environ. Sci. Technol.* 39, 3401–3408. doi: 10.1021/es048563o

101. Rabaey, K., Boon, N., Siciliano, S. D., Verhaege, M., and Verstraete, W. (2004). Biofuel cells select for microbial consortia that self-mediate electron transfer. *Appl. Environ. Microbiol.* 70, 5373–5382. doi: 10.1128/AEM.70.9.5373-5382.2004

102. Rabaey, K., and Rozendal, R. A. (2010). Microbial electrosynthesis—revisiting the electrical route for microbial production. *Nat. Rev. Microbiol.* 8, 706–716. doi: 10.1038/nrmicro2422

103. Ragsdale, S. W., and Pierce, E. (2008). Acetogenesis and the Wood–Ljungdahl pathway of CO2 fixation. *Biochim. Biophys. Acta*1784, 1873–1898. doi: 10.1016/j.bbapap.2008.08.012

104. Reguera, G., Nevin, K. P., Nicoll, J. S., Covalla, S. F., Woodard, T. L., and Lovley, D. R. (2006). Biofilm and nanowire production leads to increased current in *Geobacter sulfurreducens* fuel cells. *Appl. Environ. Microbiol.* 72, 7345–7348. doi: 10.1128/AEM.01444-06

105. Richardson, D. J., Edwards, M. J., White, G. F., Baiden, N., Hartshorne, R. S., Fredrickson, J. K., et al. (2012). Exploring the biochemistry at the extracellular redox frontier of bacterial mineral Fe (III) respiration. *Biochem. Soc. Trans.* 40, 493–500. doi: 10.1042/BST20120018

106. Richter, H., Nevin, K. P., Jia, H., Lowy, D. A., Lovley, D. R., and Tender, L. M. (2009). Cyclic voltammetry of biofilms of wild type and mutant *Geobacter sulfurreducens* on fuel cell anodes indicates possible roles of OmcB, OmcZ, type IV pili, and protons in extracellular electron transfer. *Energy Environ. Sci.* 2, 506–516. doi: 10.1039/b816647a

107. Richter, K., Schicklberger, M., and Gescher, J. (2012). Dissimilatory reduction of extracellular electron acceptors in anaerobic respiration. *Appl. Environ. Microbiol.* 78, 913–921. doi: 10.1128/AEM.06803-11

108. Rosenbaum, M., Aulenta, F., Villano, M., and Angenent, L. T. (2011). Cathodes as electron donors for microbial metabolism: which extracellular electron transfer mechanisms are involved? *Bioresour. Technol.* 102, 324–333. doi: 10.1016/j.biortech.2010.07.008

109. Ross, D. E., Flynn, J. M., Baron, D. B., Gralnick, J. A., and Bond, D. R. (2011). Towards electrosynthesis in *Shewanella*: energetics of reversing the Mtr pathway for reductive metabolism. *PLoS ONE* 6:e16649. doi: 10.1371/journal.pone.0016649

110. Sasaki, K., Tsuge, Y., Sasaki, D., and Kondo, A. (2014). Increase in lactate yield by growing *Corynebacterium glutamicum* in a bioelectrochemical reactor. *J. Biosci. Bioeng.* 117, 598–601. doi: 10.1016/j.jbiosc.2013.10.026

111. Schiel-Bengelsdorf, B., and Dürre, P. (2012). Pathway engineering and synthetic biology using acetogens. *FEBS Lett.* 586, 2191–2198. doi: 10.1016/j.febslet.2012.04.043

112. Schröder, U., Nießen, J., and Scholz, F. (2003). A generation of microbial fuel cells with current outputs boosted by more than one order of magnitude. *Angew. Chem. Int. Ed.* 42, 2880–2883. doi: 10.1002/anie.200350918

113. Schrott, G. D., Bonanni, P. S., Robuschi, L., Esteve-Nuñez, A., and Busalmen, J. P. (2011). Electrochemical insight into the mechanism of electron transport in biofilms of *Geobacter sulfurreducens*. *Electrochim. Acta* 56, 10791–10795. doi: 10.1016/j.electacta.2011.07.001

114. Schuchmann, K., and Müller, V. (2012). A bacterial electron-bifurcating hydrogenase. *J. Biol. Chem.* 287, 31165–31171. doi: 10.1074/jbc.M112.395038

115. Schuchmann, K., and Müller, V. (2014). Autotrophy at the thermodynamic limit of life: a model for energy conservation in acetogenic bacteria. *Nat. Rev. Microbiol.* 12, 809–821. doi: 10.1038/nrmicro3365

116. Shi, L., Richardson, D. J., Wang, Z., Kerisit, S. N., Rosso, K. M., Zachara, J. M., et al. (2009). The roles of outer membrane cytochromes of *Shewanella* and *Geobacter* in extracellular electron transfer. *Environ. Microbiol. Rep.* 1, 220–227. doi: 10.1111/j.1758-2229.2009.00035.x

117. Shi, L., Rosso, K., Zachara, J., and Fredrickson, J. (2012). Mtr extracellular electron-transfer pathways in Fe (III)-reducing or Fe (II)-oxidizing bacteria: a genomic perspective. *Biochem. Soc. Trans.* 40, 1261–1267. doi: 10.1042/BST20120098

118. Shi, L., Squier, T. C., Zachara, J. M., and Fredrickson, J. K. (2007). Respiration of metal (hydr) oxides by *Shewanella* and*Geobacter*: a key role for multihaem c-type cytochromes. *Mol. Microbiol.* 65, 12–20. doi: 10.1111/j.1365-2958.2007.05783.x

119. Shin, H., Zeikus, J., and Jain, M. (2002). Electrically enhanced ethanol fermentation by *Clostridium thermocellum* and*Saccharomyces cerevisiae*. *Appl. Microbiol. Biotechnol.* 58, 476–481. doi: 10.1007/s00253-001-0923-2

120. Shirodkar, S., Reed, S., Romine, M., and Saffarini, D. (2011). The octahaem SirA catalyses dissimilatory sulfite reduction in*Shewanella oneidensis* MR-1. *Environ. Microbiol.* 13, 108–115. doi: 10.1111/j.1462-2920.2010.02313.x

121. Simpson, P. J., Richardson, D. J., and Codd, R. (2010). The periplasmic nitrate reductase in *Shewanella*: the resolution, distribution and functional implications of two NAP isoforms, NapEDABC and NapDAGHB. *Microbiology* 156, 302–312. doi: 10.1099/mic.0.034421-0

122. Steinbusch, K. J., Hamelers, H. V., Schaap, J. D., Kampman, C., and Buisman, C. J. (2009). Bioelectrochemical ethanol production through mediated acetate reduction by mixed cultures. *Environ. Sci. Technol.* 44, 513–517. doi: 10.1021/es902371e

123. Straub, K. L., Benz, M., and Schink, B. (2001). Iron metabolism in anoxic environments at near neutral pH. *FEMS Microbiol. Ecol.* 34, 181–186. doi: 10.1111/j.1574-6941.2001.tb00768.x

124. Strycharz, S. M., Glaven, R. H., Coppi, M. V., Gannon, S. M., Perpetua, L. A., Liu, A., et al. (2011). Gene expression and deletion analysis of mechanisms for electron transfer from electrodes to *Geobacter sulfurreducens*. *Bioelectrochemistry* 80, 142–150. doi: 10.1016/j.bioelechem.2010.07.005

125. Strycharz, S. M., Woodard, T. L., Johnson, J. P., Nevin, K. P., Sanford, R. A., Löffler, F. E., et al. (2008). Graphite electrode as a sole electron donor for reductive dechlorination of tetrachlorethene by *Geobacter lovleyi*. *Appl. Environ. Microbiol.* 74, 5943–5947. doi: 10.1128/AEM.00961-08

126. Strycharz-Glaven, S. M., and Tender, L. M. (2012). Reply to the "*Comment on 'On Electrical Conductivity of Microbial Nanowires and Biofilms*"by N. S. Malvankar, M. T. Tuominen, and D. R. Lovley, *Energy Environ. Sci.* 2012, 5. *Energy Environ. Sci.* 5, 6250–6255. doi: 10.1039/c2ee02613a.

127. Sydow, A., Krieg, T., Mayer, F., Schrader, J., and Holtmann, D. (2014). Electroactive bacteria—molecular mechanisms and genetic tools. *Appl. Microbiol. Biotechnol.* 98, 8481–8495. doi: 10.1007/s00253-014-6005-z

128. Tatsumi, N., and Inui, M. (2012). *Corynebacterium glutamicum: Biology and Biotechnology.* Heidelberg: Springer Science & Business Media.

129. TerAvest, M. A., Zajdel, T. J., and Ajo-Franklin, C. M. (2014). The Mtr pathway of *Shewanella oneidensis* MR-1 couples substrate utilization to current production in *Escherichia coli*. *ChemElectroChem* 1, 1874–1879. doi: 10.1002/celc.201402194

130. Thamdrup, B. (2000). Bacterial manganese and iron reduction in aquatic sediments. *Adv. Microb. Ecol.* 16, 41–84. doi: 10.1007/978-1-4615-4187-5_2

131. Thrash, J. C., and Coates, J. D. (2008). Review: direct and indirect electrical stimulation of microbial metabolism. *Environ. Sci. Technol.* 42, 3921–3931. doi: 10.1021/es702668w

132. Tracy, B. P., Jones, S. W., Fast, A. G., Indurthi, D. C., and Papoutsakis, E. T. (2012). Clostridia: the importance of their exceptional substrate and metabolite diversity for biofuel and biorefinery applications. *Curr. Opin. Biotechnol.* 23, 364–381. doi: 10.1016/j.copbio.2011.10.008

133. Tremblay, P.-L., and Zhang, T. (2015). Electrifying microbes for the production of chemicals. *Front. Microbiol.* 6:201. doi: 10.3389/fmicb.2015.00201

134. Tremblay, P.-L., Zhang, T., Dar, S. A., Leang, C., and Lovley, D. R. (2013). The Rnf complex of *Clostridium ljungdahlii* is a proton-translocating ferredoxin: NAD+ oxidoreductase essential for autotrophic growth. *MBio* 4, e00406–e00412.

135. Unden, G., and Bongaerts, J. (1997). Alternative respiratory pathways of *Escherischia coli*: energetics and transcriptional regulation in response to electron acceptors. *Biochim. Biophys. Acta* 1320, 217–234. doi: 10.1016/S0005-2728(97)00034-0

136. Vignais, P. M., and Billoud, B. (2007). Occurrence, classification, and biological function of hydrogenases: an overview. *Chem. Rev.* 107, 4206–4272. doi: 10.1021/cr050196r

137. Vijgenboom, E., Busch, J. E., and Canters, G. W. (1997). In vivo studies disprove an obligatory role of azurin in denitrification in*Pseudomonas aeruginosa* and show that azu expression is under control of RpoS and ANR. *Microbiology* 143, 2853–2863. doi: 10.1099/00221287-143-9-2853

138. Wagner, R. C., Regan, J. M., Oh, S.-E., Zuo, Y., and Logan, B. E. (2009). Hydrogen and methane production from swine wastewater using microbial electrolysis cells. *Water Res.* 43, 1480–1488. doi: 10.1016/j.

watres.2008.12.037

139. Wang, S., Huang, H., Kahnt, J., and Thauer, R. K. (2013). A reversible electron-bifurcating ferredoxin-and NAD-dependent [FeFe]-hydrogenase (HydABC) in *Moorella thermoacetica*. *J. Bacteriol.* 195, 1267–1275. doi: 10.1128/JB.02158-12

140. Welte, C., Krätzer, C., and Deppenmeier, U. (2010). Involvement of Ech hydrogenase in energy conservation of *Methanosarcina mazei*. *FEBS J.* 277, 3396–3403. doi: 10.1111/j.1742-4658.2010.07744.x

141. White, G. F., Shi, Z., Shi, L., Wang, Z., Dohnalkova, A. C., Marshall, M. J., et al. (2013). Rapid electron exchange between surface-exposed bacterial cytochromes and Fe (III) minerals. *Proc. Natl. Acad. Sci. U.S.A.* 110, 6346–6351. doi: 10.1073/pnas.1220074110

142. Williams, K. H., Nevin, K. P., Franks, A., Englert, A., Long, P. E., and Lovley, D. R. (2009). Electrode-based approach for monitoring in situ microbial activity during subsurface bioremediation. *Environ. Sci. Technol.* 44, 47–54. doi: 10.1021/es9017464

143. Wrighton, K. C., Thrash, J. C., Melnyk, R. A., Bigi, J. P., Byrne-Bailey, K. G., Remis, J. P., et al. (2011). Evidence for direct electron transfer by a gram-positive bacterium isolated from a microbial fuel cell. *Appl. Environ. Microbiol.* 77, 7633–7639. doi: 10.1128/aem.05365-11

144. Yang, Y., Ding, Y., Hu, Y., Cao, B., Rice, S. A., Kjelleberg, S., et al. (2015). Enhancing bidirectional electron transfer of*Shewanella oneidensis* by a synthetic flavin pathway. *ACS Synth. Biol.* doi: 10.1021/sb500331x [Epub ahead of print].

145. Yang, Y., Xu, M., Guo, J., and Sun, G. (2012). Bacterial extracellular electron transfer in bioelectrochemical systems. *Process Biochem.* 47, 1707–1714. doi: 10.1016/j.procbio.2012.07.032

146. Zhang, T., Cui, C., Chen, S., Ai, X., Yang, H., Shen, P., et al. (2006). A novel mediatorless microbial fuel cell based on direct biocatalysis of *Escherichia coli*. *Chem. Commun.* 21, 2257–2259. doi: 10.1039/b600876c

147. Zhang, T., Gannon, S. M., Nevin, K. P., Franks, A. E., and Lovley, D. R. (2010). Stimulating the anaerobic degradation of aromatic hydrocarbons in contaminated sediments by providing an electrode as the electron acceptor. *Environ. Microbiol.* 12, 1011–1020. doi: 10.1111/j.1462-2920.2009.02145.x

148. Zhang, T., Zeng, Y., Chen, S., Ai, X., and Yang, H. (2007). Improved performances of *E. coli* catalyzed microbial fuel cells with composite graphite/PTFE anodes. *Electrochem. Commun.* 9, 349–353. doi: 10.1016/j.elecom.2006.09.025

CHAPTER 2

ELECTRIFYING MICROBES FOR THE PRODUCTION OF CHEMICALS

Pier-Luc Tremblay and Tian Zhang

Keywords: Microbial electrosynthesis, bioelectrochemical systems, electricity, CO_2 reduction, electron transfer mechanisms

INTRODUCTION

Microbial electrosynthesis (MES) happens when a microbial catalyst reduces CO_2 into multicarbon chemical commodities with electrons derived from the cathode of a bioelectrochemical system designed primarily to perform biological reductive reactions (rBES; Rabaey and Rozendal, 2010;Rabaey et al., 2011; Lovley, 2012; Lovley and Nevin, 2013; Wang and Ren, 2013; Hallenbeck et al., 2014; Rosenbaum and Franks, 2014; Figure 1). rBES-driven processes also include electrofermentation, electrorespiration, and electromethanogenesis. Electrofermentation occurs when electrons coming from a cathode are supplied to a fermentative microbial catalyst shifting the fermentation balance toward the production of more reduced products (Rabaey and Rozendal, 2010; Kracke and Krömer, 2014). In the case of electrorespiration, a terminal electron acceptor such as fumarate is reduced by a respiratory microbial catalyst with electrons coming from a cathode (Park et al., 1999; Rabaey and Rozendal, 2010). Electromethanogenesis has similarities with MES since CO_2 is the feedstock, but in this case CO_2 will be reduced to methane by a methanogenic microbial catalyst using electrons derived from a cathode (Cheng et al., 2009;Villano et al., 2010; Kobayashi et al., 2013).

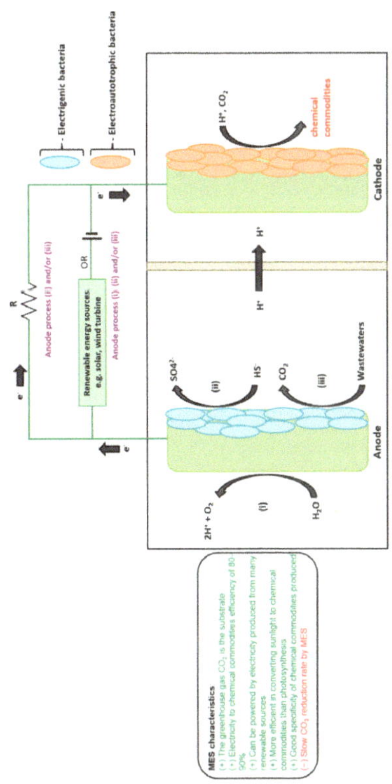

Figure 1: Principle and flexibility of MES. (i) MES can be coupled with different renewable energy sources such as wind and solar to produce a wide range of chemical commodities. MES can also be coupled to environment-friendly anodic processes such as (ii) sulfide oxidation and (iii) wastewaters treatment.

Besides its capacity of using CO_2 directly as feedstock, the two other main qualities of MES are its energetic efficiency

Citation: Tremblay P-L and Zhang T (2015) Electrifying microbes for the production of chemicals. Front. Microbiol. 6:201. doi: 10.3389/fmicb.2015.00201.

and its versatility (Figure 1). The electricity efficiency to chemical commodities of MES processes is ca. 80–90% (Nevin et al., 2010, 2011; Nie et al., 2013; Zhang et al., 2013). Many crop plants have sunlight efficiency to biomass below 3% (MacDonald, 2003), whereas common silicon solar cells are at least six times more efficient at capturing the sun energy (Green et al., 2014). Therefore, powering MES with electricity from solar cells could be a more potent strategy for storing the sun energy into the chemical bonds of multicarbon compounds (Nevin et al., 2010;Lovley and Nevin, 2011).

Microbial electrosynthesis is a versatile technology because the necessary electricity can be generated from multiple renewable sources. Apart from sun energy, MES can be powered with electricity produced by wind turbine. Renewable electricity sources are intermittent by nature and do not harmonize well with the market demand (Jürgensen et al., 2014). In this context, MES becomes a perfect technological fit making possible the direct storage of electricity surplus into value-added chemical commodities (Lovley and Nevin, 2011).

Microbial electrosynthesis can also be coupled with a bioelectrochemical system performing biological oxidation reactions (oBES). Processes driven by oBES are defined by the transfer of electrons from a microbial catalyst metabolizing a given substrate to an electrode collecting electricity. oBESs have been developed for a multitude of applications including wastewater treatments, *in situ* bioremediation, water desalination, biosensors, electrohydrogenesis, and various types of microbial fuel cells (Cheng and Logan, 2007; Cao et al., 2009; Lovley, 2009, 2012;Logan, 2010; Logan and Rabaey, 2012). Recent studies have shown that electrons coming from oBES-driven processes can be used to supply MES (Figure 1). In sulfide-driven MES, the electricity required for MES is generated by the abiotic oxidation of the toxic contaminant sulfide to sulfur and the subsequent biological oxidation of sulfur to sulfate (Gong et al., 2013). A similar process has also been developed to supply electrons for the electroproduction of methane from CO_2 (Jiang et al., 2014). Moreover, MES and electromethanogenesis have been conducted in parallel with the biorecovery of cobalt at the cathode further illustrating the versatility of this technology (Huang et al., 2014).

A vibrant illustration of the significant progress made by MES in a relatively short period of time is that multicarbon compounds production rates by MES have been increased substantially over the last 4 years. For instance, the acetate production rate has been increased 433-fold from ca. 30 mM d^{-1} m^{-2} to ca. 1.3 mM d^{-1} cm^{-2}, whereas the electron transfer rate was enhanced 521-fold from ca. 71 mA m^{-2} to ca. 3.7 mA cm^{-2} (Nevin et al., 2010; Marshall et al., 2012, 2013; Nie et al., 2013; Zhang et al., 2013; Jourdin et al., 2014; LaBelle et al., 2014; Table 1). However, the main obstacle for the development of MES as an economically viable technology is still the relatively slow microbial reduction rate of CO_2 to multicarbon compounds in scalable rBES reactors. This review will discuss the ongoing efforts to increase MES productivity, stability, long-term efficiency, and versatility by optimizing microbial catalysts and electrochemical hardware and by characterizing the electron transfer mechanisms from cathode to microbe.

Table 1: Microbial electrosynthesis systems in chronological order of publication.

Microbial catalyst	Cathode	Comments	Reference
Acidithiobacillus ferrooxidans	-Platinum	-Fe(II)-mediated	Kinsel and Umbreit (1964)
A. ferrooxidans	-Platinum mesh	-Fe(II)-mediated	Nakasono et al. (1997)
A. ferrooxidans	-Platinum mesh -0.0 V (vs. Ag/AgCl)	-Fe(II)-mediated	Matsumoto et al. (1999)
Leptospirillum ferrooxidans	-Platinum mesh - +0.1 V (vs. Ag/AgCl)	-Fe(II)-mediated	Matsumoto et al. (2000)
Sporomusa ovata	-Graphite stick - −0.4 V (vs. SHE)	-Direct electron transfer -Acetate and 2-oxobutyrate produced	Nevin et al. (2010)
A. ferrooxidans	-Graphite felt - −0.0 V (vs. SCE)	-Direct electron transfer -Current density: 5 A m^{-2}	Carbajosa et al. (2010)
Clostridium aceticum *Clostridium ljungdahlii* *Moorella thermoacetica* *Sporomusa silvacetica* *Sporomusa sphaeroides*	-Graphite stick - −0.4 V (vs. SHE)	-Direct electron transfer -Acetate, 2-oxobutyrate and formate produced	Nevin et al. (2011)
Ralstonia eutropha	-Indium foil - −1.6 V (vs. Ag/AgCl)	-Formate-mediated -Biofuels produced	Li et al. (2012)
Mixed community	-Graphite fiber brush/carbon rod/graphite plate - −0.439 V or −0.539 V (vs. SHE)	-Current density: 52 mA m^{-2} (Graphite plate) -Power density: 83 mWm^{-2} (graphite plate)	Pisciotta et al. (2012)
Mixed community	-Graphite granule - −0.59 V (vs. SHE)	-Acetate, methane, and H_2 produced -Acetate production: >4 mM d^{-1}	Marshall et al. (2012)
Nitrosomonas europaea	-Nickel, glassy carbon, or copper -Multiple potentials	-Ammonia-mediated -Multi-reactors system	Khunjar et al. (2012)
Geobacter sulfurreducens	-Stainless steel - −0.6 V (vs. Ag/AgCl)	-Direct electron transfer -Current density: 30 A m^{-2}	Soussan et al. (2013)
S. ovata	-Modified carbon cloth - −0.6 V (vs. Ag/AgCl)	-Direct electron transfer -Best cathode modification: chitosan -Current density: 475 mA m^{-2} -Acetate production: 229 mM d^{-1} m^{-2}	Zhang et al. (2013)
S. ovata	-Graphite plate	-Powered by sulfide/sulfur bioanode (0.3 V vs. SHE) -Acetate production: 49.9 mmol d^{-1} m^{-2}	Gong et al. (2013)
Mariprofundus ferrooxydans	-Graphite −0.076 V (vs. SHE)	-Direct electron transfer -Cell-normalized electrode oxidation rate: 0.075 pmol electrons $cell^{-1}$ h^{-1}	Summers et al. (2013)
Mixed community	-Carbon felt - −1.15 V (vs. Ag/AgCl)	-Methane and acetate produced -Acetate production: 94.73 mg d^{-1}	Jiang et al. (2013)

Mixed community	-Carbon fiber rod - –0.4 V (vs. SHE)	-Direct electron transfer -Acetate, ethanol, 1-butanol, propionate, butyrate, and H_2 produced -Current density: 34 mA m^{-2}	Zaybak et al. (2013)
Rhodopseudomonas palustris	-Graphite rod -+0.1 V (vs. SHE)	-Light-driven -Current density: 1.5 $\mu A\ cm^{-2}$	Bose et al. (2014)
Mixed community	-Graphite felt	-Cobalt reduction -Methane and acetate produced	Huang et al. (2014)
Mixed community	-Graphite plate modified with NanoWeb-RVC - –0.85 V (vs. SHE)	-Current density: 3.7 mA cm^{-2} -Acetate production: 1.3 mM $d^{-1}\ cm^{-2}$	Jourdin et al. (2014)
R. palustris	-Carbon cloth - –0.22 V (vs. Ag/AgCl)	-Fe(II)-mediated -Light-driven -Multi-reactor system -Current density: 7.2 $\mu A\ ml^{-1}$	Doud and Angenent (2014)
Mixed community	-Graphite granule - –0.6 to –0.8 V (vs. SHE)	-Acidic pH -Acetate, H_2, and formate produced -Acetate production: 51.6 mM d^{-1} (–0.8 V) -High H_2 production	LaBelle et al. (2014)

THE MICROBIAL CATALYSTS

Mixed Communities

Microbial electrosynthesis can be driven by two major types of microbial catalysts: mixed communities and pure cultures. In the case of mixed communities, the cathodic chamber of the MES system is inoculated with samples from wastewater, sludge, or sediment (Table 1). One of the main advantages of employing a mixed community for MES is that it eliminates the need to work under stringent sterile conditions required with pure culture-driven bioprocesses. Moreover, the MES system with the highest reported acetate production rate to date of 1.3 mM d^{-1} cm^{-2} was driven by an uncharacterized mixed community (Jourdin et al., 2014). Mixed community-driven MES systems described until now mainly produce acetate because the microbial population quickly become dominated by acetogenic bacteria like *Acetobacterium* sp. (Marshall et al., 2012, 2013;LaBelle et al., 2014) and *Eubacterium* sp. (Pisciotta et al., 2012). There is also simultaneous production of methane due to the coexistent methanogens, unless an inhibitor of methanogenesis is added to the cathode reactor (Marshall et al., 2012, 2013; Pisciotta et al., 2012; Jiang et al., 2013;Huang et al., 2014). Hydrogen and formate produced biologically or abiotically at low cathode potential are other compounds frequently found in mixed community-driven MES reactors (Marshall et al., 2012, 2013; Pisciotta et al., 2012; Zaybak et al., 2013; LaBelle et al., 2014). Furthermore, in a study for the development of a method to facilitate the start-up of autotrophic biocathodes in rBESs, the mixed community microbial catalysts were reported to produce at least six products: butanol, ethanol, hydrogen, acetate, propionate, and butyrate (Zaybak et al., 2013). This study gives a good example of the difficulty generating a single specific product when employing mixed communities to drive MES processes. Other than compromising the purity of the desired product, it also complicates the separation process and reduces the electricity conversion efficiency to a specific multicarbon compound.

Pure Cultures

Acetogenic Bacteria

Diverse autotrophic pure cultures have been employed successfully in the role of microbial catalysts for MES systems. As illustrated by several studies on mixed community-driven MES, acetogens reducing CO_2 through the Wood–Ljungdahl pathway (Drake et al., 1997, 2008; Ragsdale and Pierce, 2008) are dominating and efficient electroautotrophs. Pure cultures of Gram negative acetogens like *Sporomusa silvacetica* and *Sporomusa sphaeroides* and Gram positive acetogens like *Clostridium ljungdahlii*, *Clostridium aceticum* and the thermophile *Moorella thermoacetica*are all capable of reducing CO_2 to multicarbon compounds by MES (Nevin et al., 2011). Among all the tested acetogenic bacteria, *Sporomusa ovata* DSM-2662 was the most efficient electroautotroph with acetate production rates as high as 282 mM d^{-1} m^{-2} and with electricity conversion efficiency to acetate typically above 80% (Nevin et al., 2010; Gong et al., 2013; Nie et al., 2013; Zhang et al., 2013). The production of negligible amount of 2-oxobutyrate and formate in comparison to acetate by *S. ovata* was also reported (Nevin et al., 2010, 2011). Acetogens like *S. ovata* have high electricity conversion efficiency to chemical commodities compared to autotrophic bacteria with other types of carbon fixation metabolisms because CO_2 is the sole electron acceptor during acetogenesis (Drake et al., 1997, 2008; Ragsdale and Pierce, 2008). Thus, most of the electrons derived from the cathode will end up in reduced multicarbon products. Moreover, a study by Fast and Papousakis (2012) established that the Wood–Ljungdahl pathway is the most energetically efficient non-photosynthetic carbon fixation pathway for the electroproduction of acetate and ethanol (Fast and Papousakis, 2012).

Autotrophic Fe(II) Oxidizing Bacteria

Autotrophic Fe(II) oxidizing bacteria are also capable of reducing CO_2 by MES. The acidophilic aerobic Fe(II) oxidizer *Acidithiobacillus ferrooxidans* (Kinsel and Umbreit, 1964; Nakasono et al., 1997; Matsumoto et al., 1999) and *Leptospirillum ferrooxidans* (Matsumoto et al., 2000) were able to grow in a bioelectrochemical system by using electrons coming from electrochemically reduced Fe(II). Moreover, *A. ferrooxidans* (Carbajosa et al., 2010) and the neutrophilic aerobic Fe(II) oxidizer *Mariprofundus ferrooxydans* (Summers et al., 2013) were shown to draw current directly from a poised cathode in the absence of a redox mediator to grow with CO_2 as the source of carbon. The current draw was 5 A m^{-2} with *A. ferrooxidans* (Carbajosa et al., 2010) whereas the cell-normalized electrode oxidation rate was 0.075 pmol electrons cell^{-1} h^{-1} with *M. ferrooxydans*(Summers et al., 2013). MES systems driven by these Fe(II) oxidizing bacteria were poised at potential closer to 0 V vs. SHE compared to all the other reported MES systems (Table 1). Potential closer to 0 V at the cathode could translate in lower energy requirements (Lovley and Nevin, 2013). However, in those systems O_2 is the final electron acceptor which means that a significant amount of electrons will not be used to reduce CO_2 into multicarbon commodities but to reduce O_2 and to generate biomass.

Ammonia-Oxidizing Bacteria

Ammonia can be employed as a redox mediator in MES system to promote the growth of ammonia-oxidizing bacteria (Khunjar et al., 2012; Table 1). In the study by Khunjar et al. (2012), nitrite was reduced electrochemically to ammonia in the first reactor and then was fed to a second reactor containing the ammonia-oxidizing bacteria *Nitrosomonas europaea*. Electrons from ammonia were used by *N. europaea* to produce biomass from CO_2 and to generate nitrite that was then recycled in the first electrochemical reactor. However, like the MES system driven by autotrophic Fe(II)-oxidizer, *N. europaea* requires O_2 as an electron acceptor and thus this system has the same efficiency issue.

Recombinant Microbial Catalyst

The only example to date of a MES process driven by a genetically engineered microbial catalyst was a recombinant strain of the chemolithotrophic bacterium *Ralstonia eutropha* developed for the electroproduction of biofuels (Li et al., 2012). The recombinant strain of *R. eutropha* generated 140 mg/ml of biofuels over a period of ca. 100 h by oxidizing formate that was electrochemically produced at the surface of a poised cathode. This MES process required O_2 as the final electron acceptor and a cathode poised at very low potential of –1.6 V vs. Ag/AgCl to generate the necessary electron shuttle formate. These two factors are expected to significantly lower the overall efficiency of this MES system due to the energy loss for O_2 reduction and formate production. Nevertheless, this study demonstrates that one of the advantages of using a pure culture for MES is that it can be genetically engineered to optimize the metabolism of the microbial catalyst and to increase the range of possible products.

Geobacter sulfurreducens

Evidence suggest that *Geobacter sulfurreducens* can also reduce CO_2 into a multicarbon compound by MES (Soussan et al., 2013; Table 1). *G. sulfurreducens* is a well-characterized electrigenic bacterium generating power densities in oBESs as high as 3.9 W/m^2 (Yi et al., 2009). *G. sulfurreducens* pre-grown with acetate as an electron donor and carbon source was also shown to have the capacity to accept electrons from a cathode to reduce fumarate (Gregory et al., 2004;Dumas et al., 2008) or uranium(VI; Gregory and Lovley, 2005) after the depletion of acetate. In a study by Soussan et al. (2013), *G. sulfurreducens* first used electrons derived from the cathode to reduce the final electron acceptor fumarate into succinate. When all the fumarate was depleted, *G. sulfurreducens* appeared to start producing glycerol by a process combining CO_2 possibly reduced electrochemically to bicarbonate with succinate. Interestingly, a previous report demonstrated that*G. sulfurreducens* cannot grow with CO_2 alone (Coppi et al., 2004). Recently, the closely related species *Geobacter metallireducens* was shown to be capable to grow autotrophically with formate as the sole source of carbon (Feist et al., 2014). *G. metallireducens* genome encodes two known carbon fixation pathways, the reductive TCA cycle and the dicarboxylate–hydroxybutyrate cycle, both of which are not present in the genome of *G. sulfurreducens*. It has been suggested that formate could be assimilated by *G. sulfurreducens* in the presence of a small quantity of acetate via the pyruvate formate lyase (Speers and Reguera, 2012). However, in Soussan et al. (2013) study, the cathode is poised at a potential (–0.6 V vs. Ag/AgCl) too high for CO_2 reduction to formate, hence the biochemical mechanism for the assimilation of CO_2/bicarbonate and the production of glycerol remains unclear.

Photosynthetic Fe(II)-Oxidizing Bacteria

Microbial electrosynthesis system can also be exposed to light to provide energy for CO_2 reduction by anaerobic photosynthetic Fe(II)-oxidizing bacteria. *Rhodopseudomonas palustris* is the microbial catalyst in two reported photobiocathode-based MES studies. In the first study, *R. palustris* was accepting electrons from a poised cathode in a one-reactor system to reduce CO_2 with an electron transfer rate of 1.5 μA cm^{-2} (Bose et al., 2014). Electron uptake was stimulated by light exposure, but occurred at a significantly lower rate in the dark. However, it is not clear if *R. palustris* was accepting electrons directly from the cathode or from Fe(II) reduced electrochemically. The second study described a multi-reactor system similar to the aforementioned ammonia shuttling MES system by Khunjar et al. (2012). The redox mediator Fe(III) was first reduced abiotically by a cathode to Fe(II) in an electrochemical reactor before being fed to a second photobioreactor where *R. palustris* was growing with CO_2 as the sole carbon source, Fe(II) as the electron source, and light (Doud and Angenent, 2014).

ELECTRON TRANSFER FROM THE CATHODE TO THE MICROBIAL CATALYST

Indirect Electron Transfer

Exogenous Electron Shuttles

Understanding the electron transfer mechanisms involved in MES could lead to major breakthroughs in the effort to increase the electron transfer rate between the cathode and the microbial catalyst. Electrons can either be transferred directly or indirectly via a shuttle (Patil et al., 2012; Figure 2). H_2, formate, Fe(II) and ammonia have all been reported to function as redox mediator in MES systems (Lovley and Nevin, 2013). H_2 has been the most prevalent (Table 1) since it requires only that the cathode in the MES reactor is poised at a lower potential than –0.41 V (vs. SHE). Under this condition, significant quantities of H_2 are generated from the electrons coming from the cathode and the protons migrating from the anodic chamber. However, employing H_2 as a redox mediator for MES is not optimal because its low solubility might cause energy losses.

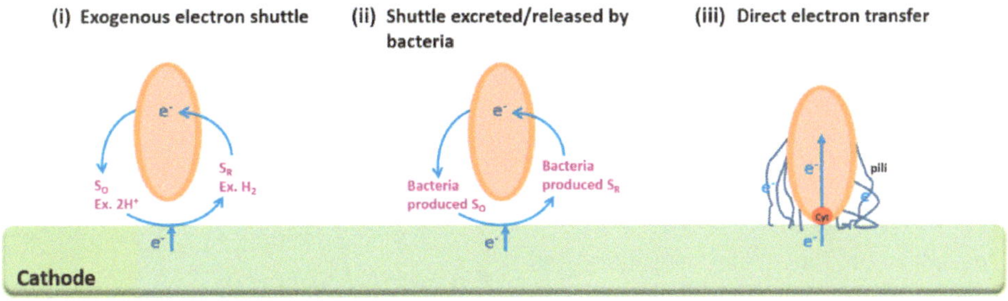

Figure 2: Possible extracellular electron transfer mechanisms from the cathode to the microbial catalyst. Indirect electron transfer via (i) an exogenous shuttle or (ii) a shuttle released/excreted by the microbial catalyst. (iii) Direct electron transfer via bacterial outer surface components such as c-type cytochromes or pili. SO is oxidized electron shuttle and SR is reduced electron shuttle.

Shuttles Excreted/Released by Bacteria

Another possible indirect electron transfer mechanism is that microbial catalysts could be producing and excreting their own soluble redox mediators to carry electrons from the cathode. Redox mediators either excreted by the bacteria such as phenazine and riboflavin or released after cell death such as vitamin B12 or DNA have all been suggested as possible components involved in extracellular electron transfer (Rosenbaum et al., 2011).

Direct Electron Transfer

Direct electron transfer requires physical contacts between extracellular components of the microbial catalyst involved in the transport of electrons and the cathode. Direct electron transfer from the cathode to the microbial catalyst in the absence of a redox mediator has been shown to occur in MES processes driven by the following electroautotrophic species: *A. ferrooxidans*(Carbajosa et al., 2010; Rodrigues and Rosenbaum, 2014), *M. ferrooxydans* (Summers et al., 2013) and a group of acetogenic bacteria (Nevin et al., 2010, 2011; Nie et al., 2013; Zhang et al., 2013). This conclusion is mainly based on the fact that all these MES systems were operated with cathodes poised at potentials too high to produce significant quantity of H_2. Moreover, driving MES systems with cathodes poised at lower potentials than the formal potential of the $2H^+/H_2$ couple does not exclude the possibility of direct electron transfer. A study by Marshall et al. (2012) reported evidence that direct electron transfer was occurring at the same time as hydrogen-mediated electron transfer. This suggests that microbial catalysts are capable of acquiring electrons from a cathode through many paths simultaneously.

Electromethanogenic Bacteria

Recently, it was demonstrated that a mutant strain of *Methanococcus maripaludis* lacking all its catabolic hydrogenases was still capable of performing electromethanogenesis (Lohner et al., 2014) thus providing supplementary significant evidence that electrotrophic bacteria can accept electrons directly from the cathode independently of the redox mediator H_2.

The Model Metal-Reducing Bacteria

Two of the most detailed studies about electron transfer mechanisms from the cathode to the microbe have been done with the two model metal-reducing bacteria *G. sulfurreducens* (Strycharz et al., 2011) and *Shewanella oneidensis* (Ross et al., 2011). In *G. sulfurreducens*, the monoheme *c*-type cytochrome GSU3274 predicted to be localized in the periplasm was found to be specifically required for electron transfer from the cathode demonstrating the importance of *c*-type cytochromes for electron uptake. Like *G. sulfurreducens*, *S. oneidensis* can do electrorespiration accepting electrons from the cathode to reduce fumarate into succinate (Ross et al., 2011). Based on functional genetic studies, Ross et al. (2011) proposed a model where the respiratory pathway normally responsible for transferring electron from the cytoplasmic metabolism to extracellular electron acceptors (Shi et al., 2012) works in reverse. Therefore, the decaheme outer membrane *c*-type cytochrome MtrC, the decaheme periplasmic *c*-type cytochrome MtrA, the porin MtrB responsible for connecting MtrC to MtrA, the tetraheme cytoplasmic membrane-associated *c*-type cytochrome CymA and the menaquinone pool are all critical components of the electron uptake pathway of *S. oneidensis*.

Electroautotrophic Bacteria

Direct electron transfer pathways are poorly characterized in electroautotrophic bacteria. For *A. ferrooxidans*, experimental evidence indicated that a Fe species excreted by the cells in the cathode biofilm could be responsible for electron uptake (Carbajosa et al., 2010). It has been speculated that*c*-type cytochromes which are critical components for the uptake of electrons from extracellular Fe(II) could also be involved in the transport of electron from the cathode (Rosenbaum et al., 2011;Sydow et al., 2014). In support of this hypothesis, metagenomics and metaproteomics of the mixed community populating a self-regenerating biocathode suggest that a member of the *Chromatiaceae*family reduced CO_2 with electrons acquired directly from the cathode via *c*-type cytochromes and other proteins associated with Fe(II) oxidation (Wang et al., 2015).

Much less information is known about how electrons are acquired by acetogens from the cathode. Recently, a genetic system developed for the Gram positive bacterium *C. ljungdahlii* (Leang et al., 2013; Banerjee et al., 2014; Ueki et al., 2014) led to the confirmation of the identity of the proton pump responsible for the generation of a proton motive force essential for growth with CO_2 as the sole source of carbon (Tremblay et al., 2013). This study provided insights on the energy conservation mechanism involved in the electroautotrophic growth of acetogens. The electron uptake mechanism of *C. ljungdahlii* is expected to be significantly different compared to other electrotrophic bacteria because it cannot synthesize *c*-type cytochromes or quinones (Köpke et al., 2010). The availability of a genetic toolbox should accelerate the characterization of the specificities of *C. ljungdahlii*'s electron transfer pathway and could provide general information about electron uptake by other Gram positive bacteria.

The genome sequence of the Gram negative and acetogenic species *S. ovata* has recently been made available (Poehlein et al., 2013). Genes coding for *c*-type cytochromes and type IV pili, two components of bacterial extracellular electron transfer mechanisms, are present in the genome. As mentioned before, *c*-type cytochromes are critical components of extracellular electron transfer pathways in both electrigenic and electrotrophic bacteria. In *Geobacter* spp., type IV pili are filaments with metallic-like conductivity facilitating long-range electron transfer (Lovley, 2011,2012; Malvankar and Lovley, 2012; Vargas et al., 2013; Malvankar et al., 2014) involved in the reduction of insoluble electron acceptors (Reguera et al., 2005; Tremblay et al., 2012). Ubiquinone (Möller et al., 1984), another critical component of electron transport pathway, was also detected in*S. ovata* and genes coding for enzymes involved in its biosynthesis were found in the genome (Poehlein et al., 2013). *S. ovata* possesses several well-characterized components of microbial extracellular transfer pathways which indicates that electron uptake by *S. ovata* could have similarities with other electrigenic or electrotrophic bacteria. Recently, an acetogenic bacteria closely related to *S. sphaeroides* was shown to do acetogenesis with metallic iron (F(0)) as the sole electron donor suggesting that direct electron transfer could be a useful strategy for Gram negative acetogens in multiple environments (Kato et al., 2015).

THE ELECTROCHEMICAL HARDWARE

Cathode Materials Tested for Current Draw

A lot of effort has been put into optimizing rBESs/MES by selecting or developing more efficient and less expensive components for bioelectrochemical reactors (Krieg et al., 2014). Until now, most of the studies published on this topic are presenting the impact of different types of biocompatible cathodes on the performance of rBESs/MES. Cathodes described in the literatures relied mainly on carbonaceous materials (Table 1). Indium foil (Li et al., 2012) and platinum (Kinsel and Umbreit, 1964; Nakasono et al., 1997; Matsumoto et al., 1999) have also been tried in rBESs but the most efficient cathode material to date for electron transfer is stainless steel with current draw as high as 30 A m^{-2} with *G. sulfurreducens* as the microbial catalyst (Soussan et al., 2013).

Conductive materials that will self-assemble in the cathodic biofilm are another approach to enhance electron transfer in rBESs. A *S. oneidensis* biofilm assembled with embodied graphene oxide could uptake electrons 74 times more efficiently (Yong et al., 2014). Oligoelectrolytes can also facilitate electron transfer from the cathode by inserting itself in the lipid membrane of bacteria to enable transmembrane charge transfer as demonstrated by Thomas et al. (2013).

Cathode Materials Tested for Chemical Production

Cathodes employed for MES are generally made of carbonaceous materials like graphite. These basic cathodes can be treated or coated with other materials resulting in modifications of their surface.

Untreated Surface

When chemical commodities production rate is not normalized to total surface area of cathode, one of the best materials is granular graphite with acetate production from a mixed community catalyst reaching 51.6 mM per day or 3.0 gL^{-1} d^{-1} (LaBelle et al., 2014). As a packed structure, the main advantage of granular graphite over other carbonaceous cathodes is the high specific area for bacterial adhesion (Wei et al., 2011).

Treated Surface

Modifications of carbonaceous cathodes have resulted in critical improvements in MES systems driven either by mixed communities or by pure cultures. Recently, Jourdin et al. (2014) coated carbon nanotubes on reticulated vitreous carbon (NanoWeb-RVC) to enhance bacterial attachment in a mixed community-driven MES system resulting in the highest normalized current density and the highest normalized acetate production rate reported for any MES systems until now (Table 1). The authors suggested that this performance improvement was due to the high surface to volume ratio of NanoWeb-RVC responsible for enhanced bacterial adhesion and effective mass transfer within the electrode-biofilm superstructure (Jourdin et al., 2014). The higher reported acetate production rate and current density compared to other MES systems relying on direct electron transfer (Nevin et al., 2010, 2011; Nie et al., 2013; Zhang et al., 2013) can also be attributed in part to the low potential of the cathode (–0.85 V vs. SHE) promoting indirect electron transfer that could be advantageous for the unknown bacterial species populating the reactor. Maintaining the cathode at this potential is possibly responsible for the slightly lower electricity to acetate efficiency of 70% observed with this MES reactor. Although this system is promising, the volumetric acetate production rate of ca. 0.025 gL^{-1} d^{-1} is 120-fold lower than that of the MES system also driven by a mixed community developed by LaBelle et al. (2014). Scaling up test for the NanoWeb-RVC biocathode of Jourdin et al. (2014) will indicate its real potential for industrial applications.

A number of modified carbonaceous cathodes have been proposed for MES systems driven by the pure culture catalyst *S. ovata* with significant success. Functionalization of carbon cloth cathodes with chitosan or other compounds conferring a positive charge to the electrode surface with the aim of increasing interactions with negatively charged bacteria like *S. ovata* (Zhang et al., 2013), resulted in higher cell density at the cathode surface (up to 9-fold increase), in better electron transfer (up to 6.7-fold increase as 475 mA m^{-2}) and in higher normalized acetate production rates (up to 7.6-fold increase as 229 mM d^{-1} m^{-2}). In a second approach, carbon cloth cathodes were treated with metal, such as gold, palladium, or nickel nanoparticles to harness their exceptional catalytic activities for MES processes. Significant increases in normalized acetate production rate by MES were recorded with all three metals in the range of 100–200 mM d^{-1} m^{-2} (Zhang et al., 2013). Modifying polyester- or cotton-based textile composite cathodes with carbon nanotubes to create a three-dimensional matrix with more surface area available for bacteria also resulted in significant increase in the productivity of MES of ca. 100 mM d^{-1} m^{-2} (Zhang et al., 2013). In a later study by the same group, using nickel nanowires anchored to graphite electrode resulted in the highest normalized acetate production rate recorded for a pure culture-driven MES system due to enhanced surface area and the generation of a porous structure (Nie et al., 2013; Table 1).

CONCLUDING REMARKS

Microbial electrosynthesis is a young technology that made significant progress in terms of productivity over the last 5 years. Large efforts have been done to optimize known microbial catalysts for MES and to screen new ones with strong electroautotrophic properties (Rodrigues and Rosenbaum, 2014). Meanwhile, cathodes fabricated with novel materials or designed with better spatial arrangement are being explored and developed. Optimization of other parts of the electrochemical hardware such as the ion-exchange membrane and the current-collecting structure for MES processes are also ongoing (Varcoe et al., 2014). For example, it has been suggested that anion-exchange membranes can be exploited in the *in situ* selective extraction of acetate produced by MES at a lower energetic cost (Andersen et al., 2014). Furthermore, understanding in detail how electrons are transferred from the cathode to the microbial catalyst will help devise better-designed strategies to improve all aspects of MES. If this rate of improvement is maintained in the future, MES could fulfill its promise as an energetically efficient, environment-friendly, and versatile bioproduction strategy.

ACKNOWLEDGMENT

This work was funded by the Novo Nordisk Foundation.

REFERENCES

1. Andersen, S. J., Hennebel, T., Gildemyn, S., Coma, M., Desloover, J., Berton, J.,et al. (2014). Electrolytic membrane extraction enables production of fine chemicals from biorefinery sidestreams. *Environ. Sci. Technol.* 48, 7135–7142. doi: 10.1021/es500483w
2. Banerjee, A., Leang, C., Ueki, T., Nevin, K. P., and Lovley, D. R. (2014). A Lactose-Inducible system for metabolic engineering of*Clostridium ljungdahlii*. *Appl. Environ. Microbiol.* 80, 2410–2416. doi: 10.1128/AEM.03666-13
3. Bose, A., Gardel, E. J., Vidoudez, C., Parra, E. A., and Girguis, P. R. (2014). Electron uptake by iron-oxidizing phototrophic bacteria. *Nat. Commun.* 5, 3391. doi: 10.1038/ncomms4391
4. Cao, X., Huang, X., Liang, P., Xiao, K., Zhou, Y., Zhang, X.,et al. (2009). A new method for water desalination using microbial desalination cells. *Environ. Sci. Technol.* 43, 7148–7152. doi: 10.1021/es901950j
5. Carbajosa, S., Malki, M., Caillard, R., Lopez, M. F., Palomares, F. J., Martin-Gago, J. A.,et al. (2010). Electrochemical growth of*Acidithiobacillus ferrooxidans* on a graphite electrode for obtaining a biocathode for direct electrocatalytic reduction of oxygen.*Biosens. Bioelectron.* 26, 877–880. doi: 10.1016/j.bios.2010.07.037
6. Cheng, S., and Logan, B. E. (2007). Sustainable and efficient biohydrogen production via electrohydrogenesis. *Proc. Natl. Acad. Sci. U.S.A.* 104, 18871–18873. doi: 10.1073/pnas.0706379104
7. Cheng, S., Xing, D., Call, D. F., and Logan, B. E. (2009). Direct biological conversion of electrical current into methane by electromethanogenesis. *Environ. Sci. Technol.* 43, 3953–3958. doi: 10.1021/es803531g
8. Coppi, M. V., O'Neil, R. A., and Lovley, D. R. (2004). Identification of an uptake hydrogenase required for hydrogen-dependent reduction of Fe(III) and other electron acceptors by *Geobacter sulfurreducens*. *J. Bacteriol.* 186,3022–3028.doi:10.1128/JB.186.10.3022-3028.2004
9. Doud, D. F. R., and Angenent, L. T. (2014). Towards electrosynthesis with uncoupled extracellular electron uptake and metabolic growth: enhancing current uptake with *Rhodopseudomonas palustris*. *Environ. Sci. Technol. Lett.* 1, 351–355. doi: 10.1021/ez500244n
10. Drake, H. L., Daniel, S. L., Kusel, K., Matthies, C., Kuhner, C., and Braus-Stromeyer, S. (1997). Acetogenic bacteria: what are the in situ consequences of their diverse metabolic versatilities? *Biofactors* 6, 13–24. doi: 10.1002/biof.5520060103
11. Drake, H. L., Gossner, A. S., and Daniel, S. L. (2008). Old acetogens, new light. *Ann. N. Y. Acad. Sci.* 1125, 100–128. doi: 10.1196/annals.1419.016
12. Dumas, C., Basseguy, R., and Bergel, A. (2008). Microbial electrocatalysis with *Geobacter sulfurreducens* biofilm on stainless steel cathodes. *Electrochim. Acta* 53, 2494–2500. doi: 10.1016/j.electacta.2007.10.018
13. Fast, A. G., and Papousakis, E. T. (2012). Stoichiometric and energetic analyses of non-photosynthetic CO_2-fixation pathways to support synthetic biology strategies for production of fuels and chemicals. *Curr. Opin. Chem. Eng.* 1, 380–395. doi: 10.1016/j.coche.2012.07.005
14. Feist, A. M., Nagarajan, H., Rotaru, A. E., Tremblay, P. L., Zhang, T., Nevin, K. P.,et al. (2014). Constraint-based modeling of carbon fixation and the energetics of electron transfer in *Geobacter metallireducens*. *PLoS Comput. Biol.* 10:e1003575. doi: 10.1371/journal.pcbi.1003575PCOMPBIOL-D-13-01735
15. Gong, Y., Ebrahim, A., Feist, A. M., Embree, M., Zhang, T., Lovley, D.,et al. (2013). Sulfide-driven microbial electrosynthesis.*Environ. Sci. Technol.* 47, 568–573. doi: 10.1021/es303837j
16. Green, M. A., Emery, K., Hishikawa, Y., Warta, W., and Dunlop, E. D. (2014). Solar cell efficiency tables (version 45). *Prog. Photovolt. Res. Appl.* 23, 1–9. doi: 10.1002/pip.2573
17. Gregory, K. B., Bond, D. R., and Lovley, D. R. (2004). Graphite electrodes as electron donors for anaerobic respiration.*Environ. Microbiol.* 6, 596–604. doi: 10.1111/j.1462-2920.2004.00593.x

18. Gregory, K. B., and Lovley, D. R. (2005). Remediation and recovery of uranium from contaminated subsurface environments with electrodes. *Environ. Sci. Technol.* 39, 8943–8947. doi: 10.1021/es050457e

19. Hallenbeck, P. C., Grogger, M., and Veverka, D. (2014). Recent advances in microbial electrocatalysis. *Electrocatalysis* 5, 319–329. doi: 10.1007/s12678-014-0198-x

20. Huang, L. P., Jiang, L., Wang, Q., Quan, X., Yang, J., and Chen, L. (2014). Cobalt recovery with simultaneous methane and acetate production in biocathode microbial electrolysis cells. *Chem. Eng. J.* 253, 281–290. doi: 10.1016/j.cej.2014.05.080

21. Jiang, Y., Su, M., and Li, D. (2014). Removal of sulfide and production of methane from carbon dioxide in microbial fuel cells-microbial electrolysis cell (MFCs-MEC) coupled system. *Appl. Biochem. Biotechnol.* 172, 2720–2731. doi: 10.1007/s12010-013-0718-9

22. Jiang, Y., Su, M., Zhang, Y., Zhan, G., Tao, Y., and Li, D. (2013). Bioelectrochemical systems for simultaneously production of methane and acetate from carbon dioxide at relatively high rate. *Int. J. Hydrogen. Ener.* 38, 3497–3502. doi:10.1016/j.ijhydene.2012.12.107

23. Jourdin, L., Freguia, S., Donose, B. C., Chen, J., Wallace, G. G., Keller, J.,et al. (2014). A novel carbon nanotube modified scaffold as an efficient biocathode material for improved microbial electrosynthesis. *J. Mater. Chem. A* 2, 13093–13102. doi: 10.1039/c4ta03101f

24. Jürgensen, L., Ehimen, E. A., Born, J., and Holm-Nielsen, J. B. (2014). Utilization of surplus electricity from wind power for dynamic biogas upgrading: Nothern Germany case study. *Biomass Bioenergy* 66, 126–132. doi: 10.1016/j.biombioe.2014.02.032

25. Kato, S., Yumoto, I., and Kamagata, Y. (2015). Isolation of acetogenic bacteria that induce biocorrosion by utilizing metallic iron as the sole electron donor. *Appl. Environ. Microbiol.* 81, 67–73. doi: 10.1128/AEM.02767-14

26. Khunjar, W. O., Sahin, A., West, A. C., Chandran, K., and Banta, S. (2012). Biomass production from electricity using ammonia as an electron carrier in a reverse microbial fuel cell. *PLoS ONE* 7:e44846. doi: 10.1371/journal.pone.0044846

27. Kinsel, N. A., and Umbreit, W. W. (1964). Method for electrolysis of culture medium to increase growth of the sulfur-oxidizing iron bacterium *Ferrobacillus sulfooxidans*. *J. Bacteriol.* 87, 1243–1244.

28. Kobayashi, H., Saito, N., Fu, Q., Kawaguchi, H., Vilcaez, J., Wakayama, T.,et al. (2013). Bio-electrochemical property and phylogenetic diversity of microbial communities associated with bioelectrodes of an electromethanogenic reactor. *J. Biosci. Bioeng.* 116, 114–117. doi: 10.1016/j.jbiosc.2013.01.001

29. Köpke, M., Held, C., Hujer, S., Liesegang, H., Wiezer, A., Wollherr, A.,et al. (2010). *Clostridium ljungdahlii* represents a microbial production platform based on syngas. *Proc. Natl. Acad. Sci. U.S.A.* 107, 13087–13092. doi: 10.1073/pnas.1004716107

30. Kracke, F., and Krömer, J. O. (2014). Identifying target processes for microbial electrosynthesis by elementary mode analysis.*BMC Bioinformatics* 15:6590. doi: 10.1186/s12859-014-0410-2

31. Krieg, T., Sydow, A., Schroder, U., Schrader, J., and Holtmann, D. (2014). Reactor concepts for bioelectrochemical syntheses and energy conversion. *Trends Biotechnol.* 32, 645–655. doi: 10.1016/j.tibtech.2014.10.004

32. LaBelle, E. V., Marshall, C. W., Gilbert, J. A., and May, H. D. (2014). Influence of acidic pH on hydrogen and acetate production by an electrosynthetic microbiome. *PLoS ONE* 9:e109935. doi: 10.1371/journal.pone.0109935

33. Leang, C., Ueki, T., Nevin, K. P., and Lovley, D. R. (2013). A genetic system for *Clostridium ljungdahlii*: a chassis for autotrophic production of biocommodities and a model homoacetogen. *Appl. Environ. Microbiol.* 79, 1102–1109. doi: 10.1128/AEM.02891-12

34. Li, H., Opgenorth, P. H., Wernick, D. G., Rogers, S., Wu, T. Y., Higashide, W.,et al. (2012). Integrated electromicrobial conversion of CO_2 to higher alcohols. *Science* 335:1596. doi: 10.1126/science.1217643

35. Logan, B. E. (2010). Scaling up microbial fuel cells and other bioelectrochemical systems. *Appl. Microbiol. Biotechnol.* 85, 1665–1671. doi: 10.1007/s00253-009-2378-9

36. Logan, B. E., and Rabaey, K. (2012). Conversion of wastes into bioelectricity and chemicals by using microbial electrochemical technologies. *Science* 337, 686–690. doi: 10.1126/science.1217412

37. Lohner, S. T., Deutzmann, J. S., Logan, B. E., Leigh, J., and Spormann, A. M. (2014). Hydrogenase-independent uptake and metabolism of electrons by the archaeon *Methanococcus maripaludis*. *ISME J.* 8, 1673–1681. doi: 10.1038/ismej.2014.82

38. Lovley, D. R. (2009). Future shock from the microbe electric. *Microb. Biotechnol.* 2, 139–141. doi: 10.1111/j.1751-7915.2009.00090_9.x

39. Lovley, D. R. (2011). Live wires: direct extracellular electron exchange for bioenergy and the bioremediation of energy-related contamination. *Energy Environ. Sci.* 4, 4896–4906. doi: 10.1039/C1ee02229f

40. Lovley, D. R. (2012). Electromicrobiology. *Annu. Rev. Microbiol.* 66, 391–409. doi: 10.1146/annurev-micro-092611-150104

41. Lovley, D. R., and Nevin, K. P. (2011). A shift in the current: new applications and concepts for microbe-electrode electron exchange. *Curr. Opin. Biotechnol.* 22, 441–448. doi: 10.1016/j.copbio.2011.01.009

42. Lovley, D. R., and Nevin, K. P. (2013). Electrobiocommodities: powering microbial production of fuels and commodity chemicals from carbon dioxide with electricity. *Curr. Opin. Biotechnol.* 24, 385–390. doi: 10.1016/j.copbio.2013.02.012

43. MacDonald, M. (2003). *Photobiology of Higher Plants*. New York, NY: Wiley.

44. Malvankar, N. S., and Lovley, D. R. (2012). Microbial nanowires: a new paradigm for biological electron transfer and bioelectronics. *ChemSusChem* 5, 1039–1046. doi: 10.1002/cssc.201100733

45. Malvankar, N. S., Yalcin, S. E., Tuominen, M. T., and Lovley, D. R. (2014). Visualization of charge propagation along individual pili proteins using ambient electrostatic force microscopy. *Nat. Nanotechnol.* 9, 1012–1017. doi: 10.1038/nnano.2014.236

46. Marshall, C. W., Ross, D. E., Fichot, E. B., Norman, R. S., and May, H. D. (2012). Electrosynthesis of commodity chemicals by an autotrophic microbial community. *Appl. Environ. Microbiol.* 78, 8412–8420. doi: 10.1128/AEM.02401-12

47. Marshall, C. W., Ross, D. E., Fichot, E. B., Norman, R. S., and May, H. D. (2013). Long-term operation of microbial electrosynthesis systems improves acetate production by autotrophic microbiomes. *Environ. Sci. Technol.* 47, 6023–6029. doi: 10.1021/es400341b

48. Matsumoto, N., Nakasono, S., Ohmura, N., and Saiki, H. (1999). Extension of logarithmic growth of *Thiobacillus ferrooxidans*by potential controlled electrochemical reduction of Fe(III). *Biotechnol. Bioeng.* 64, 716–721. doi: 10.1002/(SICI)1097-0290(19990920)64:6<716::AID-BIT11>3.0.CO;2-9

49. Matsumoto, N., Yoshinaga, H., Ohmura, N., Ando, A., and Saiki, H. (2000). High density cultivation of two strains of iron-oxidizing bacteria through reduction of ferric iron by intermittent electrolysis. *Biotechnol. Bioeng.* 70, 464–466. doi: 10.1002/1097-0290(20001120)70:4<464::AID-BIT12>3.0.CO;2-A

50. Möller, B., Ossmer, R., Howard, B. H., Gottschalk, G., and Hippe, H. (1984). *Sporomusa*, a new genus of gram-negative anaerobic-bacteria including *Sporomusa sphaeroides* spec. nov and *Sporomusa ovata* spec. nov. *Arch. Microbiol.* 139, 388–396. doi: 10.1007/Bf00408385

51. Nakasono, S., Matsumoto, N., and Saiki, H. (1997). Electrochemical cultivation of *Thiobacillus ferrooxidans* by potential control. *Bioelectrochem. Bioenerg.* 43, 61–66. doi: 10.1016/S0302-4598(97)00001-9

52. Nevin, K. P., Hensley, S. A., Franks, A. E., Summers, Z. M., Ou, J., Woodard, T. L.,et al. (2011). Electrosynthesis of organic compounds from carbon dioxide is catalyzed by a diversity of acetogenic microorganisms. *Appl. Environ. Microbiol.* 77, 2882–2886. doi: 10.1128/AEM.02642-10

53. Nevin, K. P., Woodard, T. L., Franks, A. E., Summers, Z. M., and Lovley, D. R. (2010). Microbial electrosynthesis: feeding microbes electricity to convert carbon dioxide and water to multicarbon extracellular organic compounds. *mBio* 1, e00103-10. doi: 10.1128/mBio.00103-10

54. Nie, H., Zhang, T., Cui, M., Lu, H., Lovley, D. R., and Russell, T. P. (2013). Improved cathode for high efficient microbial-catalyzed reduction in microbial electrosynthesis cells. *Phys. Chem. Chem. Phys.* 15, 14290–14294. doi: 10.1039/c3cp52697f

55. Park, D. H., Laivenieks, M., Guettler, M. V., Jain, M. K., and Zeikus, J. G. (1999). Microbial utilization of electrically reduced neutral red as the sole electron donor for growth and metabolite production. *Appl. Environ. Microbiol.* 65, 2912–2917.

56. Patil, S. A., Hägerhäll, C., and Gorton, L. (2012). Electron transfer mechanisms between microorganisms and electrodes in bioelectrochemical systems. *Bioanal. Rev.* 4, 159–192. doi: 10.1007/s12566-012-0033-x

57. Pisciotta, J. M., Zaybak, Z., Call, D. F., Nam, J. Y., and Logan, B. E. (2012). Enrichment of microbial electrolysis cell biocathodes from sediment microbial fuel cell bioanodes. *Appl. Environ. Microbiol.* 78, 5212–5219. doi: 10.1128/AEM.00480-12

58. Poehlein, A., Gottschalk, G., and Daniel, R. (2013). First insights into the genome of the gram-negative, endospore-forming organism *Sporomusa ovata* strain H1 DSM 2662. *Genome Announc.* 1, e00734-13. doi: 10.1128/genomeA.00734-13

59. Rabaey, K., Girguis, P., and Nielsen, L. K. (2011). Metabolic and practical considerations on microbial electrosynthesis. *Curr. Opin. Biotechnol.* 22, 371–377. doi: 10.1016/j.copbio.2011.01.010

60. Rabaey, K., and Rozendal, R. A. (2010). Microbial electrosynthesis - revisiting the electrical route for microbial production. *Nat. Rev. Microbiol.* 8, 706–716. doi: 10.1038/nrmicro2422

61. Ragsdale, S. W., and Pierce, E. (2008). Acetogenesis and the Wood–Ljungdahl pathway of CO_2 fixation. *Biochim. Biophys. Acta*1784, 1873–1898. doi: 10.1016/j.bbapap.2008.08.012

62. Reguera, G., Mccarthy, K. D., Mehta, T., Nicoll, J. S., Tuominen, M. T., and Lovley, D. R. (2005). Extracellular electron transfer via microbial nanowires. *Nature* 435, 1098–1101. doi: 10.1038/nature03661

63. Rodrigues, T. D., and Rosenbaum, M. A. (2014). Microbial electroreduction: screening for new cathodic biocatalysts.*Chemelectrochem* 1, 1916–1922. doi: 10.1002/celc.201402239

64. Rosenbaum, M., Aulenta, F., Villano, M., and Angenent, L. T. (2011). Cathodes as electron donors for microbial metabolism: which extracellular electron transfer mechanisms are involved? *Bioresour. Technol.* 102, 324–333. doi: 10.1016/j.biortech.2010.07.008

65. Rosenbaum, M. A., and Franks, A. E. (2014). Microbial catalysis in bioelectrochemical technologies: status quo, challenges and perspectives. *Appl. Microbiol. Biotechnol.* 98, 509–518. doi: 10.1007/s00253-013-5396-6

66. Ross, D. E., Flynn, J. M., Baron, D. B., Gralnick, J. A., and Bond, D. R. (2011). Towards electrosynthesis in *Shewanella*: energetics of reversing the mtr pathway for reductive metabolism. *PLoS ONE* 6:e16649. doi: 10.1371/journal.pone.0016649

67. Shi, L., Rosso, K. M., Clarke, T. A., Richardson, D. J., Zachara, J. M., and Fredrickson, J. K. (2012). Molecular underpinnings of Fe(III) oxide reduction by *Shewanella oneidensis* MR-1. *Front. Microbiol.* 3:50. doi: 10.3389/fmicb.2012.00050

68. Soussan, L., Riess, J., Erable, B., Delia, M. L., and Bergel, A. (2013). Electrochemical reduction of CO_2 catalysed by *Geobacter sulfurreducens* grown on polarized

stainless steel cathodes. *Electrochem. Commun.* 28, 27–30. doi: 10.1016/j.elecom.2012.11.033

69. Speers, A. M., and Reguera, G. (2012). Electron donors supporting growth and electroactivity of *Geobacter sulfurreducens*anode biofilms. *Appl. Environ. Microbiol.* 78, 437–444. doi: 10.1128/AEM.06782-11

70. Strycharz, S. M., Glaven, R. H., Coppi, M. V., Gannon, S. M., Perpetua, L. A., Liu, A.,et al. (2011). Gene expression and deletion analysis of mechanisms for electron transfer from electrodes to *Geobacter sulfurreducens*. *Bioelectrochemistry* 80, 142–150. doi: 10.1016/j.bioelechem.2010.07.005

71. Summers, Z. M., Gralnick, J. A., and Bond, D. R. (2013). Cultivation of an obligate Fe(II)-oxidizing lithoautotrophic bacterium using electrodes. *mBio* 4, e00420–e00412. doi: 10.1128/mBio.00420-12

72. Sydow, A., Krieg, T., Mayer, F., Schrader, J., and Holtmann, D. (2014). Electroactive bacteria-molecular mechanisms and genetic tools. *Appl. Microbiol. Biotechnol.* 98, 8481–8495. doi: 10.1007/s00253-014-6005-z

73. Thomas, A. W., Garner, L. E., Nevin, K. P., Woodard, T. L., Franks, A. E., Lovley, D. R.,et al. (2013). A lipid membrane intercalating conjugated oligoelectrolyte enables electrode driven succinate production in *Shewanella*. *Energy Environ. Sci.* 6, 1761–1765. doi: 10.1039/C3ee00071k

74. Tremblay, P. L., Aklujkar, M., Leang, C., Nevin, K. P., and Lovley, D. (2012). A genetic system for *Geobacter metallireducens*: role of the flagellin and pilin in the reduction of Fe(III) oxide. *Environ. Microbiol. Rep.* 4, 82–88. doi: 10.1111/j.1758-2229.2011.00305.x

75. Tremblay, P. L., Zhang, T., Dar, S. A., Leang, C., and Lovley, D. R. (2013). The Rnf complex of *Clostridium ljungdahlii* is a proton-translocating ferredoxin:NAD^+ oxidoreductase essential for autotrophic growth. *mBio* 4, e00406–e00412. doi: 10.1128/mBio.00406-12

76. Ueki, T., Nevin, K. P., Woodard, T. L., and Lovley, D. R. (2014). Converting carbon dioxide to butyrate with an engineered strain of *Clostridium ljungdahlii*. *mBio* 5, e01636–e01614. doi: 10.1128/mBio.01636-14

77. Varcoe, J. R., Atanassov, P., Dekel, D. R., Herring, A. M., Hickner, M. A., Kohl, P. A.,et al. (2014). Anion-exchange membranes in electrochemical energy systems. *Energy Environ. Sci.* 7, 3135–3191. doi: 10.1039/C4ee01303d

78. Vargas, M., Malvankar, N. S., Tremblay, P. L., Leang, C., Smith, J. A., Patel, P.,et al. (2013). Aromatic amino acids required for pili conductivity and long-range extracellular electron transport in *Geobacter sulfurreducens*. *mBio* 4, e00105–e00113. doi: 10.1128/mBio.00105-13

79. Villano, M., Aulenta, F., Ciucci, C., Ferri, T., Giuliano, A., and Majone, M. (2010). Bioelectrochemical reduction of CO_2 to CH_4via direct and indirect extracellular electron transfer by a hydrogenophilic methanogenic culture. *Bioresour. Technol.* 101, 3085–3090. doi: 10.1016/j.biortech.2009.12.077

80. Wang, H., and Ren, Z. J. (2013). A comprehensive review of microbial electrochemical systems as a platform technology.*Biotechnol. Adv.* 31, 1796–1807. doi: 10.1016/j.biotechadv.2013.10.001

81. Wang, Z., Leary, D. H., Malanoski, A. P., Li, R. W., Hervey, W. J. T., Eddie, B. J.,et al. (2015). A previously uncharacterized, nonphotosynthetic member of the Chromatiaceae is the primary CO_2-fixing constituent in a self-regenerating biocathode. *Appl. Environ. Microbiol.* 81, 699–712. doi: 10.1128/AEM.02947-14

82. Wei, J., Liang, P., and Huang, X. (2011). Recent progress in electrodes for microbial fuel cells. *Bioresour. Technol.* 102, 9335–9344. doi: 10.1016/j.biortech.2011.07.019

83. Yi, H., Nevin, K. P., Kim, B. C., Franks, A. E., Klimes, A., Tender, L. M.,et al. (2009). Selection of a variant of *Geobacter sulfurreducens* with enhanced capacity for current production in microbial fuel cells. *Biosens. Bioelectron.* 24, 3498–3503. doi: 10.1016/j.bios.2009.05.004

84. Yong, Y. C., Yu, Y. Y., Zhang, X., and Song, H. (2014). Highly active bidirectional electron transfer by a self-assembled electroactive reduced-graphene-oxide-hybridized biofilm. *Angew. Chem. Int. Ed. Engl.* 53, 4480–4483. doi: 10.1002/anie.201400463

85. Zaybak, Z., Pisciotta, J. M., Tokash, J. C., and Logan, B. E. (2013). Enhanced start-up of anaerobic facultatively autotrophic biocathodes in bioelectrochemical systems. *J. Biotechnol.* 168, 478–485. doi: 10.1016/j.jbiotec.2013.10.001

86. Zhang, T., Nie, H. R., Bain, T. S., Lu, H. Y., Cui, M. M., Snoeyenbos-West, O. L.,et al. (2013). Improved cathode materials for microbial electrosynthesis. *Energy Environ. Sci.* 6, 217–224. doi: 10.1039/C2ee23350a

CHAPTER 3

EXTRACELLULAR ELECTRON TRANSFER FROM CATHODE TO MICROBES: APPLICATION FOR BIOFUEL PRODUCTION

Okkyoung Choi and ByoungIn Sang

Keywords: Bioelectrochemical synthesis, Extracellular electron transfer, Cathodic electron, Electrofuel

BACKGROUND

An eventual replacement of fossil energy source with sustainable energy system is unavoidable. Biofuels have emerged as one of the sustainable fuels sources and it is considered as alternatives to petroleum. Biomass captured the energy from sunlight and stored it as high-energy chemical bonds, which is used for biofuels. More recently, electrofuels have been studied for liquid fuels as a means for intermittent electricity storage [1] using the energy of low-potential electrons such as hydrogen gas, reduced metal, or electricity [2]. It usually uses the interaction between microbes and electrode, through extracellular electron transfer.

Bioelectrochemical synthesis (BES) uses extracellular electron transfer of microorganisms catalyzing anodic and/or cathodic reactions. BES has two categories according to the direction of electron flow, microbial fuel cells (MFC, electricity production), and microbial electrosynthesis (MES, electricity consumption). A microbial fuel cell uses extracellular electron transfer to an electrode originating from organic compounds consumed by microorganisms. Microbial electrosynthesis uses electron transfer from an electrode to microorganisms producing reduced biochemical compounds. An electrode is thus used as an electron acceptor (MFC) or an electron donor (MES). Extracellular electron transfer has been gaining wide interest in relation to microbial electrochemical synthesis [1, 3], interspecies electron transfer [4, 5], and microbial immobilization of heavy metals for bioremediation [6,7] (Table 1). In particular, biofuels or biochemicals are reduced compounds and the reducing power is needed in microbial fermentation processes [8, 9]. An external supply of electrons using electricity enhances the reducing process in microbial metabolism. Direct electron transfer is ideal in extracellular electron transfer from a cathode to microbes.

The two mostly extensively studied microorganisms for extracellular electron transfer are *Geobacter* and*Shewanella* species. *Geobacter* and *Shewanella* are metal-reducing and gram-negative bacteria. Extracellular electron transfer in microorganisms is used in the metal reduction process by the microorganism and, in this case, the metal is used as an electron acceptor. When metal (hydr)oxides that are poorly soluble in water are present as electron acceptors, extracellular electron transfer occurs using multihaem c-type cytochromes in*Geobacter* and *Shewanella* [10]. Based on this phenomenon, the microorganisms are able to extracellularly transfer electrons and this can be applied for BES.

The mode of extracellular electron transfer is broadly divided into the following: (1) direct electron transfer: nanowire [11] or direct contact [12]; (2) mediators-shuttled: endogenous, exogenous as a redox compound or a by-product [13, 14, 15]; and (3) extracellular polymeric substances (EPS) of biofilms [16] (Fig. 1).

Citation: Okkyoung Choi and ByoungIn Sang, Extracellular electron transfer from cathode to microbes: application for biofuel production, DOI 10.1186/s13068-016-0426-0.

Table 1: The application of bioelectrochemical reduction

Application	Product	Reaction conditions	Key outcomes	Ref.
Direct reduction	$Cr^{6+} \rightarrow Cr^{3+}$	*G. sulfurreducens*, −600 mV vs. Ag/AgCl	U(VI) was removed and recovered using poised electrode	[19]
		Shewanella oneidensis MR-1, −500 mV vs. Ag/AgCl	Lactate and the electrode as the electron donors for Cr(VI) reduction	[18]
	Fumarate → succinate	*G. sulfurreducens*, −500 mV vs. Ag/AgCl	Fumarate reduction dependent on current supply	[48]
		Shewanella species in biocathode of microbial fuel cell	Similar comparison under chromate reducing condition	[102]
	Nitrate reduction	Nitrifying and denitrifying microorganisms at +197 mV vs. SHE	Simultaneous occurrence of nitrification and denitrification at a biocathode	[49]
		Denitrifying microorganisms at −123 mV vs. SHE	Long-term stability, carbon-free operation	[51]
Indirect reduction	Caproate and caprylate production from acetate	Acetate fed at −0.9 V vs. NHE	In situ-produced hydrogen as electron donor, low concentration and reaction rates	[90]
	Ethanol production from acetate	−550 mV vs. NHE, artificial mediator tested	Methyl viologen increased ethanol production but inhibited butyrate and methane formation, still hydrogen was coproduced at the cathode	[81]
	Alcohol formation from glycerol	Open circuit operation	Changes in microbial community and product outcomes after current supply	[87]
	Reduction of acetate and butyrate to mainly alcohols and acetone	−820 mV vs. Ag/AgCl	Halotolerant mixed sulfate-reducing bacteria culture	[92]
	Polyhydroxyalkanoates (PHA) from glucose	512 mV, the biocathode coupled to a bioanode in an MEC	Microaerophilic microenvironment at cathode enhanced PHA synthesis as alternative pathway to re-oxidize the NADH	[94]
	Butyraldehyde to butanol	Immobilized alcohol dehydrogenase at −400 mV vs. Ag/AgCl	Reduction to alcohol by current without supplement of NADH	[88]
	Hydrogen production	−700 mV vs. Ag/AgCl	Increased cathodic hydrogen efficiency on microbial biocathode based on a naturally selected mixed culture	[103]
		500 mV, the biocathode coupled to a bioanode in an MEC	Operated for a long period with high current density but phosphate precipitation on the biocathode	[104]
		−700 mV vs. SHE	*Desulfovibrio* sp. as a dominant microorganism in the biocathode	[22]
	Methane production	−700 mV vs. Ag/AgCl	Methane production directly from current	[53]
		−550 mV vs. NHE	CO_2 reduction to CH_4, need to reduce the internal resistance	[105]
	Improved 1,3-propandiol production from glycerol	−900 mV vs. SHE	Electrical current as the driving force for a mixed population fermenting glycerol in the cathode	[93]
	Improved butanol production from glucose	+0.045 V vs. SHE	Increased alcohol production in electrofermentation with increased a ratio $NADH/NAD^+$	[24]

Electrofuel from $CO_2$and electricity	Butyrate	−800 mV vs. SHE	Production of organic compounds from CO_2 by hydrogen driven by a cathode	[100]
	Acetate	−590 mV vs. SHE	Higher acetate production than on unmodified graphite	[99]
	Acetate, 2-oxobutyrate	−400 mV vs. SHE	The production of organic acids by current consumption	[106]

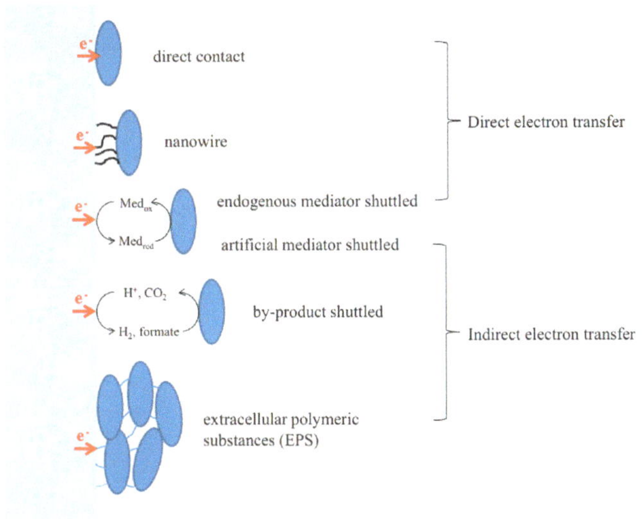

Figure 1: Cathodic electron transfer mode. Electrons from a cathode flow into a microbe directly, through direct contact, nanowire, and endogenous mediator; or indirectly, through an artificial mediator, by-product, or EPS.

Electron Transfer from a Cathode to Microbes

Multihame c-type cytochrome is a key component of the electron transfer channel in gram-negative bacteria [10]. Filamentous conductive pili are also involved in electron transfer in *Shewanella* [17] and *Geobacter* [11]. BES uses two directions, i.e., microbe → electrode (anode) in MFC and electrode (cathode) → microbe in MES, with the same or different mode. Electrons flow from an electron donor with a relatively lower redox potential to an electron acceptor with higher redox potential. In this light, in the present study we address the question that of whether it is possible to use the same electron transport chain for the opposite direction.

The redox tower in Fig. 2 shows the broad range of redox potential for MtrC (located on an extracellular site of the outer membrane), MrtA (a periplasmic c-type cytochrome), CymA (a link point between the inner membrane and the periplasm), and OmcA (anchored in the inner membrane), which were reported to play roles in electron transfer. It is proposed that reversible electron transfer within cytochrome c complex channels is feasible and the same electron transport chain can be used for the opposite direction. Extensively studied in MFC as iron-reducing bacteria, *Shewanella oneidensis* MR-1 [18] and *Geobacter* spp. [19] were reported to reduce the highly toxic hexavalent chromium (Cr(VI)) using a cathode. This indicates that both directions for electron transfer would be possible in current-producing bacteria, i.e., microbe → anode and cathode → microbe. However, it was reported that *Shewanella* showed a reversed Mtr pathway [20] but*Geobacter* used a different mode in the opposite direction [21].

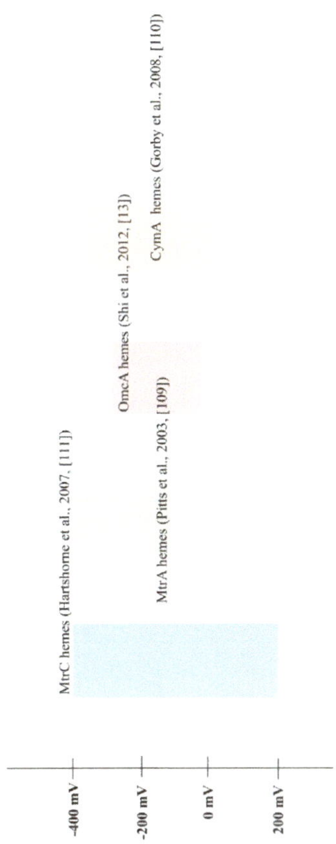

Figure 2: The broad range of redox potential in c-type cytochromes. Considering the possibility of bidirectional electron flow (cathodic, anodic), the broad redox potential suggests the direction of electron flow would be flexible in an electron transfer channel consisting of cytochrome C. The number in a *square bracket* represents the citation number in the reference list.

Direct electron transfer from a cathode to microbes has been observed in a biocathode for microbial communities including betaproteobacteria [22, 23] and firmicutes [22], in addition to *Shewanella* and*Geobacter*. The presence of other electroactive, electron endergonic strains thus should be possible.*Clostridium pasteurianum* increased butanol production using cathode electron transfer without any mediator [24]. Nevertheless, the precise electron transfer channel for acceptance of extracellular electrons has not been verified; the redox enzyme in the membrane, however, may be involved in electrochemical reduction. Ferredoxin extracted from *C. pasteurianum* previously showed direct electrochemical reduction [25], but there is still no evidence of direct electron transfer through ferredoxin in whole cells. Also, several other redox proteins could be candidates for extracellular electron transfer.

Predicted electron transfer proteins involved in extracellular electron transfer For direct electron transfer, a membrane-bound redox protein is needed. However, there has been no study of redox proteins involved in direct electron transfer except periplasmic c-type cytochrome. Several studies have reported the possibility of direct electron transfer by microorganisms in the absence of c-type cytochrome, and here we present some possible redox proteins involved in electron transfer channels including cytochromes (Table 2).

Table 2: Predicted electron transfer proteins associated in extracellular electron transfer

Predicted electron transfer proteins	Active sites	Candidate microorganism associated in extracellular electron transfer	Microorganism used in bioelectrochemical system
Cytochrome C	Heme protein	Metal-reducing bacteria (*Geobacter,Shewanella*)	*G. sulfurreducens* [19, 48], *Shewanella oneidensis* [18]
Ferredoxin	Fe-S protein	Clostridia, acetogens, methanogens	*Clostridium pasteurianum* [24], *Clostridium ljungdahlii, Clostridium aceticum, Moorella thermoacetica* [106], *Sporomusa ovata* [106,107], *Methanothermobacter thermautotrophicus* [108],*Methanobacterium palustre, Methanococcus maripaludis* [53, 54]
Rubredoxin	Fe-S protein without acid–labile sulfur	Sulfate-reducing bacteria	*Desulfovibrio* sp. [22, 70]
Hydrogenase	[NiFe] or [FeFe] or [Fe]-only	Sulfate-reducing bacteria, methanogen	*Desulfovibrio* sp. [22, 70], methanogens [52, 53]
Formate dehydrogenase	Molybdenum or tungsten	Sulfate-reducing bacteria	*Desulfovibrio* sp. [22, 70]

Cytochrome

The heme in cytochrome participates in electron transfer processes. Cell surface-localized cytochromes (OmcE and OmcS in *Geobacter sulfurreducens*, MtrC and OmcA in *Shewanella oneidensis* MR-1) are important components for electron transfer [26]. A microarray analysis of *G. sulfurreducens* gene transcript abundance showed the c-type cytochrome was highly expressed in current-producing biofilms [21]. However, it was suggested that the mechanism of two opposite directions, i.e., microbe → electrode and electrode → microbe, would be significantly different in *G. sulfurreducens.* Recently, cytochrome PccH with a unusually low redox potential for cytochrome (−24 mV at pH 7) located in the periplasm was proposed as a candidate to provide electron transfer in *G. sulfurreducens*, even though PccH could not be involved in the first step of accepting electrons [27, 28]. It is meanwhile known that *S. oneidensis* has a similar mechanism in both directions mainly using flavins (flavin mononucleotide and riboflavin) with cytochrome C [29].

Ferredoxin: Membrane-Bound Complex

Rnf complexes (a membrane-bound NADH:ferredoxin oxidoreductase) are redox-driven ion pumps and have a membrane-bound, proton-translocating ferredoxin: NAD^+ oxidoreductase contributing to ATP synthesis (energy conservation) in acetogens such as *Clostridium ljungdahlii.* RnF is a multifunctional device with nitrogen fixation, proton translocation, and electron transport capabilities [30]. It is four flavin-containing cytoplasmatic multienzyme complexes from clostridia, acetogens, and methanogens [31] and so involved in flavin-based electron bifurcation (FBEB), which is regarded as a third mode of energy conservation in addition to substrate-level phosphorylation (SLP) and electron transport phosphorylation (ETP) [32]. Not all acetogens have rnf genes.

An energy-conserving hydrogenase (Ech) also plays a role in reducing ferredoxin with proton motive force [33]. It involves a coupling mechanism: an exergonic process attributes to coupled endergonic process; ferredoxin reduction with low potential as an exergonic reaction is coupled with H_2 or NADH, a high-potential acceptor as endergonic reduction [34]. In methanogens, in the absence of cytochromes, methyltransferase is involved in a exergonic reaction to drive the extrusion of ions (Na^+ or H^+) across the membrane [35]. In the context energy conservation in a bioelectrochemical system, electron supplementation from cathode would lead to FBEB. Electron bifurcating ferredoxin reduction H^+ gradient (for *C. ljungdahlii*) or Na^+ (for *Acetobacterium woodii*) via membrane-bound Rnf complex was supposed as key components in electron transport chain [36].

Rubredoxin

Rubredoxin (Rub) is also an electron transfer protein having a Fe-S cluster with relatively small molecules (about 55 amino acids) [37, 38]. It is one of the electron transfer components of sulfate-reducing bacteria (SRB) [39] and is also detected in *Clostridium pasteurianum* [37]. In *Desulfovibrio vulgaris,* Rub reduces hydrogen peroxide and superoxide [40]. Rub showed an electrochemical response with electrodes [41, 42]. Detailed roles of Rub in microorganisms have not been found but it is expected to be involved in an electron transfer channel.

Hydrogenase and Formate Dehydrogenase

It was recently reported that a hydrogenase and formate dehydrogenase, which are released from cells, are adsorbed onto electrodes to accept electrons in biocorrosion and bioelectrosynthesis [43]. Methyl viologen-mediated electron transfer to hydrogenase from cathodes and mediatorless H_2 production using cathodic electron transfer were previously suggested as electron transfer modes [44]. Formate dehydrogenase also showed direct electron transfer from cathodes [43, 45]. The periplasmic formate dehydrogenase transfers electrons to cytochrome C in *D. desulfuricans* [46]. The combination of periplasmic enzyme and c-type cytochrome likely provides the electrical wiring [44]. Several membrane-bound enzymes such as fumarate reductase [47, 48] and a denitrification enzyme [49, 50, 51] led to bioelectrochemical reduction. Therefore, a periplasmic enzyme could be involved in an electron transfer channel in bioelectrochemical systems.

ELECTROACTIVE MICROORGANISMS

Methanogens and Acetogens

The conversion of CO_2 to CH_4 was reported in a biocathode consisting of a methanogen via direct or indirect (H_2 mediator) channels [52, 53, 54]. The electron donor for methanogenesis is H_2 for autotrophic methanogens or acetate for acetoclastic methanogens. It is supposed that, as in metal-reducing bacteria, the specific electron transfer channel in methanogens plays a role in extracellular electron transfer. Abiotically produced hydrogen is also used by methanogens in indirect electromethanogenesis, instead of direct cathodic electrons [55]. While no electron transfer channel involved in electron transfer from a cathode in methanogens has been identified, energy conservation by bifurcated electron transfer in methanogens could still potentially be found [56].

The study of enzyme purification and protein identification using mass spectroscopy in an acetotrophic methanogen, *Methanosarcina acetivorans*, showed that ferredoxin reduced membrane-associated multi-heme cytochrome c in Rnf [57, 58]. Methanogens have membrane-associated hydrogenases using ferredoxin or methanophenazine as redox partners [59]. It was reported that hydrogenase and formate dehydrogenase released out of cells mediate electron transfer between a cathode and *Methanococcus maripaludis* [43]. Also, interspecies electron transfer was shown through flagellum-like appendages between *Pelotomaculum thermopropionicum* and *Methanothermobacter thermoautotrophicus* in the form of aggregates [5].

Several acetogenic bacteria (acetate production from CO_2 and H_2) including *Sporomusa ovata, Sporomusa silvacetica, Sporomusa sphaeroides, Clostridium ljungdahlii, Clostridium aceticum*, and *Moorella thermoacetica* consumed electrons from a cathode to reduce CO_2 to acetate [60]. Recently, an acetogen closely related with *Sporomusa sphaeroides* was isolated and showed acetogenic growth using Fe(O) as a sole electron donor [61]. The acetogens *Moorella thermoacetica* and *C. formicoaceticum* reduced CO_2 to formate, consuming electricity at the cathode compartment [62]. Although the mode of electron transfer to an electroactive acetogen from a cathode is still not known, the membrane-bound cytochromes and cobalt-containing corrinoids were suggested as candidates for an electron transfer channels [63]. Also, cytochrome-b enzymes (membrane-integral b-type cytochromes, −0.215 V vs. SHE) were suggested to be involved in the electron transfer process of acetogens [64].

Metal-Oxidizing Bacteria and Sulfur-Utilizing Bacteria

The ability of iron-reducing bacteria to give electrons anodes gave rise to the hypothesis that iron-oxidizing bacteria (FeOB) could accept electrons from cathodes in two FeOB, *Mariprofundus ferrooxydans* and*Rhodopseudomonas palustris*, in recent studies [65, 66, 67] (Fig. 3). The marine isolate *Mariprofundus ferrooxydans* PV-1 used a cathode as a sole electron donor, generated ATP, and fixed CO_2 [67]. *Rhodopseudomonas palustris* TIE-1 accepted electrons from a cathode, independent of photosynthesis. The dark current indicated extracellular electron uptake uncoupled from the cyclic photosynthetic apparatus and the *pioABC* operon influenced electron uptake [65]. *Rhodopseudomonas palustris* TIE-1 increased electron uptake rate 56-fold with unlimited Fe(II) supplementation in a photobioelectrochemical system [67].

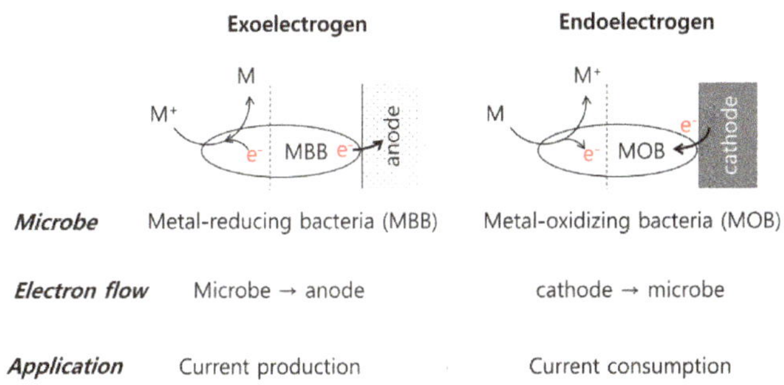

Figure 3: Analogy between metal-utilizing bacteria and direct extracellular electron transfer in a bioelectrochemical system. The *left side* of the *dotted line* shows the electron flow with metals in metal-utilizing bacteria and the *right side*indicates interaction with the electrode.

It was reported that isolated marine corrosive delta proteobacterial SRB used elemental iron as the sole electron donor and reduced sulfate, showing the possibility of extracellular electron transfer [68]. Recently, a sulfide-oxidizing bacteria, *Desulfobulbaceae*, was reported to reduce oxygen in the upper layers of marine sediments using centimeter-long filaments [69]. For removal of H_2S, the product of sulfate reduction and a toxic gas to oxygen-consuming organisms, sulfide-oxidizing bacteria used oxygen as an electron acceptor using filaments as electrical cables for H_2S oxidation to S [69].

The SRB reduced acetate, butyrate to ethanol, butanol, respectively, using electrons through direct electron transfer from a cathode [70]. It was suggested that the direct electron transfer could take place via a redox enzyme such as cytochrome-b in SRB [70]. The SRB *Desulfopila corrodens* strain IS4 showed direct electron transfer affecting iron corrosion. Electrochemical and infrared spectroelectrochemical analyses indicated c-type cytochromes were involved in electron transfer [71]. *Acidithiobacillus ferrooxidans, Desulfosporosinus orientis, Thiobacillus denitrificans, Sulfurimonas denitrificans,* and *Desulfovibrio piger* also showed electroactivity to accept electrons from a cathode in pure cultures [72].

Cathode Modification for Enhanced Performance of Bioelectrochemical Reduction

Efforts to improve the efficiency of electron transfer between a cathode and microorganisms have focused on increasing of the interfacial area and interfacial interactions. Nanoparticle attachment on a cathode was attempted with nano-nickel [73], carbon nanotubes [74, 75], conjugated oligoelectrolytes (COEs) [76], and carbon nanotubes on reticulated vitreous carbon (NanoWeb-RVC) [74, 77]. Also, a graphene-modified biocathode enhanced bioelectrochemical production of hydrogen in a MES system [78].

Another attempt involved positively charged surface modification. Extracellular electron transfer from a cathode to a microbe was increased using a positively charged functional group on the surface of a cathode [74]. Negatively charged *S. ovate* preferred to attach on a cathode and enhanced acceptance of electrons from the cathode for the reduction of CO_2 to acetate [74]. The positively charged anode led to an enriched biofilm on an anode but the negatively charged cathode has a repulsive interaction with microorganisms because the cell walls of most bacteria have an overall net negative charge. Therefore, attachment with microorganisms on a cathode has a charge barrier and one study showed that both the zeta potential and the hydrophobicity of cells increased in a current-consuming biofilm [24]. Modification should be tried according to the changes of cell surface characteristics on a cathode, in contrast with on an anode.

Application for Valuable Biofuel Production

A study of the life cycle assessment (LCA) showed MFCs do not give environmental benefit relative to the conventional anaerobic treatment [79]. The development of the MEC system connected with valuable product formation was suggested for positive energy gain [79, 80]. Thus, the product developments using bioelectrochemical reaction between microbe-cathode are promising research directions.

Metabolic Shift to Reduced Compound Production (Electrofermentation)

Electron transfer via an artificial mediator from a cathode has been applied in several studies and it showed an increase of reduced compound production [15, 81, 82, 83, 84, 85]. The distribution of final products would be determined by the electron and carbon flow in the fermentation process. Therefore, it is important to control the electron/carbon flow accordingly for production of the targeted bioproduct. Recently, an increase of butanol production in *C. pasteurianum* in a bioelectrochemical system showed the reducing power from a cathode could shift the metabolic pathway to solvent production [24]. The supplement of electrons via the cathode into a microbe led to enhanced reduction reaction directly (working on surface-associated redox enzymes, such as hydrogenases and presumably dehydrogenases [43]) or indirectly (increasing a reduced cofactor such as NADH, Fig. 4). The direct reduction process was studied in fumarate reduction to succinate [47], nitrate reduction to nitrite [48], nitrobenzene reduction to aniline [86], and hexavalent chromium reduction [18]. The indirect reduction process includes ethanol production from acetate [81], alcohol formation from glycerol [87], and butyraldehyde to butanol [88].

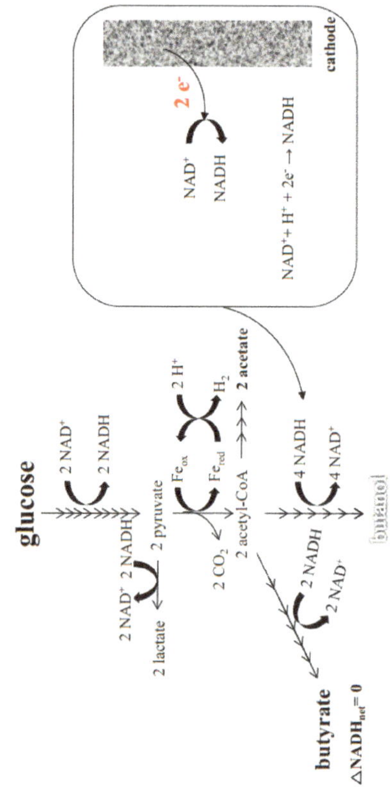

Figure 4: The microbial metabolic pathway of NADH-consuming compound production. One *arrow* indicates one step of reaction. The butyrate is NADH-balanced and generally produced more than butanol. The NADH reduction (inset) by electricity increases the flux of electron for butanol formation, more NADH-consuming pathway. ΔNADHnet(per one mol of product)=NADH production−NADH consumption,ΔNADHnet(per one mol of product)=NADH production−NADH consumption, *Fe* $_{ox}$ oxidized form of ferredoxin, *Fe* $_{red}$ reduced form of ferredoxin.

Reduction for Value-Added Bioproducts: Chain Elongation

The interaction between a cathode and microbes led to reverse β oxidation [89] and reduced propionate (C3) to valerate (C5) in a glycerol-fed bioelectrochemical system [87]. Without fermentable substrates, the reduction of acetate (C2) to caproate (C6) and caprylate (C8) took place in a *Clostridium kluyveri*-predominant mixed culture in a bioelectrochemical system at −0.9 V vs. NHE cathode potential using in-situ produced hydrogen as an electron donor [90]. The reduction of acetate (C2) and butyrate (C4) into alcohols (C1 ~ C4), acetone (C2) and caproate (C6) occurred in a mixed culture of SRB at a potential of −0.85 V vs. Ag/AgCl via direct electron transfer [70].

The application of a cathode for additional reducing power can improve low-grade chemicals to valuable biofuels with energy supplement through the reduction process of an acid to alcohol or by chain elongation. In particular, landfill leachate, which contains acetate, propionate, and butyrate as main components [91], could be used as feed stocks in bioelectrochemical systems to upgrade waste to value-added biofuels, for examples, acetate to butanol [92] (Fig. 5), glycerol to 1,3-propandiol [93], glucose to polyhydroxyalkanoates (PHA) [92].

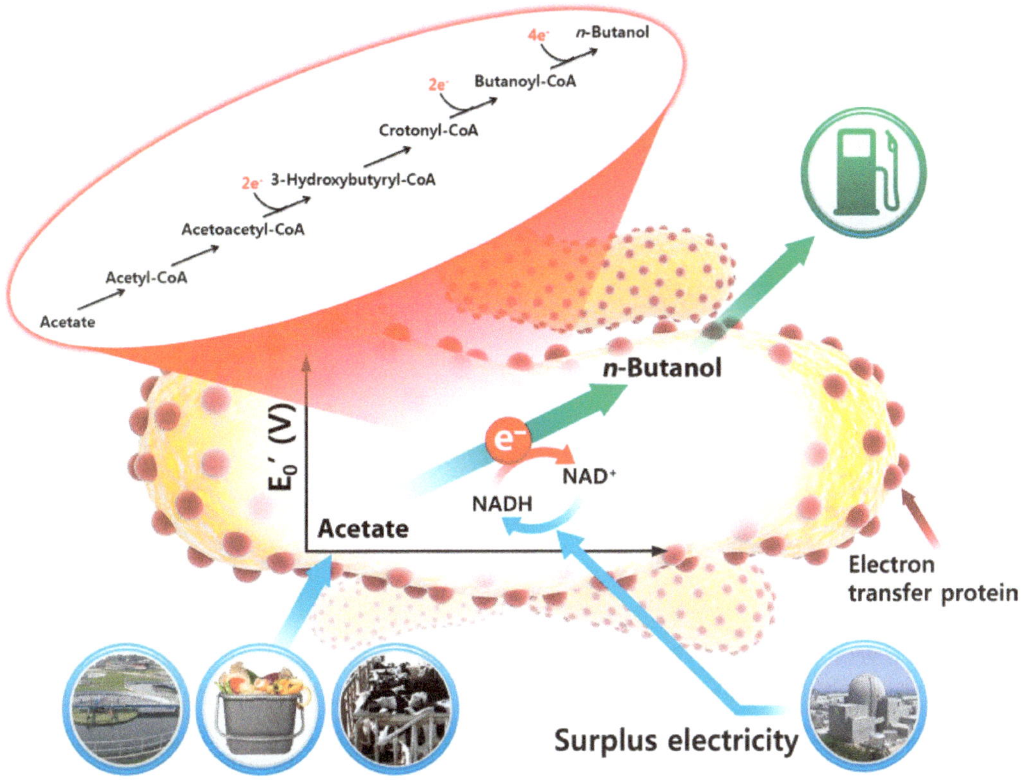

Figure 5: The upgrade of waste into value-added biofuels. The acetate from wastes, such as waste activated sludge, food waste, and animal manure, was feed stocks for biofuel production by electroactive microorganisms. The extracellular electron transfer from cathode to microbe via electron transfer protein could be used for the reduction of acetate to butanol. NADH the reduced form, NAD^+ the oxidized form of nicotinamide adenine dinucleotide, respectively.

Electrofuel Production (CO_2 Fixation)

Various electron sources can be used as electron donors (organic compounds, H_2, H_2O, etc.) or acceptors (O_2, metal, CO_2, etc.) by microbial organisms, whereas humans only use organic carbon as an electron donor and O_2 as an electron acceptor. BES uses an electrode as an electron donor (cathode) or an electron acceptor (anode). In particular, electrofuel is a carbon fixation process using a cathode as an electron donor and CO_2 as a carbon source, and this process mimics natural photosynthesis in plants [95, 96, 97].

$$n\mathrm{CO_2} + (6n+2)\left(e^{-1} + \mathrm{H}^+\right) \rightarrow \mathrm{C}_n\mathrm{H}_{2n+2} + 2n\mathrm{H_2O}$$

Electrofuel has several advantages: (1) the CO_2 greenhouse gas can be used as a substrate, and the efficiency of the electricity to chemical commodities is relatively high (80 ~ 90 %), i.e., more efficient than photosynthesis; (2) the electricity can be from many renewable sources; and (3) it has good specificity to produce desired chemical commodities. However, research in this ara is an early stage and the final titer is low and the CO_2reduction rate is slow [98].

An acetogen used an electrode as an electron source to produce 2-oxobutyrate as well as acetate [60]. The long-term operation of a bioelectrochemical system with CO_2 produced acetate at a level of 10.5 g/L over 20 days [99]. However, the concentration of other carbon compounds was still small, such as butyrate 35 mg/L [100], isobutanol 846 mg/L, and 3-methyl-a-butanol 570 mg/L [13]. In the absence of direct electron transfer, hydrogen led the reduction process with a hybrid microbial–water-splitting catalyst system [13, 101]. The hydrogen from water splitting was used to reduce carbon dioxide to produce liquid fuels and engineered *Ralstonia eutropha* produced isopropanol up to 216 mg/L [101]. Fermentative hydrogen production enhanced at −0.6 V vs. SHE led to increased 1,3-propandiol production [93]. Electrochemical generation of formate also mediated electron supplementation to microbes from a cathode in BES [13].

CONCLUSIONS

The cathodic reaction in BES is of increasing concern in the context of producing alternative fuels. Beginning with metal-utilizing bacteria, several electroactive bacteria were found and applied for the conversion of electrical to chemical energy as biofuels or biotransformation (Fig. 6). Nonetheless, many technical challenges must still be addressed and the titer of final product is also low. However, research is still in an early stage and efforts such as cell membrane modification and cathode surface modification would enhance the efficiency of BES, as shown in previous studies on MFC.

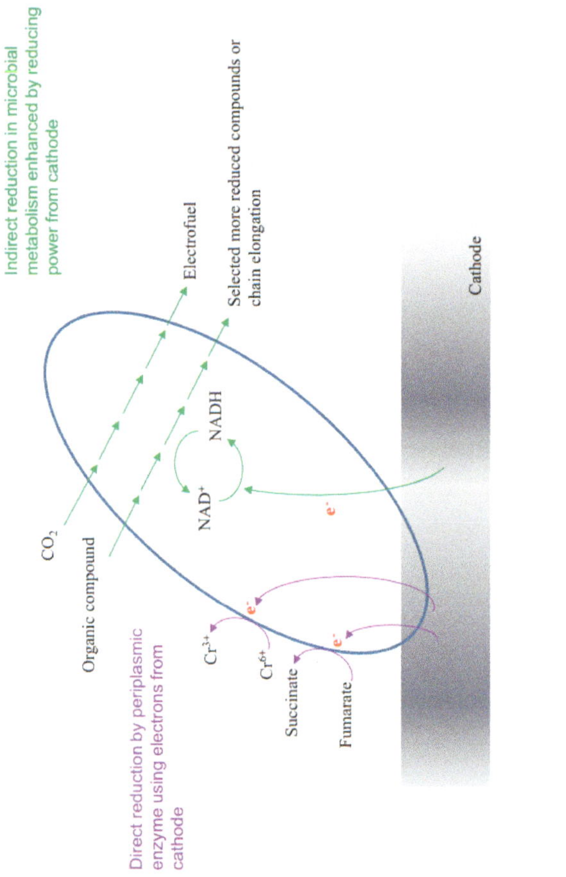

Figure 6: The application of bioelectrochemical reduction for cathodic electron transfer from a cathode to a microbe.

AUTHORS' CONTRIBUTIONS

OC and BIS conceptualized the manuscript. OC was responsible for literature review, data acquisition and analysis, and initial writing. BIS contributed with review and editing. Both authors read and approved the final manuscript.

ACKNOWLEDGEMENTS

This work was supported by the research fund of the Korean Ministry of Environment as "Converging Technology Project (202–101–006)" and the Korea Government Ministry of Trade, Industry and Energy as "the New & Renewable Energy of the Korea Institute of Energy Technology Evaluation and Planning (KETEP, No. 20133030000300)."

REFERENCES

1. Li H, Liao JC. Biological conversion of carbon dioxide to photosynthetic fuels and electrofuels. Energ Environ Sci. 2013;6(10):2892–9. doi:10.1039/C3EE41847B.
2. Hawkins AS, Han Y, Lian H, Loder AJ, Menon AL, Iwuchukwu IJ, et al. Extremely thermophilic routes to microbial electrofuels. Acs Catalysis. 2011;1(9):1043–50.
3. Schroder U, Harnisch F, Angenent LT. Microbial electrochemistry and technology: terminology and classification. Energ Environ Sci. 2015;8(2):513–9. doi:10.1039/C4EE03359K.
4. Smith JA, Nevin KP, Lovley DR. Syntrophic growth via quinone-mediated interspecies electron transfer. Front in Microbiol. 2015;6:121. doi:10.3389/fmicb.2015.00121.
5. Si Ishii, Kosaka T, Hori K, Hotta Y, Watanabe K. Coaggregation facilitates interspecies hydrogen transfer between*Pelotomaculum thermopropionicum* and *Methanothermobacter thermautotrophicus*. Appl Environ Microbiol. 2005;71(12):7838–45. doi:10.1128/aem.71.12.7838-7845.2005.
6. Pous N, Casentini B, Rossetti S, Fazi S, Puig S, Aulenta F. Anaerobic arsenite oxidation with an electrode serving as the sole electron acceptor: a novel approach to the bioremediation of arsenic-polluted groundwater. J Hazard Mater. 2015;283:617–22. doi:10.1016/j.jhazmat.2014.10.014.
7. Stams AJM, De Bok FAM, Plugge CM, Van Eekert MHA, Dolfing J, Schraa G. Exocellular electron transfer in anaerobic microbial communities. Environ Microbiol. 2006;8(3):371–82. doi:10.1111/j.1462-2920.2006.00989.x.
8. Kim T, Kim B. Electron flow shift in *Clostridium acetobutylicum* fermentation by electrochemically introduced reducing equivalent. Biotechnol Lett. 1988;10(2):123–8. doi:10.1007/BF01024638.
9. Vasconcelos I, Girbal L, Soucaille P. Regulation of carbon and electron flow in *Clostridium acetobutylicum* grown in chemostat culture at neutral pH on mixtures of glucose and glycerol. J Bacteriol. 1994;176(5):1443–50. doi:10.1128/JB.183.5.1748-1754.2001.
10. Shi L, Squier TC, Zachara JM, Fredrickson JK. Respiration of metal (hydr)oxides by *Shewanella* and *Geobacter*: a key role for multihaem c-type cytochromes. Mol Microbiol. 2007;65(1):12–20. doi:10.1111/j.1365-2958.2007.05783.x.
11. Reguera G, McCarthy KD, Mehta T, Nicoll JS, Tuominen MT, Lovley DR. Extracellular electron transfer via microbial nanowires. Nature. 2005;435(7045):1098–101. doi:10.1038/nature03661.
12. Myers CR, Myers JM. Localization of cytochromes to the outer membrane of anaerobically grown *Shewanella putrefaciens* MR-1. J Bacteriol. 1992;174(11):3429–38.
13. Shi L, Rosso KM, Clarke TA, Richardson DJ, Zachara JM, Fredrickson JK. Molecular Underpinnings of Fe(III) Oxide Reduction by *Shewanella Oneidensis* MR-1. Front Microbiol. 2012;3:50. doi:10.3389/fmicb.2012.00050.
14. Marsili E, Baron DB, Shikhare ID, Coursolle D, Gralnick JA, Bond DR. *Shewanella* secretes flavins that mediate extracellular electron transfer. Proc Natl Acad Sci. 2008;105(10):3968–73. doi:10.1073/pnas.0710525105.
15. Choi O, Um Y, Sang B-I. Butyrate production enhancement by *Clostridium tyrobutyricum* using electron mediators and a cathodic electron donor. Biotechnol Bioeng. 2012;109(10):2494–502. doi:10.1002/bit.24520.
16. Cao B, Shi L, Brown RN, Xiong Y, Fredrickson JK, Romine MF, et al. Extracellular polymeric substances from *Shewanella*sp. HRCR-1 biofilms: characterization by infrared spectroscopy and proteomics. Environ Microbiol. 2011;13(4):1018–31. doi:10.1111/j.1462-2920.2010.02407.x.

17. Gorby YA, Yanina S, McLean JS, Rosso KM, Moyles D, Dohnalkova A, et al. Electrically conductive bacterial nanowires produced by *Shewanella oneidensis strain* MR-1 and other microorganisms. Proc Natl Acad Sci. 2006;103(30):11358–63. doi:10.1073/pnas.0604517103.

18. Xafenias N, Zhang Y, Banks CJ. Enhanced performance of hexavalent chromium reducing cathodes in the presence of*Shewanella oneidensis* MR-1 and lactate. Environ Sci Technol. 2013;47(9):4512–20. doi:10.1021/es304606u.

19. Gregory KB, Lovley DR. Remediation and recovery of uranium from contaminated subsurface environments with electrodes. Environ Sci Technol. 2005;39(22):8943–7. doi:10.1021/es050457e.

20. Ross DE, Flynn JM, Baron DB, Gralnick JA, Bond DR. Towards electrosynthesis in *Shewanella*: energetics of reversing the Mtr pathway for reductive metabolism. PLoS One. 2011;6(2):e16649. doi:10.1371/journal.pone.0016649.

21. Strycharz SM, Glaven RH, Coppi MV, Gannon SM, Perpetua LA, Liu A, et al. Gene expression and deletion analysis of mechanisms for electron transfer from electrodes to *Geobacter sulfurreducens*. Bioelectrochemistry. 2011;80(2):142–50. doi:10.1016/j.bioelechem.2010.07.005.

22. Croese E, Pereira M, Euverink G-J, Stams AM, Geelhoed J. Analysis of the microbial community of the biocathode of a hydrogen-producing microbial electrolysis cell. Appl Microbiol Biotechnol. 2011;92(5):1083–93. doi:10.1007/s00253-011-3583-x.

23. Chen G-W, Choi S-J, Lee T-H, Lee G-Y, Cha J-H, Kim C-W. Application of biocathode in microbial fuel cells: cell performance and microbial community. Appl Microbiol Biotechnol. 2008;79(3):379–88. doi:10.1007/s00253-008-1451-0.

24. Choi O, Kim T, Woo HM, Um Y. Electricity-driven metabolic shift through direct electron uptake by electroactive heterotroph *Clostridium pasteurianum*. Sci Rep. 2014;. doi:10.1038/srep06961.

25. Armstrong FA, Hill HAO, Walton NJ. Direct electrochemical reduction of ferredoxin promoted by Mg2+. FEBS Lett. 1982;145(2):241–4. doi:10.1016/0014-5793(82)80175-0.

26. Shi L, Richardson DJ, Wang Z, Kerisit SN, Rosso KM, Zachara JM, et al. The roles of outer membrane cytochromes of*Shewanella* and *Geobacter* in extracellular electron transfer. Environ Microbiol Rep. 2009;1(4):220–7. doi:10.1111/j.1758-2229.2009.00035.x.

27. Dantas JM, Campelo LM, Duke NEC, Salgueiro CA, Pokkuluri PR. The structure of PccH from Geobacter sulfurreducens: A novel low reduction potential monoheme cytochrome essential for accepting electrons from an electrode. FEBS J. 2015:n/a-n/a. doi:10.1111/febs.13269.

28. Dantas JM, Tomaz DM, Morgado L, Salgueiro CA. Functional characterization of PccH, a key cytochrome for electron transfer from electrodes to the bacterium *Geobacter sulfurreducens*. FEBS Lett. 2013;587(16):2662–8. doi:10.1016/j.febslet.2013.07.003.

29. Sydow A, Krieg T, Mayer F, Schrader J, Holtmann D. Electroactive bacteria—molecular mechanisms and genetic tools. Appl Microbiol Biotechnol. 2014;98(20):8481–95. doi:10.1007/s00253-014-6005-z.

30. Tremblay PL, Zhang T, Dar SA, Leang C, Lovley DR. The Rnf complex of *Clostridium ljungdahlii* is a proton-translocating ferredoxin:nAD$^+$ oxidoreductase essential for autotrophic growth. MBio. 2012;4(1):e00406–12. doi:10.1128/mBio.00406-12.

31. Chowdhury NP, Mowafy AM, Demmer JK, Upadhyay V, Koelzer S, Jayamani E, et al. Studies on the mechanism of electron bifurcation catalyzed by electron transferring flavoprotein (Etf) and butyryl-CoA dehydrogenase (Bcd) of*Acidaminococcus fermentans*. J Biol Chem. 2014;289(8):5145–57. doi:10.1074/jbc.M113.521013.

32. Buckel W, Thauer RK. Energy conservation via electron bifurcating ferredoxin reduction and proton/Na(+) translocating ferredoxin oxidation. Biochim Biophys Acta. 2013;1827(2):94–113. doi:10.1016/j.bbabio.2012.07.002.

33. Hess V, Poehlein A, Weghoff MC, Daniel R, Muller V. A genome-guided analysis of energy conservation in the thermophilic, cytochrome-free acetogenic bacterium *Thermoanaerobacter kivui*. BMC Genom. 2014;15:1139. doi:10.1186/1471-2164-15-1139.

34. Lane N, Martin William F. The origin of membrane bioenergetics. Cell. 2012;151(7):1406–16. doi:10.1016/j.cell.2012.11.050.

35. Bertsch J, Oppinger C, Hess V, Langer JD, Muller V. Heterotrimeric NADH-oxidizing methylenetetrahydrofolate reductase from the acetogenic bacterium *Acetobacterium woodii*. J Bacteriol. 2015;197(9):1681–9. doi:10.1128/jb.00048-15.

36. Kracke F, Vassilev I, Krömer JO. Microbial electron transport and energy conservation—the foundation for optimizing bioelectrochemical systems. Front Microbiol. 2015. doi:10.3389/fmicb.2015.00575.

37. Lovenberg W, Sobel BE. Rubredoxin: a new electron transfer protein from *Clostridium pasteurianum*. P Natl Acad Sci USA. 1965;54(1):193–9.

38. Park I, Youn B, Harley J, Eidsness M, Smith E, Ichiye T, et al. The unique hydrogen bonded water in the reduced form of*Clostridium pasteurianum* rubredoxin and its possible role in electron transfer. JBIC, J Biol Inorg Chem. 2004;9(4):423–8. doi:10.1007/s00775-004-0542-3.

39. Bruschi M, Hatchikian CE, Golovleva LA, Gall JL. Purification and characterization of cytochrome c3, ferredoxin, and rubredoxin isolated from *Desulfovibrio desulfuricans* Norway. J Bacteriol. 1977;129(1):30–8.

40. Coulter ED, Kurtz DM Jr. A role for rubredoxin in oxidative stress protection in *Desulfovibrio vulgaris*: catalytic electron transfer to rubrerythrin and two-iron superoxide reductase. Arch Biochem Biophys. 2001;394(1):76–86. doi:10.1006/abbi.2001.2531.

41. Correia dos Santos MM, Paes de Sousa PM, Simões Gonçalves ML, Krippahl L, Moura JJG, Lojou É, et

al. Electrochemical studies on small electron transfer proteins using membrane electrodes. J Electroanal Chem. 2003;541:153–62. doi:10.1016/S0022-0728(02)01427-4.

42. dos Correia Santos MM, de Paes Sousa PM, Simões Gonçalves ML, Ascenso C, Moura I, Moura JJG. Electrochemical studies of rubredoxin from *Desulfovibrio vulgaris* at modified electrodes. J Electroanal Chem. 2001;501(1–2):173–9. doi:10.1016/S0022-0728(00)00521-0.

43. Deutzmann JS, Sahin M, Spormann AM. Extracellular enzymes facilitate electron uptake in biocorrosion and bioelectrosynthesis. mBio. 2015;6(2). doi:10.1128/mBio.00496-15.

44. Rosenbaum M, Aulenta F, Villano M, Angenent LT. Cathodes as electron donors for microbial metabolism: which extracellular electron transfer mechanisms are involved? Biores Technol. 2011;102(1):324–33. doi:10.1016/j.biortech.2010.07.008.

45. Srikanth S, Maesen M, Dominguez-Benetton X, Vanbroekhoven K, Pant D. Enzymatic electrosynthesis of formate through CO_2 sequestration/reduction in a bioelectrochemical system (BES). Biores Technol. 2014;165:350–4. doi:10.1016/j.biortech.2014.01.129.

46. da Silva SM, Pacheco I, Pereira IA. Electron transfer between periplasmic formate dehydrogenase and cytochromes c in*Desulfovibrio desulfuricans* ATCC 27774. J Biol Inorg Chem. 2012;17(5):831–8. doi:10.1007/s00775-012-0900-5.

47. DumasC, BasseguyR, BergelA. Microbial electrocatalysis with *Geobacter sulfurreducens* biofilm on stainless steel cathodes. Electrochim Acta. 2008;53(5):2494–500. doi:10.1016/j.electacta.2007.10.018.

48. Gregory KB, Bond DR, Lovley DR. Graphite electrodes as electron donors for anaerobic respiration. Environ Microbiol. 2004;6(6):596–604. doi:10.1111/j.1462-2920.2004.00593.x.

49. Virdis B, Read ST, Rabaey K, Rozendal RA, Yuan Z, Keller J. Biofilm stratification during simultaneous nitrification and denitrification (SND) at a biocathode. Biores Technol. 2011;102(1):334–41. doi:10.1016/j.biortech.2010.06.155.

50. Desloover J, Puig S, Virdis B, Clauwaert P, Boeckx P, Verstraete W, et al. Biocathodic nitrous oxide removal in bioelectrochemical systems. Environ Sci Technol. 2011;45(24):10557–66. doi:10.1021/es202047x.

51. Pous N, Puig S, Dolors Balaguer M, Colprim J. Cathode potential and anode electron donor evaluation for a suitable treatment of nitrate-contaminated groundwater in bioelectrochemical systems. Chem. Eng J. 2015;263:151–9. doi:10.1016/j.cej.2014.11.002.

52. Villano M, Aulenta F, Ciucci C, Ferri T, Giuliano A, Majone M. Bioelectrochemical reduction of CO_2 to CH_4 via direct and indirect extracellular electron transfer by a hydrogenophilic methanogenic culture. Biores Technol. 2010;101(9):3085–90. doi:10.1016/j.biortech.2009.12.077.

53. Cheng S, Xing D, Call DF, Logan BE. Direct biological conversion of electrical current into methane by electromethanogenesis. Environ Sci Technol. 2009;43(10):3953–8. doi:10.1021/es803531g.

54. Lohner ST, Deutzmann JS, Logan BE, Leigh J, Spormann AM. Hydrogenase-independent uptake and metabolism of electrons by the archaeon Methanococcus maripaludis. ISME J. 2014;8(8):1673–81. doi:10.1038/ismej.2014.82.

55. Hara M, Onaka Y, Kobayashi H, Fu Q, Kawaguchi H, Vilcaez J, et al. Mechanism of electromethanogenic reduction of $CO_2$by a thermophilic methanogen. Energy Procedia. 2013;37:7021–8. doi:10.1016/j.egypro.2013.06.637.

56. Costa KC, Wong PM, Wang T, Lie TJ, Dodsworth JA, Swanson I, et al. Protein complexing in a methanogen suggests electron bifurcation and electron delivery from formate to heterodisulfide reductase. Proc Natl Acad Sci. 2010;107(24):11050–5. doi:10.1073/pnas.1003653107.

57. Wang M, Tomb J-F, Ferry J. Electron transport in acetate-grown *Methanosarcina acetivorans*. BMC Microbiol. 2011;11(1):165. doi:10.1186/1471-2180-11-165.

58. Schlegel K, Welte C, Deppenmeier U, Müller V. Electron transport during aceticlastic methanogenesis by *Methanosarcinaacetivorans* involves a sodium-translocating Rnf complex. FEBS J. 2012;279(24):4444–52. doi:10.1111/febs.12031.

59. Thauer RK, Kaster A-K, Goenrich M, Schick M, Hiromoto T, Shima S. Hydrogenases from methanogenic Archaea, nickel, a novel cofactor, and H_2 Storage. Annu Rev Biochem. 2010;79(1):507–36. doi:10.1146/annurev.biochem.030508.152103.

60. Nevin KP, Hensley SA, Franks AE, Summers ZM, Ou J, Woodard TL, et al. Electrosynthesis of organic compounds from carbon dioxide is catalyzed by a diversity of acetogenic microorganisms. Appl Environ Microbiol. 2011;77(9):2882–6. doi:10.1128/aem.02642-10.

61. Kato S, Yumoto I, Kamagata Y. Isolation of acetogenic bacteria that induce biocorrosion by utilizing metallic iron as the sole electron donor. Appl Environ Microbiol. 2015;81(1):67–73. doi:10.1128/aem.02767-14.

62. Song J, Kim Y, Lim M, Lee H, Lee JI, Shin W. Microbes as electrochemical CO_2 conversion catalysts. ChemSusChem. 2011;4(5):587–90. doi:10.1002/cssc.201100107.

63. Xafenias N, Mapelli V. Performance and bacterial enrichment of bioelectrochemical systems during methane and acetate production. Int J Hydrogen Energ. 2014;39(36):21864–75. doi:10.1016/j.ijhydene.2014.05.038.

64. Müller V. Energy conservation in acetogenic bacteria. Appl Environ Microbiol. 2003;69(11):6345–53. doi:10.1128/aem.69.11.6345-6353.2003.

65. Bose A, Gardel EJ, Vidoudez C, Parra EA, Girguis PR. Electron uptake by iron-oxidizing phototrophic bacteria. Nat Commun. 2014. doi:10.1038/ncomms4391.

66. Doud DFR, Angenent LT. Toward electrosynthesis with uncoupled extracellular electron uptake and metabolic growth: enhancing current uptake with

Rhodopseudomonas palustris. Environ Sci Technol Lett. 2014;1(9):351–5. doi:10.1021/ez500244n.

67. Summers ZM, Gralnick JA, Bond DR. Cultivation of an obligate fe(ii)-oxidizing lithoautotrophic bacterium using electrodes. mBio. 2013;4(1). doi:10.1128/mBio.00420-12.

68. Dinh HT, Kuever J, Muszmann M, Hassel AW, Stratmann M, Widdel F. Iron corrosion by novel anaerobic microorganisms. Nature. 2004;427(6977):829–32. doi:10.1038/nature02321.

69. Pfeffer C, Larsen S, Song J, Dong M, Besenbacher F, Meyer RL, et al. Filamentous bacteria transport electrons over centimetre distances. Nature. 2012;491(7423):218–21. doi:10.1038/nature11586.

70. Sharma M, Aryal N, Sarma PM, Vanbroekhoven K, Lal B, Benetton XD, et al. Bioelectrocatalyzed reduction of acetic and butyric acids via direct electron transfer using a mixed culture of sulfate-reducers drives electrosynthesis of alcohols and acetone. Chem Commun. 2013;49(58):6495–7. doi:10.1039/C3CC42570C.

71. Beese-Vasbender PF, Nayak S, Erbe A, Stratmann M, Mayrhofer KJJ. Electrochemical characterization of direct electron uptake in electrical microbially influenced corrosion of iron by the lithoautotrophic SRB *Desulfopila corrodens* strain IS4. Electrochim Acta. 2015;167:321–9. doi:10.1016/j.electacta.2015.03.184.

72. Rodrigues TdC, Rosenbaum MA. Microbial electroreduction: screening for new cathodic biocatalysts. ChemElectroChem. 2014;1(11):1916–22. doi:10.1002/celc.201402239.

73. Nie H, Zhang T, Cui M, Lu H, Lovley DR, Russell TP. Improved cathode for high efficient microbial-catalyzed reduction in microbial electrosynthesis cells. Phys Chem Chem Phys. 2013;15(34):14290–4. doi:10.1039/C3CP52697F.

74. Jourdin L, Freguia S, Donose BC, Chen J, Wallace GG, Keller J, et al. A novel carbon nanotube modified scaffold as an efficient biocathode material for improved microbial electrosynthesis. J Mater Chem A. 2014;2(32):13093–102. doi:10.1039/C4TA03101F.

75. Flexer V, Chen J, Donose BC, Sherrell P, Wallace GG, Keller J. The nanostructure of three-dimensional scaffolds enhances the current density of microbial bioelectrochemical systems. Energ Environ Sci. 2013;6(4):1291–8. doi:10.1039/C3EE00052D.

76. Yan H, Catania C, Bazan GC. Membrane-intercalating conjugated oligoelectrolytes: Impact on bioelectrochemical systems. Adv Mater. 2015:n/a-n/a. doi:10.1002/adma.201500487.

77. Jourdin L, Grieger T, Monetti J, Flexer V, Freguia S, Lu Y, et al. High acetic acid production rate obtained by microbial electrosynthesis from carbon dioxide. Environ Sci Technol. 2015. doi:10.1021/acs.est.5b03821.

78. Su M, Wei L, Qiu Z, Jia Q, Shen J. A graphene modified biocathode for enhancing hydrogen production. RSC Advances. 2015;5(41):32609–14. doi:10.1039/C5RA02695D.

79. Foley JM, Rozendal RA, Hertle CK, Lant PA, Rabaey K. Life cycle assessment of high-rate anaerobic treatment, microbial fuel cells, and microbial electrolysis cells. Environ Sci Technol. 2010;44(9):3629–37. doi:10.1021/es100125h.

80. Pant D, Singh A, Van Bogaert G, Gallego YA, Diels L, Vanbroekhoven K. An introduction to the life cycle assessment (LCA) of bioelectrochemical systems (BES) for sustainable energy and product generation: relevance and key aspects. Renew Sust Energ Rev. 2011;15(2):1305–13. doi:10.1016/j.rser.2010.10.005.

81. Steinbusch KJJ, Hamelers HVM, Schaap JD, Kampman C, Buisman CJN. Bioelectrochemical ethanol production through mediated acetate reduction by mixed cultures. Environ Sci Technol. 2010;44(1):513–7. doi:10.1021/es902371e.

82. Park DH, Zeikus JG. Utilization of electrically reduced neutral red by *Actinobacillus succinogenes:* physiological function of neutral red in membrane-driven fumarate reduction and energy conservation. J Bacteriol. 1999;181(8):2403–10.

83. Park DH, Laivenieks M, Guettler MV, Jain MK, Zeikus JG. Microbial utilization of electrically reduced neutral red as the sole electron donor for growth and metabolite production. Appl Environ Microbiol. 1999;65(7):2912–7.

84. Emde R, Schink B. Enhanced propionate formation by *Propionibacterium freudenreichii subsp. freudenreichii* in a three-electrode amperometric culture system. Appl Environ Microbiol. 1990;56(9):2771–6.

85. Kracke F, Krömer JO. Identifying target processes for microbial electrosynthesis by elementary mode analysis. BMC Bioinform. 2014;15(1):410. doi:10.1186/s12859-014-0410-2.

86. Wang A-J, Cheng H-Y, Liang B, Ren N-Q, Cui D, Lin N, et al. Efficient reduction of nitrobenzene to aniline with a biocatalyzed cathode. Environ Sci Technol. 2011;45(23):10186–93. doi:10.1021/es202356w.

87. Dennis PG, Harnisch F, Yeoh YK, Tyson GW, Rabaey K. Dynamics of cathode-associated microbial communities and metabolite profiles in a glycerol-fed bioelectrochemical system. Appl Environ Microbiol. 2013;79(13):4008–14. doi:10.1128/aem.00569-13.

88. Schlager S, Neugebauer H, Haberbauer M, Hinterberger G, Sariciftci NS. Direct electrochemical addressing of immobilized alcohol dehydrogenase for the heterogeneous bioelectrocatalytic reduction of butyraldehyde to butanol. ChemCatChem. 2015;7(6):967–71. doi:10.1002/cctc.201402932.

89. Dellomonaco C, Clomburg JM, Miller EN, Gonzalez R. Engineered reversal of the β-oxidation cycle for the synthesis of fuels and chemicals. Nature. 2011;476(7360):355–9. doi:10.1038/nature10333.

90. Van Eerten-Jansen MCAA, Ter Heijne A, Grootscholten TIM, Steinbusch KJJ, Sleutels THJA, Hamelers HVM, et al. Bioelectrochemical production of caproate and caprylate from acetate by mixed cultures. ACS Sustain Chem Eng. 2013;1(5):513–8. doi:10.1021/sc300168z.

91. Lozecznik S, Sparling R, Clark SP, VanGulck JF, Oleszkiewicz JA. Acetate and propionate impact on the

methanogenesis of landfill leachate and the reduction of clogging components. Biores Technol. 2012;104:37–43. doi:10.1016/j.biortech.2011.09.123.

92. Sharma M, Aryal N, Sarma PM, Vanbroekhoven K, Lal B, Benetton XD, et al. Bioelectrocatalyzed reduction of acetic and butyric acids via direct electron transfer using a mixed culture of sulfate-reducers drives electrosynthesis of alcohols and acetone. Chem Commun. 2013;49(58):6495–7. doi:10.1039/C3CC42570C.

93. Zhou M, Chen J, Freguia S, Rabaey K, Keller J. Carbon and electron fluxes during the electricity driven 1,3-propanediol biosynthesis from glycerol. Environ Sci Technol. 2013;47(19):11199–205. doi:10.1021/es402132r.

94. Srikanth S, Venkateswar Reddy M, Venkata Mohan S. Microaerophilic microenvironment at biocathode enhances electrogenesis with simultaneous synthesis of polyhydroxyalkanoates (PHA) in bioelectrochemical system (BES). Biores Technol. 2012;125:291–9. doi:10.1016/j.biortech.2012.08.060.

95. Hawkins AS, Han Y, Lian H, Loder AJ, Menon AL, Iwuchukwu IJ, et al. Extremely thermophilic routes to microbial electrofuels. ACS Catal. 2011;1(9):1043–50. doi:10.1021/cs2003017.

96. Conrado R, Haynes C, Haendler B, Toone E. Electrofuels: a new paradigm for renewable fuels. Advanced Biofuels and Bioproducts. New York: Springer; 2013.

97. Hawkins AS, McTernan PM, Lian H, Kelly RM, Adams MWW. Biological conversion of carbon dioxide and hydrogen into liquid fuels and industrial chemicals. Curr Opin Biotech. 2013;24(3):376–84. doi:10.1016/j.copbio.2013.02.017.

98. Tremblay P-L, Zhang T. Electrifying microbes for the production of chemicals. Front Microbiol. 2015. doi:10.3389/fmicb.2015.00201.

99. Marshall CW, Ross DE, Fichot EB, Norman RS, May HD. Long-term operation of microbial electrosynthesis systems improves acetate production by autotrophic microbiomes. Environ Sci Technol. 2013;47(11):6023–9. doi:10.1021/es400341b.

100. Ganigue R, Puig S, Batlle-Vilanova P, Balaguer MD, Colprim J. Microbial electrosynthesis of butyrate from carbon dioxide. Chem Commun. 2015;51(15):3235–8. doi:10.1039/C4CC10121A.

101. Torella JP, Gagliardi CJ, Chen JS, Bediako DK, Colón B, Way JC, et al. Efficient solar-to-fuels production from a hybrid microbial–water-splitting catalyst system. Proc Natl Acad Sci. 2015;112(8):2337–42. doi:10.1073/pnas.1424872112.

102. Hsu L, Masuda SA, Nealson KH, Pirbazari M. Evaluation of microbial fuel cell Shewanella biocathodes for treatment of chromate contamination. RSC Adv. 2012;2(13):5844–55. doi:10.1039/C2RA20478A.

103. Rozendal RA, Jeremiasse AW, Hamelers HVM, Buisman CJN. Hydrogen production with a microbial biocathode. Environ Sci Technol. 2008;42(2):629–34. doi:10.1021/es071720+.

104. Jeremiasse AW, Hamelers HVM, Buisman CJN. Microbial electrolysis cell with a microbial biocathode. Bioelectrochemistry. 2010;78(1):39–43. doi:10.1016/j.bioelechem.2009.05.005.

105. Van Eerten-Jansen MCAA, Heijne AT, Buisman CJN, Hamelers HVM. Microbial electrolysis cells for production of methane from CO_2: long-term performance and perspectives. Int J Energy Res. 2012;36(6):809–19. doi:10.1002/er.1954.

106. Nevin KP, Hensley SA, Franks AE, Summers ZM, Ou J, Woodard TL, et al. Electrosynthesis of organic compounds from carbon dioxide is catalyzed by a diversity of acetogenic microorganisms. Appl Environ Microb. 2011;77(9):2882–6. doi:10.1128/aem.02642-10.

107. Nevin KP, Woodard TL, Franks AE, Summers ZM, Lovley DR. Microbial electrosynthesis: Feeding microbes electricity to convert carbon dioxide and water to multicarbon extracellular organic compounds. mBio. 2010;1(2). doi:10.1128/mBio.00103-10.

108. Sato K, Kawaguchi H, Kobayashi H. Bio-electrochemical conversion of carbon dioxide to methane in geological storage reservoirs. Energy Convers Manag. 2013;66:343–50. doi:10.1016/j.enconman.2012.12.008.

109. Pitts KE, Dobbin PS, Reyes-Ramirez F, Thomson AJ, Richardson DJ, Seward HE. Characterization of the *Shewanella oneidensis* MR-1 decaheme cytochrome MtrA: expression in *Escherichia coli* confers the ability to reduce soluble Fe(III) chelates. J Biol Chem. 2003;278(30):27758–65. doi:10.1074/jbc.M302582200.

110. Gorby Y, McLean J, Korenevsky A, Rosso K, El-Naggar MY, Beveridge TJ. Redox-reactive membrane vesicles produced by*Shewanella*. Geobiology. 2008;6(3):232–41. doi:10.1111/j.1472-4669.2008.00158.x.

111. Hartshorne R, Jepson B, Clarke T, Field S, Fredrickson J, Zachara J, et al. Characterization of *Shewanella oneidensis* MtrC: a cell-surface decaheme cytochrome involved in respiratory electron transport to extracellular electron acceptors. JBIC, J Biol Inorg Chem. 2007;12(7):1083–94. doi:10.1007/s00775-007-0278-y.

CHAPTER 4

MICROBIAL NETWORK FOR WASTE ACTIVATED SLUDGE CASCADE UTILIZATION IN AN INTEGRATED SYSTEM OF MICROBIAL ELECTROLYSIS AND ANAEROBIC FERMENTATION

Wenzong Liu, Zhangwei He, Chunxue Yang, Aijuan Zhou, Zechong Guo, Bin Liang, Cristiano Varrone and Ai-Jie Wang

Keywords: Anaerobic digestion, Microbial electrolysis cell, Biogas, Waste activated sludge

BACKGROUND

The consideration of wastewater and waste sludge as a source of renewable energy is growing, together with the public concerns for future shortage of fossil fuels and the impact on climate change [1]. Anaerobic wastewater treatment processes offer the possibility of a net energy production, rather than aerobic processes that consume energy [2]. Especially, waste activated sludge (WAS), with its abundance of nutrients, has a great potential as an alternative resource to extract value-added products [3, 4, 5], even though a long operation time is always required for anaerobic digestion (AD) to achieve an effective carbon removal and energy yield [6].

Recently, a number of studies have suggested that AD can be enhanced by inserting bioelectrochemical systems (BESs) [7, 8] in the anaerobic processes, to enhance acidogenesis [9], hydrogen production [10], or methane production [11, 12, 13]. It is well known that microorganisms capable of extracellular electron transfer have a major impact on the natural cycling of carbon and nutrients [14]. Therefore, BESs proved to be highly versatile in terms of potential application for WAS cascade utilization, ranging from energy recovery from organic substrates to product generation and specific environmental niche creation [15]. Characterization of microbial community structure in anaerobic digesters has attracted interest of engineers and researchers, because understanding microbial behavior and interactions is essential to improve the fermentation process [16, 17]. In fact, anaerobic treatment generally requires multiple groups of microorganisms working together to transform primary substrates to energy products such as hydrogen [18] or methane [19]. However, a thorough work in relation to the bacterial community shifts, and especially the impact of fermentation communities to electrode communities, is lacking.

Hydrolysis has always been considered the rate limiting stage in the fermentation of waste solids, therefore most of the studies focused on accelerating sludge hydrolysis using efficient pretreatment strategies, including chemical (such as alkaline [20]) and physical methods (such as ultrasonic [21] or freeze/thaw pretreatment, typically in cold areas [4]). Clearly, the characteristics of different pretreated sludge can also substantially influence the fermentation efficiency and the cascade utilization in the coupled BESs [22, 23, 24]. Chemical or physical pretreatments can directly impact the granular size of WAS, leading to the production of various organic compounds [22]. Consequently, digestate from degradation process in digesters presents prime importance for microbes in the recycling of carbon and nutrients [25]. Moreover, the effective diffusion and the biochemical characteristics of carbons and metabolic products can have a significant impact on the electrode biofilm matrix (for example, faster biocatalytic rates were observed under fatty acid-fed conditions) [26]. On the other hand, the efficiency in a BES will also be influenced by the initial anaerobic fermentation community, which will interact with the electrode community, after connecting the BES to a conventional AD system. Therefore, it is very important to

Citation: Wenzong Liu, Zhangwei He, Chunxue Yang, Aijuan Zhou, Zechong Guo, Bin Liang, Cristiano Varrone and Ai-Jie Wang, Microbial network for waste activated sludge cascade utilization in an integrated system of microbial electrolysis and anaerobic fermentation, DOI: 10.1186/s13068-016-0493-2.

evaluate the synergistic effects between fermentative microbes and electrode respiring microbes and the influence of carbon source characteristics. It is also worth noting that metabolic networks and interactions can determine the electrode respiring community development [27].

It is well known that a higher performance of various BESs can be achieved using mixed culture rather than pure culture [28, 29, 30]. Though exoelectrogens like *Shewanella* and *Geobacter* are proved to conduct the electron generation and transport in the mixed culture [31], they would not be absolutely dominant compared to other functional members (like *Anaerolinea*, *Bacteroida*, and *Clostridia*) in electrode biofilm. It has been found that interaction between microbes can improve system performance and energy recovery efficiency i.e., when combining *Brevibacillus* with *Pseudomonas* [32], hinting that microbes worked together and contributed to carbon recycling and electron transfer. For example, *Acetoanaerobium* sp. and *Acetobacterium* sp. were reported to be enriched on the bio-electrode of microbial electrolysis cells (MECs) and acetogenesis occurred at a limited degree [33]. *Fusibacter* sp. represented the enriched anaerobic fermentation community, able to utilize carbohydrates and produce acetate and butyrate as end products [34], which represent a favorable substrate to exoelectrogens. In a previous study on anode communities, a complex interaction on carbon degradation was revealed by functional genes (ranging from labile to recalcitrant carbon) [35]. However, synergistic and interactive effects of various communities have been insufficiently examined. In some cases, an unexpected reduction of efficiency and targeted products occured when applying microbial electrolysis to wastewater/sludge treatment [36], though a high energy harvest can easily be achieved from artificial wastewater with pure carbon sources [37, 38].

The relationship between BES and AD (for instance the understanding of their mutual benefits, or system stability enhancement, etc.) has become a debated issue [15], but clearly the effects of newly introduced communities on BES function has to be taken into consideration for further potential application [39]. So far, there were very few reports investigating the characteristics of cascade bioconversion, related to electron recovery, in mixed carbon sources to reveal microbial community interactions between digestate and BES biofilm. Thus, the present study wanted to explore the hypothesis that (i) a proper pretreatment of waste activated sludge can achieve a satisfactory efficiency in terms of final carbon utilization and energy recovery; (ii) sludge cascade utilization can be achieved for hydrogen production in MECs connecting anaerobic fermentation; (iii) fermentative communities formed in AD can have an impact on anode respiring bacteria in batch operation of integrated systems.

RESULTS AND DISCUSSION

Characteristic Change of WAS Fermentation using Different Pretreatment

The particle size primarily changed through different pretreatment methods, which substantially influenced the subsequent release and conversion of various organic matter in the WAS. Ultrasonic treatment played the most noticeable effect on sludge structure break and scatter, leading to an average particle size distribution of 29.5 μm. Alkaline treatment slightly improved the sludge particle scatter, with an average of 56.3 μm in comparison to 60.8 μm of the control sludge without pretreatment. On the other hand, an obvious increase of particle size up to 387.5 μm was obtained by the freeze/thaw treatment, because flocks were produced after freezing. Consequently, the lysis ratio of increased SCOD to TCOD was 21, 6, and 11 % after alkaline, freeze/thaw, and ultrasonic pretreatment, respectively. The freeze/thaw pretreatment was not as effective as other methods on SCOD release, indicating that the flocks of larger particle size were not broken into smaller fragments in a short reaction time [40]. In another report on the effect of sludge pretreatment on sludge characteristics, the disruption of sludge flocks led to the release of intracellular and extracellular materials [41]. Moreover, alkaline (pH 10–12) treatment is known to further enhance the organic release during the pretreatment [42] and favor VFA production in the subsequent fermentation [20, 43]. In our study, an increased amount of soluble organics was released after pretreatment, mainly in the form of carbohydrates, proteins, and volatile fatty acids (VFAs). After 3 d fermentation, SCOD increased from 147 mg/L of the raw sludge to 452 mg/L of the control, 7690 mg/L of the alkaline pretreatment, 1760 mg/L of the freeze/thaw pretreatment, and 3461 mg/L of the ultrasonic pretreatment. The VFAs were mostly produced by the alkaline pretreatment, which accumulated up to 5300 mg COD/L, accounting for 69 % of total SCOD. The same occurred with proteins, which reached 1749 mg/L, with a 24-fold increase compared to the control sludge without any pretreatment.

In our view, WAS pretreatment initially changed the particle characteristics, which played a major role in the studied process, influencing organic release and fermentative communities during pretreatment, and subsequently affecting the fermentation and VFAs production. Although all sludge pretreatments successfully improved hydrolysis and organic release, the subsequently generated short-chain fatty acids differed in terms of content and concentration, which are known to be important factors affecting the conversion rate and efficiency in bioelectrochemical systems [38]. Clearly, soluble organics play an important role, reducing the accessibility of substrates to bacterial disintegration, or stated differently, the initial particle size can affect the contact surface area, for subsequent bacterial action [44]. Our results showed that ultrasound led to the smallest particle size, followed by a higher acetate production (>80 % of total fermentative products). Alkaline pretreatment could increase the total production of short-chain fatty acids with a high conductivity fermentative liquid. Therefore, it seems to be one of the substantial factors to interact with electrode biofilm communities.

Furthermore, organics and conductivity of WAS fermentative liquid were the two key factors to the bioelectrochemical communities. To evaluate the influence of pretreatment on COD contribution in different sludge structures, COD was divided into four parts: soluble SCOD, loosely bound extracellular polymeric substances (LB-EPS), tightly bound extracellular polymeric substances (TB-EPS), and residual particles (Fig. 1). The alkaline pretreatment effectively released ~25 % particle organics (compared to the control) into SCOD (~21 %). A small part of particles (4 %) with reduced TB-EPS (~3 %) were converted into LB COD (~7 %). However, COD contribution was reduced to less than half of SCOD (6–11 %) from particles, when using the freeze/thaw or ultrasonic pretreatment. The SCOD of VFAs reached the peak accumulation during fermentation before methane production started, under the conditions of this study [45, 46]. The release of soluble matter also increased conductivity of fermentation solution. Even during fermentation without any pretreatment, there was a slight increase from 1.2 to 1.4 mS/cm in sludge fermentation liquid (SFL). Conductivity was further increased to 1.96–2.63 mS/cm by the freeze/thaw and ultrasonic treatment respectively, which matched the increasing trend of SCOD and inorganic ion release. The alkaline addition, using NaOH, highly enhanced the conductivity, reaching up to 6.23 mS/cm, which was almost close to 50 mM PBS (Phosphate buffer solution, pH 7.0) used for MEC reactor setup [35]. A high conductivity is to be considered potentially beneficial to electron transport in the following bioelectrochemical process [47]. Besides the additional alkaline contribution, organics and ion release from WAS improves during the pretreatment and is further enhanced during the fermentation. A previous study showed that the limiting factors, at the anodic biofilm, change from potential limitations at low conductivity, to dual potential and carbon source transfer limitations at a moderate conductivity, and to only mass transfer limitations at high conductivity [48]. A low conductivity (<1 mS/cm) was observed in common AD effluent after organic removal and biological treatment, moreover, a higher external voltage was required when connecting BES after AD to achieve biofuels [49]. In this respect, pretreatment is an important and flexible tool to regulate the performance of BES and AD integrated process, which would determine the total efficiency on waste treatment and biofuel recovery.

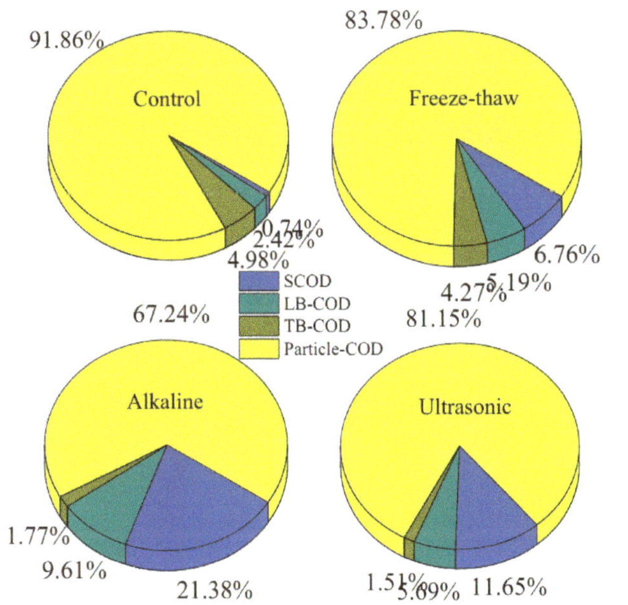

Figure 1: Analysis of main COD distributions in sludge after different pretreatments. The analysis of SCOD, LB-COD, TB-COD and the residual COD in particles after different pretreatments.

Pretreated SFL Utilization and Hydrogen Production in MECs

The setup performance of the 15 MEC replicate reactors, before fueling SFL. The average coulombic efficiency was steadily around 92.2 ± 6.5 % in all replicates, with an average peak current of 3.75 ± 0.22 mA. The COD removal efficiency of acetate reached 86.8 ± 2.1 %. The 15 MEC reactors showed similar conversion efficiencies to hydrogen, with 3.3 ± 0.5 mol H_2/mol acetate and a hydrogen production rate of 1.36 ± 0.26 mL/mg COD (1.46 ± 0.28 mL H_2/mL reactor/d). Twelve reactors were randomly divided into four groups (three replicates each), to be fed with SFL obtained from the different pretreatment methods.

The pretreated sludge properties determined the subsequent fermentation process, leading to various levels of acidification, organic contents, and production rates. The highest amount of VFAs was produced during the 3rd day fermentation of alkaline pretreated WAS, containing 2225.81 mgCOD/L and accounting for 42 % in total VFAs. There was 1077.25 mgCOD/L acetate produced in ultrasonic pretreated WAS, while still accounting for 41 % of total VFAs. The lowest amount of VFAs was observed with the freeze/thaw-pretreated WAS, though still showing a 3.8-fold increase compared to WAS without pretreatment. In all pretreated SFL, more VFAs were firstly utilized in MECs, showing a similar removal of around 70 %. Over 95 % of acetate and butyrate were utilized in alkaline SFL and ultrasonic SFL, while only ~85 % acetate and butyrate were removed in freeze/thaw SFL. Differently, a much higher percentage of propionate (removal amount was really low) was removed in freeze/thaw SFL than others at the same time. As a result, hydrogen production rate differed in MEC reactors, based on acid types and concentrations that were produced [38]. Previous results showed that pretreatment methods are very important to release organics and enhance degradation of various carbon sources from WAS [46, 50]. Probably, the cascade utilization of SFL could be regulated according to composition in VFAs, proteins, and polysaccharide [22, 51], while the energy recovery changed when the suitable organic compounds were degraded. It is therefore likely that the alkaline treatment performed best energy gains (Fig. 2) thanks to the high conductivity (increased from 2.1 ± 0.2 mS/cm for raw sludge to 3.5 ± 0.3 mS/cm for alkaline pretreatment) [51, 52] and SCOD, as well as high COD removal.

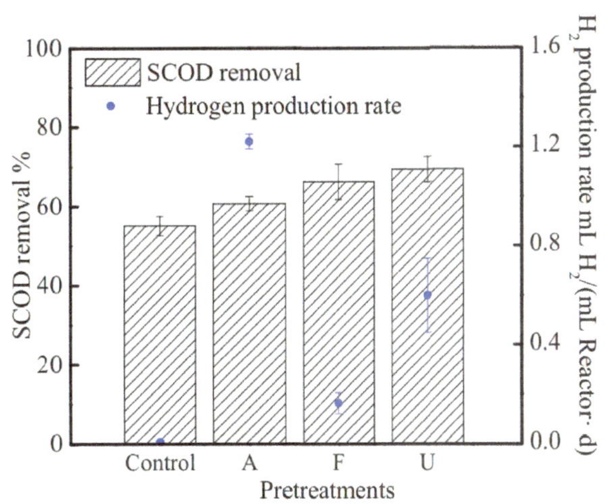

Figure 2: SCOD removal and hydrogen production rate in MECs fed with different sludge fermentative liquids. *Control* sludge without treatment, *A* Alkaline, *F* freeze/thaw, *U* ultrasonic pretreatment.

The current generation varied among different pretreated and fermented sludge. The highest peak current reached ~3.7 mA, when feeding with the alkaline pretreated SFL, which showed the highest acetate production of 2200 mg COD/L, as well as an enhanced conductivity. The peak current dropped to 2.5 mA for ultrasonic and 1.8 mA for freeze/thaw condition. The lowest current was only 1.0 mA, using SFL produced from raw WAS without pretreatment. The SCOD removal was slightly different in different SFL, with 61 ± 2 % for alkaline, 66 ± 5 % for freeze/thaw, and 69 ± 3 % for ultrasonic pretreatment in MECs (Fig. 2). However, the hydrogen production rate was quite different. The alkaline pretreated SFL achieved the highest hydrogen production of 1.22 ± 0.03 mL H_2/mL reactor/d (compared to 1.46 ± 0.28 mL H_2/mL reactor/d before fueling with SFL). The ultrasonic pretreated SFL was converted into hydrogen with a rate of 0.60 ± 0.15 mL H_2/mL reactor/d. The freeze/thaw pretreatment, instead, was not able to effectively improve the cascade utilization of WAS, compared to raw sludge.

Clearly, organic products were the result of metabolic activities of the microbial community, which was characterized by different composition and structure during fermentation. Previous studies showed that with the proper enrichment of microbial communities, anaerobic processes can be improved and perform more efficiently [53]. In this study, the solid granular sludge changed based on different pretreatment methods. Although SCOD was increased and consequently converted to more hydrogen in MECs, the hydrogen yield was reduced by ~16 % when influent was changed from artificial wastewater (acetate, ~1140 mg COD/L) to SFL (acetate, ~2225 mg COD/L accounting for 29 % of SCOD in the alkaline pretreatment) with fermentative communities. Moreover, hydrogen yield was reduced by ~59 % when feeding the ultrasonic pretreatment SFL (acetate, ~1080 mg COD/L accounting for 31 % of SCOD). It has been pointed out that further increases in organic loading do not vary hydrogen production significantly [54, 55]. Therefore, it is likely that MEC performances changed in relation to anodic community structure, which interacted with dominant fermentative communities and organic compounds produced [56].

Methane Production and Archaea Community Change in Integrated System

Methane production was detected in all MEC reactors after 2 weeks feeding SFL, however, the methane production rate was fluctuating, not being comparable among different pretreatments over all batch operations (data not shown). Although methane production was not substantially increased over 1 month (as we evaluated previously [57]), the MECs feeding SFL without pretreatment presented the highest methane production over all other conditions, together with the highest amount of acetotrophic methanogens (*Methanosaeta*), both in control SFL and MEC biofilms fed with control SFL (Fig. 3). It was interesting that the lowest amount of archaea were detected under ultrasonic pretreatment, leading to the lowest growth in MECs as a result. Compared to methanogens in initial biofilms fed with acetate (MEC sample), it seem that acetotrophic methanogens were substantially enriched to anode biofilm in all conditions. But hydrogenotrophic methanogens were further enriched with higher amounts than in SFL feeding, as shown in freeze/thaw and ultrasonic pretreatment, including *Methanocorpusculum*, *Methanosphaerula*,*Methanoregula*, *Methanospirillum*, *Methanobacterium*, and *Methanobrevibacter*. The extra hydrogen generation from MECs can favor hydrogenotrophic methanogens in anaerobic condition [57]. The H_2produced in a single chamber MEC can be lost through methanogenesis, which causes energy loss in the system [22, 51, 52].

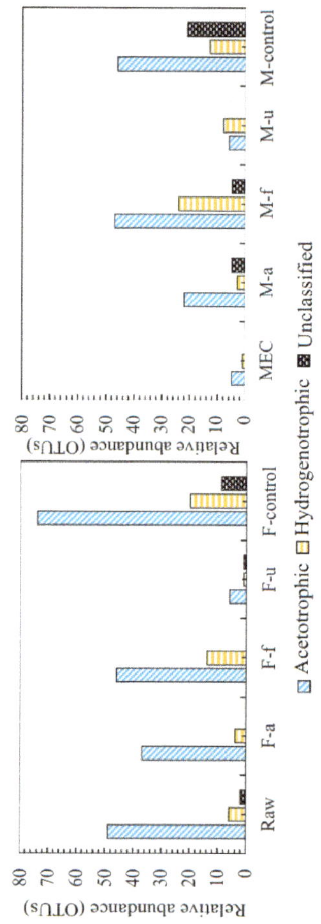

Figure 3: Relative abundance of Archaea communities based on OTUs. *F samples* fermentative liquid after treatment, *M samples* biofilm in MEC reactors (detailed sample legends see Fig. 2).

Microbial Community Structure and Anodic Biofilm Community Shift in Integrated Process

A total of 244,761 raw sequences were analyzed over all community samples. Operational taxonomic units (OTUs) at 3 % distance were the most detected ones in raw WAS (5002), with the highest diversity (Shannon index 6.96), while being the least detected in the startup anode biofilm using acetate (2341), showing a reduced diversity (Shannon index 5.39). Similar results were observed from ACE (abundance-based coverage estimator) and Chao1 indices. Interestingly, microbial community diversity in the SFL decreased, indicating that specific fermentation bacteria were enriched and became dominant. On the other hand, an increase of diversity was detected in anodic biofilm communities, after initial MECs were connected to sludge fermentation (with or without pretreatment) for several days, showing an interactive effect of fermentative communities on initial anodic communities, thus leading to subsequent changes in MEC reactor performances.

After fermentation, the unpretreated sludge showed similar community structure to raw sludge. The most abundant phylum in SFL was *Proteobacteria*, accounting for 36.7 %

in the control, 40.0 % in the alkaline pretreatment, 28.7 % in the freeze/thaw pretreatment, and 54.8 % in the ultrasonic pretreatment, over all microbial communities. Seemingly, sludge fermentation after ultrasonic pretreatment mostly increased *Gammaproteobacteria*. In comparison to the control sludge, *Firmicutes* (*Bacilli*sp. and *Clostridia* sp.) were all increased in SFL of the pretreated sludge. *Bacteroidetes* was the third most abundant community in the SFL.

When MECs were connected to the fermentation process, anodic biofilm composition obviously changed, compared to the original communities established using acetate (Fig. 4). *Desulfovibrio* [58] and *Geobacter* [14] (*Deltaproteobacteria*), responsible for electron transfer between bacteria and electrode, represented the key functional community. *Geobacter* was the most detected genus of the anode biofilm, in the case of reactors fed with acetate (startup MECs), and further increased after feeding with alkaline pretreated SFL, as well as the ultrasonic pretreated SFL (with a corresponding high energy conversion achieved in these reactors). The MEC fed with the freeze/thaw-pretreated SFL, instead, showed low abundances of *Desulfovibrio* and *Geobacter,*which was similar to the control SFL. On the other hand, compared to other treatments, freezing-thaw SFL led to an increased abundance of *Pseudomonas* in the anodic community. Moreover, large particles of organics in SFL led to enrichment of fermentative communities in the anode biofilm, including *Anaerolinea* (*Levilinea* and*Longlinea*), *Bacteroida* (*Paludibacter* and *Parabacteroides*), and *Clostridia* (*Proteinilclasticum*, *Proteocatella*, and *Sedimentibacter*). Compared to the original anode biofilm, four genera of the class *Clostridia* (namely*Acetoanaerobium*, *Acetobacterium*, *Anaerovorax*, and *Fusibacter*) decreased in all SFL-fed MECs.

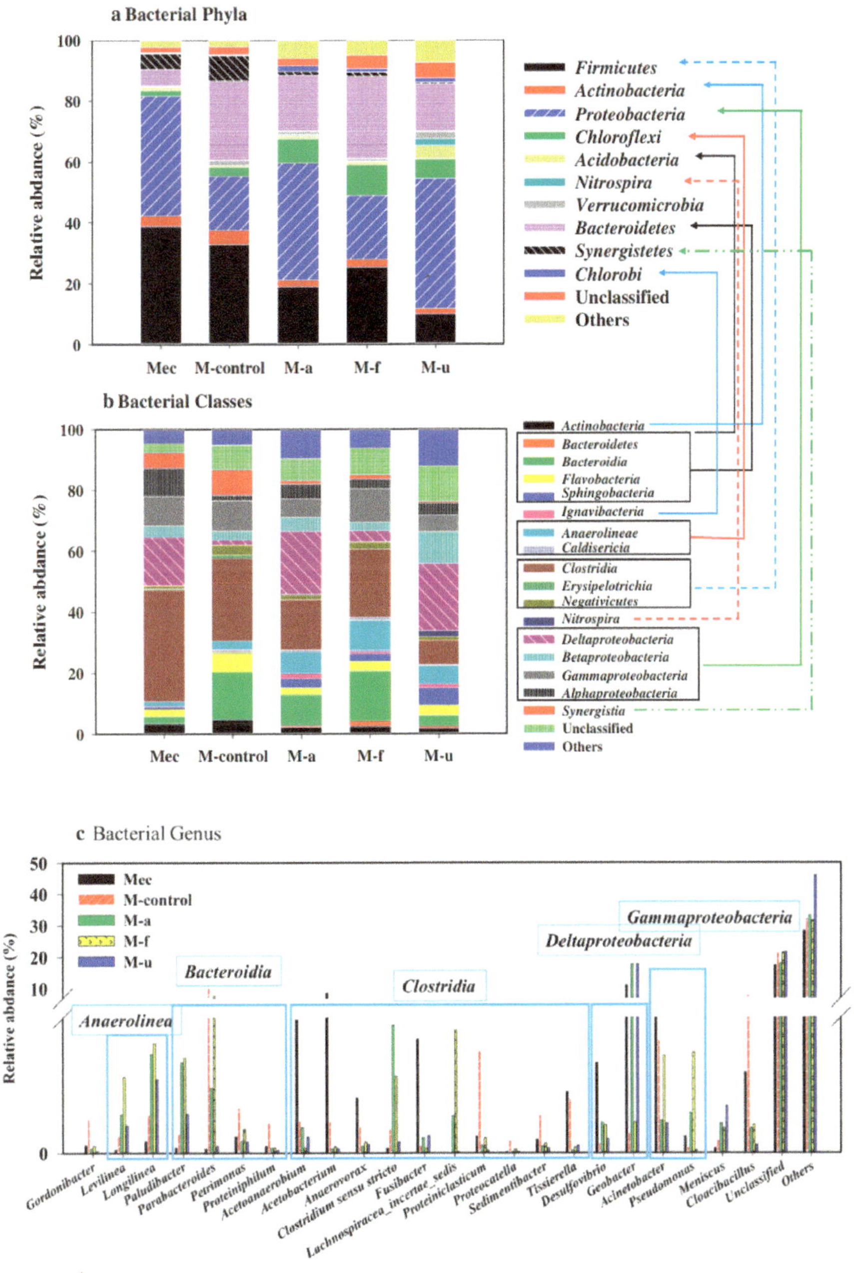

Figure 4: Taxonomic classification of 16S rRNA gene sequences of bacterial communities of anode biofilm at the phylum (a), class (b), and genus (c) levels. Relative abundance was defined as the number of sequences per sample. *Mec* MEC initially fed with acetate, *M-control* SFL-fed MEC without treatment (control treatment), *M-a* MEC fed with alkaline treated SFL, *M-f* MEC fed with freeze/thaw treated SFL, *M-u* MEC fed with ultrasonic treated SFL.

Moreover, hierarchical cluster analysis clearly showed that SFL communities varied, depending on the different treatment method (Fig. 5, F-samples). The bacterial community structure of the control SFL (F-control) without pretreatment changed less, compared to the raw sludge (Raw), while the ultrasonic pretreatment led to the greatest difference in community structure in the SFL (F-f). Figure 5 (M-samples) showed how various SFL communities had different impacts on the change of the anodic biofilm communities, after combining fermentation and microbial electrolysis process. Interestingly, anodic biofilm communities were similarly grouped among the original MEC (Mec), MECs fed with the alkaline pretreated SFL (M-a), as well as MECs fed with ultrasonic pretreated SFL (M-u). Regarding the gas production, they performed much higher hydrogen yield than the SFL-fed control or the freeze/thaw pretreatment, which had a low hydrogen conversion, below 0.2 mL/mg COD, and peak current below 1.0 mA.

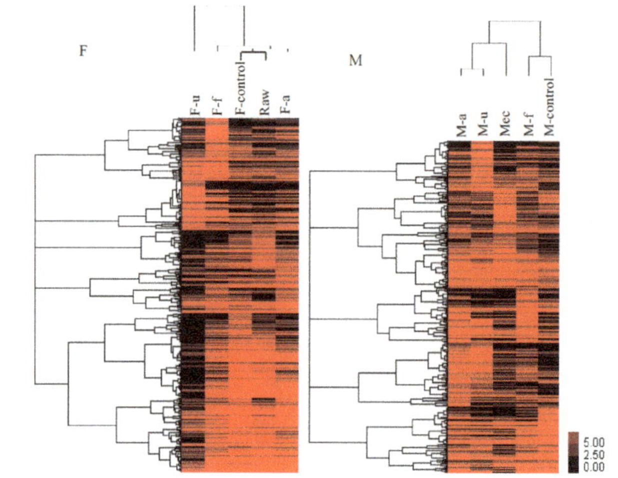

Figure 5: Hierarchical cluster analysis of bacterial communities from sludge fermentative liquid (*F*) and MEC biofilm (*M*). The OTUs of *y-axis* were ordered by phylum (3 % distance). Sample communities were clustered based on complete linkage method. The *color* intensity of scale indicates relative abundance of each OTU read (sample legends see Fig.2).

Bioelectrochemical communities were highly enriched in dominant functional groups related to*Proteobacteria* (63 %) and *Firmicutes* (25 %) when feeding acetate during reactor setup, inoculated with activated sludge (Fig. 6). They played the primary function of electron transfer and substrate degradation, with great potential on complex carbon utilization, as already suggested from a functional genes' perspective [35]. The most abundant genera for extracellular electron transfer were *Geobacter*, *Desulfovibrio*, and*Acinetobacter*, belonging to *Proteobacteria*. *Geobacter* species are considered as the most efficient exoelectrogens in bioelectrochemical systems [59]. *Desulfovibrio* and *Acinetobacter* species are dissimilatory metal-reducing bacteria involved in contaminant degradation and metal reduction, outside the cell membrane [60, 61]. In *Firmicutes*, three genera of *Clostridia* were detected, namely *Acetoanaerobium* (5.9 %),*Acetobacterium* (8.5 %), and *Fusibacter* (5.0 %). They are supposed to play an important role in carbon recycling for anode respiring bacteria, as previous studies have shown that some *Firmicutes* may closely live with anode respiring bacteria, when fed with fermentative substrates [62].

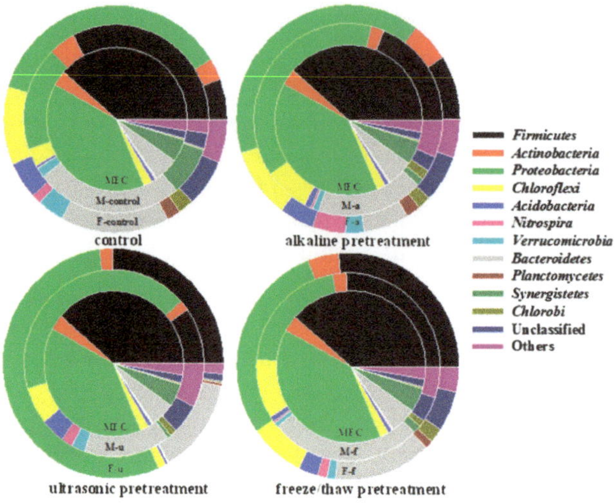

Figure 6: Community change and linkage during sludge cascade utilization (from fermentation to MEC) after different sludge treatments. Bacterial communities based on OUT (3 % distance), and the taxonomic identification of OTUs at phylum level, during sludge cascade utilization of alkaline treated SFL; anode biofilm communities (MEC) changed into a new biofilm community (M-a) after feeding SFL (communities of F-control).

Microbial Community Network on Exoelectrogenic and Fermentative Communities

Based on the discussion above, a network representing the community change and linkage was constructed (Fig. 7), taking alkaline treatment samples as example. MEC biofilm was inoculated from raw sludge (Raw), then raw sludge was pretreated to produce fermentation liquid as feedstock for the MECs. During the cascade process, MEC biofilm interacted with fermentative communities to form a new MEC biofilm. The SFL primarily led to an increase in abundance of *Bacteroidetes* and *Chloroflexi* in anodic communities, thus reducing the abundance of *Proteobacteria*. Even though the phylum *Bacteroidetes* was commonly detected in bioelectrochemical system communities, few studies pointed out their negative impacts on electron transfer efficiency [33]. On the other hand, *Bacteroidetes* can be further enriched (over *Proteobacteria*) in an open-circuit BES to convert substrates, thus competing with anode respiring bacteria for power output [63]. Lately, it was highlighted that *Bacteroidetes* can be easily enriched in BES when supplemented with other electron acceptors (NO_3^-) [64], thus potentially enhancing an electron flow that is separated from the energy yield in MECs. *Chloroflexi* also represented an enriched phylum in the open circuit, and it is usually predominant in anaerobic digester sludge. *Bacteroidetes* increased to 25.9 %, compared to 5.6 % in the initial setup MECs, with high coulombic efficiency and H_2 yield. *Proteobacteria,* on the contrary, decreased to 17.8 % in SFL control MECs, although they increased to 26.8 % by feeding with freeze/thaw-pretreated SFL. More specifically,*G ammaproteobacteria* and *Deltaproteobacteria* were partially decreased, while *Betaproteobacteria* partially increased, thus leading, as a result, to a lower efficiency of electron transfer and hydrogen recovery.

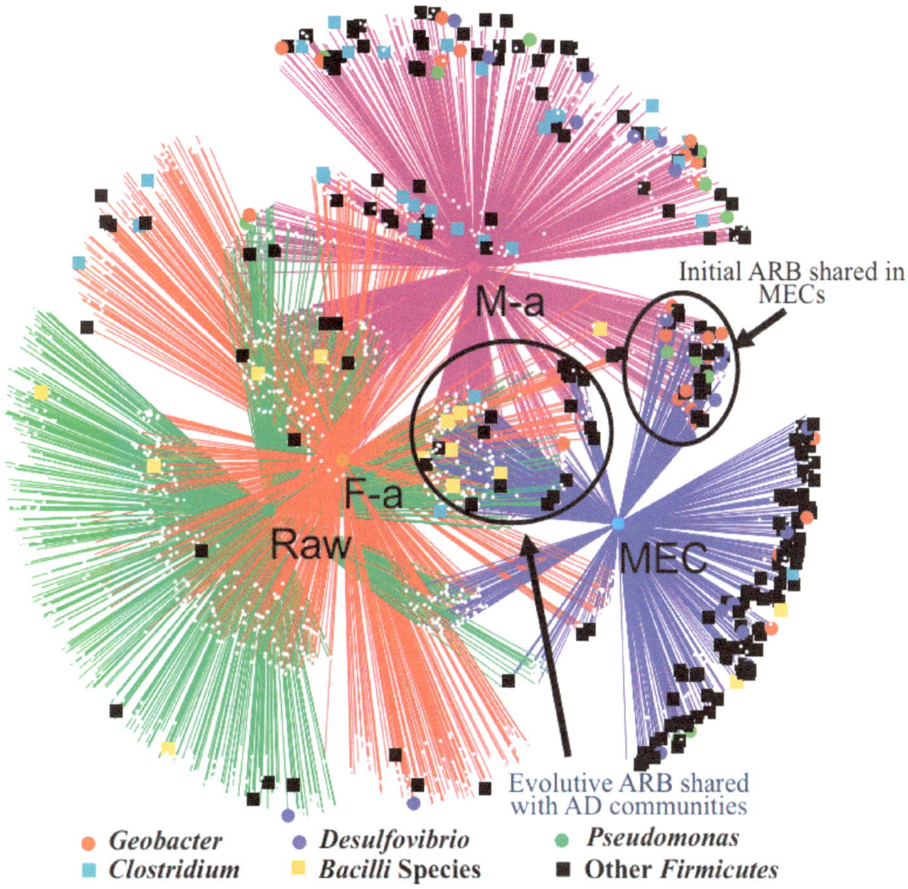

Figure 7: Network of communities based on OTUs in sludge cascade utilization, through alkaline pretreatment. Electron transfer in integrated system was conducted by part of the initial anode respiring bacteria (ARB) and part of the evolutive ARB, interacting with fermentative communities. *M-a* MEC fed with alkaline treated SFL, *F-a* Fermentative solution of alkaline pretreatment.

MECs fed with alkaline pretreated SFL showed the best energy yield and shared the largest community of*Geobacter*, *Desulfovibrio*, *Pseudomonas*, and *Clostridium,* with initial MECs (Fig. 7). Simultaneously, *Firmicutes*substantially decreased, including*Acetoanaerobium* (1.1 %), *Acetobacterium* (0.2 %), and *Fusibacter* (0.3 %). Some of the anaerobes (>20 %), i.e., the class *Bacilli* (*Pasteuria* and *Lactococcus*) and *Clostridia* (*Fusibacter*,*Anaerovorax*, and *Proteiniclasticum*), were clearly enriched on electrode biofilm by SFL feeding, suggesting a synergistic effect with exoelectrogens to degrade complex organic matter [35] (like *Lactococcus* producing soluble electron shuttles to promote electron transfer between cells and the electrode surface [65]). Some genera belonging to *Clostridia* (*Acetobacterium* and *Acetoanaerobium*) (<10 %) were probably enriched in MECs by the availability of products such as hydrogen (electron). However, their function on carbon/electron recycling seems to be very limited in various BES systems [33], which would suggest only a limited hydrogen loss (or over 100 % coulombic efficiency) in MECs fed with acetate [66], carbohydrate [67], or fermentation liquid [51]. It is worth noting that a high protein or polysaccharide but low fatty acid content would lead to the dominance of *Proteiniclasticum* and *Parabacteroides* (increased by >10 %), which are able to produce VFAs as end products from fermentation [68, 69]. Thus, part of the microbial communities did not function on extracellular electron transfer; however, they were maintained in the fermentation niche of electrode biofilm, where they could provide labile products for electrode respiring bacteria. A substantial reduction in current and hydrogen recovery (Alkaline vs. Ultrasonic) was observed when introducing SFL with an increased abundance in *Proteiniclasticum* .*Proteiniclasticum* reached 11.4 % in the ultrasonic pretreated SFL, but only 0.4 % in the alkaline pretreated SFL. In fact, MECs presented a similar COD removal (61 and 69 %) but coulombs were reduced by 31 % and hydrogen production by 50 %.

Alkaline pretreatment could provide more short-chain fatty acids and higher conductivities than other pretreatments. These two aspects are known to favor mass transfer in anodic biofilm [48]. *Geobacter*increased in both, alkaline and ultrasonic pretreated SFL, in which accumulated VFAs were higher than 2500 mgCOD/L (with acetate reaching >1000 mgCOD/L). It appeared that some species, belonging to*Parabacteroides*, *Clostridium*, and *Pseudomonas*, were potentially enriched more in alkaline SFL with higher SCOD and conductivity, compared to ultrasonic SFL. But a delayed fermentation process on raw WAS, as well as freeze/thaw-pretreated WAS, substantially led to different fermentative communities in the anode biofilm (such as *Parabacteroides* and *Proteinilclasticum)*, which produced little VFAs for anode respiring bacteria. In this situation, long time would be required to WAS cascade utilization, with a slow fermentation and inefficient electron generation in BES.

CONCLUSIONS

WAS, rich in organic carbon, can be used as an alternative bioresource, and energy recovery can be potentially improved when combining proper pretreatment method and novel AD technology. During the cascade utilization of WAS, organic removal could be improved by various pretreatment methods, while energy recovery of MEC was impacted by fermentation liquid and microbial community composition. A smaller particle size of WAS after pretreatment substantially favored the fermentation process and formation of suitable byproducts for BES. Alkaline pretreatment gave the highest VFAs production, achieving the most effective energy recovery in the MECs with abundance of *Geobacter* >10 %. Energy yield of electron transfer in integrated system was influenced, on one side, by the initially formed exoelectrogen community, but also by the newly formed one, after interacting with fermentative communities. A high protein or polysaccharide (but low fatty acids) content led to the dominance of *Proteiniclasticum* and *Parabacteroides*, which had a delayed contribution to extracellular electron transfer.

METHODS

Pretreatment of WAS and Short Fermentation Process

The WAS was collected from the secondary sedimentation tank of Wenchang municipal WWTP in Harbin, China. The sludge was concentrated by settling for 24 h and moving the water layer away, then stored in the refrigerator at 4 °C. The large particles were separated through the 40 mesh sieve. In order to facilitate comparison among different treatments, the volatile suspended solid (VSS) content of the WAS used in this study was adjusted to 14 g/L. Three kinds of pretreatment methods were used, prior to fermentation: (a) alkaline treatment, in which to the pH was adjusted to 10 (by NaOH) and the sludge reacted for 20 min at 81 °C [45]; (b) freeze/thaw treatment, by freezing the sludge at −20 °C for 72 h and then thawing it at room temperature, before use [70]; (c) ultrasonic treatment, in which the sludge was treated for 10 min with 0.5 kW/L energy density, using 40 kHz bi-frequency ultrasound; (d) the control, without any treatment of the raw sludge [46].

After pretreatment, the particle size distribution was analyzed with a particle analyzer (Mastersizer 2000 Malvern) with a Hydro 2000MU dispersing unit and detected by means of laser diffractometry (within approximately 30 s). The following parameters were set: refractive index (RI) values for particles and basis solution (water, 20 °C) were 1.52 and 1.33, respectively and the measurement cycle is 10 s. The analysis was repeated three times, and the average readings were obtained by Mastersizer 2000 software. The output of the measurements is depicted in a graph of volume (%) against particle size (μm) within a range from 0.02 to 2000 μm. The stirrer and pump speed were kept at 600 rpm, which is the minimum pump/stirrer speed of the instrument, to eliminate the potential damage to the sludge flocks [71]. The WAS fermentation took place in sealed glass bottles (effective volume of 500 mL, Sichuan Shubo CO., China). Each bottle contained 350 mL WAS. All bottles were flushed with nitrogen gas for 10 min to remove oxygen. The bottles were stirred in an air-bath shaker at 35 ± 2 °C for 3 days to achieve a good accumulation of VFAs [45]. The supernatant was taken from SFL after settling overnight (12 h), and used as carbon source for MECs.

MECs Reactor Setup and Operational Conditions

Fifteen single chamber MEC reactors were set up. The reactors were made of polycarbonate. The effective volume was 40 mL, including a 28 mL chamber (3 cm diameter × 4 cm length) and a 10 mL injection syringe as a gas collection tube (valid volume 12 mL). The anode was a graphite brush (1.6 cm diameter × 8 cm; 0.22 m^2 surface area). The cathode was made of carbon cloth (YW-50, YiBang Technology Co., Ltd., China), covered with a Pt catalyst layer (0.5 mg Pt/cm^2 in one side) [51].

The reactors were firstly inoculated with the sludge from the Wenchang municipal WWTP and were fed with acetate (1500 mg/L) as carbon source, in 50 mM phosphate buffer solution (PBS, containing NH_4Cl 0.31 g/L, KCl 0.13 g/L, $NaH_2PO_4{\cdot}2H_2O$ 5.618 g/L, $Na_2HPO_4{\cdot}12H_2O$ 6.155 g/L, pH 7.0 ± 0.5, 7.3 ± 0.3 mS/cm) [35, 51]. The external voltage was 0.80 ± 0.01 V. All the reactors kept running stably, in 24 h batch operation, for at least 10 cycles till all replicates performed similar current, COD removal, and gas production. Twelve reactors were picked out of the 15, and then every three reactors were set as replicates, which were randomly divided into four groups for the test of different sludge fermentative liquid, obtained from different treatment, as described above. The SFL for MECs was discharged and refilled at the end of every 3 d batch cycle over 1 month. A flow schematic representation of experimental methodology and reactor setup is shown in.

Analysis and Calculation Methods

The currents were automatically monitored (Acquisition system; Keithley Instrument, US) through a 10 Ω resister. The gas was collected in a gas bag (50 mL; Cali5-Bond; Calibrated Instrument Inc, US). Gas composition was analyzed by a gas chromatograph (Fuli, GC9790; Zhenjiang Instrument Inc, China), with a packed column [51] (TDX-01; 2 m length) and equipped with a TCD detector. The volume of gas was measured by a glass syringe.

LB-EPSs and TB-EPSs were extracted according to previous studies [72, 73], and were modified appropriately. The specific method was: firstly, 10 mL sample was centrifuged at 4000*g* for 10 min, the supernatant was filtered with 0.45 μm cellulose nitrate membrane filters, and the filtrate was serviced as dissolved organic matters (DOMs). Secondly, the residue in centrifuge tube was treated according to the EPS extraction method for LB-EPSs [72]. The filtrate was processed as the LB-EPSs. Finally, after LB-EPSs extraction, the residue in centrifuge tube was treated according to the EPSs extraction method for TB-EPSs and filtrate was regarded as the TB-EPSs.

The short-chain fatty acids (SCFAs) were recorded as the sum of acetic (HAc), propionic (HPr), *n*-butyric (*n*-HBu),*iso*-butyric (*iso*-HBu), *n*-valeric (*n*-HVa), and *iso*-valeric acids (*iso*-HVa). SCFAs were analyzed by a gas chromatograph (Agilent 4890; J&W Scientific, USA) equipped with a FID detector and a capillary column (19095N-123HP-INNOWAX; 30 × 0.530 mm × 1.00 μm; J&W Scientific, USA) [45]. The samples were centrifuged at 10,000 rpm and filtered through a

0.45 μm membrane filter, before analysis. Soluble carbohydrate and protein of filtrate samples were analyzed immediately.

The energy and coulombic efficiency were calculated to characterize the performance of MEC reactor. Columbic efficiency indicated the electron recovery from substrates, which was defined by the ratio of coulomb recovery to the total coulombs in the substrate. The coulomb recovery can be calculated by the equation $Q = \int i \cdot \Delta t$, where i is the current of the external circuit. The total coulombs can be calculated by the equation $Q_t=(COD_{in}-COD_{eff})\cdot V\cdot F\cdot b/$, where F represents the Faraday constant, 96,485 C/mol; MO_2 is the molar mass of oxygen, 32 g/mol; b is the complete oxidation requirement of electron per mole oxygen and b is 4 mol-e^-/mol; COD is measured from influence and effluence of SFL.

DNA Extraction and Illumina Sequencing

Microbial community samples were collected from fermentation liquid on the 3rd day of fermentation and anode biofilm at the end of MEC operations. PowerSoil DNA Isolation Kit (Mo Bio Laboratories, Carlsbad, CA, US) was used to extract the genomic DNA of the samples, according to the manufacturer's instructions. The quantity and quality of the extracted DNA were checked by measuring its absorbance at 260 and 280 nm, using a spectrophotometer. Amplicon was constructed for Illumina sequencing, using bacterial fused primers 341F (5′-CCTACACGACGCTCTTCCGATCTN (barcode) CCTACGG-GNGGCWGCAG-3′) and 805R (5′-GACTGGAGTTCCTTGGCACCCGAGAATTCCA (barcode) GACTACHVGGGTATCTAATCC-3′) for the V3–V4 region of the 16S rRNA gene [74]. PCRs were performed in a total volume of 50 μL, containing 1 × PCR buffer, 1 mM dNTPs, 5 μM each primer, 1 U Plantium Taq, and 10 ng of template DNA. The PCR amplification program contained an initial denaturation at 94 °C for 3 min, followed by 5 denaturing cycles at 94 °C for 30 s, annealing at 45 °C for 20 s, and extension at 65 °C for 30 s, then followed by 20 cycles of denaturing at 94 °C for 20 s, annealing at 55 °C for 20 s, and extension at 72 °C for 30 s, finally followed by a final extension at 72 °C for 5 min.

Before sequencing, PCR products of different samples were normalized in equimolar amounts in the final mixture, which was used to construct the PCR amplicon libraries. Sequencing was carried out on an Illumina HiSeq 2000, and the raw sequences have been deposited in the NCBI Short Read Archive (SRA) database, under the accession number SRR1532554. With similarity set at 97 % and a confidence threshold of 95 %, the obtained sequences were phylogenetically allocated down to the phylum, class, and genus level with the MOTHUR program (http://www.mothur.org/wiki/Main_Page). To define the relative abundance of a given phylogenetic group, the number of sequences affiliated to that group were divided by the total number of obtained sequences. The results were used for the analysis and comparison of microbial community structure differences. The Cytoscape network layout used in this work was made by Cytoscape 3.2.1, using detected OUTs as node connectivity to illustrate groups and inter-group relationships [75, 76].

AUTHORS' CONTRIBUTIONS

WL and ZH designed and carried out the experiments, performed the data analysis, and drafted the manuscript. CY, AZ, and ZG carried out the pretreatment method, participated in the setup operation and maintenance of the bioreactors. BL and CV participated in data analysis for pyrosequencing results and revised the manuscript. WL and AW conceived of the study, put forward the hypothesis, and gave the final approval of publication. All authors read and approved the final manuscript.

ACKNOWLEDGEMENTS

This research was supported by National Science Foundation for Distinguished Young Scholars (Grant No. 51225802), by National Natural Science Foundation of China (No. 51578534), by "Hundred Talents Program" of the Chinese Academy of Sciences, and by Project 135 of Chinese Academy of Sciences (No. YSW2013B06), and by Sino-EU International S&T cooperation program (No. S2015GR1012).

REFERENCES

1. Foresti E, Zaiat M, Vallero M. Anaerobic processes as the core technology for sustainable domestic wastewater treatment: consolidated Applications, new trends, perspectives, and challenges. Rev Environ Sci Bio. 2006;5:3–19.
2. McCarty PL, Bae J, Kim J. Domestic wastewater treatment as a net energy producer-can this be achieved? Environ Sci Technol. 2011;45:7100–6.
3. Jin LY, Zhang GM, Tian HF. Current state of sewage treatment in China. Water Res. 2014;66:85–98.
4. Barnabe S, Brar SK, Tyagi RD, Beauchesne I, Surampalli RY. Pre-treatment and bioconversion of wastewater sludge to value-added products-Fate of endocrine disrupting compounds. Sci Total Environ. 2009;407:1471–88.
5. Guo ZC, Zhou AJ, Yang CX, Liang B, Sangeetha T, He ZW, Wang L, Cai WW, Wang AJ, Liu WZ. Enhanced short chain fatty acids production from waste activated sludge conditioning with typical agricultural residues: carbon source composition regulates community functions. Biotechnol Biofuels. 2015;8:1.
6. Appels L, Baeyens J, Degreve J, Dewil R. Principles and potential of the anaerobic digestion of waste-activated sludge. Prog Energ Combust. 2008;34:755–81.
7. Borole AP, Reguera G, Ringeisen B, Wang ZW, Feng YJ, Kim BH. Electroactive biofilms: current status and future research needs. Energ Environ Sci. 2011;4:4813–34.
8. Li WW, Yu HQ, He Z. Towards sustainable wastewater treatment by using microbial fuel cells-centered technologies. Energ Environ Sci. 2014;7:911–24.
9. Zhang JX, Zhang YB, Quan X, Chen S. Enhancement of anaerobic acidogenesis by integrating an electrochemical system into an acidogenic reactor: effect of hydraulic retention times (HRT) and role of bacteria and acidophilic methanogenic Archaea. Bioresour Technol. 2015;179:43–9.
10. Vijayaraghavan K, Sagar GK. Anaerobic digestion and in situ electrohydrolysis of dairy bio-sludge. Biotechnol Bioproc Eng. 2010;15:520–6.

11. Bo T, Zhu X, Zhang L, Tao Y, He X, Li D, Yan Z. A new upgraded biogas production process: coupling microbial electrolysis cell and anaerobic digestion in single-chamber, barrel-shape stainless steel reactor. Electrochem Commun. 2014;45:67–70.
12. Tartakovsky B, Mehta P, Bourque JS, Guiot SR. Electrolysis-enhanced anaerobic digestion of wastewater. Bioresource Technol. 2011;102:5685–91.
13. Sasaki K, Sasaki D, Morita M, Hirano S, Matsumoto N, Ohmura N, Igarashi Y. Bioelectrochemical system stabilizes methane fermentation from garbage slurry. Bioresour Technol. 2010;101:3415–22.
14. Lovley DR. Live wires: direct extracellular electron exchange for bioenergy and the bioremediation of energy-related contamination. Energ Environ Sci. 2011;4:4896–906.
15. De Vrieze J, Gildemyn S, Arends JBA, Vanwonterghem I, Verbeken K, Boon N, Verstraete W, Tyson GW, Hennebel T, Rabaey K. Biomass retention on electrodes rather than electrical current enhances stability in anaerobic digestion. Water Res. 2014;54:211–21.
16. Lee C, Kim J, Shin SG, Hwang S. Monitoring bacterial and archaeal community shifts in a mesophilic anaerobic batch reactor treating a high-strength organic wastewater. FEMS Microbiol Ecol. 2008;65:544–54.
17. Shin SG, Han G, Lim J, Lee C, Hwang S. A comprehensive microbial insight into two-stage anaerobic digestion of food waste-recycling wastewater. Water Res. 2010;44:4838–49.
18. Ren NQ, Guo WQ, Liu BF, Cao GL, Ding J. Biological hydrogen production by dark fermentation: challenges and prospects towards scaled-up production. Curr Opin Biotechnol. 2011;22:365–70.
19. McCurry DL, Bear SE, Bae J, Sedlak DL, McCarty PL, Mitch WA. Superior removal of disinfection byproduct precursors and pharmaceuticals from wastewater in a staged anaerobic fluidized membrane bioreactor compared to activated sludge. Environ Sci Technol Lett. 2014;1:459–64.
20. Hu YB, Zhang CJ, Zhang C, Tan XJ, Zhu HG, Zhou Q. Effect of alkaline pre-treatment on waste activated sludge solubilization and anaerobic digestion. 3rd International Conference on Bioinformatics and Biomedical Engineering, Vols 1–11. 2009: 6359–62.
21. Tiehm A, Nickel K, Zellhorn M, Neis U. Ultrasonic waste activated sludge disintegration for improving anaerobic stabilization. Water Res. 2001;35:2003–9.
22. Wang L, Liu WZ, Kang LL, Yang CX, Zhou AJ, Wang AJ. Enhanced biohydrogen production from waste activated sludge in combined strategy of chemical pretreatment and microbial electrolysis. Int J Hydrog Energ. 2014;39:11913–9.
23. Lu L, Xing D, Ren N. Pyrosequencing reveals highly diverse microbial communities in microbial electrolysis cells involved in enhanced H_2 production from waste activated sludge. Water Res. 2012;46:2425–34.
24. Sun R, Zhou AJ, Jia JN, Liang Q, Liu Q, Xing DF, Ren NQ. Characterization of methane production and microbial community shifts during waste activated sludge degradation in microbial electrolysis cells. Bioresour Technol. 2015;175:68–74.
25. Monlau F, Sambusiti C, Ficara E, Aboulkas A, Barakat A, Carrere H. New opportunities for agricultural digestate valorization: current situation and perspectives. Energ Environ Sci. 2015;8:2600–21.
26. Renslow RS, Babauta JT, Majors PD, Beyenal H. Diffusion in biofilms respiring on electrodes. Energ Environ Sci. 2013;6:595–607.
27. Ishii S, Suzuki S, Norden-Krichmar TM, Phan T, Wanger G, Nealson KH, Sekiguchi Y, Gorby YA, Bretschger O. Microbial population and functional dynamics associated with surface potential and carbon metabolism. ISME J. 2014;8:963–78.
28. Rabaey K, Boon N, Siciliano SD, Verhaege M, Verstraete W. Biofuel cells select for microbial consortia that self-mediate electron transfer. Appl Environ Microbiol. 2004;70:5373–82.
29. Ishii S, Watanabe K, Yabuki S, Logan BE, Sekiguchi Y. Comparison of electrode reduction activities of geobacter sulfurreducens and an enriched consortium in an air-cathode microbial fuel cell. Appl Environ Microbiol. 2008;74:7348–55.
30. Watson VJ, Logan BE. Power production in MFCs inoculated with *Shewanella oneidensis* MR-1 or mixed cultures. Biotechnol Bioeng. 2010;105:489–98.
31. Logan BE. Exoelectrogenic bacteria that power microbial fuel cells. Nat Rev Microbiol. 2009;7:375–81.
32. Rabaey K, Rodriguez J, Blackall LL, Keller J, Gross P, Batstone D, Verstraete W, Nealson KH. Microbial ecology meets electrochemistry: electricity-driven and driving communities. ISME J. 2007;1:9–18.
33. Xafenias N, Mapelli V. Performance and bacterial enrichment of bioelectrochemical systems during methane and acetate production. Int J Hydrog Energ. 2014;39:21864–75.
34. Ravot G, Magot M, Fardeau ML, Patel BKC, Thomas P, Garcia JL, Ollivier B. Fusibacter paucivorans gen. nov., sp nov., an anaerobic, thiosulfate-reducing bacterium from an oil-producing well. Int J Syst Bacteriol. 1999;49:1141–7.
35. Liu W, Wang A, Cheng S, Logan BE, Yu H, Deng Y, Nostrand JD, Wu L, He Z, Zhou J. Geochip-based functional gene analysis of anodophilic communities in microbial electrolysis cells under different operational modes. Environ Sci Technol. 2010;44:7729–35.
36. Arends JB, Verstraete W. 100 years of microbial electricity production: three concepts for the future. Microb Biotechnol. 2012;5:333–46.
37. Call D, Logan BE. Hydrogen production in a single chamber microbial electrolysis cell lacking a membrane. Environ Sci Technol. 2008;42:3401–6.
38. Cheng S, Logan BE. Sustainable and efficient biohydrogen production via electrohydrogenesis. P Natl Acad Sci USA. 2007;104:18871–3.
39. Liu W, Cai W, Guo Z, Wang L, Yang C, Varrone C, Wang A. Microbial electrolysis contribution to anaerobic

digestion of waste activated sludge, leading to accelerated methane production. Renew Energ. 2016;91:334–9.

40. Yu Q, Lei HY, Li Z, Li HL, Chen K, Zhang XH, Liang RL. Physical and chemical properties of waste-activated sludge after microwave treatment. Water Res. 2010;44:2841–9.
41. Chang CJ, Tyagi VK, Lo SL. Effects of microwave and alkali induced pretreatment on sludge solubilization and subsequent aerobic digestion. Bioresour Technol. 2011;102:7633–40.
42. Dogan I, Sanin FD. Alkaline solubilization and microwave irradiation as a combined sludge disintegration and minimization method. Water Res. 2009;43:2139–48.
43. Tong J, Chen Y. Recovery of nitrogen and phosphorus from alkaline fermentation liquid of waste activated sludge and application of the fermentation liquid to promote biological municipal wastewater treatment. Water Res. 2009;43:2969.
44. Kavitha S, Brindha GMJ, Gloriana AS, Rajashankar K, Yeom IT, Banu JR. Enhancement of aerobic biodegradability potential of municipal waste activated sludge by ultrasonic aided bacterial disintegration. Bioresour Technol. 2016;200:161–9.
45. Zhou A, Yang C, Guo Z, Hou Y, Liu W, Wang A. Volatile fatty acids accumulation and rhamnolipid generation in situ from waste activated sludge fermentation stimulated by external rhamnolipid addition. Biochem Eng J. 2013;77:240–5.
46. Zhou A, Du J, Varrone C, Wang Y, Wang A, Liu W. VFAs bioproduction from waste activated sludge by coupling pretreatments with Agaricus bisporus substrates conditioning. Process Biochem. 2014;49:283–9.
47. Nam JY, Kim HW, Lim KH, Shin HS, Logan BE. Variation of power generation at different buffer types and conductivities in single chamber microbial fuel cells. Biosens Bioelectron. 2010;25:1155–9.
48. Recio-Garrido D, Perrier M, Tartakovsky B. Modeling, optimization and control of bioelectrochemical systems. Chem Eng J. 2016;289:180–90.
49. Kondaveeti S, Min B. Bioelectrochemical reduction of volatile fatty acids in anaerobic digestion effluent for the production of biofuels. Water Res. 2015;87:137–44.
50. Zhou A, Yang C, Kong F, Liu D, Chen Z, Ren N, Wang A. Improving the short-chain fatty acids production of waste activated sludge stimulated by a bi-frequency ultrasonic pretreatment. J Environ Biol. 2013;34:381–9.
51. Liu W, Huang S, Zhou A, Zhou G, Ren N, Wang A, Zhuang G. Hydrogen generation in microbial electrolysis cell feeding with fermentation liquid of waste activated sludge. Int J Hydrog Energ. 2012;37:13859–64.
52. Xu LJ, Liu WZ, Wu YN, Wang AJ, Li S, Ji W. Optimizing external voltage for enhanced energy recovery from sludge fermentation liquid in microbial electrolysis cell. Int J Hydrog Energ. 2013;38:15801–6.
53. Liu WT, Chan OC, Fang HHP. Microbial community dynamics during start-up of acidogenic anaerobic reactors. Water Res. 2002;36:3203–10.
54. Escapa A, Gil-Carrera L, Garcia V, Moran A. Performance of a continuous flow microbial electrolysis cell (MEC) fed with domestic wastewater. Bioresour Technol. 2012;117:55–62.
55. Cusick RD, Bryan B, Parker DS, Merrill MD, Mehanna M, Kiely PD, Liu GL, Logan BE. Performance of a pilot-scale continuous flow microbial electrolysis cell fed winery wastewater. Appl Microbiol Biot. 2011;89:2053–63.
56. Zhou JZ, Liu WZ, Deng Y, Jiang YH, Xue K, He ZL, Van Nostrand JD, Wu LY, Yang YF, Wang AJ. Stochastic assembly leads to alternative communities with distinct functions in a bioreactor microbial community. Mbio. 2013;4:e00584.
57. Wang A, Liu W, Cheng S, Xing D, Zhou J, Logan BE. Source of methane and methods to control its formation in single chamber microbial electrolysis cells. Int J Hydrog Energ. 2009;34:3653–8.
58. Koller KB, Hawkridge FM, Fauque G, Legall J. Direct electron-transfer reactions of cytochrome-C553 from *Desulfovibrio-Vulgaris* Hildenborough at indium oxide electrodes. Biochem Biophys Res Commun. 1987;145:619–24.
59. Commault AS, Lear G, Weld RJ. Maintenance of geobacter-dominated biofilms in microbial fuel cells treating synthetic wastewater. Bioelectrochemistry. 2015;2015:9.
60. Wang HM, Ren ZJ. Bioelectrochemical metal recovery from wastewater: a review. Water Res. 2014;66:219–32.
61. Liu YJ, Nikolausz M, Wang XC. Biodegradation and detoxication of phenol by using free and immobilized cells of*Acinetobacter* sp XA05 and *Sphingomonas* sp FG03. J Environ Sci Health A. 2009;44:130–6.
62. Zhang YF, Min B, Huang LP, Angelidaki I. Electricity generation and microbial community response to substrate changes in microbial fuel cell. Bioresour Technol. 2011;102:1166–73.
63. Huang J, Wang ZW, Zhu CW, Ma JX, Zhang XR, Wu ZC. Identification of microbial communities in open and closed circuit bioelectrochemical MBRs by high-throughput 454 pyrosequencing. PloS ONE. 2014;9:e93842.
64. Kondaveeti S, Lee SH, Park HD, Min B. Bacterial communities in a bioelectrochemical denitrification system: the effects of supplemental electron acceptors. Water Res. 2014;51:25–36.
65. Inoue K, Leang C, Franks AE, Woodard TL, Nevin KP, Lovley DR. Specific localization of the c-type cytochrome OmcZ at the anode surface in current-producing biofilms of *Geobacter sulfurreducens*. Env Microbiol Rep. 2011;3:211–7.
66. Lee H-S, Rittmann BE. Significance of biological hydrogen oxidation in a continuous single-chamber microbial electrolysis cell. Environ Sci Technol. 2010;44:948–54.
67. Selembo PA, Perez JM, Lloyd WA, Logan BE. High hydrogen production from glycerol or glucose by electrohydrogenesis using microbial electrolysis cells. Int J Hydrog Energ. 2009;34:5373–81.

68. Yang CX, Liu WZ, He ZW, Thangavel S, Wang L, Zhou AJ, Wang AJ. Freezing/thawing pretreatment coupled with biological process of thermophilic *Geobacillus* sp. G1: acceleration on waste activated sludge hydrolysis and acidification. Bioresour Technol. 2015;175:509–16.

69. Zhang KG, Song L, Dong XZ. Proteiniclasticum ruminis gen. nov., sp. nov, a strictly anaerobic proteolytic bacterium isolated from yak rumen. Int J Syst Evol Micr. 2010;60:2221–5.

70. Gao W. Freezing as a combined wastewater sludge pretreatment and conditioning method. Desalination. 2011;268:170–3.

71. Alagoz BA, Yenigun O, Erdincler A. Enhancement of anaerobic digestion efficiency of wastewater sludge and olive waste: synergistic effect of co-digestion and ultrasonic/microwave sludge pre-treatment. Waste Manag. 2015;46:182–8.

72. Li XY, Yang SF. Influence of loosely bound extracellular polymeric substances (EPS) on the flocculation, sedimentation and dewaterability of activated sludge. Water Res. 2007;41:1022–30.

73. Yan LL, Liu Y, Wen Y, Ren Y, Hao GX, Zhang Y. Role and significance of extracellular polymeric substances from granular sludge for simultaneous removal of organic matter and ammonia nitrogen. Bioresour Technol. 2015;179:460–6.

74. Hamady M, Lozupone C, Knight R. Fast UniFrac: facilitating high-throughput phylogenetic analyses of microbial communities including analysis of pyrosequencing and PhyloChip data. ISME J. 2009;4:17–27.

75. Cline MS, Smoot M, Cerami E, Kuchinsky A, Landys N, Workman C, Christmas R, Avila-Campilo I, Creech M, Gross B, Hanspers K, Isserlin R, Kelley R, Killcoyne S, Lotia S, Maere S, Morris J, Ono K, Pavlovic V, Pico AR, Vailaya A, Wang PL, Adler A, Conklin BR, Hood L, Kuiper M, Sander C, Schmulevich I, Schwikowski B, Warner GJ, Ideker T, Bader GD. Integration of biological networks and gene expression data using Cytoscape. Nat Protoc. 2007;2:2366–82.

76. Herrgard MJ, Swainston N, Dobson P, Dunn WB, Arga KY, Arvas M, Bluthgen N, Borger S, Costenoble R, Heinemann M, Hucka M, Le Novere N, Li P, Liebermeister W, Mo ML, Oliveira AP, Petranovic D, Pettifer S, Simeonidis E, Smallbone K, Spasic I, Weichart D, Brent R, Broomhead DS, Westerhoff HV, Kirdar B, Penttila M, Klipp E, Palsson BO, Sauer U, Oliver SG, Mendes P, Nielsen J, Kell DB. A consensus yeast metabolic network reconstruction obtained from a community approach to systems biology. Nat Biotechnol. 2008;26:1155–60.

CHAPTER 5 BIOELECTROCHEMICAL PRODUCTION OF HYDROGEN IN AN INNOVATIVE PRESSURE-RETARDED OSMOSIS/MICROBIAL ELECTROLYSIS CELL SYSTEM

Heyang Yuan, Yaobin Lu, Ibrahim M Abu-Reesh and Zhen He

BACKGROUND

Microbial electrolysis cells (MECs) is an attractive technology that can simultaneously remove organics and produce hydrogen gas. In MECs, exoelectrogens growing on the anode respire by releasing electrons extracellularly; driven by an external voltage >0.2 V, those electrons flow to the cathode to reduce protons into hydrogen gas [1]. MECs are of strong interests because its energy consumption could be significantly lower than that of conventional methods such as water-splitting and steam reforming [2, 3]. Life cycle assessment suggested that MECs might outperform the prevailing wastewater treatment methods (i.e. activated sludge process and anaerobic digestion) in terms of energy requirement, greenhouse gas effect and other environmental impacts [4, 5]. However, the requirement of additional energy, which is mostly from fossil fuels, should be further addressed to make MECs more sustainable.

Researchers have explored alternative energy sources to drive hydrogen production in MECs. For example, a microbial fuel cell (MFC) was used to replace external power supply and provided a voltage to achieve a hydrogen production rate of 0.015 $m^3\ m^{-3}\ d^{-1}$ in an MFC-MEC coupled system [6]. It was found that the hydrogen yield was relatively low and also instable in this coupled system, possibly due to cell voltage reversal between the two bioelectrochemical systems (BES). To avoid voltage reversal, a capacitor circuit was installed between MFCs and an MEC, and helped achieve 38 % higher hydrogen production rate compared to the directly coupled system [7]. In another study, solar energy was harvested by a dye-sensitized solar cell and then applied on an MEC, which achieved a hydrogen production rate of 0.07 $m^3\ m^{-3}\ d^{-1}$ [8]. The entropic energy stored in a salinity gradient between seawater and fresh water is estimated to be 0.8 kW m^{-3} [9, 10], which could be captured by reverse electrodialysis and then used to drive MECs [11–13].

Unlike salinity energy that relies on salt movement, osmotic energy can be produced through water interaction between saline water and freshwater, and can be harvested by using pressure-retarded osmosis (PRO). In a PRO system, water is driven by the salinity gradient and diffuses from a low-salinity solution (feed) to a high-salinity solution (draw) through a semi-permeable membrane; consequently, electrical energy is generated by pressurizing the diluted draw (whose volume becomes larger after water extraction) through a hydroturbine [14]. As a forward osmosis (FO) process, PRO shows low fouling propensity compared to reverse osmosis because of relatively low water flux and hydraulic pressure [15]. The recent advances in membrane technology could greatly reduce the capital cost of PRO and thus make it competitive with other renewable energy technologies [16]. The highest power density produced by PRO was reported to be 10.6 W m^{-2} with 1 M NaCl solution as draw [17].

We have previously integrated osmotic processes with BES to assist with bioelectricity generation in MFCs or desalination in microbial desalination cells (MDCs) [18]. For example, the first osmotic microbial fuel cell (OsMFC) was developed by

Citation: Heyang Yuan, Yaobin Lu, Ibrahim M AbuReesh and Zhen He, Bioelectrochemical production of hydrogen in an innovative pressure-retarded osmosis/microbial electrolysis cell system, DOI 10.1186/s13068-015-0305-0.

replacing the ion exchange membrane with a FO membrane, and accomplished simultaneous wastewater treatment, water extraction, and bioenergy production [19]. The performance of an OsMFC was affected by its draw solution, membrane condition, anode substrates, and cathode reactions [20–22]. The osmotic process was integrated with MDCs in several ways: replacement of anion exchange membrane with an FO membrane to create an osmotic MDC, linking an OsMFC with an MDC, or connecting an FO to an MDC [23–25]. Those prior studies have demonstrated the synergy between BES and FO, and provided a foundation for the present study.

Given the fact that PRO can generate electric energy and MECs need additional energy for hydrogen production, a PRO-MEC system is proposed here: wastewater (feed) and seawater (draw) flow into a PRO unit for water extraction and energy generation; then, the energy is applied to an MEC for organic removal and hydrogen production, with the PRO feed and draw effluents serving as the anolyte and the catholyte of the MEC, respectively (Fig. 1). This is the first time that the PRO energy (osmotic-to-electricity) is studied to drive the conversion of organic compounds to hydrogen in an MEC; in addition to electrical interaction between the two components, this system also contains hydraulic linkage between the two for wastewater treatment and reuse and saline water desalination. After further desalination, the produced water (diluted draw solution) could be useful for non-potable applications such as agricultural irrigation and landscaping. The system may have application niches in a location with seawater and a high demand for wastewater reuse.

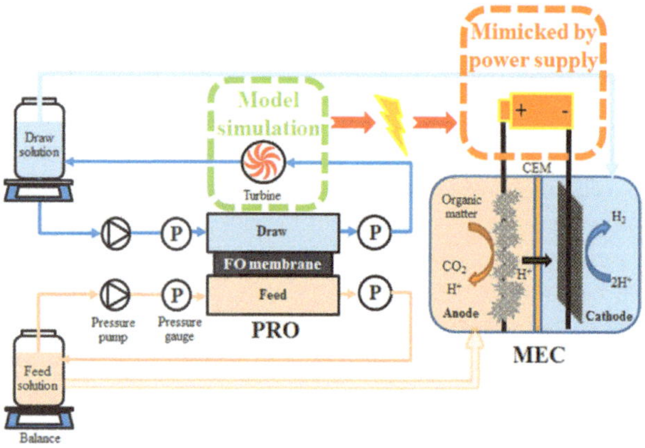

Figure 1: Schematic of the proposed system consisting of a PRO unit and an MEC through hydraulic connection.

To examine the proposed system, we first developed a time-dependent PRO model and a batch-mode MEC model. According to the simulated water flux obtained from the PRO model, the anolyte and the catholyte were prepared for an MEC that was operated experimentally using a power supply to mimic the energy supply process. The system feasibility was demonstrated by comparing the simulated PRO energy production and the experimental MEC energy consumption, together with other parameters such as water extraction, organics removal and hydrogen production. Finally, the PRO and MEC models were used to predict the system performance at varied influent volume, draw concentration and external voltage.

RESULTS AND DISCUSSION

Model Validation

Mathematical modeling can help understand the key factors in the PRO-MEC system and predict the behaviors that cannot be easily investigated by experiments. Previous PRO studies focused on achieving instantaneous maximum energy production, and it was not necessary to take the concentration change of the feed and the draw into account because of the constant replenishment of fresh solutions to maintain high osmotic pressure difference [26, 27]. However, that is not applicable to a PRO-MEC system, because the relatively fast PRO process will generate excessive amount of effluent that can exceed the treating capacity of bench-scale MECs. In this regard, the PRO unit must be operated under a batch mode and the time variable must be introduced in the PRO model to predict the water flux and energy production. To examine that hypothesis, the PRO unit was operated for 6.0 h at different hydraulic pressures (1, 2 and 3 bar) with the initial conditions listed in Table 1. At 1 bar, the feed solution decreased from 600 mL to 138 mL (462 mL of water recovery) with significant decrease in water flux (Additional file 1). The experiments at 2 and 3 bar showed similar trends, but with slightly lower water recovery due to the higher hydraulic pressure (Additional file 1). With the PRO parameters listed in the Additional file 2, the time-dependent PRO model successfully predicted the volume profile with RMSE less than 2.5 % (Table 2). The experiments with different NaCl concentrations (0.5 and 2.0 M) at 1 bar were conducted to further validate the model. After 5.0-h operation, the feed volume decreased from 600 mL to 452 mL with 0.5 M (148 mL water recovery, Additional file 3), 65 % lower than that with 0.8 M at the same operation time because of the lower osmotic pressure difference between the feed and the draw. On the other hand, water recovery rapidly reached 315 mL in 2.0 h with 2.0-M NaCl solution (Additional file 3). The results collectively demonstrated that the time-dependent PRO simulation agreed well with the experimental data (Table 2).

Based on the method presented in previous studies of MDC and MEC modeling [28, 29], a batch-mode MEC model was implemented with independent experiments of varied substrate and external resistance. The introduction of $[HCO_3^-]$ and $[H^+]$ into the anode potential (Eqs. 14, 15) and the substrate change in the concentration overpotential (Eq. 17) helped achieve satisfactory agreement between the experimental and simulation results. The RMSE for the MEC model (Table 2) was relatively high mainly due to the overestimation of organics removal (99.9 % removal) at the end of the MEC process. Consequently, at 0.8 V, the model yielded slightly higher energy consumption (488 J) and hydrogen production (36.7 mL) compared to the experiments (470 J and 32.8 mL). The MEC's operation time was prolonged from 19.2 h at 357 mg L^{-1} acetate to 96.1 h at 2,007 mg L^{-1} acetate (Additional file 4). When the external voltage was increased, the current was improved from 2.7 mA at 0.6 V to 6.6 mA at 1.0 V (Additional file 4) with 60 % increase in Coulombic efficiency and 40 % increase in cathodic efficiency (Table 3). As a result, both hydrogen production and energy consumption increased by two times. The experimental results are consistent with previous studies [30]. In summary, the simulated current generation under different conditions could match the experimental data, with slight overestimation of organics removal, energy consumption and hydrogen production.

Table 1: The initial conditions for the PRO experiments and simulation in different studies

Experiment	V^{a}_{PRO} (mL)	S^{b}_{PRO} (mg L^{-1})	$NaCl^{c}_{in}$ (M)	P^{d} (bar)	V^{e}_{A} (mL)	V^{f}_{C} (mL)	S^{g}_{A} (mg L^{-1})	Voltage[h] (V)
Exp1-model validation	600	193	0.5, 0.8 and 2.0	1, 2 and 3	115	1,085	357, 1,006 and 2,007	0.6, 0.8 and 1
Exp2-system feasibility	600	193	0.8	P_t	115	1,085	1,007	0.8
Exp3-PRO influent volume	100–2,000	193	0.8	P_t	Obtained from the time-dependent PRO model			0.8
Exp4-NaCl concentration	600	193	0.1–2.0	P_t				0.8
Exp5-external voltage	600	193	0.8	P_t	115	1,085	1,007	0.5–1.1

[a] Volume of PRO feed and draw influent.
[b] Substrate concentration in the PRO feed influent.
[c] NaCl concentration in the PRO draw influent.
[d] Hydraulic pressure applied on the PRO draw chamber.
[e] Volume of MEC anolyte influent.
[f] Volume of MEC catholyte influent.
[g] Substrate concentration in the MEC anolyte influent.
[h] External voltage applied on the MEC.

Table 2: RMSE for the PRO and MEC simulation

	RMSE (%)
PRO	
1 bar-feed	1.8 ± 0.3
1 bar draw	0.8 ± 0.0
2 bar-feed	1.9 ± 0.6
2 bar draw	0.8 ± 0.0
3 bar-feed	0.9 ± 0.1
3 bar draw	2.4 ± 0.6
0.5 M feed	1.9 ± 0.2
0.5 M draw	0.9 ± 0.0
2.0 M feed	1.4 ± 0.3
2.0 M draw	0.8 ± 0.0
MEC	
0.6 V	16.5 + 5.6
0.8 V	13.0 ± 2.3
1.0 V	23.6 + 1.3
357 mg L^{-1}	12.1
2,007 mg L^{-1}	21.6

The RMSE of the PRO model was calculated using the feed volume, and that of the MEC model was calculated using electricity.
Standard deviation was calculated with triplicate experiments.

Table 3: The MEC performance at different external voltage

Voltage (V)	S^{a}_{in} (mg L^{-1})	C^{b}_{in} (mS cm^{-1})	V^{c}_{H2} (mL)	$Rate_{H2}$ (m^3 m^{-3} d^{-1})	ΔS^{d} (%)	R^{e}_{CE} (%)	R^{f}_{cat} (%)	Energy (J)	HRT (h)
0.6	1,009 ± 31	19.5 ± 0.8	18.4 ± 0.9	0.008 ± 0.001	95.6 ± 0.0	36.1 ± 0.0	42.9 ± 0.0	221 ± 7	54.4 ± 2.3
0.8	1,009 ± 8	18.3 ± 0.1	32.8 ± 0.6	0.016 ± 0.001	93.7 ± 0.0	57.5 ± 0.0	48.1 ± 0.0	470 ± 2	46.9 ± 1.4
1.0	987 ± 53	19.9 ± 1.5	40.8 ± 2.8	0.030 ± 0.005	94.7 ± 0.0	58.3 ± 0.0	60.7 ± 0.1	581 ± 22	30.9 ± 3.2

Standard deviation was calculated with triplicate experiments.
[a] Substrate concentration in the MEC influent.
[b] Conductivity of the MEC influent.
[c] Total H_2 production.
[d] Organics removal.
[e] Coulombic efficiency.
[f] Cathodic efficiency.

System Feasibility

The feasibility of the PRO-MEC system was demonstrated by estimating energy production in the PRO unit using the models and applying the energy (mimicked by a power supply) to the MEC for hydrogen production and organic removal. It should be noted that, because of the difficulty in generating real energy from bench-scale PRO units at a high hydraulic pressure, the majority of the PRO studies or publications adopted theoretical estimate of energy production [31–34]. The results showed that the PRO unit could theoretically generate sufficient energy to drive the MEC, which successfully produced hydrogen gas and removed more than 90 % of organic compounds. In more details, the PRO simulation was performed at the hydraulic pressure of $P_t = (\pi_{D,t} - \pi_{F,t})/2$ (Eq. 7, at the maximum energy mode), and was stopped when water flux dropped below 0.5 LMH (L m^{-2} h^{-1}). This was because that after 14.5 h of simulation, P_t decreased to 2.4 bar and both the volume and the energy production reached a plateau (Fig. 2a). At the end of the simulation, the volume of the feed solution decreased from the initial 600 mL to 115 mL and the draw volume increased to 1085 mL, suggesting that 485 mL of clean water was extracted from the feed solution. Meanwhile, the PRO unit could produce a feed effluent containing 1,007 mg L^{-1} acetate, a draw effluent with 0.46 M NaCl, and 579 J of energy (Table 1; Fig. 2a).

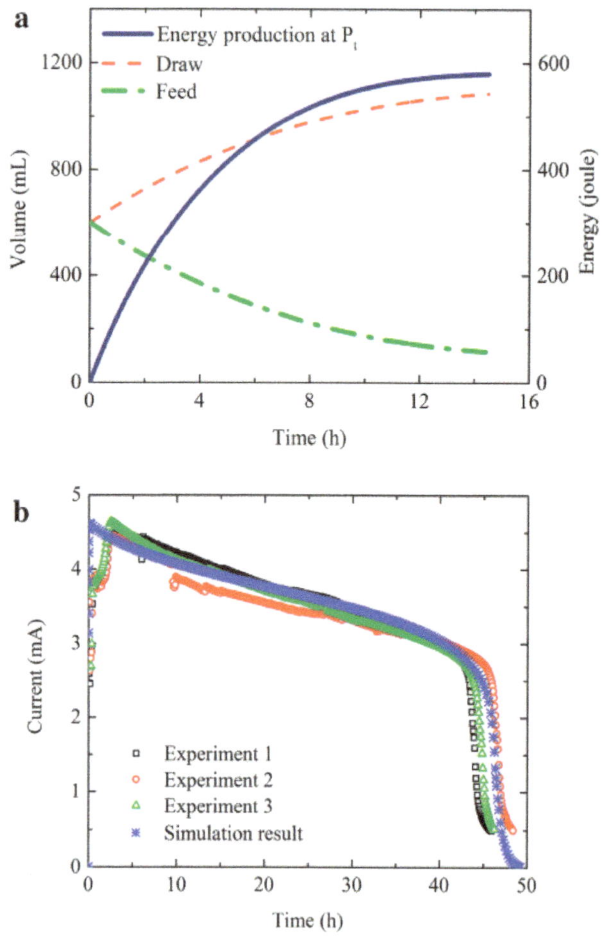

Figure 2: a Volume profile and energy production in the PRO unit at the hydraulic pressure $P_t = (\pi_{D,t} - \pi_{F,t})/2$; b experimental data and simulation result of current generation in the MEC at 0.8 V ...

Based on those results obtained from the PRO simulation, 115 mL of anolyte with 1,009 ± 8 mg L^{-1}acetate and 1,085 mL of the catholyte with the 0.46 M NaCl were prepared and supplied to the MEC (Table 1). An external voltage of 0.8 V was applied to the MEC to mimic the utilization of the energy produced by the PRO unit. The MEC operating time was determined by equating the PRO energy production with the MEC energy consumption (Eqs. 8 and 19). The MEC current decreased steadily over time (Fig. 2b), mainly due to the consumption of organic compounds. At the end of the experiment, the MEC removed 93.7 % of the organics and produced 32.8 mL of H_2 at the expense of 470 J of energy after 46.9 h (Table 3), indicating that the energy produced by the PRO unit (579 J) was sufficient to drive the MEC. The relatively low hydrogen production rate (0.016 m^3 m^{-3} d^{-1}) and cathodic efficiency (48.1 %) was probably because no buffer was added in the catholyte (the pH increased from neutral to 11.6 after 3.0 h of operation). Coulombic efficiency (57.5 %) was comparable with other MECs, likely benefited from a relatively high organics concentration and conductivity of the anolyte, which was the concentrated feed solution from the PRO unit. Similar to the substrate concentration, the conductivity in the feed was concentrated by 5.2 times due to water extraction in the PRO unit, resulting in an initial conductivity of 18.3 mS cm^{-1} in the MEC anolyte. All those results have demonstrated that it is possible to harvest osmotic energy in the PRO unit and use it to convert chemical energy into hydrogen gas in the MEC from the same organic solution. In addition, the coupling of PRO and MEC can simultaneously recover clean water and reduce the volume of the organic solution.

Model Prediction

Effects of PRO Influent Volume

In theory, the osmotic pressure difference will be less affected by the water flux at larger volumes of the feed and the draw solutions, leading to a prolonged PRO process and more energy available for driving the MEC. This hypothesis was examined with the PRO influent volume varied from 100 to 2,000 mL (simultaneous change in both the feed and the draw volumes), with other parameters listed in the Table 1 and 0.5 LMH as a simulation end-point. The PRO simulation shows that both water recovery and energy production are enhanced by increasing the PRO influent volume, while the substrate concentration in the feed effluent remains stable (Fig. 3a, b). At the PRO influent of 1,000 mL, 794 mL of clean water is recovered and 949 J of energy is produced. These values are increased by two times with 2,000 mL of PRO influent. As the initial substrate concentration in the PRO feed influent remains unchanged (193 mg L^{-1}), increasing the PRO influent volume leads to a higher total organic content in the MEC anolyte, which prolongs the MEC's HRT but does not exert effects on the maximum current (4.5 mA, Fig. 3c). As a result of the increased organics and HRT, hydrogen production is improved consistently. It is predicted that the PRO-MEC system produces 59 mL of H_2 in 77.7 h with 1,000 mL of the PRO influent, and 92 mL of H_2 in 121.2 h with 2,000 mL of the PRO influent (Fig. 3d). Interestingly, the energy consumption by the MEC increases non-linearly with increased PRO influent volume, resulting in an enlarged energy surplus between PRO unit and MEC (Fig. 3b). This is indicative that one PRO unit can drive multiple MECs at large volume loading, thereby further enhancing the overall treating capacity of the PRO-MEC system.

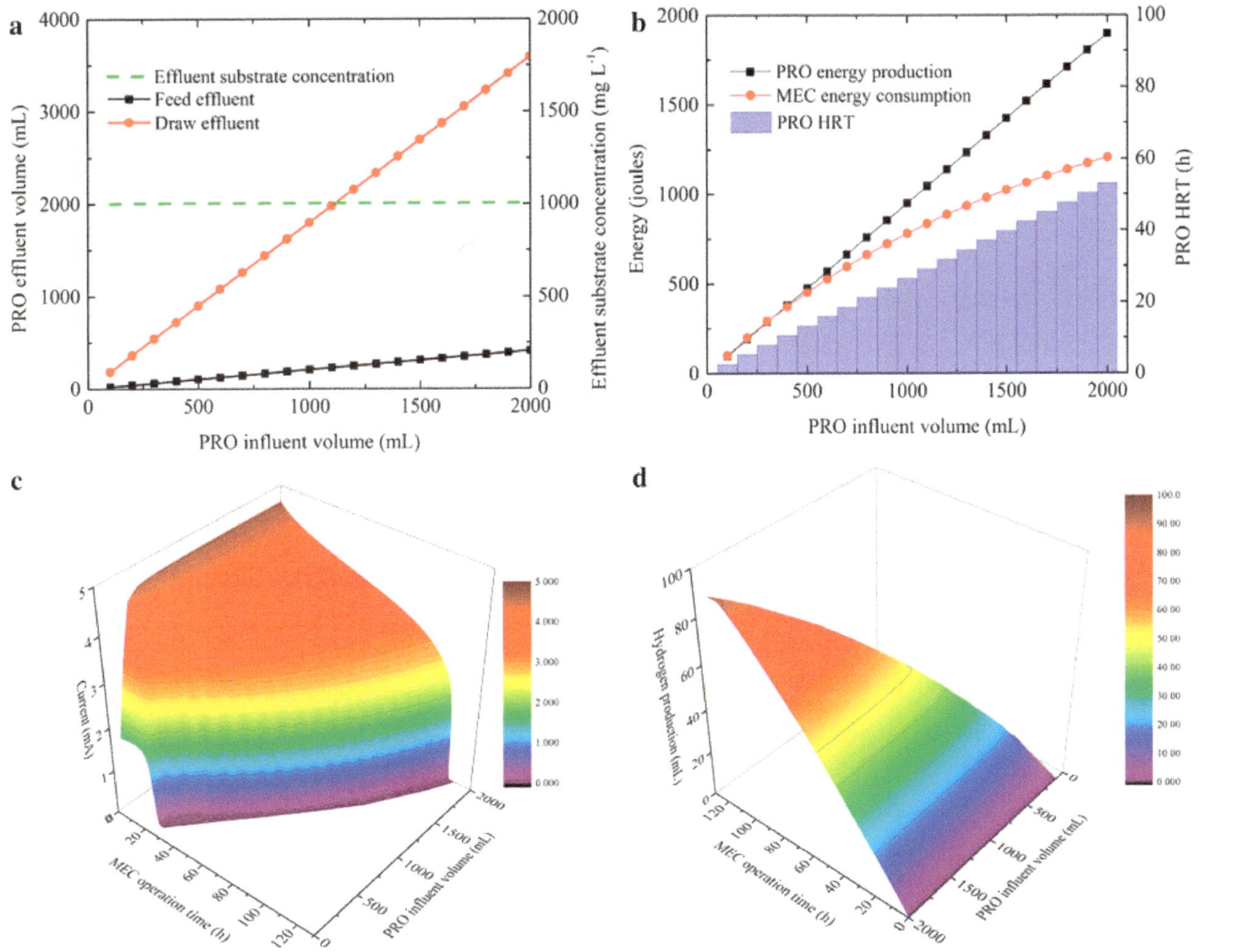

Figure 3: Effects of varied PRO influent volume on a PRO effluent volume and substrate concentration, b PRO energy production, PRO's HRT and MEC energy consumption, c current generation in the MEC and d hydrogen production in the MEC.

Effects of NaCl Concentration in the PRO Draw Influent

The NaCl concentration ranging from 0.1 to 2.0 M is input in the time-dependent PRO model with other parameters summarized in the Table 1 and 0.5 LMH as simulation end-point. It is predicted that the water recovery is enhanced significantly when the NaCl concentration increases from 0.1 to 0.7 M, and further increase in the NaCl concentration does not result in continuing improvement of water recovery (Fig. 4a). On the other hand, both the energy production and the effluent substrate concentration increase linearly with the increased NaCl concentration (Fig. 4a, b). The PRO's HRT reaches the maximum of 20.8 h at 0.4 M, and then decreases readily to 9.2 h at 2.0 M (Fig. 4b), likely because high water flux at a higher draw concentration causes faster decline in the osmotic pressure difference as described by Eqs. 5 and 6. When the NaCl concentration is lower than 0.7 M, the PRO energy production is predicted to be lower than the MEC energy consumption if the substrate is completely degraded (Fig. 4b). Therefore, the MEC performance is limited by the energy supply with low NaCl concentrations, and both the organics removal and hydrogen production cannot be performed effectively (below 93.0 % and 34.2 mL, respectively, Fig. 4c). When the NaCl concentration is higher than 0.8 M, sufficient electrical energy is generated in the PRO unit to support a complete MEC cycle with organics removal >99 %. Consequently, the MEC's operating time is prolonged to over 50.7 h, and the hydrogen production reaches 37.6 mL. Further increase in the NaCl concentration beyond 1.5 M does not noticeably enhance the hydrogen production, suggesting that the total organic content may have become a limiting factor for the MEC performance.

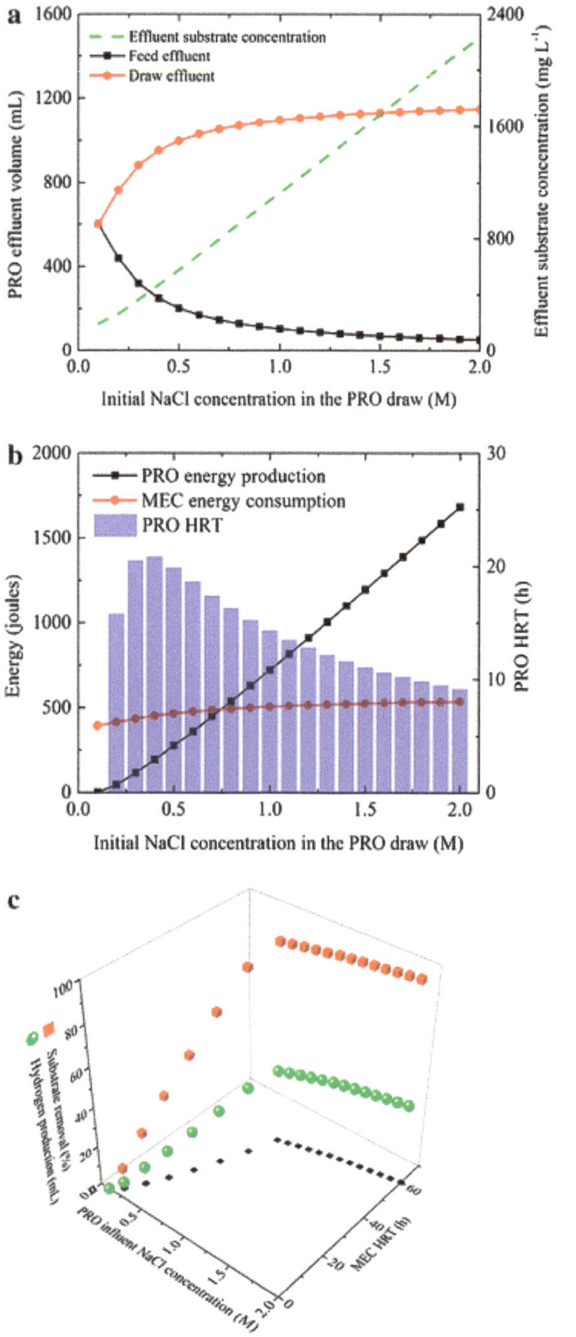

Figure 4: Effects of varied NaCl concentration on a PRO effluent volume and substrate concentration, b PRO energy production, PRO's HRT and MEC energy consumption and c substrate removal and hydrogen production in the MEC.

Effects of external voltage on the MEC

The energy consumption by the MEC is affected by the externally applied voltage according to Eqs. 18and 19. MEC simulation is therefore performed with external voltage varying from 0.5 to 1.1 V. The initial conditions for the MEC simulation are listed in Table 1. The results show that at 0.5 V, the MEC needs 89 J of energy to remove organics, but generates only 10.7 mL H_2 in 68.8 h (Fig. 5). The hydrogen production reaches the maximum value of 38.1 mL in 42.9 h at 0.9 V, and declines at a higher external voltage. That is because the MEC energy consumption exceeds the PRO energy production: the PRO energy production with given input parameters is fixed (i.e. 579 J, Fig. 5 inset, red dash line), and at 1.0 V (or 1.1 V), the MEC simulation is stopped once the energy consumption reaches 579 J; as a result, the MEC cannot accomplish a complete cycle and hydrogen production would decrease because of a shorter operating time. The organic removal is predicted to be 92.7 % at 1.0 V and only 80.3 % at 1.1 V. Meanwhile, the MEC's HRT is significantly shortened at a higher external voltage. These results indicate that the coupled system is versatile towards different purposes, and the treating capacity of the PRO and MEC can be balanced through varying external voltage.

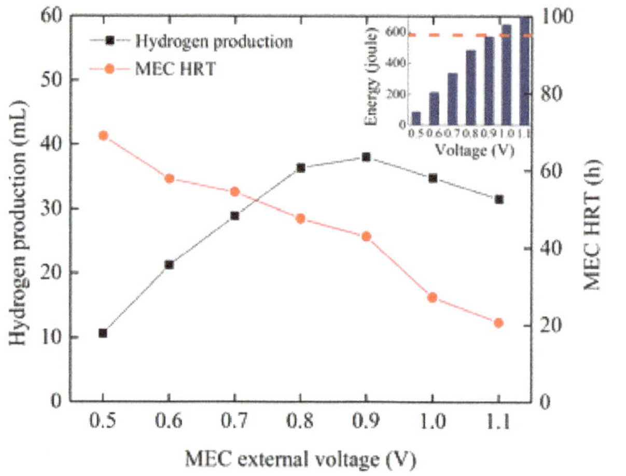

Figure 5: Hydrogen production and MEC's HRT predicted by the batch-mode MEC model with external voltage ranging from 0.5 to 1.1 V. The inset shows the MEC energy consumption at different voltage and the red line indicates the PRO energy production. ...

Perspectives

The coupled PRO-MEC system can achieve multiple benefits. Firstly, the PRO unit can serve as pre-treatment of organic solutions (e.g. wastewater), substantially reducing its volume and extracting clean water [15]. The draw solution for the PRO unit could be brine from RO desalination (91.3 % of water recovery predicted by the model) or seawater (71.7 % water recovery). Secondly, the PRO process can generate a high-conductivity feed solution as the MEC anolyte, which decreases the electrolyte resistance and thus is beneficial for bioelectrochemical processes [35, 36]. Thirdly, the osmotic energy harvested by the PRO process at certain influent volume and draw concentration is sufficient for a complete MEC cycle, making organic removal and hydrogen production more sustainable. Fourthly, compared with other approaches (e.g., MFC or solar energy) for driving hydrogen production in MECs, the present system can reduce the volume of the treated wastewater and produce a stream (diluted draw solution) for further desalination. For example, when seawater serves as a draw solution, its conductivity can be decreased from 54.7 to 32.0 mS cm^{-1} after the PRO process. Our previous study suggests that such a dilution can significantly enhance the rate of conductivity reduction in a microbial desalination cell [25].

Despite the great promise, several challenges must be addressed to move the PRO-MEC system towards practice. First, the robustness of the batch-mode MEC model should be improved by introducing endogenous respiration and more accurate pH variables [37, 38]. Because buffer was added in the feed/anolyte, the equilibrium between $[HCO_3^-]$ and $[H_2CO_3]$ as a function of pH was not taken into account in this study. When real wastewater is used as a feed/anolyte, the equilibrium constant should be introduced in Eqs. 14 and 15. Moreover, the cathode potential in Eq. 16 could be modified by relating pH change to hydrogen production and the Fick's law of diffusion. Another critical issue for practical applications is to balance the treating capacity of PRO and MEC. The simulation results with a high PRO influent volume and a high draw concentration suggest that one PRO unit is able to generate surplus energy to drive multiple MECs. In addition, the disparity in HRTs between the PRO unit and the MEC should also be addressed with caution to optimize the efficiency when one PRO unit is coupled to several MECs. Finally, the proposed system needs to be further demonstrated with actual energy generated from a PRO unit, which was reported in only a few studies [39, 40].

CONCLUSIONS

This work has demonstrated a proof-of-concept PRO-MEC system for simultaneous bioelectrochemical hydrogen production, organic removal, and water extraction driven by two forms of energy generated from (theoretically) the same liquid stream. It also presents the first attempt to introduce time variable into a PRO model, and has modified the previous single-chamber MEC model into a two-chamber batch-mode MEC model. Multiple benefits can be achieved through the synergy between PRO and MEC, including sustainable hydrogen production, clean water recovery and reduced wastewater volume. A high influent volume and a high draw concentration are predicted to enhance the performance of the coupled system, and the optimal external voltage is determined to be 0.9 V by the models. The PRO-MEC system may hold great promise in addressing water-energy nexus.

METHODS

System Setup and Operation

The experimental system consisted of a PRO unit and an MEC, as shown in Fig. 1. A forward osmosis (FO) cell (Sepa CF II Forward Osmosis Cell 316 SS, Sterlitech Corporation, USA) was used as the PRO unit, containing an FO membrane (HTI OsMem™ CTA-ES, Hydration Technology Innovations, USA) tailored to an area of 0.014 m^2. The feed solution contained (per liter of deionized water): sodium acetate, 193 mg; NaCl, 0.5 g; $MgSO_4$, 0.015 g; $CaCl_2$, 0.02 g; KH_2PO_4, 1.06 g; K_2HPO_4, 2.14 g; $NaHCO_3$, 1 g; and trace element, 1 mL. A synthetic organic solution was used because it could be better controlled for organic concentrations and other parameters. The feed solution (600 mL, $\pi_{F,0}$ = 2.4 bar) was recirculated between a reservoir and the PRO feed chamber by a peristaltic pump (Langer Instruments Corp., USA) at a cross-flow velocity of 16.4 cm s^{-1}. The NaCl solution (0.8 M, 600 mL, $\pi_{D,0}$ = 36.2 bar) was used as a draw and recirculated between a reservoir and the PRO draw chamber by a high pressure water pump (Estone 12 V DC 5 L/min 60 W, China) at a cross-flow velocity of 85.9 cm s^{-1}. Low hydraulic pressures (1, 2 and 3 bar) were applied on the draw chamber of the PRO cell using a pressure valve at the outlet and monitored with two gauges at both inlet and outlet. The PRO unit was operated for 6.0 h as a batch, and the weight change in both feed and draw solutions was monitored by using digital balances (Scort Pro, Ohaus, USA).

A two-chamber MEC was built as previously described [41], with a carbon-brush anode electrode and a carbon-cloth cathode electrode that contained 5 mg cm^{-2} of Pt/C and a surface area of 10 cm^2. The anode was inoculated with anaerobic sludge from a local wastewater treatment plant (Peppers Ferry, Radford, VA, USA). The liquid volumes of the anode compartment and the cathode compartment of the MEC were equal at 140 mL/each. Those electrolytes were recirculated between the MEC and the external reservoirs. An external voltage of 0.8 V was applied on the MEC with a power supply (3644A, Circuit Specialists, Inc., USA) to mimic the utilization of the energy from the PRO unit. The electrolytes were prepared as shown in Table 1 and purged with nitrogen gas for 15 min prior to each test. The MEC was operated under a batch mode at room temperature. The operation time of the MEC was determined by equating the PRO energy production with the MEC energy consumption by assuming no energy loss during energy conversion process. In both the PRO and MEC experiments, triplicate experiments were conducted for each test condition. The system at different PRO influent volumes, NaCl concentrations and external voltages was evaluated with the time-dependent PRO model and the batch-mode MEC model.

Measurement and Analysis

Weight change of the feed and draw solutions was recorded by an auto logging software (Scort Pro, Ohous, USA) at a time interval of 2 min. Water flux was calculated as derivative of the volume change. Conductivity was measured by using a benchtop conductivity meter (Mettler-Toledo, USA). The voltage (U) of the MEC on a resistor (R = 1 Ω, HARS-X-3_0.001, USA) was recorded by a digital multimeter (2700, Keithley Instruments, Inc., USA) at a time interval of 5 min. The current was calculated according to Ohm's law: i = U/R. Hydrogen produced by the MEC was collected using water displacement method and the volume was measured with a syringe. COD was measured using a DR/890 colorimeter (HACH Co., Ltd., USA) according to manufacturer's instruction. Organics (acetate) concentration in the effluent was calculated by dividing the COD by the empirical coefficient of 0.78. The pH was measured by a benchtop pH meter (Oakton Instruments, USA). Important MEC parameters were calculated as previously described [41]. Hydrogen production rate was calculated based on the catholyte volume, which was obtained from the PRO simulation.

Time-Dependent PRO Model

The time-dependent water flux was simulated based on the PRO model developed by Yip et al. [27]:

$$J_t = A\left\{\frac{\pi_{D,t}\exp\left(-\frac{J_t}{k}\right) - \pi_{F,t}\exp\left(\frac{J_tS}{D}\right)}{1+\frac{B}{J_t}\left[\exp\left(\frac{J_tS}{D}\right)-\exp\left(-\frac{J_t}{k}\right)\right]} - P_t\right\} \quad (1)$$

where J_t (m^3 m^{-2} s^{-1}) is the instantaneous water flux at time t, $\pi_{D,t}$ and $\pi_{F,t}$ (bar) are the osmotic pressure of the bulk draw and feed solution at time t, respectively, A (m^3 m^{-2} s^{-1} bar^{-1}) is the permeability coefficient, B (m^3 m^{-2} s^{-1}) is the salt permeability coefficient, S (m) is the support layer structural parameter, D (m^2 s^{-1}) is the diffusion coefficient of NaCl in the membrane substrate, k(m^3 m^{-2} s^{-1}) is the mass transfer coefficient, and P_t (bar)is the hydraulic pressure applied on the draw chamber at time t. The real-time volume of the feed ($V_{F,t}$) and draw ($V_{D,t}$) are thus:

$$V_{F,t} = V_{F,0} - \int J_t dt \tag{2}$$

$$V_{D,t} = V_{D,0} + \int J_t dt \tag{3}$$

where $V_{F,0}$ and $V_{D,0}$ (L) are the initial feed and draw volume. The reverse salt flux $J_{S,t}$ (mol m^{-2} s^{-1}) can be approximated as [42]:

$$J_{S,t} = \frac{BJ_t}{AiRT} \tag{4}$$

where i is the number of dissolved species (2 for NaCl), R (8.3145 J mol^{-1} K^{-1}) is the ideal gas constant, and T (298.15 K) is the absolute temperature. According to the van't Hoff equation: π= iRTC, where C (M) is the molar concentration, the real-time osmotic pressure is obtained by combining Eqs. 1–4:

$$\pi_{F,t} = iRT\left(\frac{V_{F,0}C_{F,0} + a\int J_{S,t}dt}{V_{F,t}}\right) \tag{5}$$

$$\pi_{D,t} = iRT\left(\frac{V_{D,0}C_{D,0} - a\int J_{S,t}dt)}{V_{D,t}}\right) \tag{6}$$

where a (0.014 m^2) is the membrane area, $C_{F,0}$ (M) is the initial salt concentration in the feed solution obtained by multiplying the conductivity (mS cm^{-1}) by the empirical coefficient of 0.64 ppm and dividing the result by the molecular weight of NaCl (58.5 g mol^{-1}), and $C_{D,0}$ (M) is the initial NaCl concentration. It can be derived from the expression of power density and ideal water flux that the maximum power is generated when P = (π_D − π_F)/2 [16]. In order to maintain the maximum energy production Q_{PRO} (joule) during the PRO process, the instantaneous hydraulic pressure applied on the draw chamber in Eq. 1 is calculated as:

$$P_t = \frac{(\pi_{D,t} - \pi_{F,t})}{2} \tag{7}$$

$$Q_{PRO} = a\int P_t J_t dt \tag{8}$$

Batch-Mode MEC Model

The batch-mode MEC model was modified based on previously reported multi-population MEC model and MDC model [28, 29, 43]. Because acetate was used as substrate and the cathode chamber is abiotic, it was assumed that the anode biofilm was composed of acetoclastic methanogens and exoelectrogens. Moreover, the MEC was operated under a batch mode and thus the dilution rate used in the previous study is not applicable [28]. The mass balance for the substrate and the growth of the organisms can be written:

$$\frac{dS}{dt} = -q_{e,max}\frac{S}{K_e + S}\frac{M_{OX}}{K_M + M_{OX}}x_e - q_{m,max}\frac{S}{K_m + S}x_m \tag{9}$$

$$\frac{dx_e}{dt} = -\mu_{e,max}\frac{S}{K_e + S}\frac{M_{OX}}{K_M + M_{OX}}x_e - d_e x_e \tag{10}$$

$$\frac{dx_m}{dt} = -\mu_{m,max}\frac{S}{K_m + S}x_m - d_m x_m \tag{11}$$

$$\frac{dM_{OX}}{dt} = -Y_M\frac{S}{K_e + S}\frac{M_{OX}}{K_M + M_{OX}} + \frac{\gamma}{V_a x_e}\frac{I}{nF} \tag{12}$$

where S (mg-S L^{-1}) is the acetate concentration, x_e and x_m (mg-x L^{-1}) are the concentrations of exoelectrogens and methanogens, respectively, $q_{e,max}$ and $q_{m,max}$ (mg-S mg-x^{-1} day^{-1}) are the maximum substrate consumption rates by organisms, K_e, K_m, and K_M (mg-S L^{-1}, mg-S L^{-1}, and mg-M mg-x^{-1}) are the half saturation concentrations for exoelectrogens, methanogens, and the redox mediators, respectively, $\mu_{e,max}$ and $\mu_{m,max}$ (day^{-1}) are the maximum growth rates by the organisms, d_e and d_m (day^{-1}) are the decay rates of the organisms, M_{OX} (mg-M mg-x^{-1}) is the fraction of oxidized mediators per exoelectrogen, Y_M (mg-M mg-S^{-1}) is the mediator yield, γ (mg-M mole-M^{-1}) is the mediator molar mass, I (A) is the current through the circuit of the MEC, F (96,485 C mol^{-1}) is the Faraday constant, n (2) is number of electrons transferred per mole of mediator, V_a (L) is the anolyte volume. The anode and cathode potential E_A and E_C (V) are calculated with the Nernst equation [44]:

$$E_A = E_A^0 - \frac{RT}{8F}In\frac{S}{[HCO_3^-]^2[H^+]^9} \tag{13}$$

$$[HCO_3^-] = [HCO_3^-]_0\frac{V_{F,0}}{V_F} + \frac{2(S_0 - S)}{m_S} \tag{14}$$

$$\frac{d[H^+]}{dt} = \frac{\beta}{m_S}\frac{dS}{dt}(9 - 8 \cdot CE \cdot Y_{HZ}) \tag{15}$$

$$E_C = E_C^0 - \frac{RT}{2F}In\frac{1}{(10^{-11})^2} \tag{16}$$

where E0A (0.187 V) is the standard reduction potential of HCO_3^-/CH_3COOH^- and E0C (0 V) is the standard reduction potential of H^+/H_2, $[HCO_3^-]_0$ (0.00012 M) is the initial bicarbonate concentration in the feed influent, $V_{F,0}$ and V_F (L) are the initial and final volume of the PRO feed, S_0 and S (mg L^{-1}) are the initial and final acetate concentration in the PRO feed, m_S (82 g mol^{-1}) is the molar weight of acetate, β (%) is the buffer efficiency of the anolyte, CE (%) is the coulombic efficiency and Y_{H2} (%) is the cathodic efficiency. The rationale of Eq. 15 is that every mole of the acetate produces 9 mol of protons and 8 mol of electrons. With 8 CE moles of electrons being transferred to the cathode, 8 CE Y_{H2} mol of protons migrate and diffuse to the cathode and are reduced to H_2, leaving the residual protons to affect the anolyte pH. The cathode potential is assumed to be stable and calculated with a pH value of 11 (10^{-11} in Eq. 16 as proton concentration) because no buffer is added in the draw solution/catholyte, and the pH increased from neutral to 11.64 ± 0.02 in the first 3 h and remained stable throughout the MEC experiments. At batch mode, the concentration overpotential η_{con} (V) in the anode is determined by both the mediator concentration and the substrate concentration, and thus can be written:

$$\eta_{con} = \frac{RT}{F} In \frac{M_{Total}}{M_{Total} - M_{OX}} \frac{S_0}{S} \qquad 17$$

where M_{Total} (mg-M mg-x^{-1}) is the total mediator fraction per exoelectrogen. Combining Eqs. 13–17yields the MEC current, the energy consumption Q_{MEC} (joule) and the hydrogen production V_{H2} (mL):

$$I = \frac{(E_C - E_A) + E_{ext} - \eta_{con}}{R_{ext} + R_{in}} \frac{M_{Total} - M_{OX}}{\varepsilon + M_{Total} - M_{OX}} \qquad 18$$

$$Q_{MEC} = E_{ext} \int Idt \qquad 19$$

$$V_{H2} = Y_{H2} \frac{I}{2F} \frac{RT}{P} \qquad 20$$

where E_{ext}(V) is the external voltage applied on the MEC, R_{ext}(Ω) is the external resistance, ε (0.0001 mg-M mg-x^{-1}) is a constant from a previous study [43], P (1 atm) is the air pressure in the cathode and R_{in} (Ω) is the internal resistance [43]:

$$R_{in} = R_{min} + (R_{max} - R_{min}) e^{-K_R x_e} \qquad 21$$

where R_{min} (Ω) is the lowest observed internal resistance, R_{max} (Ω) is the highest observed internal resistance, and K_R (L mg-x^{-1}) is the constant that determines the curve steepness.

Parameter Estimation

The parameters were estimated as previously described and listed in the Additional file 2 [28]. The relative root-mean square error (RMSE) as a measure of the discrepancy between the experimental data and the simulation results was calculated:

$$RMSE = \frac{\sqrt{\frac{\sum_{i=1}^{N} (y_i - \hat{y}_i)^2}{N}}}{\hat{y}_{i,max}} \qquad 22$$

where N is the total sampling time points in the simulation; $\hat{y}_i$ and y_i are experimental data and model predicted values at t, respectively; and $\hat{y}_{i,max}$ is the maximum value of the experimental data. The RMSE of the PRO model was calculated using the feed volume, and that of the MEC model was calculated using electricity.

AUTHOR'S CONTRIBUTIONS

ZH, HY, and YL designed and conceived this work. HY and YL performed the experiment. HY, YL, ZH, and IA performed the data analysis. HY, YL, ZH, and IA wrote the manuscript. All authors read and approved the final manuscript.

ACKNOWLEDGEMENTS

The authors would like to thank Ms. Qingyun Ping (Virginia Tech) for her help with mathematical modeling. This work was made possible by NPRP grant # 6-289-2-125 from the Qatar National Research Fund (a member of Qatar Foundation). The statements made herein are solely the responsibility of the authors.

REFERENCES

1. Liu H, Grot S, Logan BE. Electrochemically assisted microbial production of hydrogen from acetate. Environ Sci Technol. 2005;39:4317–4320. doi: 10.1021/es050244p.
2. Ni M, Leung MKH, Leung DYC, Sumathy K. A review and recent developments in photocatalytic water-splitting using for hydrogen production. Renew Sustain Energy Rev. 2007;11:401–425. doi: 10.1016/j.rser.2005.01.009. [Cross Ref]
3. Haryanto A, Fernando S, Murali N, Adhikari S. Current status of hydrogen production techniques by steam reforming of ethanol: a review. Energ Fuels. 2005;19:2098–2106. doi: 10.1021/ef0500538.
4. Foley JM, Rozendal RA, Hertle CK, Lant PA, Rabaey K. Life cycle assessment of high-rate Anaerobic treatment, microbial fuel cells, and microbial electrolysis cells. Environ Sci Technol.2010;44:3629–3637. doi: 10.1021/es100125h.
5. Heidrich ES, Dolfing J, Scott K, Edwards SR, Jones C, Curtis TP. Production of hydrogen from domestic wastewater in a pilot-scale microbial electrolysis cell. Appl Microbiol Biotechnol.2013;97:6979–6989. doi: 10.1007/s00253-012-4456-7.
6. Sun M, Sheng G-P, Zhang L, Xia C-R, Mu Z-X, Liu X-W, et al. An MEC-MFC-coupled system for biohydrogen production from acetate. Environ Sci Technol. 2008;42:8095–8100. doi: 10.1021/es801513c.
7. Hatzell MC, Kim Y, Logan BE. Powering microbial electrolysis cells by capacitor circuits charged using microbial fuel cell. J Power Sour. 2013;229:198–202. doi: 10.1016/j.jpowsour.2012.12.006.
8. Chae K-J, Choi M-J, Kim K-Y, Ajayi FF, Chang I-S, Kim IS. A solar-powered microbial electrolysis cell with a platinum catalyst-free cathode to produce hydrogen. Environ Sci Technol. 2009;43:9525–9530. doi: 10.1021/es9022317.
9. Yip NY, Elimelech M. Thermodynamic and energy efficiency analysis of power generation from natural salinity gradients by pressure retarded osmosis. Environ Sci Technol. 2012;46:5230–5239. doi: 10.1021/es300060m.
10. Logan BE, Elimelech M. Membrane-based processes for sustainable power generation using water.

Nature. 2012;488:313–319. doi: 10.1038/nature11477.

11. Kim Y, Logan BE. Hydrogen production from inexhaustible supplies of fresh and salt water using microbial reverse-electrodialysis electrolysis cells. Proc Natl Acad Sci USA. 2011;108:16176–16181. doi: 10.1073/pnas.1106335108. [PMC free article]

12. Cusick RD, Kim Y, Logan BE. Energy capture from thermolytic solutions in microbial reverse-electrodialysis cells. Science. 2012;335:1474–1477. doi: 10.1126/science.1219330. [PubMed]

13. Nam J-Y, Cusick RD, Kim Y, Logan BE. Hydrogen generation in microbial reverse-electrodialysis electrolysis cells using a heat-regenerated salt solution. Environ Sci Technol. 2012;46:5240–5246. doi: 10.1021/es300228m.

14. Achilli A, Childress AE. Pressure retarded osmosis: from the vision of Sidney Loeb to the first prototype installation—Review. Desalination. 2010;261:205–211. doi: 10.1016/j.desal.2010.06.017.

15. Lutchmiah K, Verliefde ARD, Roest K, Rietveld LC, Cornelissen ER. Forward osmosis for application in wastewater treatment: a review. Water Res. 2014;58:179–197. doi: 10.1016/j.watres.2014.03.045.

16. Helfer F, Lemckert C, Anissimov YG. Osmotic power with pressure retarded osmosis: theory, performance and trends—a review. J Membr Sci. 2014;453:337–358. doi: 10.1016/j.memsci.2013.10.053.

17. Chou S, Wang R, Shi L, She Q, Tang C, Fane AG. Thin-film composite hollow fiber membranes for pressure retarded osmosis (PRO) process with high power density. J Membr Sci. 2012;389:25–33. doi: 10.1016/j.memsci.2011.10.002.

18. Lu Y, Qin M, Yuan H, Abu-Reesh I, He Z. When bioelectrochemical systems meet forward osmosis: accomplishing wastewater treatment and reuse through synergy. Water. 2015;7:38–50. doi: 10.3390/w7010038.

19. Zhang F, Brastad KS, He Z. Integrating forward osmosis into microbial fuel cells for wastewater treatment, water extraction and bioelectricity generation. Environ Sci Technol. 2011;45:6690–6696. doi: 10.1021/es201505t.

20. Werner CM, Logan BE, Saikaly PE, Amy GL. Wastewater treatment, energy recovery and desalination using a forward osmosis membrane in an air-cathode microbial osmotic fuel cell. J Membr Sci. 2013;428:116–122. doi: 10.1016/j.memsci.2012.10.031.

21. Ge Z, He Z. Effects of draw solutions and membrane conditions on electricity generation and water flux in osmotic microbial fuel cells. Bioresour Technol. 2012;109:70–76. doi: 10.1016/j.biortech.2012.01.044.

22. Ge Z, Ping Q, Xiao L, He Z. Reducing effluent discharge and recovering bioenergy in an osmotic microbial fuel cell treating domestic wastewater. Desalination. 2013;312:52–59. doi: 10.1016/j.desal.2012.08.036.

23. Zhang B, He Z. Integrated salinity reduction and water recovery in an osmotic microbial desalination cell. RSC Adv. 2012;2:3265–3269. doi: 10.1039/c2ra20193c.

24. Zhang B, He Z. Improving water desalination by hydraulically coupling an osmotic microbial fuel cell with a microbial desalination cell. J Membr Sci. 2013;441:18–24. doi: 10.1016/j.memsci.2013.04.005.

25. Yuan H, Abu-Reesh IM, He Z. Enhancing desalination and wastewater treatment by coupling microbial desalination cells with forward osmosis. Chem Eng J. 2015;270:437–443. doi: 10.1016/j.cej.2015.02.059.

26. She Q, Jin X, Tang CY. Osmotic power production from salinity gradient resource by pressure retarded osmosis: effects of operating conditions and reverse solute diffusion. J Membr Sci. 2012;401–402:262–273. doi: 10.1016/j.memsci.2012.02.014.

27. Yip NY, Tiraferri A, Phillip WA, Schiffman JD, Hoover LA, Kim YC, et al. Thin-film composite pressure retarded osmosis membranes for sustainable power generation from salinity gradients.Environ Sci Technol. 2011;45:4360–4369. doi: 10.1021/es104325z.

28. Ping Q, Zhang C, Chen X, Zhang B, Huang Z, He Z. Mathematical model of dynamic behavior of microbial desalination cells for simultaneous wastewater treatment and water desalination. Environ Sci Technol. 2014;48:13010–13019. doi: 10.1021/es504089x.

29. Pinto RP, Srinivasan B, Escapa A, Tartakovsky B. Multi-population model of a microbial electrolysis cell. Environ Sci Technol. 2011;45:5039–5046. doi: 10.1021/es104268g. [PubMed]

30. Tartakovsky B, Manuel MF, Wang H, Guiot SR. High rate membrane-less microbial electrolysis cell for continuous hydrogen production. Int J Hydro Energ. 2009;34:672–677. doi: 10.1016/j.ijhydene.2008.11.003.

31. Song X, Liu Z, Sun DD. Energy recovery from concentrated seawater brine by thin-film nanofiber composite pressure retarded osmosis membranes with high power density. Energy Environ Sci.2013;6:1199–1210. doi: 10.1039/c3ee23349a.

32. Yip NY, Elimelech M. Comparison of energy efficiency and power density in pressure retarded osmosis and reverse electrodialysis. Environ Sci Technol. 2014;48:11002–11012. doi: 10.1021/es5029316.

33. Straub AP, Lin S, Elimelech M. Module-scale analysis of pressure retarded osmosis: performance limitations and implications for full-scale operation. Environ Sci Technol. 2014;48:12435–12444. doi: 10.1021/es503790k.

34. Prante JL, Ruskowitz JA, Childress AE, Achilli A. RO-PRO desalination: an integrated low-energy approach to seawater desalination. Appl Energy. 2014;120:104–114. doi: 10.1016/j.apenergy.2014.01.013.

35. Yossan S, Xiao L, Prasertsan P, He Z. Hydrogen production in microbial electrolysis cells: choice of catholyte. Int J Hydrogen Energy. 2013;38:9619–9624. doi: 10.1016/j.ijhydene.2013.05.094.

36. Rozendal RA, Hamelers HV, Rabaey K, Keller J, Buisman CJ. Towards practical implementation of bioelectrochemical wastewater treatment. Trends Biotechnol. 2008;26:450–459. doi: 10.1016/j.tibtech.2008.04.008.

37. Torres C, Kato Marcus A, Rittmann B. Kinetics of consumption of fermentation products by anode-respiring bacteria. Appl Microbiol Biotechnol. 2007;77:689–697. doi: 10.1007/s00253-007-1198-z.

38. Popat SC, Ki D, Young MN, Rittmann BE, Torres CI (2014) Buffer pKa and transport govern the concentration overpotential in electrochemical oxygen reduction at neutral pH. Chem Electro Chem n/a-n/a

39. Achilli A, Prante JL, Hancock NT, Maxwell EB, Childress AE. Experimental results from RO-PRO: a next generation system for low-energy desalination. Environ Sci Technol. 2014;48:6437–6443. doi: 10.1021/es405556s.

40. Straub AP, Yip NY, Elimelech M. Raising the Bar: Increased Hydraulic Pressure Allows Unprecedented High Power Densities in Pressure-Retarded Osmosis. Environ Sci Technol Lett.2014;1:55–59. doi: 10.1021/ez400117d.

41. Yuan H, Li J, Yuan C, He Z. Facile synthesis of MoS2@CNT as an effective catalyst for hydrogen production in microbial electrolysis cells. Chem Electro Chem. 2014;1:1828–1833.

42. Phillip WA, Yong JS, Elimelech M. Reverse draw solute permeation in forward osmosis: modeling and experiments. Environ Sci Technol. 2010;44:5170–5176. doi: 10.1021/es100901n. [PubMed]

43. Pinto RP, Srinivasan B, Manuel MF, Tartakovsky B. A two-population bio-electrochemical model of a microbial fuel cell. Bioresour Technol. 2010;101:5256–5265. doi: 10.1016/j.biortech.2010.01.122.

44. Logan BE, Hamelers B, Rozendal R, Schröder U, Keller J, Freguia S, et al. Microbial fuel cells: methodology and technology. Environ Sci Technol. 2006;40:5181–5192. doi: 10.1021/es0605016.

CHAPTER 6

MICROBIAL EXTRACELLULAR ELECTRON TRANSFER AND ITS RELEVANCE TO IRON CORROSION

Souichiro Kato

INTRODUCTION

Acquisition of energy is an indispensable activity for all living organisms. Most organisms, including human beings, conserve energy through respiration, with organic compounds and oxygen gas as the electron donor and acceptor respectively. In contrast, many microorganisms have the ability to utilize diverse inorganic compounds as the substrates for respiration. Furthermore, some particular microorganisms have the ability to acquire energy through transferring electrons to or from extracellular solid compounds. This microbial metabolism is specifically termed as 'extracellular electron transfer (EET)' (Gralnick and Newman, 2007; Richter *et al.*, 2012). In addition to naturally occurring metal minerals, microorganisms harbouring EET abilities can utilize artificial conductive materials, including graphite and metal electrodes, as the electron donor or acceptor (Bond and Lovley,2003; Gregory *et al.*, 2004). This microbial activity has received considerable attention for biotechnological applications, including bioremediation of toxic metals and diverse bioelectrochemical systems (Arends and Verstraete, 2012; Logan and Rabaey, 2012; Kato, 2015). Besides, recent studies have disclosed that some microorganisms utilize zero-valent metallic iron as their electron donor via EET metabolisms, which stimulates iron corrosion under anoxic conditions. This review introduces the basic knowledge in microbial EET metabolisms and the recent research progresses on the relevance of microbial EET to iron corrosion.

Citation: Souichiro Kato, Microbial extracellular electron transfer and its relevance to iron corrosion, doi: 10.1111/1751-7915.12340.

MICROBIAL EET

As for commonly characterized energy metabolisms, microorganisms first must incorporate substrates for respiration into their cells, where oxidation and reduction of substrates proceed. In contrast, special molecular mechanisms are required for EET-based energy metabolisms, since microorganisms cannot incorporate solid materials into their cells. Mechanisms of microbial EET can be categorized into either direct or indirect manners. In the direct EET, microorganisms attach to solid surfaces, to or from which they directly transfer electrons (Fig. 1A). The molecular mechanisms for the direct EET have been intensively investigated for some iron-reducing and iron-oxidizing bacteria, including *Geobacter sulfurreducens*, *Shewanella oneidensis* and *Acidithiobacillus ferrooxidans* (Stams *et al.*, 2006; Weber*et al.*, 2006; Shi *et al.*, 2007; Castelle *et al.*, 2008). These microorganisms utilize metal-containing redox proteins, such as *c*-type cytochromes and rusticyanin, to electrically connect intracellular respiratory chains and extracellular solid materials. Furthermore, *G. sulfurreducens* and *S. oneidensis*were reported to produce conductive filamentous apparatus (pili and outer membrane extensions, respectively) that are specifically termed as 'nanowires' (Reguera *et al.*, 2005; Gorby *et al.*, 2006; Pirbadian *et al.*, 2014). These microorganisms have the ability to transfer electrons to or from distantly located solid materials using the filaments as 'electric wire' (Fig. 1B).

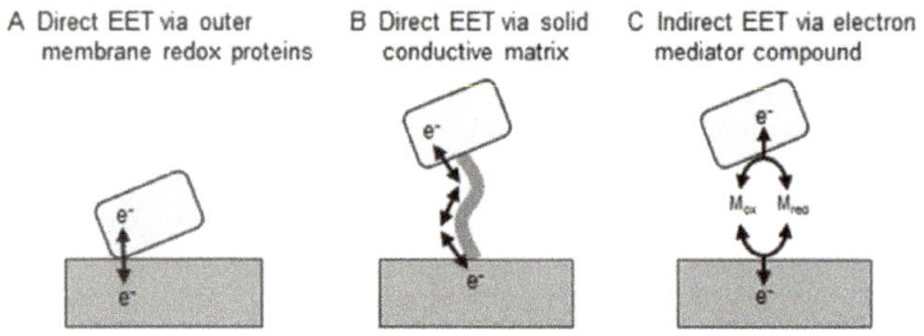

Figure 1: The schematic images of three microbial EET mechanisms.

In contrast, some microorganisms indirectly transfer electrons to or from solid compounds using diffusible redox chemicals, referred to as electron mediators (Watanabe *et al.*, 2009). Microorganisms reduce (or oxidize) intracellular electron mediators, after which the reduced (or oxidized) electron mediators diffuse out of the cell to solid surfaces and donate (or accept) electrons, and then the oxidized (or reduced) mediators return back into the cells and are again utilized as respiratory substrates (Fig. 1C). Some microorganisms have the ability to synthesize low-molecular-weight organic compounds that work as electron mediators, including phenazine compounds and flavin derivatives (Rabaey *et al.*, 2004; Marsili *et al.*, 2008). Naturally occurring (e.g. humic substances) and artificial (e.g. quinone derivatives, ferrocene derivatives) redox chemicals can also work as electron mediators to propel microbial EET (Jiang and Kappler, 2008; Watanabe *et al.*, 2009; Nishio *et al.*,2013). The reader is referred to several excellent review articles on the mechanisms of direct and indirect EET (Hernandez and Newman, 2001; Shi *et al.*, 2007; Lovley, 2008; Thrash and Coates, 2008; Watanabe *et al.*, 2009; Richter *et al.*, 2012).

Table 1: Isolated strains identified as EMIC-inducing microorganisms

Metabolic groupStrain name	Isolation sources	Phylogeny: Phylum/class	Phylogeny: Family	References
Sulfate-reducing bacteria				
'*Desulfopila corrodens*' IS4	Marine sediment	*Deltaproteobacteria*	*Desulfobulbaceae*	Dinh and colleagues (2004)
'*Desulfovibrio ferrophilus*' IS5	Marine sediment	*Deltaproteobacteria*	*Desulfovibrionaceae*	Dinh and colleagues (2004)
Methanogenic archaea				
Methanobacterium sp. IM1	Marine sediment	*Euryarchaeota*	*Methanobacteriaceae*	Dinh and colleagues (2004)
Methanococcus maripaludis KA1	Crude-oil storage tank	*Euryarchaeota*	*Methanococcaceae*	Uchiyama and colleagues (2010)
Methanococcus maripaludis Mic1c10	Crude-oil storage tank	*Euryarchaeota*	*Methanococcaceae*	Mori and colleagues (2010)
Acetogenic bacteria				
Sporomusa sp. GT1	Rice paddy field	*Firmicutes*	*Veillonellaceae*	Kato and colleagues (2015)
Nitrate-reducing bacteria				
Bacillus licheniformis ATCC 14580	Soil	*Firmicutes*	*Bacillaceae*	Xu and colleagues (2013)
Prolixibacter sp. MIC1-1	Crude-oil well	*Bacteroidetes*	*Prolixibacteraceae*	Iino and colleagues (2015)

MICROORGANISMS AND CORROSION OF IRON

Corrosion of iron is an electrochemical process consisting of oxidation of metallic iron to ferrous ion (anodic reaction, Eq. (1)) and reduction of electron acceptor compounds (cathodic reaction):

$$Fe^{2+} + 2e^- \leftrightarrow Fe\,(0);\ E_0{}' = -0.47\ V, \quad (1)$$

where −0.47 V references the standard hydrogen electrode (the same applies hereafter). In oxic environments, the cathodic reaction is reduction of oxygen (Eq. (2)) and corrosion rate is relatively high (Fig. 2A):

$$O_2 + 4e^- + 2H_2O \leftrightarrow 4OH^-;\ E_0{}' = +0.82\ V. \quad (2)$$

In contrast, the most common electron consuming reaction under anoxic condition is proton reduction to hydrogen gas (Eq. (3)) (Fig. 2B):

$$2H^+ + 2e^- \leftrightarrow H_2;\ E_0{}' = -0.41\ V. \quad (3)$$

Theoretically, iron corrosion in anoxic environments does not become a serious problem, since the proton reduction reaction on iron surfaces is usually a particularly slow reaction. However, iron corrosion under anoxic conditions has often been reported, and in most cases it is thought to be mediated by microbial metabolic activities (Lee *et al.*, 1995; Hamilton, 2003; Beech and Sunner, 2004; Videla and Herrera, 2005).

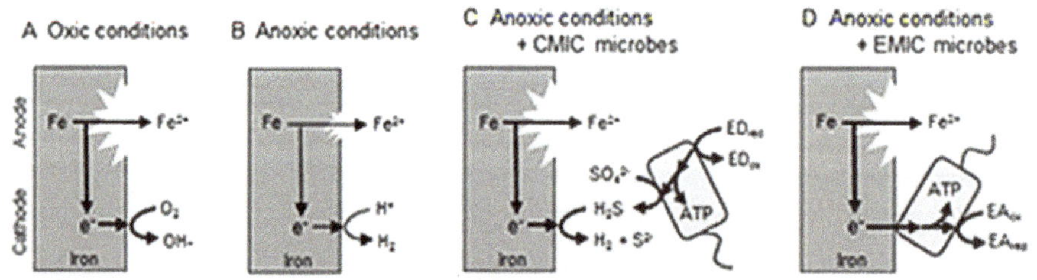

Figure 2: The schematic images of the iron corrosion mechanisms. Spontaneous, oxidative dissolution of metallic iron into ferrous iron (the anodic reaction) is the first step of corrosion in all of the cases.

Corrosion of iron caused by microbial metabolism is specifically termed as microbiologically influenced corrosion (MIC) or biocorrosion. Microbiologically influenced corrosion under anaerobic conditions often damages energy infrastructures, including crude oil reservoir tanks and underground oil and gas pipelines, causing enormous economic losses. The microorganisms first identified as MIC-inducers were sulfate-reducing bacteria (SRB) (Gaines, 1910). Sulfate-reducing bacteria produce hydrogen sulfide through reduction of sulfate (Eq. (4)) with various organic compounds and often H_2as the source of reducing equivalents:

$$SO_4^{2-} + 10H^+ + 8e^- \leftrightarrow H_2S + 4H_2O;\ E_0' = -0.22\ V. \tag{4}$$

Hydrogen sulfide is a highly oxidative chemical and is known to react with metallic iron to generate iron sulfide (Eq. (5)):

$$H_2S + Fe^0 \rightarrow FeS + H_2. \tag{5}$$

This process enhances corrosion of iron (Fig. 2C). Similarly, other microorganisms that secrete corrosive chemicals such as organic acids also stimulate corrosion of iron (Little *et al.*, 2001; Usher*et al.*, 2014). This manner of MIC, i.e. stimulation of the cathodic reaction by microbial metabolic end-products, is referred to as chemical MIC (CMIC; Enning *et al.*, 2012).

In contrast, acceleration of cathodic reactions via consumption of cathodic electrons, often referred to as cathodic depolarization, had been believed as one of the most prominent causes of MIC (Wolzogen Kühr, 1961). Microorganisms consuming H_2 abiotically generated on iron surfaces (Eq. (3)) coupled with sulfate reduction (Eq. (4)) or methanogenesis (Eq. (6)) were considered to induce cathodic depolarization (King and Miller, 1971; Daniels *et al.*, 1987):

$$HCO_3^- + 9H^+ + 8e^- \leftrightarrow CH_4 + 3H_2O;\ E_0' = -0.24\ V. \tag{6}$$

However, stimulation of anaerobic corrosion via H_2 consumption has been viewed critically (Hardy,1983), and recent studies revealed that H_2 consumption itself is not sufficient to induce fatal iron corrosion as often observed in actual fields (Dinh *et al.*, 2004; Mori *et al.*, 2010). These studies implied that only microorganisms with EET ability, which enables efficient uptake of electrons from metallic iron, can stimulate the cathodic reaction and hence iron corrosion (Fig. 2D). This manner of MIC, i.e. stimulation of the cathodic reaction by consuming cathodic electrons as the metabolic energy sources, is referred to as electrical MIC (EMIC) (Enning *et al.*, 2012). In the following sections, specific examples of microorganisms causing EMIC are introduced and their putative EET mechanisms are discussed.

MICROORGANISMS INDUCING EMIC

So far microorganisms in four different metabolic groups (SRB, methanogenic archaea, acetogenic bacteria and nitrate-reducing bacteria) are elucidated as EMIC-inducers (Table 1). In this section, specific examples of the EMIC-inducing microorganisms are introduced.

SRB

As introduced in the previous section, SRB is known as one of the most prominent bacteria to cause anaerobic MIC, mainly via production of corrosive chemical H_2S. In addition to the indirect mechanisms, recent research disclosed that some SRB strains stimulate corrosion via more direct manners. Dinh and colleagues (2004) reported enrichment and isolation of two novel SRB strains from marine sediment using metallic iron as the sole electron donor. These isolated strains, subsequently denominated as '*Desulfopila corrodens*' strain IS4 and '*Desulfovibrio ferrophilus*' strain IS5 (Enning*et al.*, 2012), reduced sulfate with concomitant oxidation of metallic iron much faster than abiotic H_2generation in an organic matter-free medium. In contrast, the closely related SRB with efficient H_2-utilizing abilities did not show such activities (Dinh *et al.*, 2004; Gittel *et al.*, 2010). So far, EMIC-inducing abilities in SRB have been identified in only a limited number of strains in the deltaproteobacterial families *Desulfovibrionaceae* and *Desulfobulbaceae* (Enning and Garrelfs, 2014). These reports suggest that the EMIC-inducing SRBs have special molecular mechanisms for electron uptake from metallic iron, and the ability is a species and/or strains-specific trait rather than universal in SRB. Under sulfate-reducing conditions, black mineral crust mainly consisting of iron sulfide minerals is generally formed as corrosion products (Enning *et al.*, 2012) (Fig. 3).

Figure 3: Corrosion products formed by EMIC-inducing microorganisms.

Methanogenic Archaea

Methane production by methanogenic archaea (Eq. (6)) is a major microbial metabolism in anoxic environments, especially in sulfate-depleted zones. Participation of microbial methanogenesis to cathodic depolarization and anaerobic MIC was proposed in the 1980s (Daniels *et al.*, 1987). The first direct evidence for an EMIC-inducing methanogen was reported by Dinh and colleagues (2004). A new methanogenic strain, *Methanobacterium* sp. strain IM1, isolated from marine sediment, was shown to grow and produce methane with metallic iron as the sole electron source, and stimulate corrosion of iron (Dinh *et al.*, 2004). Another research group in Japan also isolated EMIC-inducing methanogenic strains, namely *Methanococcus maripaludis* strain KA1 and strain Mic1c10, from crude oil reservoir tanks (Mori *et al.*, 2010; Uchiyama *et al.*, 2010). Interestingly, authentic hydrogenotrophic methanogens closely related to these EMIC-inducing methanogens did not grow well on and produce methane from metallic iron, indicating that the corrosive ability is a limited trait for certain methanogenic archaea.

Acetogenic Bacteria

Acetogenic bacteria have the abilities to conserve energy via the reductive acetyl-CoA pathway, by which acetate is produced through reduction of carbon dioxide (Eq. (7)):

$$2HCO_3^- + 9H^+ + 8e^- \leftrightarrow CH_3COO^- + 4H_2O;\ E_0' = -0.28\ V. \quad (7)$$

Acetogenic bacteria generally utilize hydrogen gas or certain organic compounds as the source of reducing equivalents. While engagement of microbial acetogenesis for anaerobic MIC in low-sulfate environments was postulated (Suflita *et al.*, 2008), the direct evidence had not been demonstrated. Recently, Mand and colleagues (2014) demonstrated acetate production in microbial enrichment cultures with metallic iron as the sole electron source, from which bacteria closely related to known acetogenic bacterium (*Acetobacterium* sp.) were detected. Our research group also enriched acetogenic microbial communities with metallic iron as the sole energy source (Kato *et al.*, 2015). The enriched communities produced acetate coupled with oxidation of metallic iron and produced significantly larger amount of ferrous iron than the abiotic controls. We eventually isolated *Sporomusa* sp. strain GT1 from the enrichments and demonstrated that the isolate acetogenetically grew on metallic iron and enhanced iron corrosion, while other well-known acetogenic bacteria did not. In addition to the direct effects, the metabolism of acetogenic bacteria should cause another peripheral effects on iron corrosion. Most SRB cannot grow autotrophically and require organic compounds such as acetate as the carbon source. Hence, it is postulated that acetate generated by acetogenic bacteria indirectly enhances iron corrosion via stimulation of growth of CMIC- and/or EMIC-inducing SRB (Mand *et al.*,2014).

Nitrate-Reducing Bacteria (NRB)

Reduction of nitrate (Eq. (8)) is also a major microbial metabolism under anoxic conditions:

$$NO_3^- + 2H^+ + 2e^- \leftrightarrow NO_2^- + H_2O;\ E_0' = +0.44\ V. \quad (8)$$

Recent studies revealed that particular NRB also causes EMIC. For instance, Xu and colleagues (2013) demonstrated that *Bacillus licheniformis* ATCC 14580, a facultative NRB strain, stimulates corrosion of carbon steel under nitrate-reducing conditions. Iino and colleagues (2015) isolated another EMIC-causing NRB, *Prolixibacter* sp. strain MIC1-1, which is the first corrosive representative belonging to the phylum *Bacteroidetes*. In addition to the ability of *Prolixibacter* sp. strain MIC1-1 to utilize cathodic electrons in metallic iron as the electron donor, the metabolic end-product, nitrite, which is known as a strong corrosive compound (Lin *et al.*, 2008), appears to induce severe corrosion (Iino*et al.*, 2015). Nitrate injection into oil and gas reservoirs has been used for mitigation of souring and MIC caused by SRB, since it promotes growth of NRB and in turn suppresses growth of SRB (Gieg*et al.*, 2011). However, the discovery of MIC-inducing NRB led to the revelation that supplementation of nitrate may risk inducing NRB-assisted corrosion. In contrast to sulfate-reducing conditions, greyish mineral crust mainly containing iron carbonate and iron phosphate is generally formed as corrosion products under methanogenic, acetogenic and nitrate-reducing conditions (Uchiyama *et al.*, 2010; Iino*et al.*, 2015) (Fig. 3).

How do Corrosive Microorganisms Uptake Electrons from Metallic Iron?

The EMIC-inducing microorganisms should have either direct or indirect EET abilities to efficiently uptake electrons from metallic iron. Physiological and electrochemical analyses have disclosed that the corrosive SRB strains (*Desulfopila corrodens* strain IS4 and *Desulfovibrio ferrophilus* strain IS5) appear to directly uptake electrons from metallic iron (Enning *et al.*, 2012; Venzlaff *et al.*, 2013). Production of the mineral crust (containing FeS, $FeCO_3$ and Mg/$CaCO_3$) with semiconductor-like properties by the EMIC-inducing SRB also supports this assumption. Deng and colleagues (2015) electrochemically demonstrated that the corrosive SRB *Desulfovibrio ferrophilus* strain IS5 is capable of extracting electrons from an inert electrode without consuming H_2 as an electron carrier. Similarly, Beese-Vasbender and colleagues (2015) reported that the corrosive methanogen *Methanobacterium* sp. strain IM1 generates cathodic current and produces methane at an electrode potential of −0.4 V, which is much more positive than the potential required for H_2 production on a graphite cathode. Furthermore, Iino and colleagues (2015) confirmed that the *Prolixibacter* sp. strain MIC1-1 does not have ability to oxidize H_2, while this strain showed high EMIC activities coupled with nitrate reduction. These studies indicate that at least a part of EMIC-causing microorganisms have the abilities for direct EET, while the molecular mechanisms remain unknown.

There are some reports on the relevance of indirect EET to EMIC. In the indirect EET, electron mediator molecules are required to achieve electron transfer from metallic iron to microbial cells. Zhang and colleagues (2015) demonstrated that exogenous addition of electron mediators [riboflavin and flavin adenine dinucleotide (FAD)] accelerates corrosion of stainless steel by *Desulfovibrio vulgaris*. Some EET-harbouring microorganisms such as *Shewanella* spp. were reported to produce and excrete riboflavin and FAD to facilitate their EET activities (Marsili *et al.*, 2008; Brutinel and Gralnick, 2012). These findings suggest that electron mediators secreted by microorganisms themselves facilitate EMIC, although there has been no direct evidence yet. Naturally occurring inorganic chemicals are another candidate of electron mediators facilitating EMIC. Manganese-oxidizing bacteria were reported to conduct this type of corrosion (Dickinson *et al.*, 1997; Ashassi-Sorkhabi *et al.*, 2012). In this case, oxidized forms of manganese [Mn(III/IV)] consume cathodic electrons in metallic iron and is reduced to Mn(II), which is oxidized back to Mn(III/IV) by manganese-oxidizing bacteria. Iodide-oxidizing bacteria isolated from brine in an iodide production facility stimulate iron corrosion in a similar manner with redox coupling of I^-/I_2 (Wakai *et al.*, 2014).

Hydrogen gas is also a possible candidate of an electron mediator molecule. As discussed above, however, a number of studies showed that microorganisms that simply consume H_2 do not induce significant corrosion of iron. Recently, Deutzmann and colleagues (2015) demonstrated a novel mechanism for electron uptake from metallic iron with hydrogen gas (and possibly formate) as an important intermediate. They discovered that cell-free spent culture medium of a corrosive methanogen*Methanococcus maripaludis* strain MM901 accelerated H_2 (and formate) generation from metallic iron, while that of a hydrogenase-deficient mutant did not. This study suggests that surface-associated redox enzymes, such as hydrogenases and formate dehydrogenase, are sufficient to mediate an apparent electron uptake that enhances iron corrosion.

The EET mechanisms for EMIC remains controversial, and it may be highly possible that each EMIC-inducing microorganism utilizes completely different EET strategies. Also in the research field of bioelectrochemistry, the molecular mechanisms of microbial electron uptake from cathodes are much more enigmatic rather than the opposite reaction, i.e. microbial electron injection to anodes (Rosenbaum *et al.*, 2011). Further research with multiple viewpoints, including microbial physiology, genomics, molecular biology and electrochemistry, is required to understand how EMIC-inducers efficiently uptake electrons from metallic iron.

CONCLUDING REMARKS

The studies on MIC in the last decades have demonstrated that physiologically and phylogenetically diverse microorganisms, including SRB (*Proteobacteria*), methanogens (*Euryarchaeota*), acetogens (*Firmicutes*) and NRB (*Firmicutes* and *Bacteroidetes*), stimulate iron corrosion via their EET metabolisms. Further investigation on these microorganisms, especially on their EET mechanisms, and also identification of novel EMIC-causing microorganisms will lead to technological development for MIC mitigation. In addition, EMIC-causing microorganisms can potentially be utilized as novel biocatalysts. Microbial electron uptake from solid materials is being pursued as biocatalysts for cathode reactions in diverse microbial electrochemical systems (Rabaey and Rozendal, 2010; Erable*et al.*, 2012; Kim *et al.*, 2015). Based on this idea, cathodic methane production by a corrosive methanogen has already been examined (Beese-Vasbender *et al.*, 2015). Close relatives of a corrosive acetogen are also being investigated as biocatalysts for conversion on electrical energy into liquid fuels (Nevin *et al.*, 2011).

Considering that almost all metallic iron has been recently introduced into environments by human activities, EMIC reaction, namely utilization of metallic iron as an electron donor of respiration, is an enigmatic metabolism from the viewpoint of evolution. One plausible explanation is that EMIC reaction is due to the promiscuous usage of other analogous metabolisms. Our group recently demonstrated that electrically conductive metal minerals, which abundantly exist in natural environments (Hochella *et al.*, 2008; Nakamura *et al.*, 2010a), serve as an electron source and sink for some EET-harbouring microorganisms (Nakamura *et al.*, 2009; 2010b; Kato *et al.*, 2010, 2013). Furthermore, a novel microbial symbiotic metabolism, in which electrons released by one microorganism are transferred to another via electric current flowing through biological and mineralogical solid materials, has recently been demonstrated (Summers *et al.*, 2010; Kato *et al.*,2012a). Diverse kinds of microorganisms, including NRB, ferric iron reducers, methanogens and dehalorespirators, have been reported to participate in the electron-consuming part of the symbiotic process (Summers *et al.*, 2010; Kato *et al.*, 2012a,b; Aulenta *et al.*, 2013). Further studies on EMIC will also give novel insights for ecological and evolutional aspects of microbial EET.

ACKNOWLEDGEMENT

I thank Dr. Mia Terashima for critical reading of the manuscript.

REFERENCES

1. Arends J.B., and Verstraete W. (2012) 100 years of microbial electricity production: three concepts for the future. Microb Biotechnol 5: 333–346.
2. Ashassi-Sorkhabi H., Moradi-Haghighi M., and Zarrini G. (2012) The effect of*Pseudoxanthomonas* sp. as manganese oxidizing bacterium on the corrosion behavior of carbon steel. Mater Sci Eng C 32: 303–309.
3. Aulenta F., Rossetti S., Amalfitano S., Majone M., and Tandoi V. (2013) Conductive magnetite nanoparticles accelerate the microbial reductive dechlorination of trichloroethene by promoting interspecies electron transfer processes. ChemSusChem 6: 433–436.
4. Beech I.B., and Sunner J. (2004) Biocorrosion: towards understanding interactions between biofilms and metals. Curr Opin Biotechnol 15: 181–186.
5. Beese-Vasbender P.F., Grote J.P., Garrelfs J., Stratmann M., and Mayrhofer K.J. (2015) Selective microbial electrosynthesis of methane by a pure culture of a marine lithoautotrophic archaeon.Bioelectrochemistry 102: 50–55.
6. Bond D.R., and Lovley D.R. (2003) Electricity production by *Geobacter sulfurreducens* attached to electrodes. Appl Environ Microbiol 69: 1548–1555.
7. Brutinel E.D., and Gralnick J.A. (2012) Shuttling happens: soluble flavin mediators of extracellular electron transfer in Shewanella. Appl Microbiol Biotechnol 93: 41–48.
8. Castelle C., Guiral M., Malarte G., Ledgham F., Leroy G., Brugna M., and Giudici-Orticoni M.T. (2008) A new iron-oxidizing/O_2-reducing supercomplex spanning both inner and outer membranes, isolated from the extreme acidophile *Acidithiobacillus ferrooxidans* . J Biol Chem283: 25803–25811.
9. Daniels L., Belay N., Rajagopal B.S., and Weimer P.J. (1987) Bacterial methanogenesis and growth from CO_2 with elemental iron as the sole source of electrons. Science 237: 509–511.
10. Deng X., Nakamura R., Hashimoto K., and Okamoto A. (2015) Electron extraction from an extracellular electrode by *Desulfovibrio ferrophilus* strain IS5 without using hydrogen as an electron carrier. Electrochemistry 83: 529–531.
11. Deutzmann J.S., Sahin M., and Spormann A.M. (2015) Extracellular enzymes facilitate electron uptake in biocorrosion and bioelectrosynthesis. mBio 6: e00496–15.
12. Dickinson W.H., Caccavo F., Olesen B., and Lewandowski Z. (1997) Ennoblement of stainless steel by the manganese-depositing bacterium *Leptothrix discophora* . Appl Environ Microbiol63: 2502–2506.
13. Dinh H.T., Kuever J., Mussmann M., Hassel A.W., Stratmann M., and Widdel F. (2004) Iron corrosion by novel anaerobic microorganisms. Nature 427: 829–832.
14. Enning D., and Garrelfs J. (2014) Corrosion of iron by sulfate-reducing bacteria: new views of an old problem. Appl Environ Microbiol 80: 1226–1236.
15. Enning D., Venzlaff H., Garrelfs J., Dinh H.T.,

Meyer V., Mayrhofer K., *et al* (2012) Marine sulfate-reducing bacteria cause serious corrosion of iron under electroconductive biogenic mineral crust. Environ Microbiol 14: 1772–1787.

16. Erable B., Féron D., and Bergel A. (2012) Microbial catalysis of the oxygen reduction reaction for microbial fuel cells: a review. ChemSusChem 5: 975–987.
17. Gaines R. (1910) Bacterial activity as a corrosive influence in the soil. J Ind Eng Chem 2: 128–130.
18. Gieg L.M., Jack T.R., and Foght J.M. (2011) Biological souring and mitigation in oil reservoirs.Appl Microbiol Biotechnol 92: 263–282.
19. Gittel A., Seidel M., Kuever J., Galushko A.S., Cypionka H., and Könneke M. (2010)*Desulfopila inferna* sp. nov., a sulfate-reducing bacterium isolated from the subsurface of a tidal sand-flat. Int J Syst Evol Microbiol 60: 1626–1630.
20. Gorby Y.A., Yanina S., McLean J.S., Rosso K.M., Moyles D., Dohnalkova A., *et al* (2006)Electrically conductive bacterial nanowires produced by *Shewanella oneidensis* strain MR-1 and other microorganisms. Proc Natl Acad Sci USA 103: 11358–11363.
21. Gralnick J.A., and Newman D.K. (2007) Extracellular respiration. Mol Microbiol 65: 1–11.
22. Gregory K.B., Bond D.R., and Lovley D.R. (2004) Graphite electrodes as electron donors for anaerobic respiration. Environ Microbiol 6: 596–604.
23. Hamilton W.A. (2003) Microbially influenced corrosion as a model system for the study of metal microbe interactions: A unifying electron transfer hypothesis. Biofouling 19: 65–76.
24. Hardy J.A. (1983) Utilisation of cathodic hydrogen by sulphate-reducing bacteria. Br Corros J18: 190–193.
25. Hernandez M.E., and Newman D.K. (2001) Extracellular electron transfer. Cell Mol Life Sci 58: 1562–1571.
26. Hochella M.F. Jr, Lower S.K., Maurice P.A., Penn R.L., Sahai N., Sparks D.L., and Twining B.S. (2008) Nanominerals, mineral nanoparticles, and Earth systems. Science 319: 1631–1635.
27. Iino T., Ito K., Wakai S., Tsurumaru H., Ohkuma M., and Harayama S. (2015) Iron corrosion induced by nonhydrogenotrophic nitrate-reducing *Prolixibacter* sp. strain MIC1-1. Appl Environ Microbiol 81: 1839–1846.
28. Jiang J., and Kappler A. (2008) Kinetics of microbial and chemical reduction of humic substances: implications for electron shuttling. Environ Sci Technol 42: 3563–3569.
29. Kato S. (2015) Biotechnological aspects of microbial extracellular electron transfer. Microbes Environ 30: 133–139.
30. Kato S., Kai F., Nakamura R., Watanabe K., and Hashimoto K. (2010) Respiratory interactions of soil bacteria with (semi)conductive iron-oxide minerals. Environ Microbiol 12: 3114–3123.
31. Kato S., Hashimoto K., and Watanabe K. (2012a) Microbial interspecies electron transfer via electric currents through conductive minerals. Proc Natl Acad Sci USA 109: 10042–10046.
32. Kato S., Hashimoto K., and Watanabe K. (2012b) Methanogenesis facilitated by electric syntrophy via (semi)conductive iron-oxide minerals. Environ Microbiol 14: 1646–1654.
33. Kato S., Hashimoto K., and Watanabe K. (2013) Iron-oxide minerals affect extracellular electron-transfer paths of *Geobacter* spp. Microbes Environ 28: 141–148.
34. Kato S., Yumoto I., and Kamagata Y. (2015) Isolation of acetogenic bacteria that induce biocorrosion by utilizing metallic iron as the sole electron donor. Appl Environ Microbiol 81: 67–73.
35. Kim B.H., Lim S.S., Daud W.R., Gadd G.M., and Chang I.S. (2015) The biocathode of microbial electrochemical systems and microbially-influenced corrosion. Bioresour Technol 190: 390–401.
36. King R.A., and Miller J.D.A. (1971) Corrosion by the sulphate-reducing bacteria. Nature 233: 491–492.
37. Lee W., Lewandowski Z., Nielsen P.H., and Hamilton W.A. (1995) Role of sulfate-reducing bacteria in corrosion of mild steel: A review. Biofouling 8: 165–194.
38. Lin K.S., Chang N.B., and Chuang T.D. (2008) Fine structure characterization of zero-valent iron nanoparticles for decontamination of nitrites and nitrates in wastewater and groundwater.Sci Technol Adv Mater 9: 025015.
39. Little B., Staehle R., and Davis R. (2001) Fungal influenced corrosion of post-tensioned cables.Int Biodeterior Biodegrad 47: 71–77.
40. Logan B.E., and Rabaey K. (2012) Conversion of wastes into bioelectricity and chemicals by using microbial electrochemical technologies. Science 337: 686–690.
41. Lovley D.R. (2008) The microbe electric: conversion of organic matter to electricity. Curr Opin Biotechnol 19: 564–571.
42. Mand J., Park H.S., Jack T.R., and Voordouw G. (2014) The role of acetogens in microbially influenced corrosion of steel. Front Microbiol 5: 268.
43. Marsili E., Baron D.B., Shikhare I.D., Coursolle D., Gralnick J.A., and Bond D.R. (2008)*Shewanella* secretes flavins that mediate extracellular electron transfer. Proc Natl Acad Sci USA105: 3968–3973.
44. Mori K., Tsurumaru H., and Harayama S. (2010) Iron corrosion activity of anaerobic hydrogen-consuming microorganisms isolated from oil facilities. J Biosci Bioeng 110: 426–430.
45. Nakamura R., Kai F., Okamoto A., Newton G.J., and Hashimoto K. (2009) Self-constructed electrically conductive bacterial networks. Angew Chem Int Ed 48: 508–511.
46. Nakamura R., Takashima T., Kato S., Takai K., Yamamoto M., and Hashimoto K. (2010a)Electrical current generation across a black smoker chimney. Angew Chem Int Ed 49: 7692–7694.
47. Nakamura R., Okamoto A., Tajima N., Newton G.J., Kai F., Takashima T., and Hashimoto K. (2010b) Biological iron-monosulfide production for efficient electricity harvesting from a deep-sea metal-reducing bacterium. Chembiochem 11: 643–645.

48. Nevin K.P., Hensley S.A., Franks A.E., Summers Z.M., Ou J., Woodard T.L., *et al* (2011)Electrosynthesis of organic compounds from carbon dioxide is catalyzed by a diversity of acetogenic microorganisms. Appl Environ Microbiol 77: 2882–2886.

49. Nishio K., Nakamura R., Lin X., Konno T., Ishihara K., Nakanishi S., and Hashimoto K. (2013)Extracellular electron transfer across bacterial cell membranes via a cytocompatible redox-active polymer. Chemphyschem 14: 2159–2163.

50. Pirbadian S., Barchinger S.E., Leung K.M., Byun H.S., Jangir Y., Bouhenni R.A., *et al* (2014)*Shewanella oneidensis* MR-1 nanowires are outer membrane and periplasmic extensions of the extracellular electron transport components. Proc Natl Acad Sci USA 111: 12883–12888.

51. Rabaey K., and Rozendal R.A. (2010) Microbial electrosynthesis – revisiting the electrical route for microbial production. Nat Rev Microbiol 8: 706–716.

52. Rabaey K., Boon N., Siciliano S.D., Verhaege M., and Verstraete W. (2004) Biofuel cells select for microbial consortia that self-mediate electron transfer. Appl Environ Microbiol 70: 5373–5382.

53. Reguera G., McCarthy K.D., Mehta T., Nicoll J.S., Tuominen M.T., and Lovley D.R. (2005)Extracellular electron transfer via microbial nanowires. Nature 435: 1098–1101.

54. Richter K., Schicklberger M., and Gescher J. (2012) Dissimilatory reduction of extracellular electron acceptors in anaerobic respiration. Appl Environ Microbiol 78: 913–921.

55. Rosenbaum M., Aulenta F., Villano M., and Angenent L.T. (2011) Cathodes as electron donors for microbial metabolism: which extracellular electron transfer mechanisms are involved?Bioresour Technol 102: 324–333.

56. Shi L., Squier T.C., Zachara J.M., and Fredrickson J.K. (2007) Respiration of metal (hydr)oxides by *Shewanella* and *Geobacter*: A key role for multihaem *c*-type cytochromes. Mol Microbiol 65: 12–20.

57. Stams A.J., de Bok F.A., Plugge C.M., van Eekert M.H., Dolfing J., and Schraa G. (2006)Exocellular electron transfer in anaerobic microbial communities. Environ Microbiol 8: 371–382.

58. Suflita J.M., Phelps T.J., and Little B. (2008) Carbon dioxide corrosion and acetate: a hypothesis on the influence of microorganisms. Corros Sci 64: 854–859.

59. Summers Z.M., Fogarty H.E., Leang C., Franks A.E., Malvankar N.S., and Lovley D.R. (2010)Direct exchange of electrons within aggregates of an evolved syntrophic coculture of anaerobic bacteria. Science 330: 1413–1415.

60. Thrash J.C., and Coates J.D. (2008) Review: direct and indirect electrical stimulation of microbial metabolism. Environ Sci Technol 42: 3921–3931.

61. Uchiyama T., Ito K., Mori K., Tsurumaru H., and Harayama S. (2010) Iron-corroding methanogen isolated from a crude-oil storage tank. Appl Environ Microbiol 76: 1783–1788.

62. Usher K.M., Kaksonen A.H., and MacLeod I.D. (2014) Marine rust tubercles harbour iron corroding archaea and sulphate reducing bacteria. Corros Sci 83: 189–197.

63. Venzlaff H., Enning D., Srinivasan J., Mayrhofer K., Hassel A.W., Widdel F., and Stratmann M. (2013) Accelerated cathodic reaction in microbial corrosion of iron due to direct electron uptake by sulfate-reducing bacteria. Corros Sci 66: 88–96.

64. Videla H.A., and Herrera L.K. (2005) Microbiologically influenced corrosion: looking to the future. Int Microbiol 8: 169–180.

65. Wakai S., Ito K., Iino T., Tomoe Y., Mori K., and Harayama S. (2014) Corrosion of iron by iodide-oxidizing bacteria isolated from brine in an iodine production facility. Microb Ecol 68: 519–527.

66. Watanabe K., Manefield M., Lee M., and Kouzuma A. (2009) Electron shuttles in biotechnology.Curr Opin Biotechnol 20: 633–641.

67. Weber K.A., Achenbach L.A., and Coates J.D. (2006) Microorganisms pumping iron: anaerobic microbial iron oxidation and reduction. Nat Rev Microbiol 4: 752–764.

68. Wolzogen Kühr C.A.H. (1961) Unity of anaerobic and aerobic iron corrosion process in the soil.Corrosion 17: 293–299.

69. Xu D., Li Y., Song F., and Gu T. (2013) Laboratory investigation of microbiologically influenced corrosion of C1018 carbon steel by nitrate reducing bacterium *Bacillus licheniformis* . Corros Sci77: 385–390.

70. Zhang P., Xu D., Li Y., Yang K., and Gu T. (2015) Electron mediators accelerate the microbiologically influenced corrosion of 304 stainless steel by the *Desulfovibrio vulgaris*biofilm. Bioelectrochemistry 101: 14–21.

CHAPTER 7

TEMPERATURE AND HUMIDITY SENSOR POWERED BY AN INDIVIDUAL MICROBIAL FUEL CELL IN A POWER MANAGEMENT SYSTEM

Qi Zheng, Lei Xiong, Bing Mo, Weihong Lu, Suki Kim and Zhenyu Wang

Keywords: Microbial fuel cells (MFCs), power management system (PMS), sensors, charge pump, microcontroller unit (MCU), boost converter, DC-DC

INTRODUCTION

With increasing attention being paid to global warming and the depletion of fossil fuels, substantial investment and effort has been aimed toward developing renewable energy technologies. Microbial fuel cells (MFCs) represent a novel alternative power source, utilizing organic energy for the production of electricity. Energy generation based on MFCs offers bright prospects as an environmentally friendly energy source because of their mild reaction conditions and the wide availability of suitable fuel [1]. Although reactors based on MFCs can develop an open circuit voltage of approximately 1.1 V [2], a more typical voltage is 0.5 V, which is insufficient to drive most practical sensors and electrical components. Nevertheless, some researchers have designed energy harvesting circuits that can be powered by the lower output of MFCs [3,4]. Gong *et al.* have use super capacitors and a boost converter to power the oxygen and temperature sensor; however, this system requires an additional battery [5]. Hatzell *et al.* have used multiple MFCs connected in series to increase the output voltage [6]; however, this method is rather complex for practical applications because multiple cascaded MFCs are difficult to construct and may result in voltage reversal [7,8,9]. Therefore, the study presented in this paper used an individual MFC as the power source to drive the load with the help of the designed circuit.

In this paper, a power management system (PMS) is designed to harvest energy based on a charge pump with the help of a supercapacitor as an energy storage element. When the PMS starts operating, the supercapacitor accumulates energy in small increments during the charge period, and discharges it to a DC-DC boost circuit during the discharge period. The DC-DC boost circuit generates a 5 V voltage to simultaneously power a temperature and humidity sensor. With the assistance of the cyclical operation of this PMS, an external load can be powered by an individual MFC.

For this study, a wireless sensor is added at the load end of an nRF24L01 data transmission module. A PMS is needed to export the 5 V output voltage from an individual MFC to enable the sensors to capture the temperature and humidity data. Microcontroller units (MCUs) obtain the data and transmit them using a wireless module. Computers can acquire the information from the sensors at the receiving end using serial ports. The computers use a serial port software based on a virtual instrument, enabling the display and storage of data and the communication between the MCUs and computers.

This study demonstrates that the designed circuit has realized the function of gathering the steadily accumulating energy produced from an individual MFC and using this energy to drive the sensor and telemetry system. To confirm functionality, a remote sensing of temperature and humidity experiment was conducted using a DC-DC boost circuit with the aim of increasing the voltage and power availability. The circuit shown in this article will allow wider adoption of MFCs and should find wide application in the unmanned environmental monitoring field.

Citation: Qi Zheng, Lei Xiong, Bing Mo, Weihong Lu, Suki Kim and Zhenyu Wang, Temperature and Humidity Sensor Powered by an Individual Microbial Fuel Cell in a Power Management System, doi:10.3390/s150923126.

MATERIALS AND METHODS

Section 2.1, Section 2.2, Section 2.3 and Section 2.4 describe the operational mechanism of MFCs, the proposed PMS, the sensor and telemetry system used to test the functionality of the PMS, and the sensor's program design, respectively.

Operational Mechanism of MFCs

Mechanism of Electricity Production by MFCs

Dual-chamber MFCs are commonly used in the laboratory; they consist of an anode chamber and a cathode chamber separated by a proton exchange membrane (PEM). The anode chamber provides an anaerobic environment that is beneficial to the growth of microorganisms. The metabolism of the microorganisms oxidizes the organic substance and transfers electrons to the anode; these electrons then traverse the external load to the cathode resulting in the production of a current. At the same time, protons from which water is produced from the reaction of oxygen with the protons pass through the PEM to the cathode. As has been found in many surveys, the electrode materials, the space between the electrodes, and the density of the substrate all affect the efficiency of electricity production by MFCs [10,11,12,13,14,15,16].

MFC Setup and Operation

For this study, the dual-chamber cubic MFC was used, with each chamber having an empty volume of 27 mL. Both the anode and cathode employed a carbon felt material, each with a projected surface area of 9 cm^2 (3 cm × 3 cm), as electrode materials upon which microorganisms can easily adhere, thus producing more electrons with more rapid migrations to the electrodes. The reason for selecting the large area of the electrodes is that the large area allows a more convenient transfer of electrons, and the internal resistance is reduced accordingly.

The MFC was inoculated with an anaerobic mixed culture collected from a wastewater treatment plant. The anode chamber was filled with artificial wastewater containing CH_3COONa (with density of 0.823 g/L), $Na_2HPO_4{\cdot}12H_2O$ (with density of 11.46048 g/L), and $NaH_2PO4{\cdot}2H_2O$ (with density of 2.80818 g/L). A phosphate-buffered $K_3[Fe(CN)_6]$ solution containing $K_2HPO_4{\cdot}3H_2O$ (with density of 7.3 g/L), KH_2PO_4 (with density of 2.44 g/ L), and $K_3[Fe(CN)_6]$ (with density of 16.46 g/L) was used as a catholyte to minimize the cathode effects on system performance.

To start the MFC, 5 mL of inoculum and 20 mL of artificial wastewater was injected into the anode chamber using a syringe. Originally, the reactor is run in a batch-fed mode and is manipulated until repeatable voltages are achieved at the external resistor of 1 kΩ, for the MFC at room temperature. Subsequently, the reactor is run in a continuous-flow operation by exchanging 20 mL of fresh artificial wastewater with 20 mL of depleted anolyte in each MFC.

Design of the PMS

The proposed PMS shown in Figure 1 is composed of a charge pump, a supercapacitor, a switch, and a DC-DC converter. The charge pump is used to store the energy produced by the MFCs in the supercapacitor until sufficient electricity has been accumulated to provide an initial voltage for the boost circuit. The supercapacitor was employed as the storage component because it can store a higher energy density than a conventional capacitor over the short charging time. When the voltage increases to a set threshold value, a control signal is generated from the charge pump that closes the switch between the supercapacitor and the DC-DC converter, and the supercapacitor begins to discharge. The switch remains off until the voltage of the supercapacitor drops below another set value. In this way, the charge pump acts as a switch to connect and disconnect the supercapacitor with the load. The equivalent input resistance of the charge pump should be proper to make the MFCs achieve reasonable electricity production performance. There are some reports on harvesting energy at the MFC's maximum power point [17,18,19]. However, all the reported maximum power point tracking circuits require batteries as an additional power supply. To drive sensors with an individual MFC without any additional power, this study pursues the appropriate power point by using a fixed value supercapacitor rather than the maximum power point tracking circuits. This study uses the 3 F value as the best value of the supercapacitor for achieving the desired result, based on our previous testing results. The proposed system aims to accumulate the energy generated from the MFCs, and then use the accumulated energy to power the external circuit through the DC-DC converter.

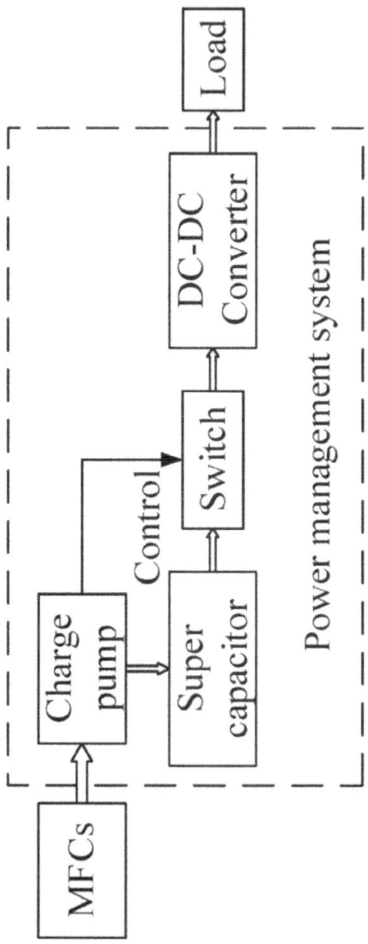

Figure 1: Schematic of the PMS.

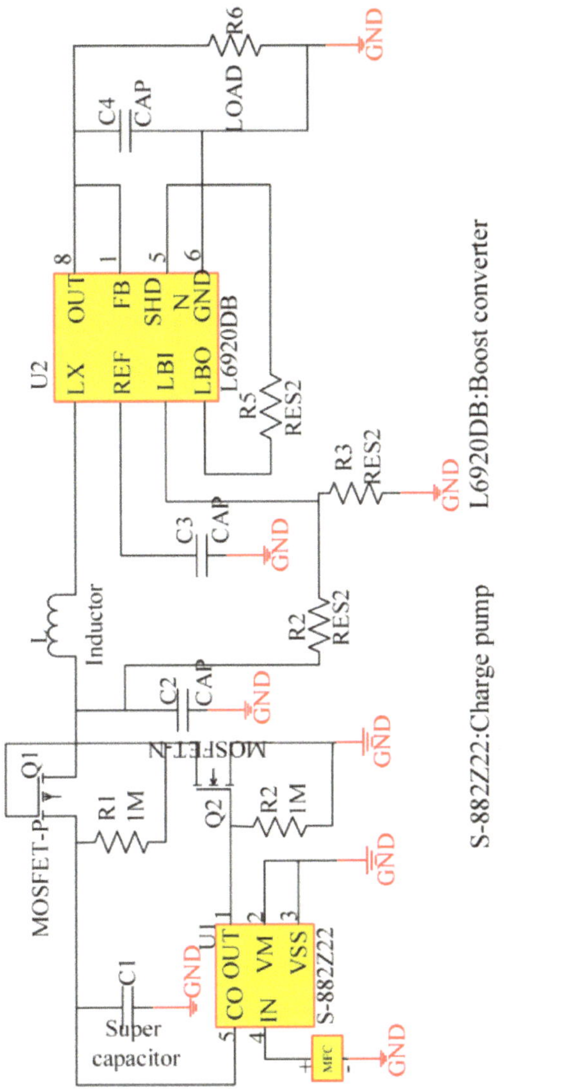

Figure 2: Energy harvesting circuit based on the charge pump and supercapacitor.

The energy harvesting circuit has been implemented using an individual MFC as the power source. As shown in Figure 2, the positive input of the MFC can be attached to the charge pump (S-882Z22). A 3 F super capacitor (C1 in Figure 2) is capable of storing energy to accumulate electricity and control the switch. The N-channel MOSFET (Q2 in Figure 2) and P-channel MOSFET (Q1 in Figure 2) are used as the two switches. When the voltage of the supercapacitor's positive electrode reaches the value of 2.2 V, the charge pump exports 2.2 V at the OUT node, and then closes the Q2 switch when the voltage of Q2 drain node has been drained to almost 0 V. Because the Q1 gate node is connected to the Q2 drain node, switch Q1 is also closed at this time. Therefore, the C1 supercapacitor discharges to the DC-DC boost circuit (L6920DB), which can increase the input voltage of 2.2 V to an output voltage of 5 V. When the voltage of C1's positive electrode drops below 1.6 V, the charge pump outputs a low level, which opens switch Q2, resulting in the disconnect of the Q2 drain node from the ground node and a sudden increase in the voltage to almost 1.6 V. Subsequently, the switch Q1 opens and the MFC again begins to charge the C1 supercapacitor.

A low input voltage can satisfy the charge pump's drive demand when using S-882Z22, which can operate normally with an input of 0.3 V, 0.5 mA. The functions of the charge pump's pins are described in Table 1 [20]. The schematic circuit diagram of S-882Z22 is shown in Figure 3 [20].

Table 1: Functions of the charge pump's pins [20]

Pin No	Pin Name	Pin Description
1	OUT	Output pin (step-up DC-DC converter connection pin)
2	VSS	GND pin
3	VM	Step-up DC-DC converter output voltage monitor pin
4	VIN	Power supply input pin
5	CPOUT	Startup capacitor connection pin

The schematic circuit diagram of S-882Z22 is shown in Figure 3. The power supply is connected to S-882Z22 through the pin Vin, and the external supercapacitor C_{CPOUT} is used as the energy storage element [20]. In this study, discharge start voltage and stop voltage are 2.2 V and 1.6 V respectively, the VM pin is connected to the VSS pin to ensure that the charge pump circuit operates all the time, preventing malfunction of the PMS.

The L6920DB is a high efficiency monolithic step-up switching converter integrated circuit (IC) especially designed for battery-powered applications. The start-up voltage of the L6920DB is 0.8 V, and its input voltage can be as high as 5.5 V, with current as high as 750 mA. The output voltage can be adjusted, ranging from 1.8 to 5 V. The pin description for the L6920DB is shown in Table 2 [21].

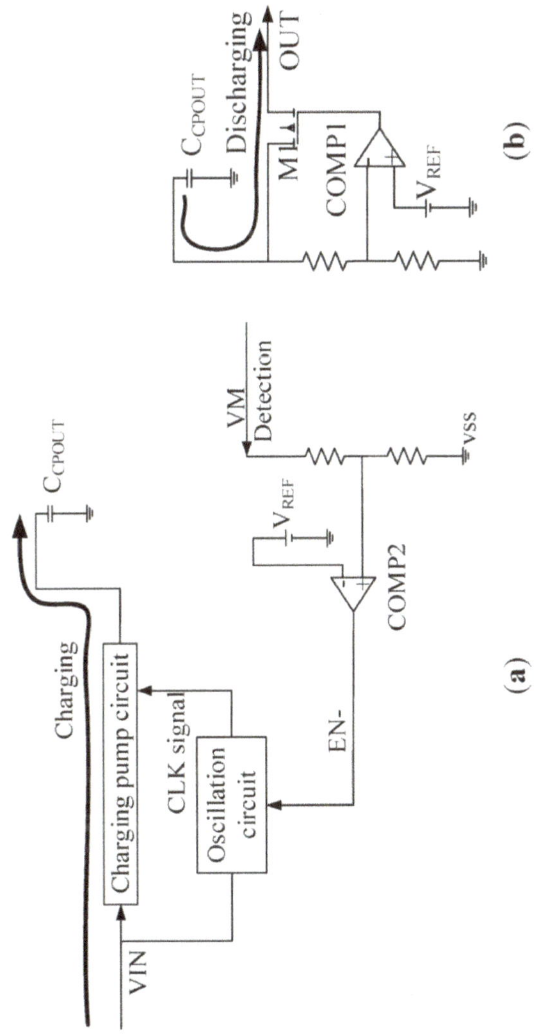

Figure 3: Schematic of S-882Z22. (a) Principle of charging mode; (b) Principle of discharging mode.

Table 2: Pins description for L6920DB [21]

Pin No.	Pin Name	Pin Description
1	FB	Output voltage selector. Connect FB to GND for Vout = 5 V or to OUT for Vout = 3.3 V. Connect FB to an external resistor divider for adjustable output voltage
2	LBI	Battery low voltage detector input. The internal threshold is set to 1.23 V. A resistor divider is needed to adjust the desired low battery threshold.
3	LBO	Battery low voltage detector output. If the voltage at the LBI pin drops below the internal threshold typ. 1.23 V, LBO goes low. The LBO is an open drain output and so a pull-up resistor (about 200 KΩ) has to be added for correct output setting
4	REF	1.23 V reference voltage. Bypass this output to GND with a 100 nF capacitor for filtering high frequency noise. No capacitor is required for stability
5	SHDN	Shutdown pin. When pin 5 is below 0.2 V the device is in shutdown, when pin 5 is above 0.6 V the device is operating.
6	GND	Ground pin
7	LX	Step-up inductor connection
8	OUT	Power OUTPUT pin

The switches used in the circuit are N-channel MOSFET (Si3460) and P-channel MOSFET (Si3499), with thresholds between 0 and 0.6 V. The working threshold changes with the temperature, with a threshold of approximately 0.6 V at a room temperature of 25 °C.

Sensor and Telemetry System

One transmitter and one receiver, shown in Figure 4, constitute the sensor and telemetry system. The energy produced from the MFCs has been administered by the PMS so as to increase a low potential to a high potential with a value of 3.3 or 5 V, which can drive the transmission system. The temperature and humidity sensors and the microcontroller unit (MCU) require a working voltage of 5 V, whereas the wireless transmitting module nRF24L01 requires 3.3 V. The function of the MCU is to obtain and transmit the data from the sensor. It is necessary for the receiver to use serial ports, through which the data can appear on a computer monitor while also storing the data in the computer.

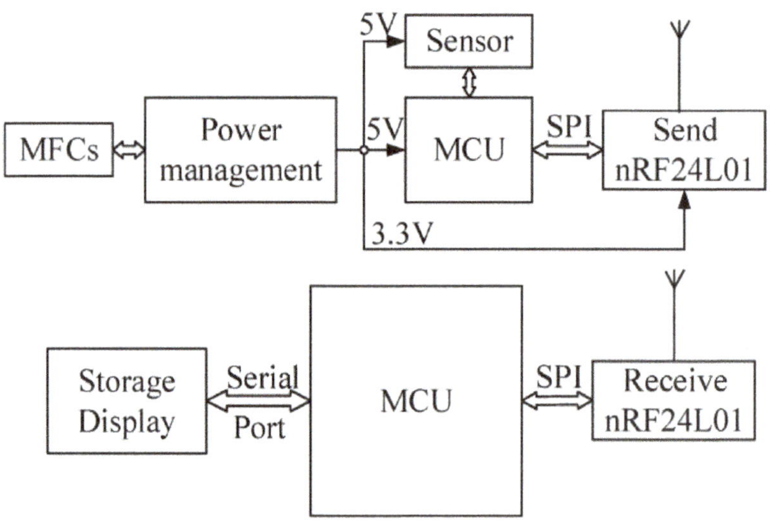

Figure 4: Schematic of the sensor and telemetry system driven by the MFCs.

Transmitter of the Sensor and Telemetry System

The hardware of the transmitter consisted of sensors, a microcontroller unit (MCU), a wireless module, and the connections at communication interfaces among the different kinds of circuit modules.

In this study, an 8-bit low power MCU (STC89C52) was used to obtain and transmit the data from the temperature and humidity sensor. The DHT11 temperature and humidity sensor used in this study can be connected to the MCU with only one data line, which conserves the interface resources. The pins description of DHT11 is shown in Table 3 [22]. The circuit scheme is shown in Figure 5.

Table 3: The pins description of DHT11 [22]

Pin No.	Pin Name	Pin Description
1	VDD	3 V to 5 V, powered directly
2	DATA	Serial data lines, signal bus
3	NC	Untapped pin, suspended
4	GND	Power negative ground

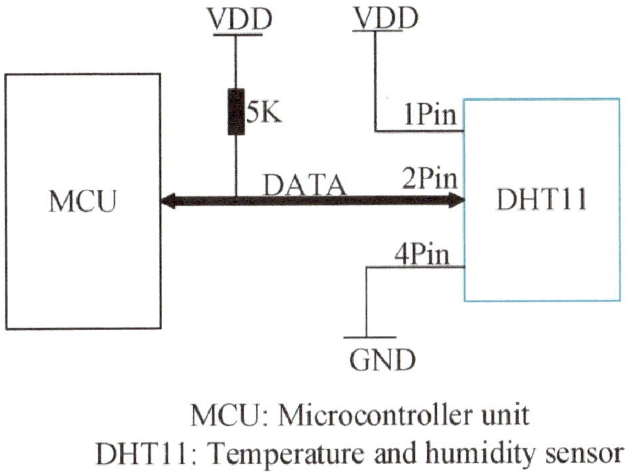

Figure 5: Application circuit of the temperature and humidity sensor.

This work used a wireless transmission module (nRF24L01) as the transceiver, which works in the Industrial Scientific Medical (ISM) band from 2.4 to 2.5 GHz. The current consumption of the transmission module is relatively low at 11.3 mA at 0 dBm output power. If the system remains in standby and power down mode, the current consumption of the wireless transmission module is even lower at 900 nA in power down mode, 26 μA in standby-I mode, and 320 μA in standby-II mode [23]. Therefore, this wireless transmission module is appropriate for low power wireless applications. The pins description of nRF24L01 is shown in Table 4 [23].

Table 4: Pins description of nRF24L01 [23]

Pin No.	Pin Name	Pin Description
1	CE	Chip Enable Activates RX or TX mode
2	CSN	Serial Peripheral Interface (SPI) chip select
3	SCK	SPI Clock
4	MOSI	SPI Slave Data Input
5	MISO	SPI Slave Data Output
6	IRQ	Interrupt Pin
7	VDD	Power Supply (+3 V DC)
8	VSS	Ground(0 V)

Sensor DHT11 can capture the temperature and humidity data from the ambient environment, and the MCU detects the data for transmission. A serial peripheral interface (SPI) is used to transmit data with the wireless module nRF24L01. Because there is no hardware SPI in the MCU STC89C52, the IO (input and output) port of the MCU was used to simulate the timing sequence of the SPI. In this way, the data communication between the MCU and wireless module can be achieved.

Receiver of the Sensor and Telemetry System

The hardware of the receiver consisted of a MCU, wireless receiver module, voltage converter circuit, and the connections among communication circuits between the MCU and computer. In this study, conversion between USB and the serial port is implemented by a conversion module CH340T. CH340T is a full speed USB interface device, the peripheral components of which require only capacitors and a crystal oscillator. CH340T is also a full duplex communication serial interface with a built-in transceiver buffer. The communication Baud rate ranges from 50 bps to 2 Mbps. The power supply voltage is either 5 V or 3.3 V. The pins description of CH340T is shown in Table 5 [24].

Table 5: The pins description of CH340T [24]

Pin No.	Pin Name	Pin Description
3	TXD	Transmitting data
4	RXD	Receiving data
6	UD+	USB data line D+
7	UD-	USB data line D-
9	XI	Linking with 12M crystal oscillator
10	XO	Linking with 12M crystal oscillator

The supply voltage of the wireless module is 3.3 V; however, the voltage generated from an individual MFC is increased by the DC-DC booster converter (AMS117) to 5 V. AMS117 is a low dropout voltage regulator that can implement the conversion between 5 and 3.3 V.

Program Design of the Wireless Sensor

The software design of the MFC energy harvesting system also consists of transmitting and receiving parts. In the transmitter part, the sensor DHT11 detects the temperature and humidity data and sends the data through the wireless module nRF24L01 to the MCU, which reads the data from the sensor. In the receiver part, the data can be stored in a buffer prepared for the communication between the MCU and computer. The flow diagram of the program is described in Figure 6.

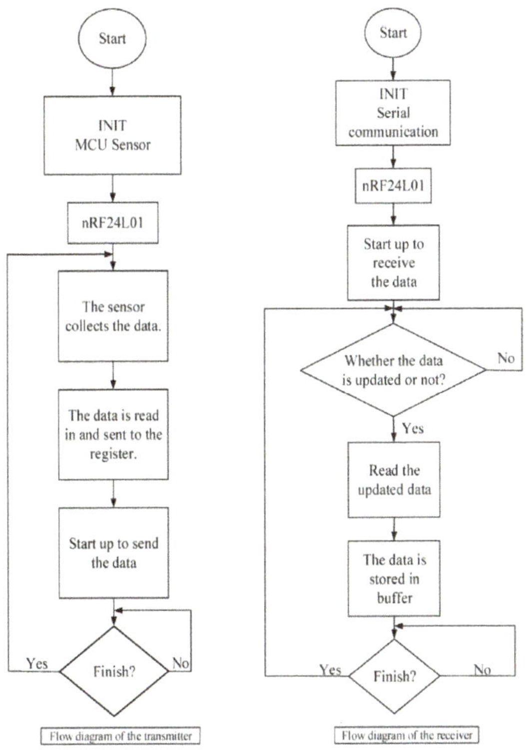

Figure 6: Flow diagram of the program.

Program Design of the Sensor Collecting Data

The sensor DHT11 uses the signal bus with a bidirectional data-flow transmission format. The data line is used to synchronize and communicate between the sensors and the MCU. The data consists of integer and decimal parts with a total volume of 32 bit in the following format: 8 bit integer part of the temperature plus 8 bit decimal part of the temperature plus 8 bit integer part of the humidity plus 8 bit decimal part of the humidity. Firstly, the MCU sends a start signal. The sensor DHT11 changes from a lower power consumption mode into a high-speed operation mode. When the initial signal from the MCU ends, the sensor DHT11 transmits the response signal, which is a complete 32 bit signal.

The users can read part of the data selectively. If the sensor cannot receive the initial signal transmitted from the MCU, the sensor cannot detect the temperature and humidity data. The sensor will revert to the low-speed working mode after data collection.

The Program Design of the Wireless Module

The nRF24L01 can be configured in four main modes of operation: TX, RX, power down, and standby. The mode of operation depends on the registers CE, PWR_UP, and PRIM_RX, as shown in Table 6. By setting the PWR_UP bit in the CONFIG register to 1, the device enters standby-I mode. Standby-I mode is used to minimize average current consumption while maintaining short start up times. In this mode, only part of the crystal oscillator is active. Change to active modes only occurs if CE is set high; when CE is set low, the nRF24L01 returns to standby-I mode in both the TX and RX modes. In standby-II mode, additional clock buffers are active and more current is used compared to standby-I mode. The nRF24L01 enters standby-II mode if CE is maintained high on a PTX device with an empty TX FIFO (first in first out). If a new packet is uploaded to the TX FIFO, the PLL (Phase Locked Loop) immediately begins operation, and the packet is transmitted after the normal PLL settling delay of 130 µs. The RX mode is an active mode in which the nRF24L01+ radio is used as a receiver. To enter this mode, the nRF24L01 must have the PWR_UP bit, PRIM_RX bit, and tCE pin set high. The TX mode is an active mode for transmitting packets. To enter this mode, the nRF24L01 must have the PWR_UP bit set high, PRIM_RX bit set low, a payload in the TX FIFO, and a high pulse on the CE for more than 10 µs. Therefore, reasonable adoption of these different modes can reduce the power consumption of the whole circuit.

Table 6: Operation modes configuration [24]

Mode	CE	PWR_UP	PRIM_RX	Register FIFO (First In First Out)
TX mode	1	1	0	Data in TX FIFOs. Will empty all levels in TX FIFOs [a]
TX mode	Minimum 10 µs high pulse	1	0	Data in TX FIFOs. Will empty one level in TX FIFOs [b]
RX mode	1	1	1	/
Standby-I	0	1	/	No ongoing packet transmission
Standby-II	1	1	0	TX FIFO empty
Power down	/	0	/	/

[a] in this operating mode, if the CE is held high, all TX FIFOs are emptied and all necessary ACK and possible retransmits are conducted. The transmission continues as long as the TX FIFO is refilled. If the TX FIFO is empty and the CE is still high, nRF24L01+ enters standby-II mode. In this mode the transmission of a packet is initiated as soon as the CSN is set high after an upload (UL) of a packet to TX FIFO; [b] In this operating mode, the CE pulses high for at least 10 µs. This is the normal operating mode, which allows one packet to be transmitted. After the packet is transmitted, the nRF24L01+ enters standby-I mode.

There are two methods for handling the data packet from nRF24L01: ShockBurst mode and enhanced ShockBurst mode. ShockBurst mode is typically used for simplex communications. Enhanced ShockBurst mode has the advantage of offering two-way transmission. The format of the data packet used in this study is shown as follows:

Preamble 1 byte	Address 3–5 byte	Data 1–32 byte	C R C 1–2 byte

The preamble is a bit sequence used to detect 0 and 1 levels in the receiver. An address ensures that the correct packet is detected by the receiver. The width of the cyclic redundancy check (CRC) can be set to 0, 1, or 2 bytes. The 1-byte CRC method has a higher data transmission efficiency. However, the 2-byte CRC method has better data integrity. In general, the data transmission speed via the wireless module is different from the MCU. The data handling speed of the MCU STC89C52 is less than the data transmission speed of the wireless module nRF24L01. Therefore, the transmitted signals from the lower MCU must be written in the wireless module FIFO to achieve the transportation of the data between the low-speed devices and the high-speed devices.

The entire configuration of nRF24L01 is in a register configured as a SPI, which is a full-duplex communication interface that transmits the data synchronously. The maximum transfer speed of a standard SPI is 10 Mbps. The SPI enables serial communication among various pieces of peripheral equipment and the MCU. In general, the interface uses four lines: Serial Clock (SCK), Master Input/Slave Output (MISO), Master Output/Slave Input (MOSI), and Slave Selected (NSS). The internal hardware of the SPI used in this study consisted of two shifting registers, and the SPI can transmit the 8-bit data. There are two working modes for the SPI: master mode and slave mode. The communication method between the master device and the slave device is shown in Figure 7. This study used the MCU as the master device and the wireless nRF24L01 as the slave device. In the master mode, the SPI requires data transmission through the NSS and SCK lines, and the slave device receives data in the NSS and SCK lines. The register should be set to the initial value before the SPI begins data transmission.

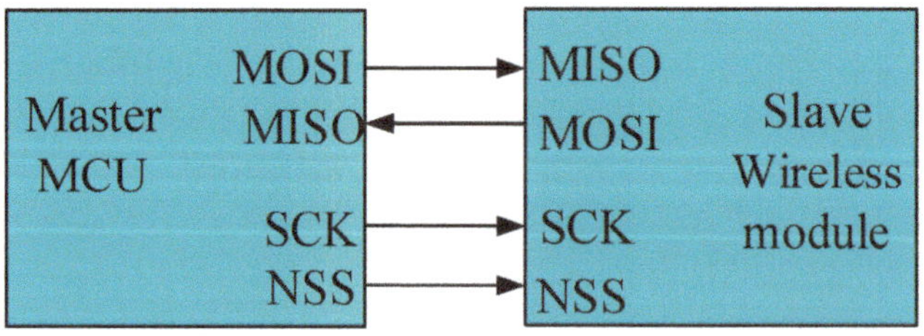

Figure 7: Communication method between the master device and the slave device.

In the nRF24L01 data transmission program, the wireless module remains in the transmitting mode when the CE is set to a high level and maintained for at least 10 s. The pins of the nRF24L01 data lines are set as input. The humidity data detected by the sensors is written through MOSI. Subsequently, the data is written in the TX_FIFO register of nRF24L01 and the CE setting changes from the high level to a low level. The wireless module goes into the ShockBurst mode of transmitting the data. The nRF24L01 data transmission process is shown in Figure 8.

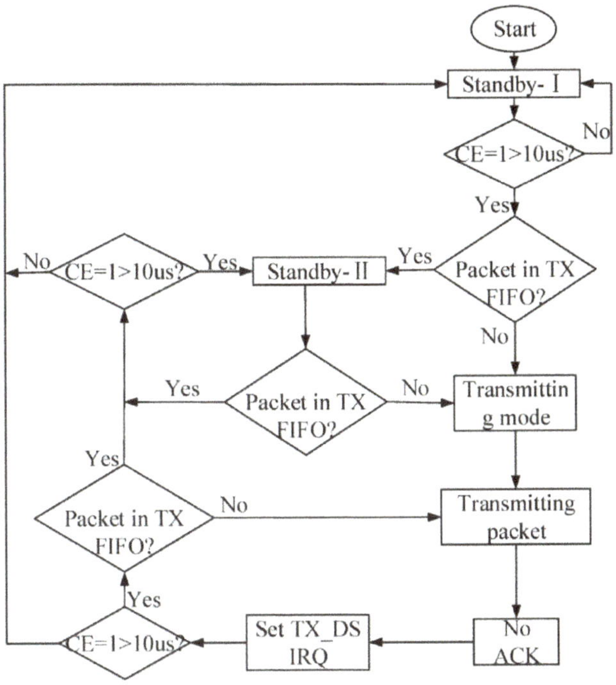

Figure 8: The nRF24L01 data transmission process.

In the nRF24L01 receiver program, CE was set to 1, and PWR_UP was also set to 1. The configured CRC, the width of the address, the channel, and the transfer speed of the receiver should all be equal to those of the transmitter. The nRF24L01 data reception process is shown in Figure 9.

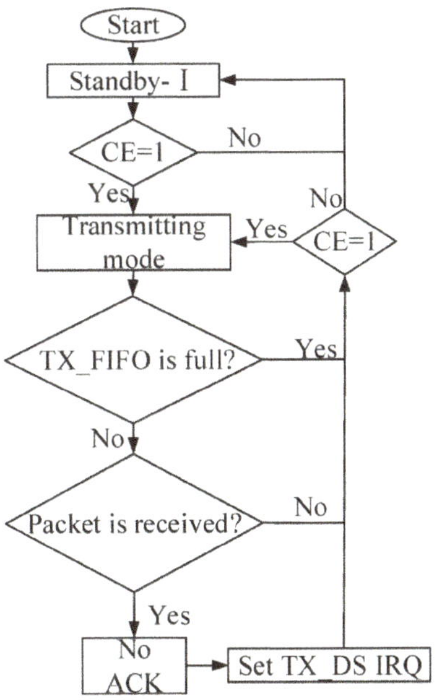

Figure 9: The nRF24L01 data reception process.

RESULTS AND DISCUSSION

Characterization of the MFCs

In general, after starting the MFC, the open-circuit voltage (OCV) was measured by a digital multimeter (Keithley Instruments Inc., Cleveland, OH, USA). The OCV could reach 0.733 V. Subsequently, a 1 kΩ resistor was linked to the MFC. The beginning of a first period was defined as the point at which the output voltage approached 0.6 V; from this point, three stable periods were recorded by the same digital multimeter, as shown in Figure 10. From this procedure, it could be concluded that the voltage of the resistor was approximately 0.59 V, and each period lasted approximately 6.5 h. The steep decline in the voltage curve after the stable period of 6.5 h results from the exhaustion of the organic substance by the microorganisms. If wastewater in the anode chamber is refreshed, the output voltage could increase up to the stable value within a few minutes. In this study, 15 mL of wastewater is replenished at the end of each exhaustion period to the anode chamber, the empty volume of which is 27 mL; subsequently, approximately 0.59 V of output voltage from the MFC could be achieved in a few minutes. With 0.59 V as the output voltage and 6.5 h as the average time of every period, the current, power, and energy can be calculated as follows:

- Current: $I = U/R = 0.59\ V/1\ k\Omega = 0.59\ mA$
- Power: $P = U^2/R = (0.59\ V)^2/1000\ \Omega = 0.35\ mW$
- Energy: $W = P \times t = 0.35\ mW \times 6.5 \times 3600\ s = 8.19\ J$

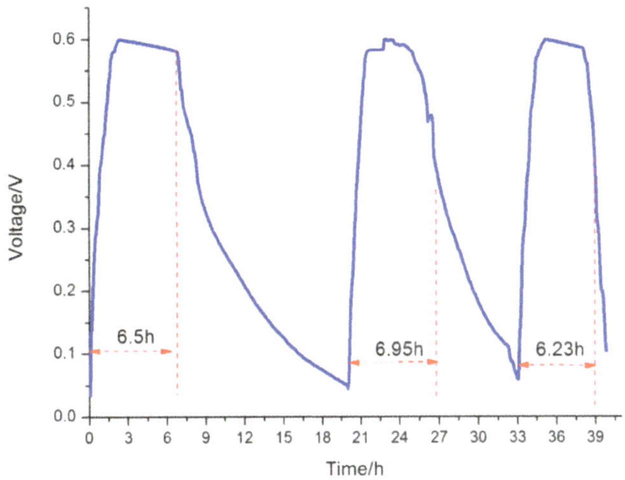

Figure 10: Three continuous flow cycles showing voltage variation at the external resistance of 1 kΩ.

Energy Harvesting Test Results

Figure 11 shows the voltage and current input through the charge pump. As can be seen from the figure, the voltage remained at approximately 0.658 V after an initial drop. The current maintained a steady value of 0.395 mA. Using the charge pump, this energy was transferred from the individual MFC where it was generated to the supercapacitor for storage.

Figure 12 shows that it required approximately 49.5 hours to increase the voltage from 0 to 2.2 V in the first period. The second period started at 1.6 V and charged to 2.2 V in 23 h, close to the time required during the first period to charge the voltage from 1.6 to 2.2 V. The discharging time required to reduce the voltage from 2.2 to 1.6 V is indicated in Figure 12 by the red dotted circle; this is when the voltage was boosted to 5 V to drive the load. The average current of MFC during charge period is about 0.2 mA. Supposing that Q1 is electrical charge produced from an individual MFC, Q2 is electrical charge stored in the supercapacitor, and E is the energy efficiency of the circuit, these values could be calculated as follows:

- $Q1 = I \times t = 0.2 \times 10^{-3} \times 49.5 \times 3600\ s = 35.64\ C$
- $Q2 = C \times U = 3\ F \times 2.2\ V = 6.6\ C$
- $E = Q2 / Q1 = 6.6 / 35.64 = 18.518\%$

The efficiency of the boost converter is 85%–90% according to the datasheet of the boost converter [21]. So that the energy efficiency of designed PMS is about 15.7%–16.7%, higher than 5.33% reported on reference [25]. To illustrate this cyclical charging and discharging process in detail, the turning points between the charging and discharging of the supercapacitor, and *vice versa*, are shown in Figure 13. When the supercapacitor discharged, the curve fell immediately. On the contrary, the voltage rose quickly when the supercapacitor was in the charging state.

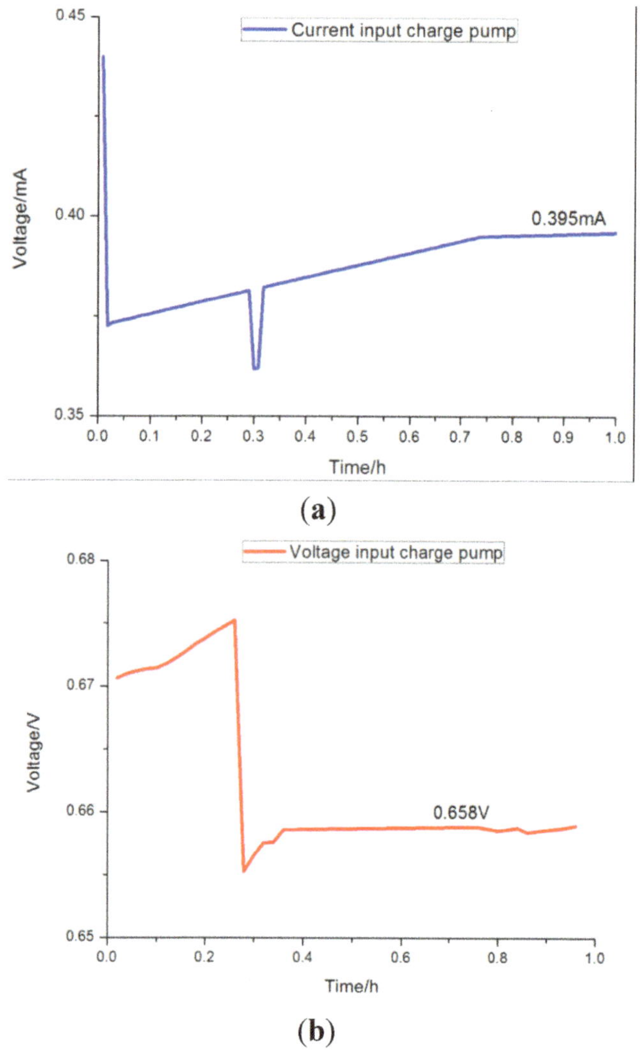

Figure 11: Voltage and current input through the charge pump. (a) Current input of charge pump; (b) Voltage input of charge pump.

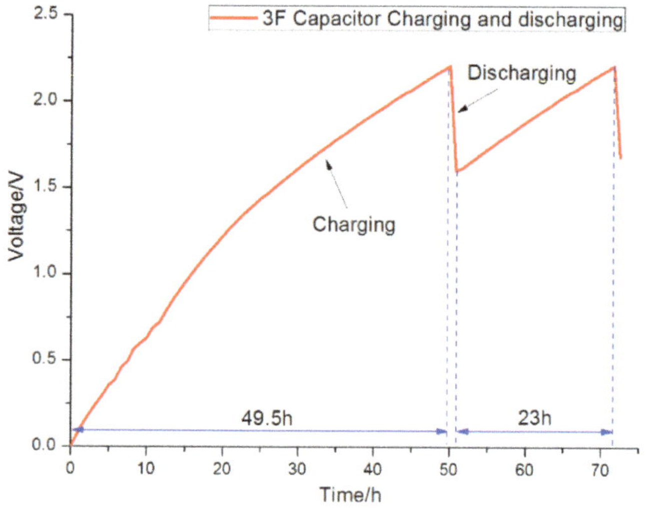

Figure 12: Charging and discharging curves of the 3 F supercapacitor.

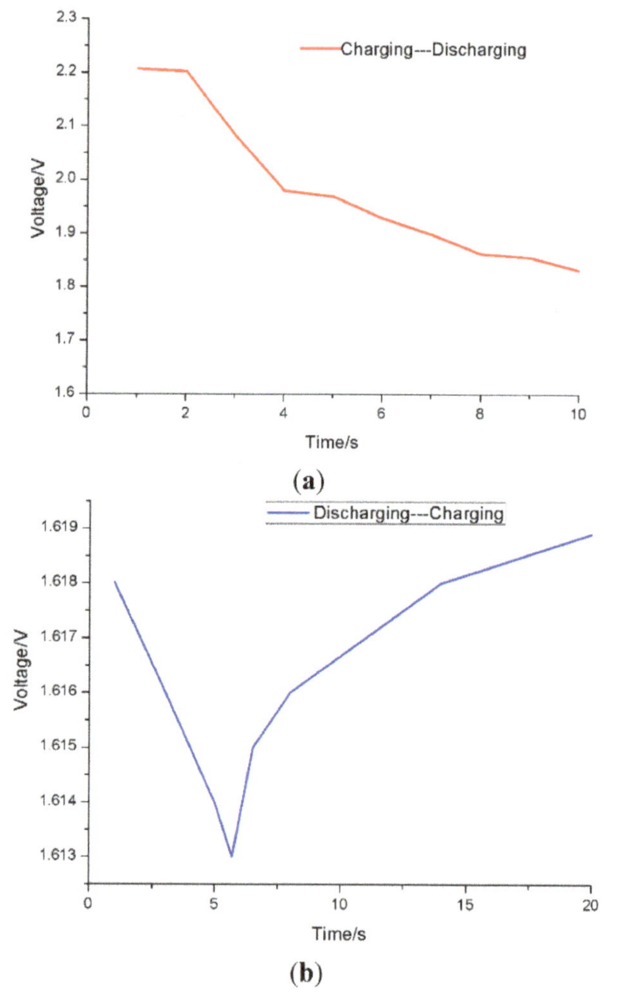

Figure 13: Turning points between charging and discharging and vice versa. (a) Turning points between charging and discharging; (b) Turning points between discharging and charging.

Application of the Designed PMS to Drive the Sensor and Telemetry System

The entire experimental setup consisted of an individual

MFC, a collecting and storing energy circuit, and the wireless sensing transmitter. The 3 F supercapacitor stored the energy generated from the MFCs by means of the proposed PMS. The elevated voltage (5 V) discharged from the supercapacitor could be conducted as the power to the wireless sensor. The current and voltage inputting the MCU were 21.9 mA and 5.2 V, respectively; these input curves are shown in Figure 14. The operation time of the sensor depended on the storage capacity of the supercapacitor. In this case, the 3 F supercapacitor could power the sensor for 5 s.

At the end of the receiver, the data emitted from the sensor could be accepted through serial ports. During the 5 s of the sensor's operation, 30 data points were detected from the transmitter, and the results are shown in Figure 15. The temperature was 28 °C, and the relative humidity was 0.58, relative to vapor saturation in the air. At the same time, the system was shown to be capable of storing and displaying the data points using the serial port of the virtual instrument.

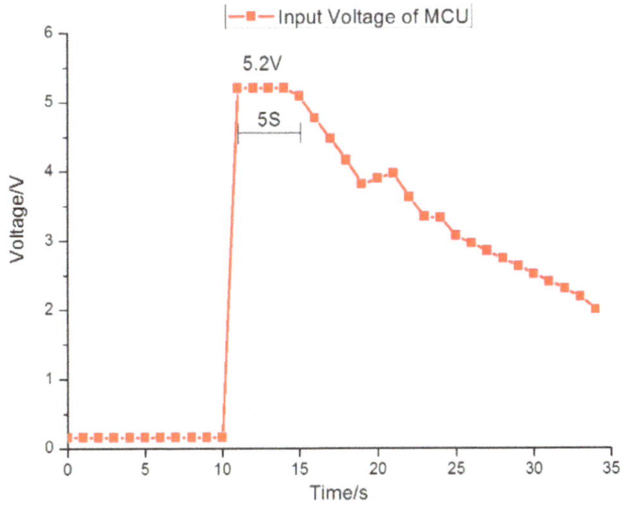

(a)

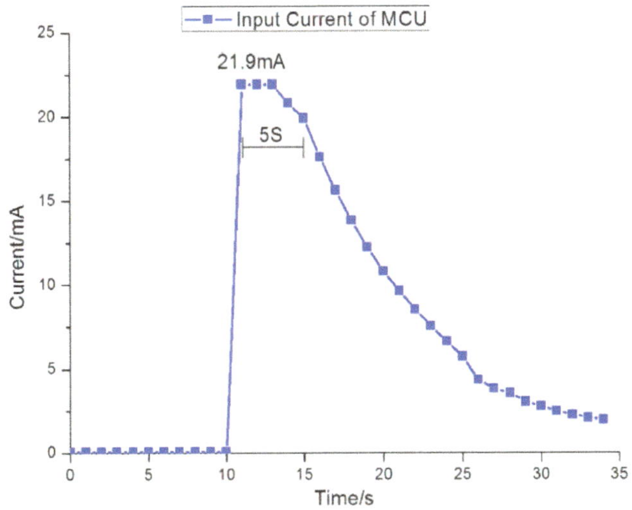

(b)

Figure 14: Voltage and current inputting the MCU. (a) Voltage inputting the MCU; (b) Current inputting the MCU.

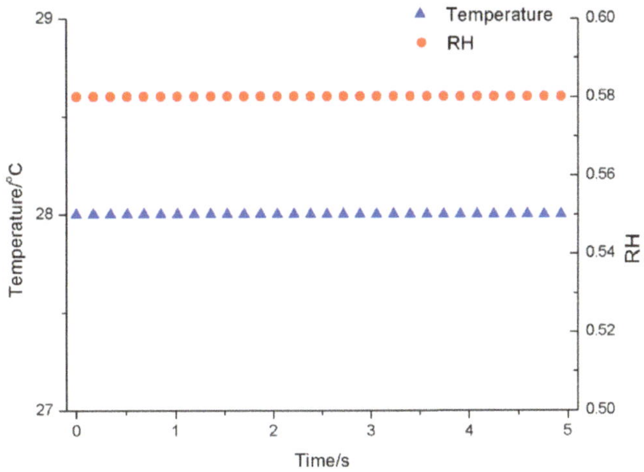

Figure 15: Received temperature and relative humidity data.

From the test, it can be concluded that the application of the method–first for storing energy, and then for boosting voltage—enabled the actuation of the low-power devices with the intermittent output energy from the MFCs.

CONCLUSIONS

An energy harvesting circuit based on an individual MFC, a charge pump, and a supercapacitor has been implemented in this paper. The proposed PMS uses a supercapacitor to store energy to accumulate electricity. When the voltage at the supercapacitor's node reaches 2.2 V, the charge pump produces a control signal allowing the switch to be on, and the supercapacitor starts to discharge. The discharge can be increased to 5 V with the help of a DC-DC boost circuit. The transmitter of the wireless sensing system can operate while capturing temperature and humidity data. The computer at the receiving end can store and display the data using serial ports.

According to the testing performed in this study, the energy efficiency of designed PMS is about 15.7%–16.7%. The time interval between two samples is approximately 23 h. With the 3 F supercapacitor as the energy harvester, the sensor and transmitter system could work normally for 5 s, and 30 samples could be detected and transmitted during each sampling period. The time interval can be shortened if fewer samples are required from each sampling period. In theory, if only one sample was needed per period, the time interval could be reduced to approximately 50 min by using the 3 F supercapacitor as the energy harvester.

MFCs are a promising technology that may contribute toward sustainable energy production and energy savings in wastewater treatment. However, the low power and voltage produced by an individual MFC cannot satisfy the demands of practical devices. Although multiple MFCs can be directly linked in series, the overall voltage of these cascaded MFCs cannot be sustainably increased because of voltage reversal, which decreases the stack voltage. This study certifies that it is feasible to use an individual MFC along with the designed PMS, as the power supply for a sensor and telemetry system that can collect one sample per hour. This sample period is useful for environmental monitoring where time based changes to temperature and humidly are small. In addition, with minor improvements to efficiency and cell capacity, this sample rate could easily approach 4–5 samples per hour.

ACKNOWLEDGMENTS

The authors would like to show their gratitude to Huaqiao University and Chinese Academy of Sciences; This research was supported by "the National Natural Science Foundation of China," Grant No. 61204122.

AUTHOR CONTRIBUTIONS

All authors collaborated to carry out the work presented here. Zhenyu Wang and Weihong Lu defined the research topic. Qi Zheng, Lei Xiong and Bing Mo carried out the experiments, set up the system and performed the experiments. Qi Zheng and Lei Xiong wrote the paper. Bing Mo and Suki Kim reviewed and edited the manuscript. All authors read and approved the manuscript.

REFERENCES

1. Logan, B.E.; Murano, C.; Scott, K. Electricity generation from cysteine in a microbial fuel cell. *Water Res.* 2005, *39*, 942–952.

2. Aelterman, P.; Rabaey, K.; Pham, H.T.; Boon, N.; Verstrrete, W. Continuous Electricity Generation at High Voltages and Currents Using Stacked Microbial Fuel Cells. *Environ. Sci. Technol.* 2006, *40*, 3388–3394.

3. Wu, P.K.; Biffinger, J.C.; Fitzgerald, L.A.; Ringeisen, B.R. A Low Power DC/DC Booster Circuit Designed for Microbial Fuel Cells. *Process Biochemis.* 2012, *47*, 1620–1626.

4. Nicolas, D.; François, B. Self-Starting DC-DC Boost Converter for Low-Power and Low-Voltage Microbial Electric Generators. In Proceedings of the 2011 IEEE Energy Conversion Congress and Exposition (ECCE), Phoenix, AZ, USA, 17–22 September 2011; pp. 889–896.

5. Gong, Y.; Sage, E. Benthic Microbial Fuel Cell as Direct Power Source for an Acoustic Modem and Seawater Oxygen/Temperature Sensor System. *Environ. Sci. Technol.* 2011, *45*, 5047–5053.

6. Kim, Y.; Marta, C.H. Capturing power at higher voltages from arrays of microbial fuel cells without voltage reversal.*Energy Environ. Sci.* 2011, *4*, 4662–4667.

7. Oh, S.-E.; Logan, B.E. Voltage reversal during microbial fuel cell stack operation. *Sci. Direct.* 2007, *167*, 11–17.

8. Ledezma, P.; Greenman, J.; Ieropoulos, I. MFC-cascade stacks maximize COD reduction and avoid voltage reversal under adverse conditions. *Bioresour. Technol.* 2013, *134*, 158–165.

9. Gurung, A.; Kim, J.; Jung, S.; Jeon, B.H.; Yang, J.E.; Oh, S.-E. Effects of Substrate Concentrations on Performance of Serially Connected Microbial Fuel Cells (MFCs) Operated in a Continuous Mode. *Biotechnol. Lett.* 2012, *34*, 1833–1839.

10. Gurung, A.; Oh, S.-E. The Performance of Serially and Parallelly Connected Microbial Fuel Cells. *Energy Sour.* 2012,*34*, 1591–1598.

11. Cheng, S.; Liu, H.; Logan, B.E. Power densities using different cathode catalysts (Pt and CoTMPP) and polymer binders (Nafion and PTFE) in single microbial fuel cells. *Environ. Sci. Technol.* 2006, *40*, 364–369.

12. Cheng, S.; Liu, H.; Logan, B.E. Increased performance of single-chamber microbial fuel cells using an improved cathode structure. *Electrochem. Commun.* 2006, *8*, 489–494.

13. You, S.J.; Zhao, Q.L.; Zhang, J.N.; Jiang, J.Q.; Wan, C.L.; Du, M.A.; Zhao, S.Q. A graphite granule membraneless tubular air-cathode microbial fuel cell for power generation under continuously operational conditions. *J. Power Sour.*2007, *173*, 172–177.

14. Zuo, Y.; Cheng, S.; Call, D.; Logan, B.E. Tubular membrane cathodes for scalable power generation in microbial fuel cells. *Environ. Sci. Technol.* 2007, *41*, 3347–3353.

15. Logan, B.; Cheng, S.; Watson, V.; Estadt, G. Graphite fiber brush anodes for increased power production in air-cathode microbial fuel cells. *Environ. Sci. Technol.* 2007, *41*, 3341–3346.

16. Liang, P.; Huang, X.; Fan, M.Z.; Cao, X.X.; Wang, C. Composition and distribution of internal resistance in three types of microbial fuel cells. *Appl. Microbiol. Biotechnol.* 2007, *77*, 551–558.

17. Park, J.D.; Ren, Z. High efficiency energy harvesting from microbial fuel cells using a synchronous boost converter. *J. Power Sour.* 2010, *208*, 322–327.

18. Park, J.D.; Ren, Z. Hysteresis controller based maximum power point tracking energy harvesting system for microbial fuel cells. *J. Power Sour.* 2012, *205*, 151–156.

19. Degrenne, N.; Buret, F.; Allard, B.; Bevilacqua, P. Electrical energy generation from a large number of microbial fuel cells operating at maximum power point electrical load. *J. Power Sour.* 2012, *205*, 188–193.

20. S-882Z22, Datasheet. Available online: http://wenku.baidu.com/link?url=-vDPf-fkHfspDolEoMRPM-LoXYdZviuSxWj3m0shK793hUytt_TGOWT-miRn2ATTlqGdKTLWxG8iwnq8rZVYBw8EB-1m7qUq5Qwfln0zTCbT0a (accessed on 5 May 2015).

21. L6920DB, Datasheet. Available online: http://www.st.com/st-web-ui/static/active/en/resource/technical/document/datasheet/CD00055960.pdf (accessed on 5 May 2015).

22. DHT11, Datasheet. Available online: http://wenku.baidu.com/link?url=6QLWNSEuDHvWgx9kExhex2W-

MOWFLzerjFY9AXphx7e0uQEX-4MZP9Z2JhBW5Dudx3SA99zYpFnidC-98Suyxx8Z3x6KNQX42OeivNbGjJDqK (accessed on 5 May 2015).

23. nRF24L01, Datasheet. Available online: http://www.nordicsemi.com/eng/Products/2.4GHz-RF/nRF24L01/ (accessed on 5 May 2015).

24. CH340T, Datasheet. Available online: http://www.winchiphead.com/download/CH340/CH340DS1.PDF (accessed on 5 May 2015).

25. Zhang, D.; Yang, F.; Shimotori, T.; Wang, K.C.; Huang, Y. Performance evaluation of power management systems in microbial fuel cell-based energy harvesting applications for driving small electronic devices. *J. Power Sour.* 2012, *217*, 65–71.

CHAPTER 8 HIGH RATE COPPER AND ENERGY RECOVERY IN MICROBIAL FUEL CELLS

Pau Rodenas Motos, Annemiek ter Heijne, Renata van der Weijden, Michel Saakes, Cees J. N. Buisman and Tom H. J. A. Sleutels

Keywords: Microbial fuel cells, bioelectrochemical systems, metal recovery, copper

INTRODUCTION

Heavy metals pose a serious problem when they are released into the environment due to their toxicity for humans and its negative effect on biodiversity (Norgate and Rankin, 2002). Therefore, removal, and recovery of heavy metals, remediation of polluted sites, and decontamination of waste streams is needed in heavy metal and mining industries. Today, the conventional method to recover metals, such as copper, is by electrowinning, which uses electric power to electrochemically reduce dissolved metals to their metallic form. As metals are becoming more scarce and expensive (Watkins and McAleer, 2006; Geman and Smith, 2013), the revenues compensate for the electricity costs. However, considerable amounts of electrical energy are still required in electrolysis cells (2.23 kW h kg^{-1} of Copper; Vegliò et al., 2003). In comparison, bioelectrochemical systems (BESs) can provide the energy by the bioelectrochemical degradation of organic compounds present in waste waters, for instance, acetate (electron donor), at a so called bioanode. By coupling this bioanode to a metal reducing cathode, waste water treatment can be coupled to the recovery of heavy metals in a clean and sustainable way. When additional energy is harvested from these two reactions, the system is known as a microbial fuel cell (MFC). It is also possible to add extra electrical energy to drive and accelerate a reduction reaction at the cathode, in that case, the system is known as a microbial electrolysis cell (MEC) (Logan et al., 2008). The anode and the cathode compartment are generally separated by a membrane to simultaneously avoid mixing of the electrolytes and in theory, these membranes allow ions to pass selectively, so charge neutrality is maintained.

ter Heijne et al. (2010) showed the proof of principle of this concept where they combined bioelectrochemical oxidation of acetate with the reduction of copper. In their system, electricity was produced while copper was removed and plated onto a solid electrode. A maximum current density of 3.2 A m^{-2} was achieved in a setup with a flat plate graphite electrodes and a bipolar membrane (BPM). Since then, other studies have reported MFCs in which copper was reduced at the cathode, reaching current densities ranging from 0.9 to 7 A m^{-2} with, for example, an increased surface area of the anode compared to the cathode (Modin et al., 2012; Zhang et al., 2012).

To make BESs practical and economically suitable to replace existing technologies, higher current densities are required. To reach high current densities, it is essential to reduce the voltage losses by reducing the internal resistance. Sleutels et al. (2012) analyzed the effect of the internal resistance on the practical applicability of BES. A resistance below 40 mΩ m^2 was considered acceptable for BES applications to achieve sufficiently high current and power densities to make BESs economic. Besides economic studies based on internal resistance, other attempts have been done to estimate feasibility of BESs using life-cycle assessment (LCA) analysis (Foley et al., 2010; Pant et al., 2011;Sleutels et al., 2012).

On the other hand, copper recovery is a highly valuable application for BES research. The recovery of valuable

Citation: Rodenas Motos P, ter Heijne A, van der Weijden R, Saakes M, Buisman CJN and Sleutels THJA (2015) High rate copper and energy recovery in microbial fuel cells. Front. Microbiol. 6:527. doi: 10.3389/fmicb.2015.00527.

products allows for higher internal resistances compared to the recovery of energy alone (Sleutels et al., 2012). Copper recovery as a cathode reaction can potentially lead to high current densities due to its low overpotential. It is therefore, a suitable cathode reaction to demonstrate that MFCs can be operated at high current densities.

The main objective of this study was to increase the current and power density of a copper reducing MFC. Our approach was to reduce the internal resistance compared to previous studies by four changes: (i) decreasing the distance between electrodes, (ii) use an anion exchange membrane (AEM), (iii) use a copper plate as cathode, and (iv) use carbon felt as anode material.

First, the distance between anode and cathode was reduced from 3 to 0.5 cm (ter Heijne et al., 2010). Second, an AEM is known to have lower internal resistance than other membranes when applied in MEC (Rozendal et al., 2007; Sleutels et al., 2009a). As the BPM resulted in high energy losses; the internal resistance was expected to be reduced by using an AEM. Third, cathode material was changed from graphite paper to copper, which is a material conventionally used in copper electroplating. Fourth, anode material was replaced from graphite paper to carbon felt to achieve high specific surface area for electroactive biofilm growth. To make this anode material available for microbial activity, the solution was forced via a perpendicular flow through this felt as shown by Sleutels et al. (2009b, 2011).

We operated this new cell design at different current densities and analyzed its performance in terms of power production and voltage losses.

MATERIALS AND METHODS

Research Set Up

The cell used in these experiments was similar to the one used by Kuntke et al. (2014). This cell was built using 10 cm × 10 cm metallic copper plates as cathode and 10 cm × 10 cm carbon felt of 1.5 mm thickness as anode. A Ralex®AEM (MEGA a.s., Stráž pod Ralskem, Czech Republic) was placed in between the anode and cathode compartment. To force the electrolyte to flow perpendicular through the electrodes, 1.2 mm spacer (64% open; PETEX 07-4000/64, Sefar BV, Goor, Netherlands) was placed between the membrane and the electrodes in both compartments (Figure1). The anolyte was forced to flow toward the membrane through the electrode (Sleutels et al., 2011). The anode, cathode, spacer material, and membrane were kept in place by bolting them in between two PMMA endplates.

The anode compartment was controlled at 30°C via a water flow through the outer wall in the anode recirculation bottle. Only the cathode temperature was measured and it oscillated between 20 and 23°C. Nitrogen was flushed continuously through anolyte and catholyte to keep anaerobic conditions in anode and cathode compartments. Anode and cathode potentials were measured against Ag/AgCl 3 M reference electrode (ProSense, Oosterhout, Netherlands; 201 mV vs. Standard Hydrogen Electrode). Voltage losses across the membrane were measured as the voltage difference between the two reference electrodes placed in catholyte and anolyte. Cell voltage and potentials of cathode, anode, and membrane were continuously recorded together with pH and temperature in both compartments using a data logger (Endress+Hauser Messtechnik GmbH+Co., Rhein, Germany).

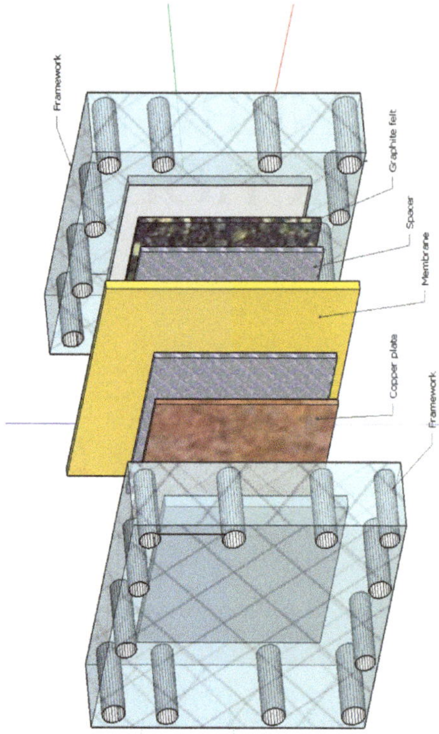

Figure 1: Cell design 3D view with copper electrode (brown), membrane (yellow), spacer (clear blue patterned), graphite felt (black with white dots), and framework (transparent).

Start-Up and Operation

At the beginning of the experiment a 2 g L^{-1} Cu^{2+} solution was prepared dissolving $CuSO_4$ $5{\cdot}H_2O$ as catholyte. The anode influent (2mL min^{-1}) consisted of synthetic waste water solution containing 20 mM acetate, phosphate buffer (10 mM, pH 7), nutrients, and vitamins as described in ter Heijne et al. (2008). The anode was fed with an excess of acetate to make sure the current production was not limited by substrate availability. Acetate was measured over time to ensure an excess of electron donor in the anode to produce current. Any competing processes at the anode for example methanogens were taken into consideration. Anolyte and catholyte had a constant recirculation rate of 200 ml min^{-1}. The total catholyte volume was 10 L. Cathode and anode pH were controlled at pH 3 and pH 7, respectively, with two dose pumps (STEPDOS 08, KNF, Germany) with 0.1 M H_2SO_4 and 0.1 M KOH. No supporting electrolyte was used. The conductivity of the solution was not measured during the experiment, only at the beginning of the experiment, and the values are 9.2 mS/cm for the catholyte and 6.2 mS/cm for the anolyte. The presence of an excess of acetate and sulfate (transported from the cathode) in the anolyte resulted in sulfate reduction (due to sulfate reducing bacteria; Hamilton, 1985) and consequently, in a consumption of protons. Hence, acid addition was required in the anode compartment. It should be noted that addition of chemicals to control the pH is not a sustainable solution for practical applications but it was used here to be able to study the maximum performance of the system for copper removal. The cell started up at a constant resistance of 1 kΩ. The resistor was switched to a lower value (stepwise using 1 kΩ, 500 Ω, 250 Ω, and 100 Ω) after the anode potential had stabilized at a value of -450 mV vs Ag/AgCl. During this growing period of 2 weeks stable conditions were ensured using buffer solution as catholyte for oxygen reduction and measuring stable voltages over the entire cell. During the measurements that are reported in the results section below, the resistances were lowered sequentially from 1000 Ω to 0.5 Ω and catholyte was changed to copper sulfate solution. For each resistance steady state values over a period of 24 h were used to calculate the polarization curve.

Analyses

Copper concentration was measured using inductively coupled plasma optical emission spectrometry (ICP-OES) Perkin Elmer Optima 5300 (Perkin Elmer, Groningen, Netherlands). The consumed acetate during the experiment was tracked every 24 h measuring TOC (Total Organic Carbon) and IC (Inorganic Carbon) that was correlated with the concentration of acetate and bicarbonate using Shimadzu TOC-L CPH combustion TOC analyzer (Shimadzu Benelux, 's-Hertogenbosch, Netherlands). The gas composition in the headspace of the cells was also examined with a gas chromatographer [Varian Inc. (Part A) – CP-4900 Micro-GC].

The deposited copper on the electrode was scratched with a knife from the surface, weighed in a test tube and dissolved in 5 ml of 33% HNO_3. Then, 5 ml of Mili-Q water was added to dilute the solution. Then the solution obtained was diluted 1000 times again and measured by ICP–OES Perkin Elmer Optima 5300 to determine the amount of copper plated on the electrode.

Calculation

The theoretical cell voltage (E_{emf} in V) can be calculated from the Gibbs free energy change of the overall reaction occurring at the cathode and anode, while the practical value can be directly measured at open cell conditions

$$E_{emf} = \frac{-\Delta G_r}{nF} \tag{1}$$

Where, ΔG_r is the Gibbs free energy of the overall reaction (kJ mol^{-1}), n is the number of electrons transferred in the reaction, F is Faraday constant (96485 C mol^{-1}).

The contribution of the partial voltage losses to the total voltage loss inside the cell can be calculated according to

$$E_{CELL} = E_{emf} - \eta_{Cath} - \eta_{Anod} - \eta_{mem} \tag{2}$$

Where E_{CELL} is the voltage measured during cell operation (V), η_{Cath} is the measured cathode overpotential (theoretical cathode potential-measured cathode potential; V), η_{Anod} is the measured anode overpotential (theoretical anode potential-measured anode potential; V), and η_{mem} is the voltage lost across the membrane (cathode ref potential-anode ref potential; V).

RESULTS AND DISCUSSION

Current and Power Output

Figure 2 shows the current and power density generated over the entire experimental period. The arrows indicate the moment at which the value of the external resistance was lowered. The current density increased from 0.5 A m^{-2} at an external resistance of 100 Ω up to a stable value of 19 A m^{-2}at an external resistance of 1 Ω. The peak current density was 23 A m^{-2} and at this maximum current density, the peak power density was 5.5 W m^{-2}. When the resistor was further decreased to 0.5 Ω, the current density did not increase further, and the power output dropped from 4.0 to 2.0 W m^{-2}. Figure 3 shows the polarization curve of the MFC, summarizing the whole experimental period, using the average values for cell voltage, current, and power density at each external resistor. The open circuit voltage (OCV) measured the first day was 485 mV, which is lower than the theoretical value of 575 mV.

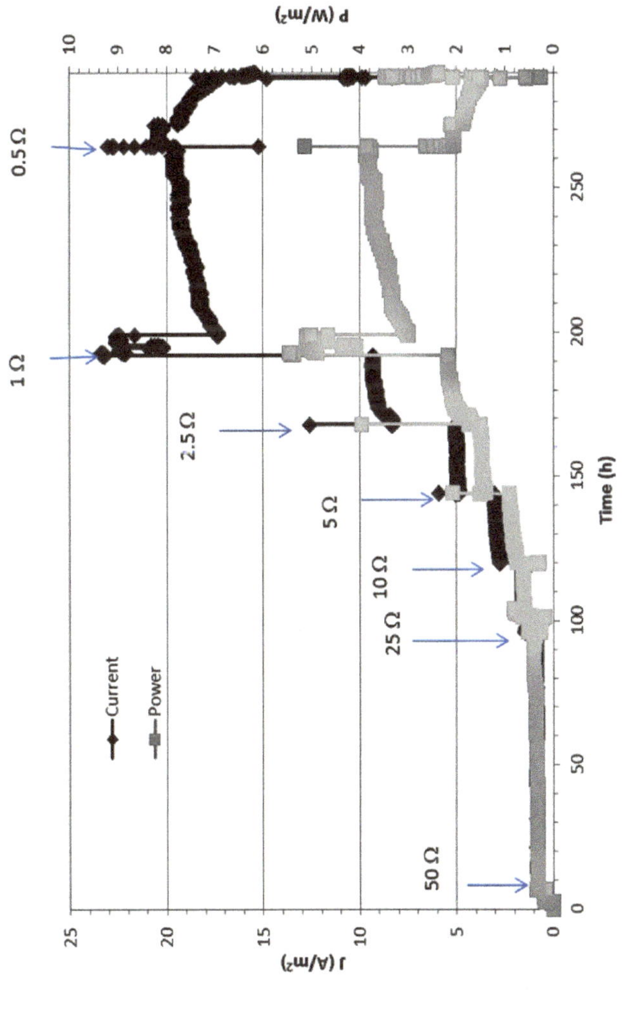

Figure 2: Evolution of current density (black) and the power density (grey) shown vs. time during the entire range of the experiment. The arrows indicate when resistances were changed.

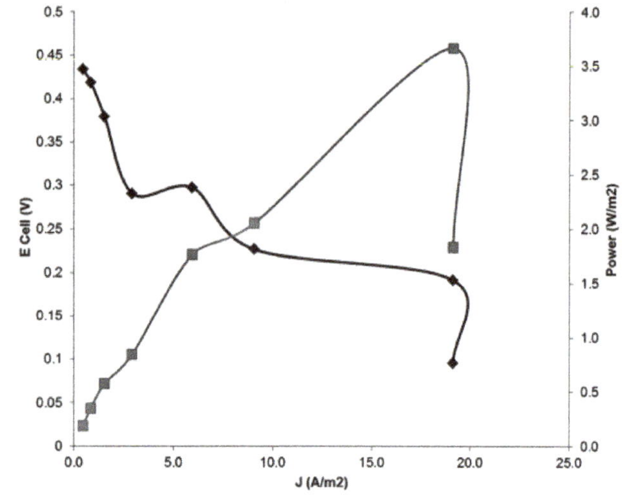

Figure 3: Polarization curve of the copper reducing microbial fuel cell (MFC).The potential is represented vs. current density (black) and the power is shown versus current density (grey).

During the whole period only 60% of the total amount of acetate was consumed by the biofilm, from this amount only the 26% was used to produce electricity. The detection of methane and sulfide in the head space of the recirculation bottle let us conclude that acetate was used for methane production and sulfate reduction by methanogens and sulfate reducers competing for the substrate with the electrogenic bacteria. The oxidation of acetate coupled to the reduction of sulfate by sulfate reducing bacteria was described by Muyzer and Stams (2008) via:

$$Acetate^{-} + SO_4^{2-} \rightarrow 2HCO_3^{-} + HS^{-} \qquad (3)$$

Analysis of the Performance

To analyze the reason for the drop in power production in more detail, we have analyzed the separate contributions of anode, cathode, and voltage losses across the membrane throughout the experimental period in line with a previous study from ter Heijne et al. (2006). Figure 4 shows the relative contribution of the voltage losses for these three main components of the system: at every external resistance. The voltage loss over the membrane was measured using the reference electrode inserted in the anode and cathode compartment. Then, the ionic resistance of the electrolyte (anode and cathode) is measured as part of the membrane voltage loss. Note that Figure4 shows the contribution of the three components, the three of them summing up to 100%. This was done to illustrate the differences with time more clearly. Obviously, at a high external resistance the produced current was lowest while at a decreasing external resistor the produced current increased (Figure 2). So overall, the internal resistance of the system decreased with time.

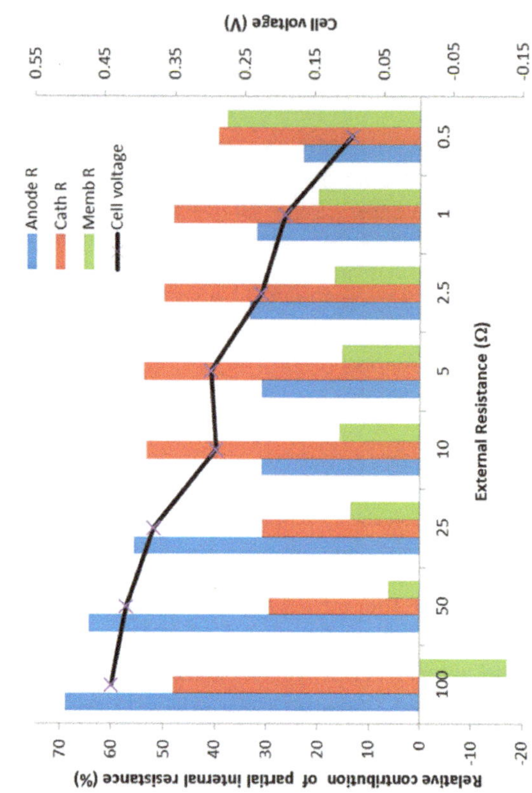

Figure 4: Contribution of the cell components (anode, cathode, and membrane) to the total voltage loss in the system at different external resistors. Also the cell voltage is shown for each external resistor.

With an increase in current density, the contribution of anode overpotential to the total voltage losses decreased considerably from 68 to 23%. We observed an increase in anode voltage from -450 to -400 mV vs Ag/AgCl for the first steps with high external resistances. However, when current reached its maximum values, the anode potential raised only to -350 mV vs Ag/AgCl. Apparently the biofilm was not affected by the high protons production rate that is associated with the current production. Instead an increase in the voltage loss over the membrane was observed. At increasing current densities voltage losses across the membrane increased from -50 mV on first day to 180 mV on ninth day. This value of -50 mV was result of the difference in chemical potential between anolyte and catholyte due to the difference in composition of these electrolytes. Sulfate was transported through the membrane from catholyte to anolyte. This chemical potential contributes positively to the cell potential. At the highest current reached, the dominant losses occurred at the cathode, approximately around 50% of total losses contribution between 10 Ω external resistance and 1 Ω external resistance. The increase in current density at lower external resistance was also reflected in an increase of the cathode overpotential. Cathode potential went down from 50 mV during first day to -60 mV vs Ag/AgCl at the end of the experiment due to the decrease in Copper concentration.

An increase in the membrane voltage losses is likely at higher current densities since more ions need to be transported in the same time period. Additionally, the increase in membrane voltage losses can also be related to scaling. At the end of the experiment, when the cell was disassembled, precipitates were found in the catholyte and on the membrane surface. Figures 5A,B show the scanning electron microscope (SEM) image of scaling and biofouling in the cathode and in the anode side respectively on top of the membrane. Analysis of these structures with electron diffraction X-ray (EDX) shows the presence of copper, sulfur, and phosphor, which lead us to believe that the precipitates are copper sulfates, copper sulfides, and copper phosphates. Scaling is more likely to occur when divalent ions (like Copper and sulfate in this system) are present. Scaling is a serious issue that may limit the transport of ionic species from one compartment to the other, and thereby limiting performance of these systems. Prevention of this scaling is an important aspect of further study.

Figure 5: (A) SEM image of crystalline scaling from cathode side on anion exchange membrane (AEM), (B) SEM image of biofilm on anode side of AEM. (C) SEM image of smooth copper deposited on flat copper surface (D) SEM image of the dendrites formed on cathode surface.

Performance Comparison to Previous Studies

To improve MFC performance the cell design was improved compared to previous study (ter Heijne et al., 2010) by modifying the electrode materials, the membrane, and the distance between electrodes and mass transport within the system. Table 1 compares the performance of the MFC described in this study to other studies that used copper reduction as cathode reaction. Main differences between this and other studies are: cell configuration, type of membrane, and different electrode materials. For example, all studies list in Table 1 except Modin et al. (2012) and present study used carbon electrodes as cathode instead metal electrodes. The maximum current density in short circuit conditions produced by Modin et al. (2012) cell, was 7 A m^{-2} for 128 cm^2 of anode felt surface using a small cathode surface of 1 cm^{2}, and using also an AEM. Even though, ter Heijne et al. (2010) achieved 3.2 A m^{-2} using graphite paper electrodes with a surface of 20 cm^2 and a BPM.Cheng et al. (2013) achieved 1.2 A m^{-2} using a carbon fiber brush anode and a carbon rod cathode. On the other hand, this study achieved a current density of 23 A m^{-2} for 100 cm^2 of anode graphite felt surface and 100 cm^2 copper cathode and using AEM.

Table 1: Microbial fuel cell (MFC) comparison between different authors and cathode configurations

Anode material and surface area	Cathode material and surface area	Membrane used	Voc (V)	Jpeak (A m^{-2})	Pmax (W m^{-2})	Reference
Graphite felt of 100 cm^2	Copper plate of 100 cm^2	AEM	0.485	23.0	5.50	This study
Graphite paper of 20 cm^2	Graphite paper of 20 cm^2	BPM	0.600	3.20	0.80	ter Heijne et al. (2010)
Carbon fiber brush	Carbon rod of 4.7 cm^2	AEM	0.590	1.20	0.28	Cheng et al. (2013)
Graphite plate/graphite felt of 147 cm^2	Graphite plate of 27 cm^2	PEM	0.480	2.00*	0.24*	Tao et al. (2011)
2 × Graphite felt of 29.6 cm^2	Titanium Wire of 1.02 cm^2	AEM	–	7.00	–	Modin et al. (2012)
4 × Graphite felt of 22.5 cm^2	Graphite plate of 22.5 cm^2	AEM	0.550	0.90	0.20	Zhang et al. (2012)

*The measurements of Tao et al. (2011) are expressed in A m^{-3} and W m^{-3} for current and power densities.

Although previous studies reported considerable current densities, the power production was not always equivalent due to, either short circuit conditions or high internal resistance. However, in this study the produced power density was 5.5 W m^{-2}, which is considerably higher than the reported before with power densities ranging between 0.2 and 0.8 W m^{-2}. This achievement is explained as a result of the low voltage losses of this MFC compared to other MFC configurations.

Copper Removal and Analysis of the Deposited Copper

Samples from catholyte were taken and analyzed five times per week. Interestingly, during the first 100 h, we found out a slight increase in copper concentration from 2.0 g/L to 2.2 g/L. In this time period, the current density was still low, below 1 A m^{-2}. Only when current density exceeds 1.0 A m^{-2}, copper concentration started to drop and showed a steady decrease to 0.2 g/L after 24 h. A reasonable explanation for the increase in copper concentration may be that the copper electrode itself was oxidized to copper (II) under acidic conditions (pH 3). In order to check that copper oxidation occurs at acidic pH, a parallel experiment at open circuit in the same acidic conditions and copper concentration was performed. Here, a loss in weight of the electrode of 1.72 g of copper was observed after 24 h. This confirmed that the copper electrode can be oxidized under acidic conditions and explains the increase in copper concentration at low current densities (below 1 A m^{-2}). Finally, it is important to mention that no copper was detected in the anode during any of the experiments.

After 1 month of operation, the cell was disassembled and the copper layer on the cathode was analyzed for purity. Figures 5C,D show the SEM images of the copper deposited on top of the copper cathode. Copper exhibited dendritic formations, which are likely a result of the high current densities. These dendrites crossed the spacer and touched the membrane surface. This is a point of attention because these structures might damage the membrane. Analysis of the Copper electrode by nitric acid leaching showed that the composition of the deposited copper on the cathode has a copper purity of 99.4 ± 1.0% with 0.3% in zinc. XRD analysis confirmed the copper purity of the copper deposited onto the electrode.

Implications

High current density and power production were achieved in this improved cell design compared to the proof of principle by ter Heijne et al. (2010). The current was enhanced from 3.2 to 23 A m^{-2}while the power density increased from 0.8 to 5.5 W m^{-2}.

State of the art technologies like solvent extraction and electrowinning SX/EW consume 2.716 kWh Kg^{-1} of Cu (Alvarado, 2002). On the contrary, this study produced a constant power of 3.7 W m^{-2}equivalent to an energy production of 0.081 kWh Kg^{-1} of Cu. It should be mentioned that, the energy required to operate the system was not taken into account here. Also, 2.056 kg of acetate is required per kilogram of copper, which can be considered as a waste product.

The current density can be improved even further by supplying an external voltage to the system. Although the produced current and removed amount of copper will be higher, this would go at the expense of energy input.

Analysis of the voltage losses showed that anode was the main contribution to the voltage losses in the system at low current densities, but when currents were higher the cathode and the membrane started to limit the system. Consequently, the transport of ionic species through the system and membrane should be studied in more detail, as well as the scaling on the membrane surface related to the ion transport are an important aspect that needs attention in the future. However, to improve the bioanode performance further, many strategies have been considered for future research like Inhibition of methanogens and sulfate reducers to improve coulombic efficiency. Recently also some studies had improved the bioanode by improving electrode material by organic and inorganic coatings on top, that improves the amount of bacteria that can be attached onto the electrode (Zhang et al., 2011, 2015).

Several challenges remain to be solved before this technology becomes practically applicable. The use of a copper electrode, which is normally used in electro-winning, may pose difficulties in MFCs. Those difficulties are typically the slow reaction rates and the fact that copper actually dissolves. Further research on copper deposition at low current densities with other metallic electrodes, such as stainless steel or titanium, would give more insights on this system performance.

ACKNOWLEDGMENTS

This work was in the TTIW-cooperation framework of Wetsus, European Centre of Excellence for sustainable water technology (www.wetsus.nl). Wetsus is founded by Dutch Ministry of Economic Affairs, the city of Leeuwarden, the Province of Fryslân, the European Union by European regional Development Fund and by EZ/KOMPAS program of the "Samenwerkingsvarband Noord Nederland." This work is also part of BioelectoMET project. Title: bioelectrochemical systems for metal recovery. Funded under the Seventh Framework Program (FP7) Research area: ENV.2011.3.1.9-1 (Eco-innovation). We thank Vinnie de Wilde for the XRD analysis.

REFERENCES

1. Alvarado, S. (2002). Long term energy-related environmental issues of copper production. *Energy* 27, 183–196. doi: 10.1016/S0360-5442(01)00067-6
2. Cheng, S.-A., Wang, B.-S., and Wang, Y.-H. (2013). Increasing efficiencies of microbial fuel cells for collaborative treatment of copper and organic wastewater by designing reactor and selecting operating parameters. *Bioresour. Technol.* 147, 332–337. doi: 10.1016/j.biortech.2013.08.040
3. Foley, J. M., Rozendal, R. A., Hertle, C. K., Lant, P. A., and Rabaey, K. (2010). *Life Cycle Assessment of High-Rate Anaerobic Treatment, Microbial Fuel Cells, and Microbial Electrolysis Cells.* Available at:http://pubs.acs.org.ezproxy.library.wur.nl/doi/full/10.1021/es100125h [accessed March 26, 2015].
4. Geman, H., and Smith, W. O. (2013). Theory of storage, inventory and volatility in the LME base metals. *Resour. Policy* 38, 18–28. doi: 10.1016/j.resourpol.2012.06.014
5. Hamilton, W. A. (1985). Sulphate-reducing bacteria and anaerobic corrosion. *Annu. Rev. Microbiol.* 39, 195–217. doi: 10.1146/annurev.mi.39.100185.001211
6. Kuntke, P., Sleutels, T., Saakes, M., and Buisman, C. J. N. (2014). Hydrogen production and ammonium recovery from urine by a Microbial Electrolysis Cell. *Int. J. Hydrogen Energy* 39, 4771–4778. doi: 10.1016/j.ijhydene.2013.10.089
7. Logan, B. E., Call, D., Cheng, S., Hamelers, H. V. M., Sleutels, T. H. J. A., Jeremiasse, A. W., et al. (2008). Microbial electrolysis cells for high yield hydrogen gas production from organic matter. *Environ. Sci. Technol.* 42, 8630–8640. doi: 10.1021/es801553z
8. Modin, O., Wang, X., Wu, X., Rauch, S., and Fedje, K. K. (2012).Bioelectrochemical recovery of Cu, Pb, Cd, and Zn from dilute solutions. *J. Hazard. Mater.* 235–236, 291–297. doi: 10.1016/j.jhazmat.2012.07.058
9. Muyzer, G., and Stams, A. J. M. (2008). The ecology

and biotechnology of sulphate-reducing bacteria. *Nat. Rev. Microbiol.* 6, 441–454. doi: 10.1038/nrmicro1892

10. Norgate, T. E., and Rankin, W. J. (2002). *The Role of Metals in Sustainable Development, 49–55.* Available at:http://www.scopus.com/inward/record.url?eid=2-s2.0-2342565850&partnerID=40&md5=87a444cdfc8c05a408550a2a5b33804b
11. Pant, D., Singh, A., Van Bogaert, G., Gallego, Y. A., Diels, L., and Vanbroekhoven, K. (2011). An introduction to the life cycle assessment (LCA) of bioelectrochemical systems (BES) for sustainable energy and product generation: relevance and key aspects. *Renew. Sustain. Energy Rev.* 15, 1305–1313. doi: 10.1016/j.rser.2010.10.005
12. Rozendal, R. A., Hamelers, H. V. M., Molenkamp, R. J., and Buisman, C. J. N. (2007). Performance of single chamber biocatalyzed electrolysis with different types of ion exchange membranes. *Water Res.* 41, 1984–1994. doi: 10.1016/j.watres.2007.01.019
13. Sleutels, T. H., Hamelers, H. V., and Buisman, C. J. (2011). Effect of mass and charge transport speed and direction in porous anodes on microbial electrolysis cell performance. *Bioresour. Technol.* 102, 399–403. doi: 10.1016/j.biortech.2010.06.018
14. Sleutels, T. H. J. A., Hamelers, H. V. M., Rozendal, R. A., and Buisman, C. J. N. (2009a). Ion transport resistance in Microbial Electrolysis Cells with anion and cation exchange membranes. *Int. J. Hydrogen Energy* 34, 3612–3620. doi: 10.1016/j.ijhydene.2009.03.004
15. Sleutels, T. H. J., Lodder, R., Hamelers, H. V. M., and Buisman, C. J. N. (2009b). Improved performance of porous bio-anodes in microbial electrolysis cells by enhancing mass and charge transport. *Int. J. Hydrogen Energy* 34, 9655–9661. doi: 10.1016/j.ijhydene.2009.09.089
16. Sleutels, T. H. J. A., Ter Heijne, A., Buisman, C. J. N., and Hamelers, H. V. M. (2012). Bioelectrochemical systems: an outlook for practical applications. *ChemSusChem* 5, 1012–1019. doi: 10.1002/cssc.201100732
17. Tao, H. C., Liang, M., Li, W., Zhang, L. J., Ni, J. R., and Wu, W. M. (2011). Removal of copper from aqueous solution by electrodeposition in cathode chamber of microbial fuel cell. *J. Hazard. Mater.* 189, 186–192. doi: 10.1016/j.jhazmat.2011.02.018
18. ter Heijne, A., Hamelers, H. V. M., de Wilde, V., Rozendal, R. A., and Buisman, C. J. N. (2006). A bipolar membrane combined with ferric iron reduction as an efficient cathode system in microbial fuel cells. *Environ. Sci. Technol.* 40, 5200–5205. doi: 10.1021/es0608545
19. ter Heijne, A., Hamelers, H. V. M., Saakes, M., and Buisman, C. J. N. (2008). Performance of non-porous graphite and titanium-based anodes in microbial fuel cells. *Electrochim. Acta* 53, 5697–5703. doi: 10.1016/j.electacta.2008.03.032
20. ter Heijne, A., Liu, F., Weijden, R. V., Weijma, J., Buisman, C. J., and Hamelers, H. V. (2010). Copper recovery combined with electricity production in a microbial fuel cell. *Environ. Sci. Technol.* 44, 4376–4381. doi: 10.1021/es100526g
21. Vegliò, F., Quaresima, R., Fornari, P., and Ubaldini, S. (2003). Recovery of valuable metals from electronic and galvanic industrial wastes by leaching and electrowinning. *Waste Manag.* 23, 245–252. doi: 10.1016/S0956-053X(02)00157-5
22. Watkins, C., and McAleer, M. J. (2006). Pricing of non-ferrous metals futures on the London Metal Exchange. *Appl. Finan. Econ.* 16, 853–880. doi: 10.1080/09603100600756514
23. Zhang, C., Liang, P., Jiang, Y., and Huang, X. (2015). Enhanced power generation of microbial fuel cell using manganese dioxide-coated anode in flow-through mode. *J. Power Sources* 273, 580–583. doi: 10.1016/j.jpowsour.2014.09.129
24. Zhang, L. J., Tao, H. C., Wei, X. Y., Lei, T., Li, J. B., Wang, A. J., et al. (2012). Bioelectrochemical recovery of ammonia-copper(II) complexes from wastewater using a dual chamber microbial fuel cell. *Chemosphere* 89, 1177–1182. doi: 10.1016/j.chemosphere.2012.08.011
25. Zhang, Y., Hu, Y., Li, S., Sun, J., and Hou, B. (2011). Manganese dioxide-coated carbon nanotubes as an improved cathodic catalyst for oxygen reduction in a microbial fuel cell. *J. Power Sources* 196, 9284–9289. doi: 10.1016/j.jpowsour.2011.07.069

CHAPTER 9 APPLICATION OF CARBON-MICROSPHERE-MODIFIED ELECTRODES FOR ELECTROCHEMISTRY OF HEMOGLOBIN AND ELECTROCATALYTIC SENSING OF TRICHLOROACETIC ACID

Wen-Cheng Wang, Li-Jun Yan, Fan Shi, Xue-Liang Niu, Guo-Lei Huang, Cai-Juan Zheng and Wei Sun

Keywords: Hemoglobin, carbon microsphere, direct electrochemistry, trichloroacetic acid

INTRODUCTION

Recently, the electrochemical behavior of proteins has roused great interest, and the results can be applied to the study of electron transfer mechanisms in biosystems and the construction of third-generation electrochemical biosensors or biofuel cells [1,2]. However, the electroactive centers are often buried in the polypeptide chains of redox proteins, which make electron transfer of proteins in a conventional biosensor more difficult [3]. Therefore, various protein-based biosensors have been fabricated for the realization of electrochemical behavior with the usage of multifarious modifiers such as polymers, surfactants and nanosized materials [4,5]. The existence of modifiers can preserve the original structure of redox proteins and their enzymatic activity, which offer a suitable microenvironment for electron transfer between electrode and proteins.

Nanomaterials with multifarious morphologies and unique properties such as excellent biocompatibility and large surface area have been applied to the electrochemistry of protein [6]. Among them, carbon nanomaterials are commonly used due to their excellent electrical conductivity. Various carbon materials have been applied to the investigation of electrochemical behavior of proteins, such as mesoporous carbon [7], carbon nanotubes (CNT) [8], and graphene (GR) [9,10]. Carbon microspheres (CMS) are a kind of carbon material that has been widely investigated. Shin et al. proposed a hydrothermal technique for the fabrication of colloidal spheres from aqueous cyclodextrin solution [11]. Sun et al. prepared core-shell structures of colloidal carbon spheres with the loading of different nobel-metal nanoparticles [12]. Chen et al. synthesized monodispersed carbon spheres from glucose for the supercapacitor [13]. Jin et al. proposed a direct hydrocarbon pyrolysis technique for the large-scale synthesis of carbon spheres [14]. The synthesized carbon spheres can be used for the loading of other nanomaterials and applied in different fields such as magnetism, catalysis and biosensor [15,16,17]. However, few reports about the application of CMS in the field of protein electrochemistry have been found.

In the present study, CMS was synthesized from glucose using a hydrothermal method and was further applied to the protein electrochemistry. CMS-hemoglobin (Hb) composite was prepared and cast on the ionic liquid modified carbon paste electrode (CILE) surface. Then, CTS were cast for the immobilization of the composite on the electrode surface. The fabrication procedure of this Hb modified electrode is shown in Scheme 1. CILE has exhibited many advantages, such as wide electrochemical windows, high conductivity, good antifouling capability and inherent electrocatalytic ability, which is reported in electroanalysis and electrochemical sensors [18,19]. CTS is an abundant natural cationic biopolymer that is composed of structural repeating units of N-acetyl-glucosamine and glucosamine, which offers a biocompatible microenvironment for the immobilized redox proteins.

Citation: Wen-Cheng Wang, Li-Jun Yan, Fan Shi, Xue-Liang Niu, Guo-Lei Huang, Cai-Juan Zheng and Wei Sun, Application of Carbon-Microsphere-Modified Electrodes for Electrochemistry of Hemoglobin and Electrocatalytic Sensing of Trichloroacetic Acid, doi:10.3390/s16010006.

Therefore, CTS-modified electrodes have been commonly used in electrochemical biosensors [20]. The synthesized CMS were checked using different techniques and exhibited large surface area with porous structure. On CTS/CMS-Hb/CILE the direct electrochemistry of Hb was carried out and the electrocatalysis of trichloroacetic acid (TCA) was achieved, demonstrating the potential applications of this electrochemical sensor.

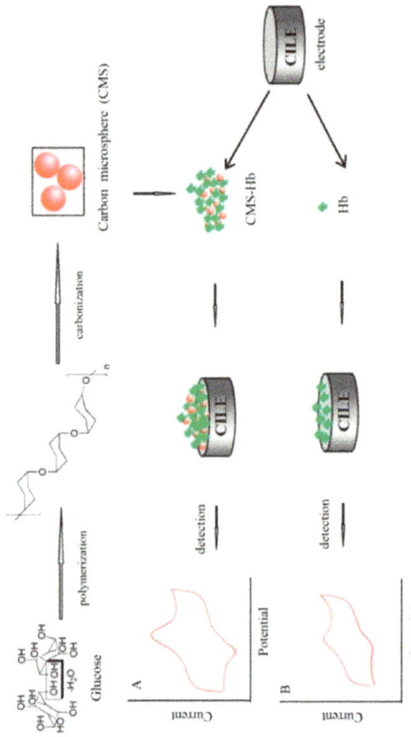

Scheme 1: The fabrication process of this Hb modified electrode.

EXPERIMENTAL

Reagents

Glucose (Beijing Chem. Reagent Factory, Beijing, China), Hb (Sinopharm. Chem. Reagent, Shanghai, China), 1-butylpyridinium hexafluorophosphate ($BPPF_6$ > 99%, Lanzhou Greenchem. CAS., Lanzhou, China), chitosan (CTS, Dalian Xindie Ltd., Dalian, China) graphite powder (30 μm, Shanghai Colloid Chem., Shanghai, China) and TCA (Tianjin Kemiou Chem. Ltd., Tianjin, China).The experiments were conducted in 0.1 mol·L^{-1} phosphate buffer saline (PBS) at room temperature (25 ± 1 °C), which was bubbled with pure N_2 for half hour before experiments to deoxygenate and keep under N_2 atmosphere during the electrochemical measurements. Ultrapure water and other chemicals (analytical reagent grade) were used in the experiments.

Apparatus

CHI 660D workstation (Shanghai CH Instrument, Shanghai, China); Nicolet 6700 FT-IR spectrophotometer (Thermo Fisher Scientific Inc., Waltham, Massachusetts, USA); TU-1901 double beam UV-Visible spectrophotometer (Beijing General Instrument Ltd. Co., Beijing, China); Renishaw InVia Raman microspectrometer using 532 nm lasers (Renishaw Plc., London, UK); JSM-7100F scanning electron microscope (Japan Electron Co., Tokyo, Japan); FEI Tecnai G2 F20 microscope (FEI, Hillsborough, Oregon, USA) with a field-emission gun operating at 200 kV; D8 advance X-ray diffractometer (Germany Bruker Co., Karlsruhe, Germany). A three-electrode system was used with a modified electrode (the working electrode), platinum wire (the counter electrode) and saturated calomel electrode (SCE, the reference electrode).

Synthesis of CMS

CMS was fabricated on the basis of the previous report [12]. In general, 8.0 g of glucose was added to 40 mL water with ultrasonic agitation for 2 min to get a colorless solution, which was positioned in teflon-sealed autoclave and kept for 4 h in 180 °C. The brown or black product was segregated by centrifugal separation, cleaned by 3 cycles of centrifugal separation/washing/re-dispersion in water and alcohol, and dehydrated at 80 °C for over 4 h to get black powder. A 1.0 mg·mL^{-1} CMS liquor was produced by redispersing CMS into the water with ultrasonic agitation for 3 h to get a homogeneous suspension solution.

Electrode Fabrication

CILE was manufactured as described previously [21]. In brief, graphite powder were mixed thoroughly with ionic liquid $BPPF_6$ at 3/1 (w/w) in a mortar, then the paste was inserted into a glass tube ($\Phi = 4.0 \times 10^{-3}$ m). Prior to use, CILE was smoothed to get a mirror-like surface.

The step of fabricating the modified electrode was as follows. A 6.0 μL of 0.5 mg·mL^{-1} CMS and 15.0 mg·mL^{-1} Hb mixture solution was directly cast on the CILE surface with a 10.0 μL microsyringe. Then, the working electrode (CTS/CMS-Hb/CILE) was fabricated by spreading 5.0 μL of 1.0 mg·mL^{-1} chitosan (in 1.0% HOAC) solution evenly onto the CMS-Hb/CILE surface. A uniform film on the modified electrode was formed by covering a beaker to alleviate the evaporated solution. The preparation processes of CTS/Hb/CILE, CTS/CMS/CILE, CTS/CILE etc., were parallel to that of CTS/CMS-Hb/CILE.

RESULTS AND DISCUSSION

Morphological and Structural Characterization

SEM images of CMS with different magnification are shown in Figure 1A,B. It can be seen that the synthesized CMS had an average diameter of 600 nm, which was in good agreement with the reference [12]. TEM images of CMS at different magnitude were shown as Figure 1C,D, which exhibited that the solid structure of the surface had many nanosized pores. These pores on the surface of CMS could be attributed to the release of unreacted organic compounds such as oligosaccharides that been washed by water and alcohol. The existence of pores could offer large surface area and more reactive species for aiding in penetration and adsorption. Based on the reference [12], this synthesis process of CMS is a completely environmental-friendly procedure without the employment of any poisonous reagents, surfactants or organic reagent. Therefore the as-prepared CMS are nontoxic, with potential applications in biosensors or bioelectrochemistry. The graphitization degrees of CMS were further checked by XRD and Raman spectroscopy. As shown in Figure 1E, two peaks at 22.5° and 42.8° appeared on XRD, which could be specified as the typical graphitic (002) and (100) planes. The broadening of these two peaks suggests that the degree of graphitization was low and the amorphous carbon was possibly subsistent [14]. Figure 1F shows the Raman spectrum of CMS, which had one significant peak at 1574.0 cm^{-1}. This peak corresponds to the G-band that is attributed to the ordered graphite structure [22].

The functional groups on CMS were further characterized by an FT-IR spectrum with the result shown in Figure 1G. The O–H bond stretching vibration led to the strong characteristic peak at 3434.7 cm^{-1}. The C–H bond stretching vibration led to the weak peaks at 3172.4 cm^{-1} and 2927.5 cm^{-1}. The absorption bands at 1398.2 cm^{-1} corresponded to C-C stretching vibration and that of 1625.7 cm^{-1} was the carboxyl groups (C=O) stretching vibration [13]. Therefore, on the surface of CMS, many functional groups were present, which resulted in the hydrophilicity and stability of CMS in aqueous solution. As shown in Figure 1H, the CMS solution could remain stable for 2 days without congregation, indicating that CMS was stable in water and could be used for bioapplications.

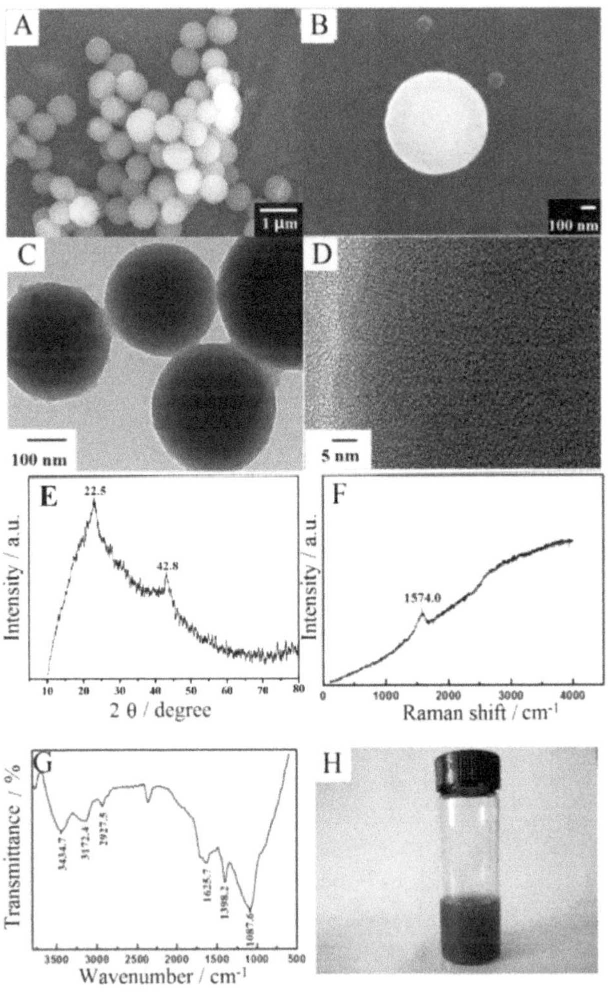

Figure 1: (A,B) SEM; (C,D) HRTEM; (E) XRD; (F) Raman spectrum; (G) FT-IR spectrum of CMS; (H) Photo of 1.0 mg·mL^{-1} CMS solution that kept for 2 days.

Spectroscopic Results

FT-IR spectroscopy was employed to examine the integrity of the structure and structural changes of the proteins [23]. Detailed data of the polypeptide chain could be determined from the shapes of amide I and II [24]. The amide I (1650.8 cm^{-1}) and II band (1542.8 cm^{-1}) of Hb after being mixed with CMS is shown in Figure 2Ab, which had less difference compared with amide I (1647.0 cm^{-1}) and II (1533.2 cm^{-1}) bands of the native Hb (Figure 2Aa). The results indicated that the original structure of Hb after being mixed with CMS was unchanged. UV-Vis adsorption spectroscopy is another useful tool to monitor the conformation change of heme proteins [25]. The Hb molecules had a characteristic band (406.0 nm) in pH 3.0 PBS (Figure 2Ba), which was the same as that of CMS-Hb (Figure 2Bb), meaning Hb molecules retained their original conformation in the CMS-Hb solution. All the spectroscopic research indicated that the biocompatibility of CMS and Hb kept the fundamental active conformation of the original structure after being mixed with CMS.

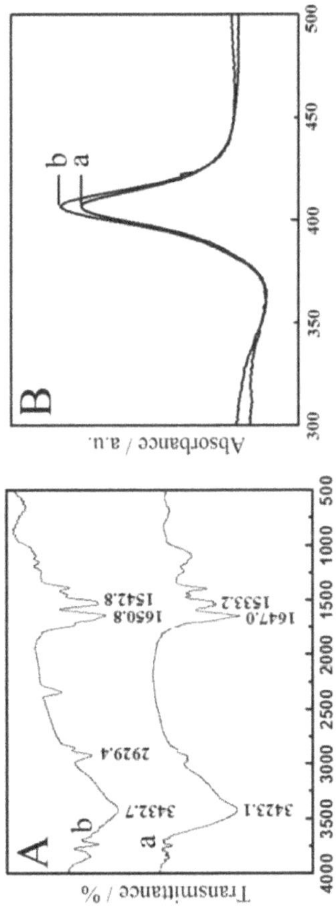

Figure 2: (A) FT-IR spectra of Hb (a) and CMS-Hb (b); (B) UV-Vis spectra of Hb (a) and CMS-Hb (b).

Electrochemical Characterization

Electrochemical behaviors were verified by cyclic voltammetry in the 0.1 mol·L^{-1} KCl and 10.0 mmol·L^{-1} $[Fe(CN)_6]^{3-/4-}$ mixture solution and the results are shown in Figure 3A,B. On CILE (Figure 3Aa) a couple of symmetric redox peaks was shown and it was the representative response of CILE. The electrochemical response of CTS/CILE (Figure 3Ab) was weaker than that of CILE, which was attributed to the existence of unconductive CTS on the surface of electrode impeding the electron transfer. The smallest redox peak currents appeared on CTS/Hb/CILE (Figure 3Bc), proving that the existence of Hb molecules on the electrode surface further impeded the electron transfer. However, on CTS/CMS-Hb/CILE (Figure 3Bd) the redox peak currents increased with the highest value, which was ascribed to the existence of CMS that promoted the electron transfer rate of $[Fe(CN)_6]^{3-/4-}$ with increased responses.

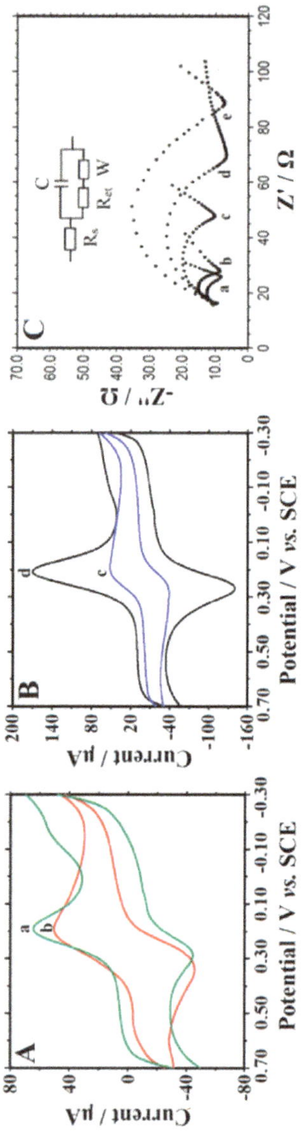

Figure 3: CV of (A) CILE (a), CTS/CILE (b); (B) CTS/Hb/CILE (c), CTS/CMS-Hb/CILE (d) in a 10.0 mmol·L^{-1} $[Fe(CN)_6]^{3-/4-}$ and 0.1 mol·L^{-1} KCl solution, scan rate:0.1 V·s^{-1}; (C) EIS of (a) CTS/CMS/CILE; (b) CTS/CMS-Hb/CILE; (c) CILE; (d) CTS/CILE and (e) CTS/Hb/CILE with the frequencies ranging from 10^5 to 10^{-1} Hz. (Inset is the Randles circuit model in the cell).

Figure 3C shows the electrochemical impedance spectroscopy (EIS) of different electrodes, which was employed to investigate the interfacial information [26]. The electron transfer resistance (R_{et}) value of CILE was found to be 37.97 Ω (curve c) and that of CTS/CILE increased to 54.75 Ω (curve d), proving that the interfacial resistance was increased with the existence of unconductive CTS film. On CTS/Hb/CILE (curve e) the R_{et} was 82.83 Ω, which could be ascribed to the existence of Hb further increasing the interfacial resistance. While on CTS/CMS/CILE (curve a) and CTS/CMS-Hb/CILE (curve b) theR_{et} decreased to 10.76 Ω and 15.25 Ω, respectively, showing that the existence of CMS decreased the resistance. CMS is a carbon material with good conductivity that can quicken the electron transfer.

Direct Electrochemistry of Hb

Figure 4 showed the electrochemical behaviors of different modified electrodes in PBS (pH 3.0). No redox peaks were found on CTS/CILE (curve a) and CTS/CMS/CILE (curve b). With a layer of IL present on the surface, CILE exhibits good conductivity and a biocompatible surface [9], and a couple of asymmetric redox peaks were found on CTS/Hb/CILE (curve c), showing that direct electron transfer between Hb and CILE was realized. On CTS/CMS-Hb/CILE the redox peaks increased greatly (curve d), which remained nearly unchanged at multi-scan cyclic voltammogram. The mixture of CMS with Hb on the electrode can form a biocomposite with good stability, which is suitable for accelerating electron transfer from the electroactive center of Hb to the electrode. The cathodic (Epc) and the anodic (Epa) peak potential were found to be −0.140 V and −0.227 V with the ΔE_p as 87 mV. The formal peak potential ($E^{0'}$) was calculated as −0.184 V (vs. SCE). The ratio of the cathodic (Ipc) and the anodic (Ipa) peak current was found to be 1.06. To explore the best CV responses of Hb on CTS/CMS-Hb/CILE, the amount of CMS cast on the electrode was optimized in the control experiments with a concentration range of 0.05 to 5.0 mg·mL^{-1}. The highest redox currents appeared at 0.5 mg·mL^{-1} CMS, which was used for electrode modification.

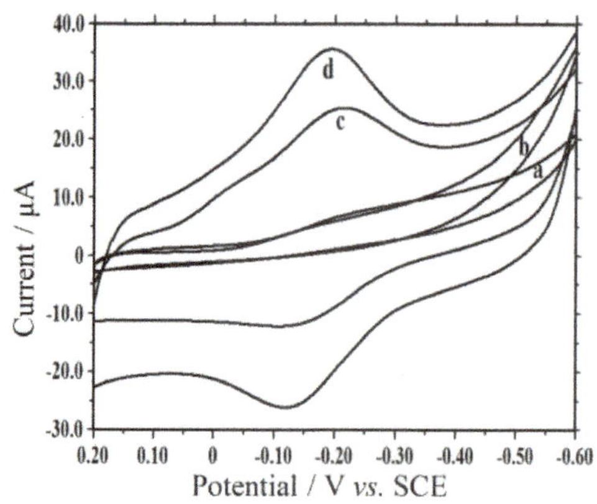

Figure 4: CV of (a) CILE; (b) CTS/CMS/CILE; (c) CTS/Hb/CILE and (d) CTS/CMS-Hb/CILE in pH 3.0 PBS, scan rate: 0.1 V·s^{-1}.

Electrochemical Investigation

As shown in Figure 5A, the influence of scan rate (υ) on voltammetric responses of CTS/CMS-Hb/CILE was checked. A couple of redox peak appeared at different scan rates with the regression equations of peak currents and υ as Ipc(μA) = 92.5·υ (V·s^{-1}) + 6.9 ($n = 10$, $\gamma = 0.999$) and Ipa(μA) = −78.0 υ·(V·s^{-1}) − 10.0 ($n = 10$, $\gamma = 0.999$) (Figure 5B), indicating a surface-controlled thin-layer electrode behavior. The integration of the redox peak in cyclic voltammograms can count the surface coverage (Γ*) of electroactive Hb by the formula ($\Gamma^* = Q/nAF$). By subtracting the background current, the peak currents of Hb were integrated with the Γ* value found to be 9.06 × 10^{-10} mol·cm^{-2}, which was larger than the theoretical monolayer coverage (1.89 × 10^{-11} mol·cm^{-2}) [27]. The fraction of electroactive Hb among the total Hb (1.20 × 10^{-8} mol·cm^{-2}) was calculated as 7.6%. As shown in Figure 5C, the increase in v resulted in the change of the redox peak potentials and the relationships of Ep with lnυ at high scan rate range were built. Two regression equations were got as

Epc(V) = −0.03 lnυ − 0.23 (n = 7, γ = 0.997) and Epa(V) = 0.02 lnυ − 0.10 (n = 7, γ = 0.993). Based on the Laviron's method [28], the kinetic of the redox reaction could be calculated with the electron transfer coefficient (α) and electron transfer rate constant (k_s) as 0.444 and 0.946 s^{-1}. The k_s value was larger than those of Nafion/Hb-graphene oxide-IL/CILE (0.92 s^{-1}) [29], GR/Fe_3O_4/Hb/GCE (0.30 s^{-1}) [30], Nafion/GR-TiO_2-Hb/CILE (0.65 s^{-1}) [31] and Hb-IL-MWCNT-CPE (0.84 s^{-1}) [32], indicating a relatively high rate of electron transfer. Also the ks value of CTS/Hb/CILE was calculated as 0.389 s^{-1} with the same method, which was less than that of CTS/CMS-Hb/CILE (0.946 s^{-1}). Therefore the presence of CMS with high conductivity provides an enhanced electron transfer reaction for Hb.

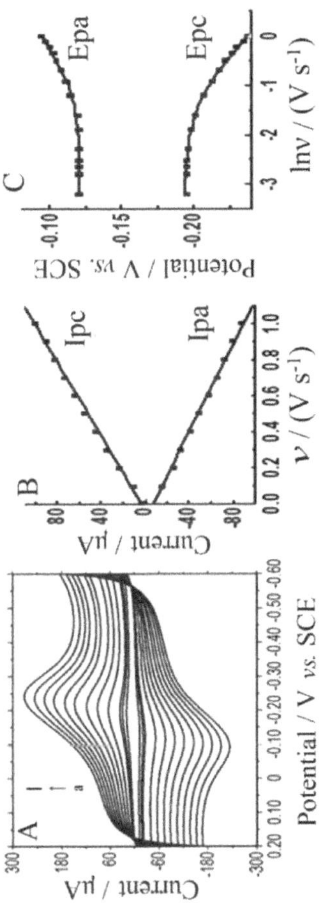

Figure 5: (A) Effect of scan rates (υ) (from a→l as 50, 80, 100, 200, 300, 400, 500, 600, 700, 800, 900, 1000 mV·s^{-1}) on electrochemical responses of CTS/CMS-Hb/CILE in pH 3.0 PBS; (B) Plot of the redox peak currents against υ; (C) plot of the redox peak potentials against lnυ.

The influence of buffer pH on electrochemical behaviors of CTS/CMS-Hb/CILE was studied in different PBS and the results are displayed in Figure 6A. The linear equation between $E^{0'}$ and pH was found to be $E^{0'}$(mV) = −44.25·pH − 54.73 (γ = 0.998). The slope value (−44.25 mV·pH^{-1}) was smaller than the theoretical value (−59.0 mV·pH^{-1}) for a one-proton and one-electron transfer process [33,34]. At pH 3.0 buffer solution, the largest redox peak currents appeared, and it was selected for the electrochemical experiments.

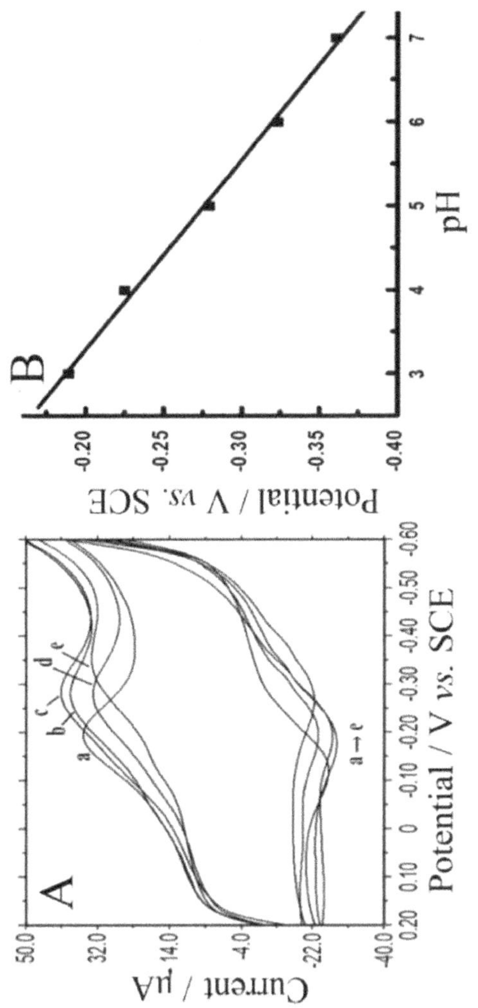

Figure 6: (A) CV of CTS/CMS-Hb/CILE in 0.1 mol·L^{-1} different pH PBS (from a→e: 3.0, 4.0, 5.0, 6.0, 7.0), scan rate: 0.1 V·s^{-1}; (B) the relationship of $E^{0'}$ with pH.

Electrocatalysis

The redox-protein-based biosensor exhibits excellent electrocatalytic ability for TCA, which is an important analytical target in biochemistry and environmental chemistry. As shown in Figure 7, when various concentrations of TCA were analyzed by CTS/CMS-Hb/CILE, the reduction peak current increased at −0.243 V without oxidation peak (curves a-h). With the further increase of the TCA concentration another reduction peak appeared at −0.515 V, indicated that di- and mono-chloroacetic acid might be dechlorinated by the formation of Hb [Hb Fe(I)] after TCA dechlorination with Hb Fe(II) [35]. On CTS/CMS/CILE direct electroreduction of TCA was studied with the potential negatively than −0.8 V (curves i, j). Therefore the reduction potential of TCA was reduced due to the existence of Hb. The reaction of electrocatalysis could be inferred from the following equations [35]:

$$HbFe(III) + e \rightarrow HbFe(II)$$
$$2HbFe(II) + H^+ + Cl_3CCOOH \rightarrow 2HbFe(III) + Cl^- + Cl_2CHCOOH$$
$$HbFe(II) + e \rightarrow HbFe(I)$$
$$2HbFe(I) + H^+ + Cl_2CHCOOH \rightarrow 2HbFe(II) + Cl^- + ClCH_2COOH$$
$$2HbFe(I) + H^+ + ClCH_2COOH \rightarrow 2Hb\,Fe(II) + Cl^- + CH_3COOH$$

The catalytic cathodic peak current at −0.243 V depended linearly on the TCA concentration with the equation of Iss(μA)

= 3.04 C (mmol·L^{-1}) − 4.89 (n = 13, γ = 0.999). The linear range and the detection limit were obtained as 2.0 ~ 70.0 mmol·L^{-1} and 0.30 mmol·L^{-1} (3σ), respectively. The cathodic peak current reached a stable value when the TCA concentration exceeded 70.0 mmol·L^{-1}, indicating a typical Michaelis-Menten kinetic process. The Lineweaver-Burk equation of $1/I_{ss} = (1/I_{max})(1 + K_M^{app}/C)$ [36] was used to calculate the apparent Michaelis-Menten constant (K_M^{app}) at a value of 1.60 mmol·L^{-1}, which was smaller than those of published values [27,37,38,39]. The low KMapp value indicated that Hb entrapped on the electrode retained its bioactivity and had a much higher biological affinity to TCA.

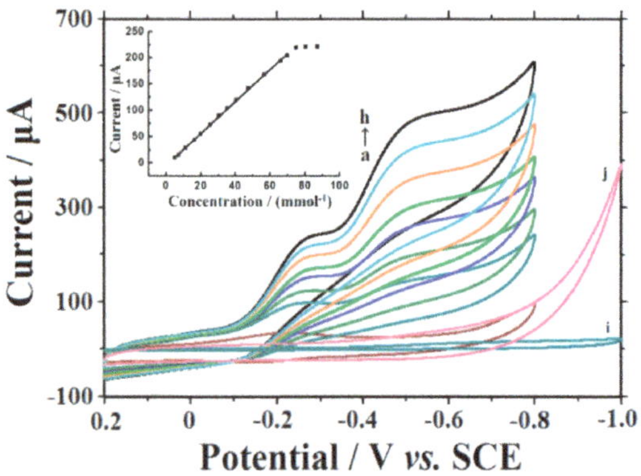

Figure 7: CV of CTS/CMS-Hb/CILE with 0, 15.0, 25.0, 35.0, 45.0, 55.0, 65.0, 70.0 mmol·L^{-1} TCA (curves a→h) and of CTS/CMS/CILE with of 0, 80.0 mmol·L^{-1} TCA (curves i and j), scan rate: 0.1 V·s^{-1} (Inset was the linearity of catalytic reduction currents and TCA concentration).

Analytical Application

To verify its application, CTS/CMS-Hb/CILE was used for determining of TCA in lab water with the standard addition method. As shown in Table 1, no TCA were found in the water specimens and the recovery was in the range of 97.00%–104.00%. Therefore this Hb-modified electrode could be used for water specimen detection.

Table 1: Detection results of TCA in the lab water specimen (n = 3)

Specimen	Found (mmol·L^{-1})	Added (mmol·L^{-1})	Total (mmol·L^{-1})	Recovery (%)	RSD (%)
water	0	2.00	2.08	104.00	3.06
		4.00	3.88	97.00	3.10
		6.00	6.12	102.00	2.98

Stability and Reproducibility

The reproducibility was studied by applying six Hb-modified electrodes to the determination of 10.0 mmol·L^{-1} TCA independently. An acceptable relative standard deviation (RSD) of 2.15% was determined. CTS/CMS-Hb/CILE was put into a 4 °C refrigerator for a certain period to check the storing stability. Every 5 days, the peak response of CTS/CMS-Hb/CILE to 10.0 mmol·L^{-1} TCA was tested, which decreased by 3.1% after 10 days and 7.5% for 25-days storage, proving the relative good stability of CTS/CMS-Hb/CILE.

CONCLUSIONS

CMS was prepared using a hydrothermal method and direct electrochemistry of Hb was carried out on CMS-modified CILE. A couple of well-defined redox peaks could be seen on CTS/CMS-Hb/CILE, indicating the acceleration of direct electron transfer of Hb due to high conductivity, large surface area and good biocompatibility of CMS. The immobilized Hb molecules kept their original structure and exerted good electrocatalytic ability for the reduction of TCA. As compared with other types of redox-protein-modified electrodes for TCA detection [29,31,37,38,39], this Hb-based electrode exhibited advantages such as favorable sensitivity, linear range, detection limit, stability and reproducibility. Hence, CMS has potential for the construction of third-generation redox-protein-based bioelectrochemical sensors.

ACKNOWLEDGMENTS

We acknowledge the financial support of the National Natural Science Foundation of China (Nos. 21365010 and 51363008), the Nature Science Foundation of Hainan Province (20152016), the International Science and Technology Cooperation Project of Hainan Province (KJHZ2015-13), the Marine Science and Technology Program of Hainan Province (2015XH06) and Graduate Student Innovation Research Projects of Hainan Normal University (Hsyx2015-47).

AUTHOR CONTRIBUTIONS

All authors contributed extensively to the work in this paper. Wen-Cheng Wang synthesized the sensing material, Li-Jun Yan preformed the electrocatalysis, Fan Shi prepared the modified electrode, Xue-liang Niu conducted the sensor analysis, Guo-Lei Huang characterized the carbon microsphere, Cai-juan Zheng conducted the characterization of protein, and Wei Sun investigated the electrochemical performances.

REFERENCES

1. Armstrong, F.A.; Hill, H.A.O.; Walton, N.J. Direct electrochemistry of redox proteins. Acc. Chem. Res. 1988, 21, 407–413.
2. Ramanavicius, A.; Ramanaviciene, A. Hemoproteins in design of biofuel cells. Fuel Cells 2009, 1, 25–36.
3. Armstrong, F.A.; Wilson, G.S. Recent developments in faradaic bioelectrochemistry. Electrochim. Acta 2000, 45, 2623–2645.
4. Prakash, P.A.; Yogeswaran, U.; Chen, S.M. A review on direct electrochemistry of catalase for electrochemical sensors.Sensors 2009, 9, 1821–1844.
5. Blanford, C.F. The birth of protein electrochemistry. Chem. Commun. 2013, 49, 11130–11132.
6. Mani, V.; Devadas, B.; Chen, S.M.; Li, Y. Immobilization of enzymes and redox proteins and their electrochemical biosensor applications. ECS Trans. 2013, 50, 35–41.
7. Sun, W.; Guo, C.X.; Zhu, Z.H.; Li, C.M. Ionic liquid/mesoporous carbon/protein composite microelectrode and its biosensing application. Electrochem. Commun. 2009, 11, 2105–2108.
8. Wang, J.; Li, M.; Shi, Z.; Li, N.; Gu, Z. Direct electrochemistry of cytochrome c at a glassy carbon electrode modified with single-wall carbon nanotubes. Anal. Chem. 2002, 74, 1993–1997.

9. Li, M.G.; Xu, S.D.; Tang, M.; Liu, L.; Gao, F.; Wang, Y.L. Direct electrochemistry of horseradish peroxidase on graphene-modified electrode for electrocatalytic reduction towards H_2O_2. Electrochim. Acta 2011, 56, 1144–1149.

10. Kang, X.H.; Wang, J.; Wu, H.; Aksay, I.A.; Liu, J.; Lin, Y.H. Glucose oxidase–graphene–chitosan modified electrode for direct electrochemistry and glucose sensing. Biosens. Bioelectron. 2009, 25, 901–905.

11. Shin, Y.S.; Wang, L.Q.; Bae, I.T.; Arey, B.W.; Exarhos, G.J. Hydrothermal syntheses of colloidal carbon spheres from cyclodextrins. J. Phys. Chem. C 2008, 112, 14236–14240.

12. Sun, X.M.; Li, Y.D. Colloidal carbon spheres and their core/shell structures with noble-metal nanoparticles. Angew. Chem. Int. Ed. 2004, 43, 597–601.

13. Chen, J.X.; Xia, N.N.; Zhou, T.X.; Tan, S.X.; Jiang, F.P.; Yuan, D.S. Mesoporous carbon spheres: Synthesis, characterization and supercapacitance. Int. J. Electrochem. Sci. 2009, 4, 1063–1073. [Google Scholar]

14. Jin, Y.Z.; Gao, C.; Hsu, W.K.; Zhu, Y.Q.; Huczko, A.; Bystrzejewski, M.; Roe, M.; Lee, C.Y.; Acquah, S.; Kroto, H.; et al. Large-scale synthesis and characterization of carbon spheres prepared by direct pyrolysis of hydrocarbons. Carbon2005, 43, 1944–1953.

15. Auer, E.; Freund, A.; Pietsch, J.; Tacke, T. Carbons as supports for industrial precious metal catalysts. Appl. Catal. A1998, 173, 259–271.

16. Wang, Y.; Su, F.; Wood, C.D.; Lee, J.Y.; Zhao, X.S. Preparation and characterization of carbon nanospheres as anode materials in lithium-ion secondary batteries. Ind. Eng. Chem. Res. 2008, 47, 2294–2300.

17. Yan, X.B.; Xu, T.; Xu, S.; Chen, G.; Liu, H.W.; Yang, S.R. Fabrication of carbon spheres on a C: H films by heat-treatment of a polymer precursor. Carbon 2004, 42, 2769 2771.

18. Shiddiky, M.J.A.; Torriero, A.A.J. Application of ionic liquids in electrochemical sensing systems. Biosens. Bioelectron.2011, 26, 1775–1787.

19. Sun, W.; Gao, R.F.; Jiao, K. Research and application of ionic liquids in analytical chemistry. Chin. J. Anal. Chem. 2007,35, 1813–1819. [Google Scholar]

20. Sheng, M.L.; Gao, Y.; Sun, J.Y.; Gao, F. Carbon nanodots–chitosan composite film: a platform for protein immobilization, direct electrochemistry and bioelectrocatalysis. Biosens. Bioelectron. 2014, 58, 351–358.

21. Sun, W.; Yang, M.X.; Jiao, K. Electrocatalytic oxidation of dopamine at an ionic liquid modified carbon paste electrode and its analytical application. Anal. Bioanal. Chem. 2007, 389, 1283–1291.

22. Xu, X.; Shi, B.X.; Zheng, W.; Xia, W.; Guang, H. Photo-responsive behaviors and structural evolution of carbon-nanotube-supported energetic materials under a photoflash. Mater. Lett. 2012, 88, 27–29.

23. Byler, D.M.; Susi, H. Examination of the secondary structure of proteins by deconvolved FTIR spectra. Biopolymers1986, 25, 469–487.

24. Kauppinen, J.K.; Moffat, D.J.; Mantsch, H.H.; Cameron, D.G. Fourier self-deconvolution: a method for resolving intrinsically overlapped ban. Appl. Spectrosc. 1981, 35, 271–276.

25. Rusling, J.F.; Nassar, A.E.F. Enhanced electron transfer for myoglobin in surfactant films on electrodes. J. Am. Chem. Soc. 1993, 115, 11891–11897.

26. Orazem, M.E.; Tribollet, B. An integrated approach to electrochemical impedance spectroscopy. Electrochim. Acta 2008,53, 7360–7366.

27. Wang, S.F.; Chen, T.; Zhang, Z.L.; Shen, X.C.; Lu, Z.X.; Pang, D.W.; Wong, K.Y. Direct electrochemistry and electrocatalysis of heme proteins entrapped in agarose hydrogel films in room-temperature ionic liquids. Langmuir2005, 21, 9260–9266.

28. Laviron, E. General expression of the linear potential sweep voltammogram in the case of diffusionless electrochemical systems. J. Electroanal. Chem. 1979, 101, 19–28.

29. Sun, W.; Gong, S.X.; Shi, F.; Cao, L.L.; Ling, L.Y.; Zheng, W.Z.; Wang, W.C. Direct electrochemistry and electrocatalysis of hemoglobin in graphene oxide and ionic liquid composite film. Mater. Sci. Eng. C 2014, 40, 235–241.

30. Wang, Y.Q.; Zhang, H.J.; Yao, D.; Pu, J.J.; Zhang, Y.; Gao, X.R.; Sun, Y.M. Direct electrochemistry of hemoglobin on graphene/Fe_3O_4 nanocomposite-modified glass carbon electrode and its sensitive detection for hydrogen peroxide. J. Solid State Electrochem. 2013, 17, 881–887.

31. Sun, W.; Guo, Y.Q.; Ju, X.M.; Zhang, Y.Y.; Wang, X.Z.; Sun, Z.F. Direct electrochemistry of hemoglobin on graphene and titanium dioxide nanorods composite modified electrode and its electrocatalysis. Biosens. Bioelectron. 2013, 42, 207–213.

32. Wei, W.; Jin, H.H.; Zhao, G.C. A reagentless nitrite biosensor based on direct electron transfer of hemoglobin on a room temperature ionic liquid/carbon nanotube-modified electrode. Microchim. Acta 2009, 164, 167–171.

33. Yamazaki, I.; Araiso, T.; Hayashi, Y.; Yamada, H.; Makino, R. Analysis of acid-base properties of peroxidase and myoglobin. Adv. Biophys. 1978, 11, 249–281. [Google Scholar]

34. Liu, H.H.; Tian, Z.Q.; Lu, Z.X.; Zhang, Z.L.; Zhang, M.; Pang, D.W. Direct electrochemistry and electrocatalysis of heme-proteins entrapped in agarose hydrogel films. Biosens. Bioelectron. 2004, 20, 294–304.

35. Fan, C.H.; Zhuang, Y.; Li, G.X.; Zhu, J.Q.; Zhu, D.X. Direct electrochemistry and enhanced catalytic activity for hemoglobin in a sodium montmorillonite film. Electroanal 2000, 12, 1156–1158.

36. Kamin, R.A.; Wilson, G.S. Rotating ring-disk enzyme electrode for biocatalysis kinetic studies and characterization of the immobilized enzyme layer. Anal. Chem. 1980, 52, 1198–1205.

37. Sun, W.; Li, X.Q.; Wang, Y.; Zhao, R.J.; Jiao, K. Electrochemistry and electrocatalysis of hemoglobin on multi-walled carbon nanotubes modified carbon

ionic liquid electrode with hydrophilic EMIMBF4 as modifier. Electrochim. Acta2009, 54, 4141–4148.

38. Zhang, Z.Q.; Wu, J.; Sun, W.; Jiao, K. Direct electrochemistry of hemoglobin and its electrocatalysis based on a carbon nanotube paste electrode. J. Chin. Chem. Soc. 2009, 56, 561–567. [Google Scholar]

39. Sun, W.; Guo, Y.Q.; Lu, Y.P.; Hu, A.H.; Shi, F.; Li, T.T.; Sun, Z.F. Electrochemical biosensor based on graphene, Mg_2Al layered double hydroxide and hemoglobin composite. Electrochim. Acta 2013, 91, 130–136.

CHAPTER 10 GRAPHENE OXIDE AND H_2 PRODUCTION FROM BIOELECTROCHEMICAL GRAPHITE OXIDATION

Lu Lu, Cuiping Zeng, Luda Wang, XiaoboYin, Song Jin, Anhuai Lu & Zhiyong Jason Ren

INTRODUCTION

Graphene and related materials have come to the forefront of research and development due to their unique electrical, thermal, or mechanical properties and wide application potentials. For large-scale graphene production, chemical oxidation of graphite to graphite oxide or graphene oxide (GO) followed by reducing GO to graphene using chemical, thermal, or electrochemical methods has been widely studied1·2. GO not only is the important precursor for mass production of graphene-based materials2, it also draws extensive attention for its own potentials in many areas, including electronics and optoelectronics3, bio-nanotechnology4, renewable energy5, membrane research6 and environmental applications7. To date, GO is mainly synthesized by chemical oxidation based on either the Hummers, Brodie, or Staudenmaier methods2. All of these chemical methods use either concentrated acids, such as sulfuric acid (H_2SO_4) and nitric acid (HNO_3), toxic reagents such as potassium dichromate ($K_2Cr_2O_7$) and potassium permanganate ($KMnO_4$), or even explosive potassium chlorate ($KClO_3$) to oxidize graphite to GO, and the production procedure can be expensive, dangerous, and non-sustainable1. Electrochemical exfoliation of graphite to GO or graphene was recently performed in ionic liquids8·9·10, aqueous acids11·12·13 and inorganic salt solution14 under a 7–20 V voltage, and the products were reported with different levels of defect in the crystal lattice and oxygen-doping. It is possible that microorganisms can oxide dispersed graphite to graphite oxide nanosheets, but external carbon sources and oxygen were needed and the reaction rate was low15. Data from our preliminary studies suggests a new method for producing GO at relatively high rate through bioelectrochemical oxidization of graphite under ambient conditions.

Here we report such new and green approach to produce bioelectrochemically exfoliated graphite oxide and graphene oxide (BEGO) sheets from solid graphite rods by using a bioelectrochemical system (BES). The BEGO production is also accompanied with the production of value-added H_2 and organic compounds. This new method eliminates the use of expensive and potentially hazardous chemicals for bulk GO production, and presents the possibility of using abundant graphite as the electron source for co-production of clean energy and chemicals. In this process, graphite rod electrode (anode) was exfoliated in a BES to produce BEGO colloidal suspension, CO_2, and electrons. The electrons derived from the anode oxidation are transferred to a biocathode, where H_2 and organics are produced by microbial reduction of protons and CO_2, respectively, a process known as microbial electrosynthesis (MES)16·17 (Fig. 1A). In addition to the electricity-driven reduction of CO_2, MES here is expanded to a more generalized process including all the microbially catalyzed synthesis of chemical compounds in an electrochemical system using direct or indirect electron transfer from the electrode to the microorganisms17·18. Bioelectrochemical system is a technology platform that uses microorganisms to catalyze electrochemical reactions for energy and chemicals production from biodegradable substrates, such as those from wastewater, soil and sediments16·17·19. A BES reactor traditionally uses an inert graphite anode, which serves as the electron acceptor

Citation: Lu, L. et al. Graphene oxide and H_2 production from bio-electrochemical graphite oxidation. Sci. Rep. 5, 16242; doi: 10.1038/srep16242 (2015).

for exoelectrogenic bacteria, also known as electricigens or anode-respiring bacteria for their anaerobic respiration. While the BES platform has been studied to produce inorganic and organic chemicals through electrolysis or electrosynthesis on the cathode20, the pure carbon based anode has not been reported sacrificial16·21·22·23, because it is not biologically available and poised at a low potential compared to abiotic electrochemical reactions. However, we found in this study that by coupling the biological and electrochemical mechanisms, the graphite anode could be oxidized and exfoliated into BEGO colloidal suspension at a faster rate than abiotic controls.

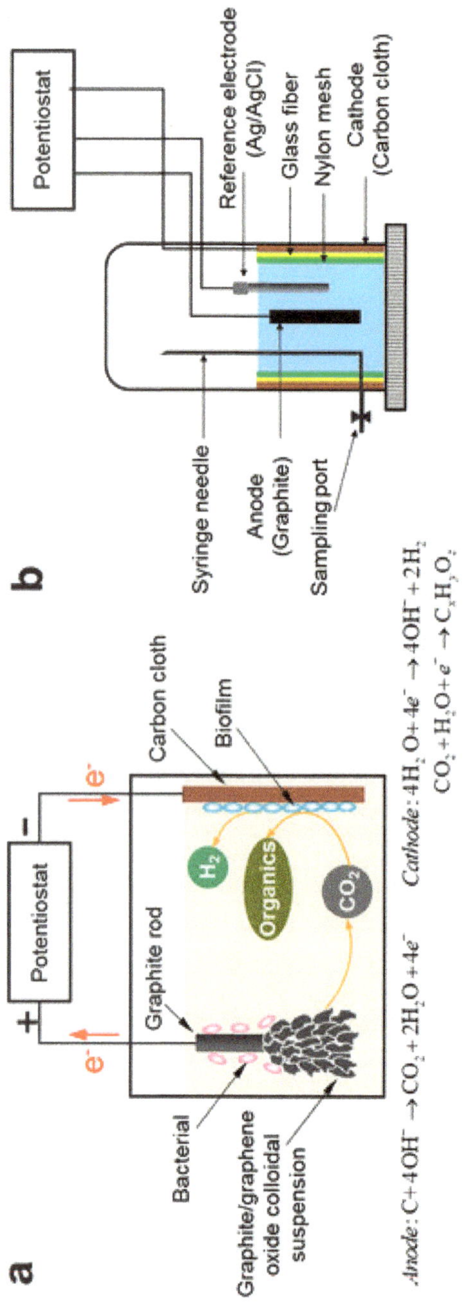

Figure 1: Schematic illustration of 3-electrode bioelectrochemical system (BES).

(a) The hypothesized reaction mechanisms in BES, including microbial catalyzed GO production on the anode and microbial electrosynthesis on the cathode. (b) The configuration of water sealed anaerobic BES reactors.

RESULTS

Establishing an Electrotrophic Biocathode for Electrosynthesis

The first stage of experiment aimed to establish a microbial community for electrosynthesis on the cathode. The active reactor was one time inoculated with anaerobic sludge and fed with 2 mM glucose as initial substrate, which was then operated at a fixed cathode potential of −0.6 V (vs. SHE). The system was flushed with 100% CO_2 to reach a constant CO_2 content of ~5.6 mmol in the headspace. The current production can be observed after 40 days and reached to 1.24 A/m^2 at the end of batch. In contrast, only 0.01 A/m^2 current was generated in the abiotic control reactor that was operated the same way but without inoculum. In the meantime, $H_2$was produced from the active reactor, suggesting the formation of biocatalysts reduced the potential required to overcome the real thermodynamic barrier of cathode reaction. Although the theoretical potential for H_2 production by reduction of proton under normal condition (298K, 1 atm and pH=7) is −0.414 V (vs. SHE), more negative potential is normally required to drive the reaction in practical due to the existence of overpotential. The slight increase of CO_2 and the generation of CH_4 should be due to fermentation of glucose or sludge and potentially hydrogenotrohic methanogenesis. The solution COD and pH were stable, suggesting no organic matters were produced through CO_2 reduction at this stage. On day 57, the media were completely replaced with mineral solution without carbon source or inoculum. CO_2 was provided again by flushing the fresh solution. Current density increased rapidly from 0.53 to 2.06 A/m^2 from day 57 to day 77, which resulted in accumulation of 18.7 mmol H_2 with average production rate of 9.4 mmol/L/day. Different from previous stages, significant CO_2 reduction was observed, with the headspace CO_2dropped from 5.4 to 2.4 mmol. In the meantime, COD concentration in the liquid increased significantly from 96 mg/L to the 4,060 mg/L and accompanied by a pH drop from 6.78 to 4.54, indicating the formation of organic acids. In contrast, the control reactor showed no obvious current or H_2 generation production, nor obvious CO_2 reduction or COD accumulation. The pH kept stable in the control reactor too. Such differences suggest that an electrotrophic biocathode capable of microbial electrosynthesis is established22. Methanogenesis was greatly inhibited presumably due to low pH.

Coupling Graphite Oxidization to BEGO with Microbial Electrosynthesis

After 76 days of operation, the solution was replaced again and flushed with CO_2. As shown in Fig. 2, repeating pattern was observed, and $H_2$was the predominant product that accumulated to 110.8 mmol over 30 days. The average production rate was 36.9 mmol/L/day. No obvious CH_4 production was observed. One key difference in this stage is the production of CO_2 from the active reactor. CO_2 accumulated to 35.2 mmol at an average production rate of 11.7 mmol/L/day. The solution COD continuously increased from 136 mg/L to 1350 mg/L, while the solution pH dropped from 6.76 to 5.87. No obvious gas production was observed in the abiotic control reactor, and the COD and pH kept constant. Along with CO_2 production, gradual physical mass loss of the graphite anode and increased turbidity and color of the solution were observed in the active reactor (Fig. 3), indicating the graphite electrode was the major carbon source for CO_2 production. Although cell decomposition and microbial metabolism also generate CO_2, they were not dominant based on electron balance. Figure 3 shows the images of the anodes at different stages, as well as the solutions containing the exfoliated BEGO sheets. The supernatant exhibited blackish brown color. By comparison, the anode of the abiotic control reactor (cathode poised at −0.6 V vs. SHE) remained intact, and the supernatant was as clear as the fresh medium.

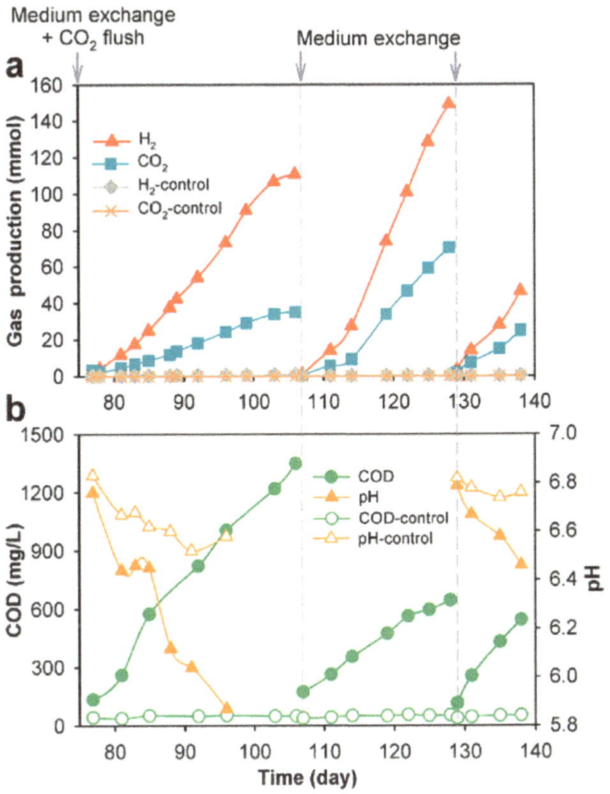

Figure 2: Products generation during BEGO production and microbial electrosynthesis.

(a) Gas (H_2 and CO_2) production in the BES. (b) Changes of COD and pH of the solution. Cathodes were poised at −0.6 V (vs. SHE). Uninoculated abiotic reactor with the cathode poised at −0.6 V vs. SHE was used as control. Data shown from day 76 to day 138, representing repeatable cycles of system performance after microbial acclimation and enrichment. Results of acclimation period from day 1 to day 76 can be found in Supplementary Information.

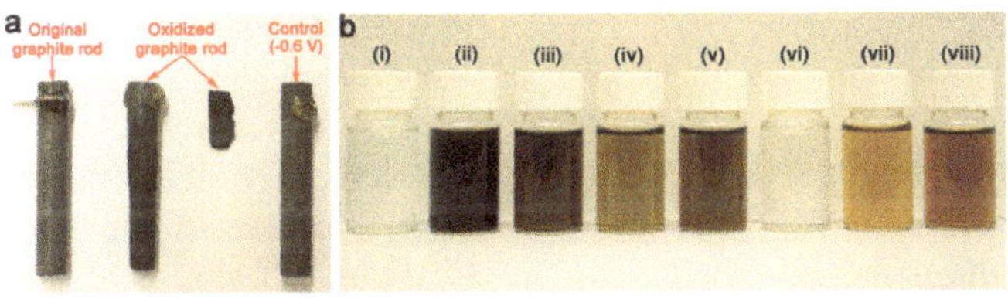

Figure 3: Digital images of (a) graphite rods before and after bioelectrochemical exfoliation, and (b) exfoliation production mixture dispersed in phosphate buffer medium for three weeks. (i) fresh phosphate medium; (ii) original sample from active reactor; (iii,iv) 2- and 4- fold dilution of sample ii, respectively; (v,vi) samples from abiotic control reactors with anode poised at +1.6 V (vs. SHE) and cathode poised at −0.6 V, respectively; (vii,viii) purchased chemical graphene oxide (CGO) dispersion in H_2O (Sigma-Aldrich) with concentrations of 0.2 and 0.4 mg/mL, respectively.

To double check the involvement of graphite in this bioelectrochemical reaction and understand the underlying mechanisms, we repeated the experiment by completely replacing the medium again on days 107 and 129, as well as replacing a new graphite anode on day 107. But this time we did not perform external CO_2 flush so carbon could be balanced.Figure 2A shows the average production rates of H_2 and CO_2 increased further compared to the previous cycle and reached 46.3–67.8 mmol/L/day (H_2) and 25.0–32.0 mmol/L/day (CO_2), respectively. The ratio of accumulated volume of H_2 to CO_2 ranged from 1.9 to 2.1. The production of COD decreased to 545–646 mg/L at the end of each cycle, indicating that planktonic microorganisms in solution were mainly responsible for microbial electrosynthesis of organics during the enrichment period (day 57–76), where they could use indirect electron donors, such as H_2, for electron transfer. Once they were washed away during medium replacement, less organic COD was synthesized by the biofilm attached on the electrodes. The carbon-based colloidal suspension was continuously produced in every batch cycle, which resulted in gradual loss of the graphite anode. The graphite anode lost almost 90% of its initial mass at the end of the batch (Fig. 3). Again, no change was found in the control reactor, with cathode poised at −0.6 V (vs. SHE).

In an attempt to quantify the Coulombic efficiency to measure the electron recovery in products from electrical current, we found that hydrogen production rate was highly correlated with the current density (Fig. 4b). Among the coulombs consumed, most coulombs ended in H_2 and a small amount was used for organic synthesis (Fig. 4a). The highest total Coulombic efficiency for H_2 and organics production using current was around 90%, indicating the energy loss due to internal resistance and cell synthesis is low. The produced H_2 energy can offset 58–63% of total electric energy input to the system. Cyclic voltammetry (CV) profiles between the active and control reactor cathodes demonstrates that catalytic current density of active cathode was almost one order of magnitude higher than that of abiotic cathode (Fig. 4c), showing much higher electrochemical activities of the biocathode.

A maximum anode potential of +1.6 V (vs. SHE) was observed in the biotic reactor during operation, while the anode potential in the abiotic control ranged from +0.71 to +0.98 V (vs. SHE) despite a same cathode potential (−0.6 V vs. SHE) was poised on both reactors. In order to characterize the potential of abiotic electrochemical exfoliation as compared with the bioelectrochemical process, another abiotic control was operated by poising the anode (working electrode) at +1.6 V (vs. SHE) for 48 h. The current production (0.30–1.13 A/m^2) was much lower than that in the biotic reactor (1.68–4.46 A/m^2) despite a higher voltage (2.67 V) was applied on the abiotic reactor compared to biotic reactor (2.2 V) (Figure S3). As a result, the exfoliation rate of graphite anode was 83 mg/L/day, much lower than the biotic active reactor, which had a rate of 388 mg-BEGO /L/day.

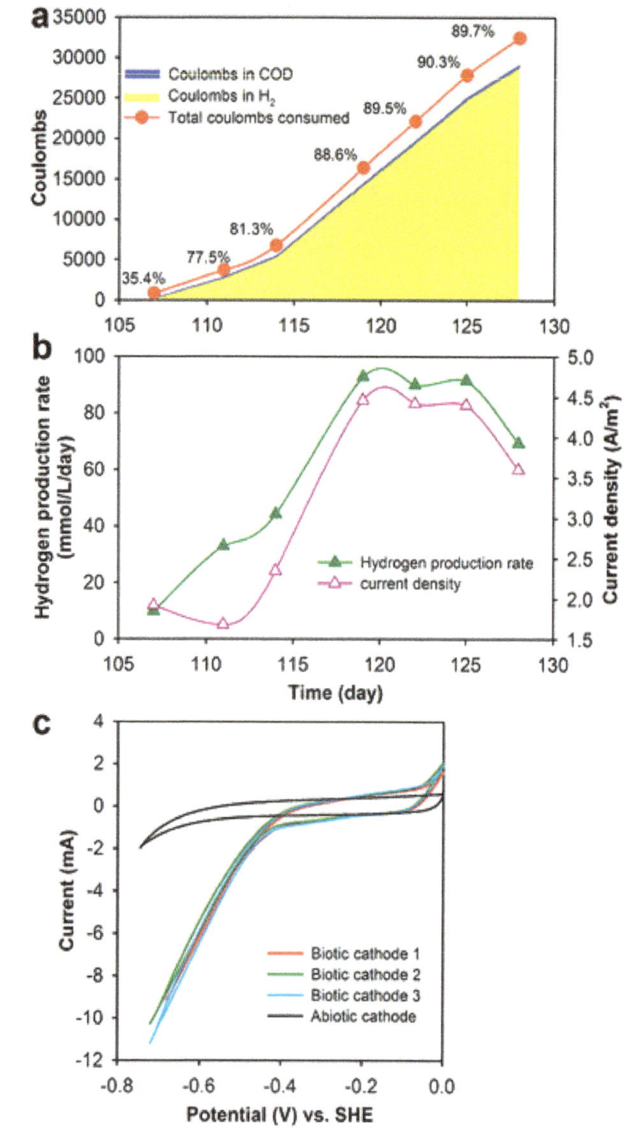

Figure 4: System electron balance and bacterial electrochemical activities.

(a) Distribution of coulombs in products (H_2 and organics) compared to total coulombs consumed. (b) The positive correlation between H_2 production rate and current density (cathode poised at −0.6 V vs. SHE). (c) Cyclic voltammograms (1 mV/s) for the biotic cathode (active reactor, three reproducible scans) and abiotic cathode (control reactor).

Characterization of BEGO

We used a suite of tools to characterize the BEGO suspension produced in the active BES reactors. The fourier transform infrared (FT-IR) spectrum of BEGO exhibited the presence of several oxygen-containing groups in the studied region but absence of any significant absorption band for graphite (Fig. 5a, S3). The groups of −OH (structural hydroxyl group, 3421–3441 cm^{-1}), −COOH (carboxyl, 1730 cm^{-1}), O–H (1385 cm^{-1}), C–O–C (epoxy, 1221–1226 cm^{-1}) and C–O (alkoxy, 1056–1060 cm^{-1}) were found in both BEGO samples and CGO24·25·26. The peak at 1622–1635 cm^{-1} has been assigned to deformation vibration of water molecules24. This observation confirms the introduction of oxygen to graphite crystal and is consistent with the now widely accepted chemical composition of graphite oxide/GO[2]. The BEGO also shows typical GO hydrophilic characteristics that disperse in water but not in dimethylformamide (DMF). The UV-vis absorption spectra (Fig. 5b) of aqueous BEGO samples exhibits two characteristic features that can be used to identify graphite oxide/GO27: a peak at 229–232 nm (CGO is 229 nm), corresponding to $\pi \rightarrow \pi^*$ transitions of aromatic C–C bonds, and a shoulder at 300 nm, corresponding to $n \rightarrow \pi^*$ transitions of C=C bonds. X-ray photoelectron spectroscopy (XPS) was used to further assess the chemical composition of the BEGO (Fig. 5c). It is unreliable to use C/O ratio for evaluation of the oxidation levels of BEGO due to impossible full dehydration of a GO sample. Instead, we first deconvoluted XPS spectra of the C 1s peak into four peaks that correspond to the functional groups of C–C (carbon sp^2, 284.5 eV), C–O (alkoxy, 285.9 eV), C=O (carbonyl, 287.4 eV) and O–C=O (carboxyl, 288.7 eV). Then, the area of each deconvoluted peak was normalized with respect to the C–C peak (Table S1). The degree of oxidation for BEGO is comparable with that indicated in microbial oxidation15, but both of them are less oxidized than chemical oxidized GO, suggesting that the regular structure of carbon sp2 domains may be easily retained during bioelectrochemical oxidation of graphite. Raman spectroscopy is a proven tool for molecular morphology characterization of carbon materials, and it reveals highly ordered graphite crystal with prominent G-band representing the planar configuration sp2 bonded carbon at 1583 cm^{-1} and the weak disorder band (D-band) at 1356 cm^{-1} as well as 2D-band (second order or overtone of the D-band) at 2696 cm^{-1} (Fig. 5d). The BEGO spectrum is almost identical to CGO with characteristics of broader G-band and D-band, higher intensity ratio of the D to G band (I_D/I_G), decrease of 2D-band and a shift of G-band to higher frequency due to amorphization of graphite15·28·29, indicating similar structure and composition. However, the I_D/I_G intensity ratios is lower in the BEGO spectra, suggesting that BEGO has less disordered carbon atoms or larger average size of the in-plane sp^2 domains, and therefore a higher crystalline quality than that generated by chemical oxidation28. The improved BEGO crystallinity is also supported by its less blue shift of G-band than CGO relative to that in graphite29. In BEGO, the G-band located at frequency of 1587 cm^{-1}, near that in graphite (1583 cm^{-1}). This indicates a less adsorption of oxygen atoms on the BEGO network.

Atomic force microscopy (AFM) images show that there are around 12 sheets (average size 300–500 nm) per $5 \times 5\ \mu m$ area based on the size of AFM image, and the height of sheet is 2.5–4 nm (Fig. 5e,f,g). Due to different interaction force of tapping mode for the GO and mica, the height difference is amplified at the interface. Thus, BEGO consists of 1 ~ 3 layers of monolayer GO (~1–1.4 nm)3. Transition electron microscopy (TEM) images show that the BEGO displays one or few-layer structure (Fig. 5h,I, S4), which is consistent with the result of AFM. We did observe overlapped sheet structure (left side) with thickness of more than a few layers, and this is due to the lack external exfoliation means, such as sonication or mechanical stirring, commonly used in processes of chemical GO production2. The selected area electron diffraction (SAED) pattern (insert of Fig. 5i) of few layers BEGO shows a typical sharp crystalline ring pattern. The bright spots corresponding to the (1100) reflections are observed on the first ring came from the (1100) plane and retained the hexagonal symmetry of the (1100) diffraction pattern30. The observation of more than six strong hexagonal spots indicates that BEGO contains unordered crystal structure arising from oxidation31.

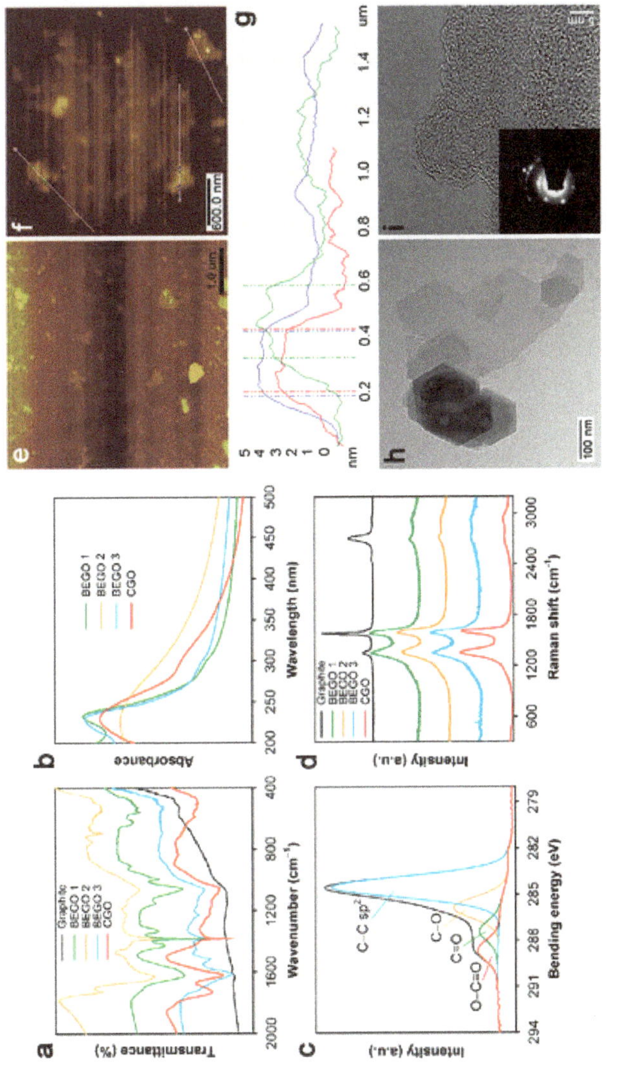

Figure 5: Characterization of the produced BEGO.

(a) FT-IR spectra of BEGO samples, graphite powder obtained from raw electrode and purchased chemical graphene oxide (CGO). (b) UV-vis spectra of BEGO and CGO in aqueous solutions. (c) XPS of C 1s spectra of BEGO. (d) Raman spectra of BEGO, graphite and CGO. (e) AFM image, (f) AFM zoomed image for the middle region, and (g) AFM cross-section height profile. (h) TEM and (i) HRTEM images. The inset of (**i**) is the selected area electron diffraction (SAED) pattern.

Bacterial Community involved in Bioelectrochemical Graphite Oxidation

Bacterial community structure analysis using high throughput sequencing showed the genus *Pseudomonas* is the dominant population on the graphite anode (74% of total reads) (Fig. 6b). Other major populations include *Rhodococcus*, *Ralstonia,* and *Propionibacterium*, which accounted for 7%, 4% and 3% of the total composition, respectively. Other 90 genera making up less than 1% of total composition were found with combined total abundance of 7%. There were 5% novel genera that had not been identified yet. Some species of*Pseudomonas*, such as *P. aeruginosa*32, *P. alcaliphila*33 and *P. putida*34, can self-excrete redox mediators to transfer electrons to the electrode. Among *Pseudomonas* population, the majority of sequences (75.3%) were closely similar to *P. syringae* (100% similarity) (Fig. 6c), which has not been reported to carry bioelectrochemical processes of extracellular electron transfer. *Rhodococcus* is typical aromatic compound degrader by oxygenating the aromatic ring. All dominant bacterial genera are aerobic microbes except aerotolerant anaerobic*Propionibacterium*, indicating the possible oxygen production on the anode. However, the system was sealed and maintained in anaerobic condition and no molecule oxygen was detected (or below the GC detection limit) during whole process. This suggests that O_2 may be depleted *in situ* around microbial cells. Because hydroxyl can also be resolved to generate molecule oxygen on the anode ($4OH^- \rightarrow 2H_2O + O_2 + 4e^-$) through water electrolysis, more studies are needed to understand what are the specific roles of anode microorganisms on graphite oxidation and the evolution and fate of potential O_2.

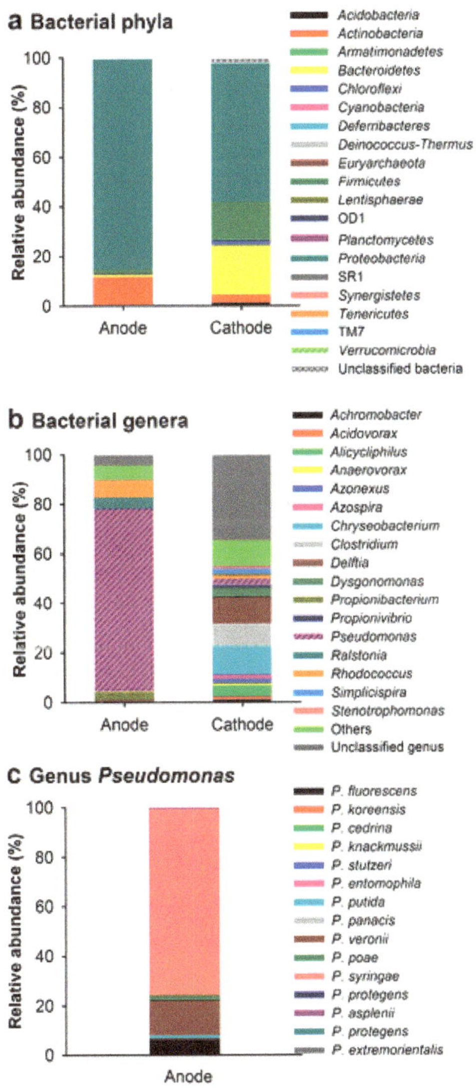

Figure 6: Bacterial community composition on the anode and cathode.

Dominant bacterial (a) phyla and (b) genera. (**c**) Distribution of specific species (identity >97%) within genus of *Pseudomonas* on the anode. Genera making up less than 1% of total composition were classified as "others".

Bacterial community on the biocathode has higher diversity than that on the anode. In addition to phylum *Proteobacteria* (anode 85%, cathode 56%), the community on the cathode was also dominated by *Fimicutes*(15%) and *Bacteroidetes* (20%), which were rarely found on the anode (Fig. 6a). *Fimicutes* was reported to be electrochemically active though the majority of such bacteria are known as *Proteobacteria*35. *Delftia*(11%), *Clostridium* (9%), *Alicycliphilus* (4%) and *Chryseobacterium* (12%) are dominant genera on the cathode. For *Clostridium*, all the sequences were closely similar to *C. carboxidivorans* (100% similarity), an anaerobic solvent-producing bacteria, which can grow autotrophically using H_2/CO_2 or CO with acetate, ethanol, butyrate and butanol as end products36, suggesting it may have been involved in MES process. Other dominant populations are facultative or aerobic bacteria, which likely function as O_2 sinkers to maintain anaerobic condition in the MES.

DISCUSSION

The graphite oxidation in bioelectrochemical system (BES) provides a new approach for GO production, and it concurrently offers a new and abundant electron source for microbial electrosynthesis, an emerging area for value-added organic fuel and chemical production17·19·37·38. The graphite anode oxidation is hypothesized similar as the anode oxidation in molten hydroxide-based direct carbon fuel cell (MH-DCFC)39·40(Equation 1).

$$C + 4OH^- \rightarrow CO_2 + 2H_2O + 4e^- \quad (1)$$

However, operation of MH-DCFC requires high temperature (500–650 °C), while the processes in this study occurred at room temperature. The CO_2 should be generated from graphite oxidation, because no other carbon source was available in the medium. Similarly, the electrical current was also primarily derived from graphite oxidation, because no apparent O_2 was detected in the system, which indicates limited possibilities of electron generation by water electrolysis. Reduction of O_2 by electrons at the cathode is not believed to be the main O_2 sink due to extreme high Coulombic efficiency (90%) for H_2 and organics production. Hydrolysis driven by bioelectrochemical reactions provided hydroxyl and proton for graphite anode oxidation and H_2 generation at the cathode, respectively (Eq. 2).

$$4H_2O + 4e^- \rightarrow 4OH^- + 2H_2 \quad (2)$$

This hypothesis is supported by the observed 2:1 production ratio between H_2 and CO_2 (Fig. 2, Eqs. 1 and 2), and the balanced consumption of both protons and hydroxyl ions described in Equations 1–2 is supported by the observation of stable pH during the whole experiment. The standard Gibbs free energy change of whole reaction $C + 2H_2O \rightarrow CO_2 + 2H_2$ is positive, indicating it is a non-spontaneous process. A theoretical minimum voltage of 0.196 V has to be applied in order for the reaction to proceed under standard biological condition (T=298.15 K, P=1 bar, pH=7) here based on thermodynamic calculation (Supplementary Information). However, much higher external energy input (>2.0 V) was needed in practice to overcome the overpotential loss at both electrodes.

In the BES reactors, microorganisms served as the biocatalysts to reduce the over-potential of reaction on the electrode and significantly enhanced current production compared to the abiotic control reactors. Microbial community structure shows that the majority of bacteria on the anode were heterotrophs. Thus, the graphite carbon was not used as carbon source, but rather the microbes utilized the organics produced from the MES. Both H_2 and organics produced during MES could provide the energy source for the growth of anode microorganisms. While the specific contribution of microbial activities on GO exfoliation needs further investigation, it was clear that with bacteria the exfoliation rate was much higher. While a cathode potential was poised (−0.6 V vs. SHE), the anode potential of the reactor increased to a maximum of +1.6 V as compared with the abiotic control (0.71–0.98 V). This change should be associated with the biofilm catalysts formation on the cathode that reduced activation energy barrier, because the current in active reactor was four times that in the abiotic control. In addition, solution chemistry such as pH and ionic composition were changed due to microbial metabolisms, which also impacts GO exfoliation. Therefore, we hypothesize the mechanisms of bioelectrochemical exfoliation of graphite for BEGO production include: (i) Hydroxyl may act as a nucleophile to attack the edge and grain boundaries of graphite14 to open the sites up. (ii) Intermediates of microbial metabolisms such as electrosynthesis organics, a small amount of salt ions, and water may intercalate into the graphitic layers and cause expansion of graphite interlayers, and (iii) CO_2 produced during process can also exert large force into the graphite layers to exfoliate them from one another.

The electrotrophic biofilm established on the cathode or planktonic cells utilized CO_2 and electrons from the cathode or H_2 for microbial electrosynthesis, which resulted in the accumulation of organic compounds as reflected by COD. The H_2 production rate (46.3–67.8 mmol/L/day) was more than twice that of the previous BES studies (17.5–25.3 mmol/L/day) using water or reduced ferricyanide as electron sources at lower cathode potential (−0.8 to −0.7 V vs. SHE)41·42 and we attribute this to the higher current density (4.5 A/m^2) by graphite oxidation compared to 1.1–2.4 A/m^2 in previous systems. Hydrogenotrophic methanogenesis was effectively inhibited despite the presence of large amount of H_2 and CO_2, which was likely due to the low pH (4.54) during acclimation stage. It is widely accepted that most methanogens prefer a neutral environment for metabolism43.

The unique properties of graphene materials have been well recognized for their application potentials, and researchers around the world are diligently developing new methods to produce lower cost, high quality, and mass amount of graphene materials. The European Commission recently committed an unprecedented €1 billion Euros investment to graphene research (http://graphene-flagship.eu/), calling it the “the largest research excellence awards in history”, and it is forecasted that the graphene material markets will grow from $20 million in 2014 to more than $390 million in 202444. The preliminary economic analysis shows the cost of GO production through bioelectrochemical route can be orders of magnitudes lower than current methods (Supplementary Information). While this study reports the first findings of such a method, more studies are needed to understand the mechanisms, characterize microbial and electrochemical functions, test different materials, and optimize system configuration and performance. Some specific tasks may include more precise carbon and electron balance studies, system characterizations under different potentials, microbial communities, effects of GO on microbial growth and metabolisms, and materials purification and system optimization.

METHODS

Reactor Construction and Operation

Cylindrical glass bottles are used as three-electrode reactor systems. Each reactor has an effective volume of 220 mL (Fig. 1B). Graphite rod (D 0.6 cm × L 5 cm, Graphite Sales, OH) was used as the anode, and carbon cloth (5 cm x 10 cm, Fuel Cell Earth, MA) served as the cathode. A layer of nonconductive permeable glass fiber and a layer of nylon mesh were placed onto cathode as separator and support45. An Ag/AgCl reference electrode (RE-5B, BASi) was placed in between the anode and the cathode (0.198 V vs SHE). The electrodes were connected using titanium wires. A potentiostat (CHI1000B, CH Instruments Ins.) was used to poise the cathode (working electrode) at −0.6 V (vs. SHE), where anode was the counter electrode. All potentials are reported versus SHE.

The active reactors were inoculated with anaerobic sludge collected from an anaerobic digester, and the initial medium include (per liter): 0.36 g of glucose, 11.55 g of $Na_2HPO_4{\cdot}12H_2O$, 2.77 g of $NaH_2PO_4{\cdot}2H_2O$, 0.31 g of NH_4Cl, 0.13 g of KCl, and trace materials and vitamins46 flushed with 100% CO_2 to reach a constant CO_2 content of ~5.6 mmol in the headspace. Carbon dioxide was used as carbon source for microbial electrosythesis of organics. When current production was observed, the liquid was replaced with the same medium but without glucose and sludge inoculum. The operation was conducted in anaerobic condition, and after each transfer, the liquid and headspace were flushed with 100% CO_2. The enrichment stage was considered complete when high current, H_2 and organic production were observed. After that at least three batch cycles were operated. To quantify carbon flow, no more CO_2 flushing was conducted during media change after the first cycle. The exfoliated anode would be replaced if necessary during this stage. Three reactors were prepared as experiment replicates but were started at different time. The three reactors showed similar operation profiles and one reprehensive time course reactor profile is reported. The BEGO1, BEGO2 and BEGO3 are samples taken from three replicated reactors, respectively. All reactors were operated at room temperature of 25±2 °C. Two abiotic control reactors were operated. One followed the same protocol as the active reactor, with the cathode poised at −0.6 V (vs. SHE). Another control was setup by poising the anode at +1.6 V (vs. SHE), which was the maximum anode potential observed in biotic BES reactor.

Chemical Analysis and Calculation

Gas production was continuously collected using a gas bag (1L, Cali-5-Bond, Calibrated Instruments Inc.) connected to the headspace through a syringe needle. H_2 and CH_4 were measured by a gas chromatograph (Model 8610C, SRI Instruments) equipped with a thermal conductivity detector with nitrogen as the carrier gas. CO_2 and O_2 were analyzed using helium as the carrier gas47. Liquid samples were filtered through 0.22 μm membrane and then centrifuged at 13,000 rpm for 1 h to remove impurities such as carbon nanoparticles and bacteria. For organic product measurement, soluble chemical oxygen demand (COD) was analyzed using a standard method (TNTplus COD Reagent; HACH Company). Cyclic voltammetry (CV) was conducted at a scan rate of 1 mV/s between −0.8 to 0 V (vs. SHE) using a separate potentiostat (PC4/300, Gamry Instruments, NJ). The scan was carried out several times and the stable CV curves were used for analysis. The pH was measured using a pH meter (HQ440d, HACH company).

Total coulombs consumed (C_T) based on current measurements were calculated by integrating the current over time. Coulombs found in the product (C_P) was calculated using equation $C_p = n{\cdot}b{\cdot}F$, where n is the moles of product, b is the number of electrons in the product (b=2 and 4 for per mole of hydrogen and COD, respectively), and F is Faraday's constant (96,485 C/mol). Coulombic efficiencies were calculated by dividing C_P by C_T. Gas production rate (mmol/L/d) and current density (A/m^2) was calculated on the basis of solution volume (100 mL) and cathode projected area (50 cm^2), respectively. The amount of energy (W_E) added to the circuit by the potentiostat over a period of time (t) was calculated by equation $W_E = \int_0^t I E_{cell}\, dt$, where I and E_{cell} are current in the circuit and the voltage over the BES, respectively. The electricity energy consumption was normalized to kilowatt-hours (kWh) via 3600 kilojoules(kJ) per kWh.

Characterization of BEGO

The exfoliation mixture was first centrifuged at 2500 rpm to remove large graphite particles and possible planktonic microorganisms, and then the supernatant was filtered through a 0.1 μm Durapore PVDF filter (Millipore Co.,). The remaining solid materials on the filter were washed with the following sequence: 18 MΩ water, 80% ethanol, 18 MΩ water, 1 N HCl, and 18 MΩ water again. The resulting materials called BEGO were re-suspended in 18 MΩ water for characterization. UV-vis absorption spectra were recorded using a Thermo GENESYS 10S UV-vis spectrophotometer. Fourier transform infrared spectroscopy (FT-IR) was performed by using a Thermo-Nicolet FT-IR Avatar 370 spectrometer. XPS analysis was carried out by a Kratos Axis His spectrometer. Raman spectra were obtained by Raman Microscope with wavelength of 532 nm and a 100 × objective. Atomic force microscopy (AFM) images were generated on a VEECO Dimension 3100. Transition electron microscopy (TEM) and electron diffraction images were obtained using a FEI Tecnai G20 electron microscope. Samples for AFM and TEM imaging were prepared by drop-casting the dispersion onto freshly cleaved mica substrates and lacey carbon TEM grid, respectively, which were then air dried under ambient lab conditions.

Bacterial Community Composition Analysis

Anode biofilm samples were scratched using a sterilized blade, and small amount of carbon cloth cathode was cut using a sterile scissor. Both of them were used for total genomic DNA extraction using a PowerSoil DNA Isolation Kit (MoBio Laboratories, Inc., CA), and bacterial 16S rRNA gene targeted at the hypervariable V1-V3 region was amplified by PCR using a 10-nucleotide barcoded forward primer 8F (5′-AGAGTTTGATCCTGGCTCAG-3′) and the reverse primer 533R (5′-TTACCGCGGCTGCTGGCAC-3′) for high-throughput 454 GS-FLX pyrosequencing. The pyrosequencing and bioinformatics analysis were carried out according to our previous description46·48. Pyrosequencing produced 6816 and 6855 qualified sequences with an average length of 450 bp for anode and cathode community, respectively. Raw sequencing data were deposited to the NCBI Sequence Read Archive (SRA) with accession No. SRP040685.

ACKNOWLEDGEMENTS

We thank Dr. Hongjun Liang at Colorado School of Mines and Dr. Dong Su at Brookhaven National Laboratory for their gracious technical support. L.L. and Z.R. were supported by the College of Engineering and Applied Sciences of University of Colorado Boulder, and C.Z. was partially supported by the National Basic Research Program of China (973 Program, Program No. 2014CB846001) to A.L.

REFERENCES

1. Zhu, Y. *et al.* Graphene and graphene oxide: synthesis, properties, and applications. *Adv. Mater.* 22, 3906–3924 (2010).

2. Park, S. & Ruoff, R. S. Chemical methods for the production of graphenes. *Nat. Nanotechnol.* 4, 217–224 (2009).

3. Eda, G. & Chhowalla, M. Chemically derived graphene oxide: Towards large-area thin-film electronics and optoelectronics.*Adv. Mater.* 22, 2392–2415 (2010).

4. Liu, Z., Robinson, J. T., Sun, X. & Dai, H. PEGylated nanographene oxide for delivery of water-insoluble cancer drugs. *J. Am. Chem. Soc.* 130, 10876–10877 (2008).

5. Wang, L. *et al.* Graphene oxide as an ideal substrate for hydrogen storage. *ACS nano* 3, 2995–3000 (2009).

6. Mi, B. Graphene Oxide Membranes for Ionic and Molecular Sieving. *Science* 343, 740–742 (2014).

7. Yang, S.-T. *et al.* Folding/aggregation of graphene oxide and its application in Cu^{2+} removal. *J. Colloid Interf. Sci.* 351, 122–127 (2010).

8. Liu, N. *et al.* One-step ionic-liquid-assisted electrochemical synthesis of ionic-liquid-functionalized graphene sheets directly from graphite. *Adv. Funct. Mater.* 18, 1518–1525 (2008).

9. Lu, J. *et al.* One-pot synthesis of fluorescent carbon nanoribbons, nanoparticles, and graphene by the exfoliation of graphite in ionic liquids. *ACS nano* 3, 2367–2375 (2009).

10. Singh, V. V. *et al.* Greener electrochemical synthesis of high quality graphene nanosheets directly from pencil and its SPR sensing application. *Adv. Funct. Mater.* 22, 2352–2362 (2012).

11. Su, C.-Y. *et al.* High-quality thin graphene films from fast electrochemical exfoliation. *ACS nano* 5, 2332–2339 (2011).

12. Parvez, K. *et al.* Electrochemically exfoliated graphene as solution-processable, highly conductive electrodes for organic electronics. *ACS nano* 7, 3598–3606 (2013).

13. Liu, J. *et al.* A green approach to the synthesis of high-quality graphene oxide flakes via electrochemical exfoliation of pencil core. *RSC Adv.* 3, 11745–11750 (2013).

14. Parvez, K. *et al.* Exfoliation of Graphite into Graphene in Aqueous Solutions of Inorganic Salts. *J. Am. Chem. Soc.* 136, 6083–6091 (2014).

15. Zhu, C., Hao, Q., Huang, Y., Yang, J. & Sun, D. Microbial oxidation of dispersed graphite by nitrifying bacteria 2011.2. *Nanoscale* 5, 8982–8985 (2013).

16. Nevin, K. P., Woodard, T. L., Franks, A. E., Summers, Z. M. &Lovley, D. R. Microbial electrosynthesis: feeding microbes electricity to convert carbon dioxide and water to multicarbon extracellular organic compounds. *mBio* 1, 1–4 (2010).

17. Rabaey, K. & Rozendal, R. A. Microbial electrosynthesis—revisiting the electrical route for microbial production. *Nat. Rev. Microbiol.*8, 706–716 (2010).

18. Desloover, J., Arends, J. B. A., Hennebel, T. & Rabaey, K.Operational and technical considerations for microbial electrosynthesis. *Biochemical Society Transactions* 40, 1233 (2012).

19. Wang, H. & Ren,Z. J. A Comprehenisve Review of Microbial Electrochemical Systems As A Platform Technology, *Biotechnol. Adv.* 31, 1796–1807 (2013).

20. Gregory, K. B., Bond, D. R. & Lovley, D. R. Graphite electrodes as electron donors for anaerobic respiration. *Environ. Microbiol.* 6, 596–604 (2004).

21. Nevin, K. P. *et al.* Electrosynthesis of organic compounds from carbon dioxide is catalyzed by a diversity of acetogenic microorganisms. *Appl. Environ. Microbiol.* 77, 2882–2886 (2011).

22. Marshall, C. W., Ross, D. E., Fichot, E. B., Norman, R. S. & May, H. D. Electrosynthesis of commodity chemicals by an autotrophic microbial community. *Appl. Environ. Microbiol.* 78, 8412–8420 (2012).

23. Lu, L., Huang, Z., Rau, G. & Ren, Z. J. Microbial Electrolytic Carbon Capture for Carbon Negative and Energy Positive Wastewater Treatment. *Environ. Sci. Technol.* 49, 8193–8201 (2015).

24. Titelman, G. I. *et al.* Characteristics and microstructure of aqueous colloidal dispersions of graphite oxide. *Carbon* 43, 641–649 (2005).

25. Kuila, T. *et al.* A green approach for the reduction of graphene oxide by wild carrot root. *Carbon* 50, 914–921 (2012).

26. Hontoria-Lucas, C., Lopez-Peinado, A., López-González, J. d. D.,Rojas-Cervantes, M. & Martin-Aranda, R. Study of oxygen-containing groups in a series of graphite oxides: physical and chemical characterization. *Carbon* 33, 1585–1592 (1995).

27. Paredes, J. I., Villar-Rodil, S., Martinez-Alonso, A. & Tascon, J. M. D. Graphene oxide dispersions in organic solvents. *Langmuir* 24, 10560–10564 (2008).

28. Jin, M. *et al.* Facile physical route to highly crystalline graphene.*Adv. Funct. Mater.* 21, 3496–3501 (2011).

29. Kudin, K. N. *et al.* Raman spectra of graphite oxide and

functionalized graphene sheets. *Nano Lett.* 8, 36–41 (2008).

30. McAllister, M. J. *et al.* Single sheet functionalized graphene by oxidation and thermal expansion of graphite. *Chem. Mater.* 19, 4396–4404 (2007).
31. Moon, I. K., Lee, J., Ruoff, R. S. & Lee, H. Reduced graphene oxide by chemical graphitization. *Nat. commun.* 1, (2010).
32. Rabaey, K., Boon, N., Siciliano, S. D., Verhaege, M. & Verstraete, W.Biofuel cells select for microbial consortia that self-mediate electron transfer. *Appl. Environ. Microbiol.* 70, 5373–5382 (2004).
33. Zhang, T. *et al.* The direct electrocatalysis of phenazine-1-carboxylic acid excreted by *Pseudomonas alcaliphila* under alkaline condition in microbial fuel cells. *Bioresource technology*102, 7099–7102 (2011).
34. Friman, H., Schechter, A., Nitzan, Y. & Cahan, R. Effect of external voltage on *Pseudomonas putida* F1 in a bio electrochemical cell using toluene as sole carbon and energy source. *Microbiology* 158, 414–423 (2012).
35. Lu, L., Xing, D. & Ren, N. Pyrosequencing reveals highly diverse microbial communities in microbial electrolysis cells involved in enhanced H_2 production from waste activated sludge. *Water Res.*46, 2425–2434 (2012).
36. Liou, J. S.-C., Balkwill, D. L., Drake, G. R. & Tanner, R. S.*Clostridium carboxidivorans* sp. nov., a solvent-producing clostridium isolated from an agricultural settling lagoon, and reclassification of the acetogen *Clostridium scatologenes* strain SL1 as *Clostridium drakei* sp. nov. *Int. J. Syst. Evol. Microbiol.* 55, 2085–2091 (2005).
37. Lovley, D. R. & Nevin, K. P. Electrobiocommodities: powering microbial production of fuels and commodity chemicals from carbon dioxide with electricity. *Curr. Opin. Biotechnol.* 24, 385–390 (2013).
38. Rabaey, K., Girguis, P. & Nielsen, L. K. Metabolic and practical considerations on microbial electrosynthesis. *Curr. Opin. Biotechnol.* 22, 371–377 (2011).
39. Gür, T. M. Critical review of carbon conversion in “carbon fuel cells”. *Chem. Rev.* 113, 6179–6206 (2013).
40. Giddey, S., Badwal, S., Kulkarni, A. & Munnings, C. A comprehensive review of direct carbon fuel cell technology. *Prog. Energy Combust. Sci.* 38, 360–399 (2012).
41. Rozendal, R. A., Jeremiasse, A. W., Hamelers, H. V. M. & Buisman, C. J. N. Hydrogen production with a microbial biocathode.*Environ. Sci. Technol.* 42, 629–634 (2008).
42. Geelhoed, J. S. & Stams, A. J. M. Electricity-assisted biological hydrogen production from acetate by *Geobacter sulfurreducens.Environ. Sci. Technol.* 45, 815–820 (2011).
43. Dworkin, M., Falkow, S., Rosenberg, E., Schleifer, K.-H. &Stackebrandt, E. *The Prokaryotes: Vol. 3: Archaea. Bacteria: Firmicutes, Actinomycetes*, Third Edition edn. Springer Science & Business Media (2006).
44. Khasha, G. *Graphene markets, technologies and opportunities 2014-2024.* (ed^(eds). IDTechEx Inc (2014).
45. Lu, L. *et al.* Enhanced Bioremediatiosn of Hydrocarbon-Contaminated Soil Using Pilot-Scale Bioelectrochemical Systems.*J. Hazard. Materials*, 274, 8–15 (2014).
46. Lu, L., Xing, D. & Ren, N. Bioreactor performance and quantitative analysis of methanogenic and bacterial community dynamics in microbial electrolysis cells during large temperature fluctuations.*Environ. Sci. Technol.* 46, 6874–6881 (2012).
47. Luo, H., Jenkins, P. E. & Ren, Z. Concurrent desalination and hydrogen generation using microbial electrolysis and desalination cells, *Environ. Sci. Technol.* 45, 340–344 (2011).
48. Lu, L., Huggins, T., Jin, S., Zuo, Y. & Ren. Z. J. Microbial Metabolism and Community Structure in Response to Bioelectrochemically Enhanced Remediation of Petroleum Hydrocarbon-contaminated Soil. *Environ. Sci. Technol.* 48, 4021–4029 (2014).

CHAPTER 11

BIOELECTROCHEMICAL SYSTEMS: ENGINEERING AND STATISTICAL ASPECTS

Shuai Luo , Hongyue Sun , Qingyun Ping, Ran Jin, and Zhen He

Keywords: Bioelectrochemical systems, data mining, differential equations, engineering models, regression, statistical models

INTRODUCTION

Bioelectrochemical systems (BES) are emerging technologies that apply microorganisms to transform chemical energy in wastewater to electrical energy through multiple microbial-electrochemical reactions [1,2,3,4]. The increasing energy demand and desire for environmental sustainability motivates the technology development for energy recovery from wastes over the past decades [5,6,7]. Among the emerged concepts in energy recovery and wastewater treatment, BES have received significant attentions and are extensively studied with different configurations and designs, including microbial fuel cells (MFCs), microbial electrolysis cells (MECs), and microbial desalination cells (MDCs), as well as the combination of up-flow constructed wetland and MFC [2,8,9,10,11,12,13,14,15,16]. The performance of BES is affected by biological, physical-chemical and electrochemical aspects (e.g., microbial community, configurations, electrode material, *etc.*) [17,18,19]. Considerable studies have been conducted to optimize the above aspects [1,7,8,20,21,22,23,24]. However, the BES development for energy recovery from wastewater treatment still has bottlenecks [18,25]. For example, BES may not replace anaerobic digesters to treat the high-strength wastewater [26], and the scale-up BES cannot achieve a similar performance in current generation as that of the lab scale systems [18,27,28]. How to improve the BES performance is still challenging, and thorough understanding of these significant aspects that limit the overall BES performance is needed for future improvement. A variety of works have been conducted to study the effect of design and operational parameters on BES performance, such as reactor configurations and scales, electrode materials, electrode surface areas and the types of electron donors [8,29,30,31,32,33,34,35,36,37]. However, because of the lack of sensing technology, some aspects are hard to be measured *in situ*, such as microbial community composition, biofilm thickness and growth rate, redox mediator transferring, and substrate utilization rate of bacteria attached on the electrode [38,39,40,41,42,43,44]. These hard-to-measure aspects are likely to be key components for understanding the mechanisms of electricity generation and improving the efficiency of BES.

Mathematical modeling can be a powerful platform to investigate the effect of above aspects on the overall BES performance [19,23,45,46,47]. The use of mathematical models may provide analytical description of these hard-to-measure aspects, and thus reduce the measurement effort. The versatility of mathematical models enables the conversion of complexly systematic phenomena into a relatively simple series of mathematical expressions to describe the effect of each component on the overall output [19,48]. In general, mathematical modeling is performed in two approaches. One approach is to derive model based on engineering/physical laws governing the system processes, and this is called engineering modeling [49,50]. In this review, most engineering models are deterministic models, because the output of the model is determined by the parameter values and the initial conditions [51]. For example, differential equations (DEs) are used as main mathematical equations in the engineering models due to their strong ability to embody the dynamics of BES [19,52,53,54,55]. The parameter values and initial conditions are set depending on specific situations to operate the model. The main equations are combined with physical, chemical and electrochemical principles to solve specific problems. For

Citation: Shuai Luo, Hongyue Sun, Qingyun Ping, Ran Jin , and Zhen He, Bioelectrochemical Systems: Engineering and Statistical Aspects, doi:10.3390/en9020111.

example, Monod-type equations are created for describing the growth curve of specific microorganisms in a certain living environment [40,56,57]. Nernst-Michaelis-Menten equations are applied to calculate the amount of Nernstian electron transferred to embody bio-anode kinetics [58,59,60], and Nernst-Planck equations are applied to represent the ion diffusion through the interface or membrane of a system [61,62,63,64]. The other approach is statistical modeling, which is developed based on the data collection from BES [49,50], and is also valuable for characterizing the system input-output relationships, especially when there is limited engineering-domain knowledge to characterize the complex mechanisms of electricity generation and organic material removal. In addition, the statistical models can better capture the system uncertainty and remedy the error originated from engineering models for better system quantification [49].

To help satisfy the increasing needs for mathematical models to achieve better development of BES for wastewater treatment and energy recovery, this review aims to summarize and discuss different modeling strategies for BES in the categories of engineering models and statistical models. It is significantly different from a recent review paper of mathematical models that focuses on overview of model studies [19]. A general modeling methodological tree is presented in Figure 1. The motivations, assumptions, advantages, limitations and future directions of various modeling strategies summarized in this paper can be a reference for scholars to choose a proper modeling strategy according to their objectives.

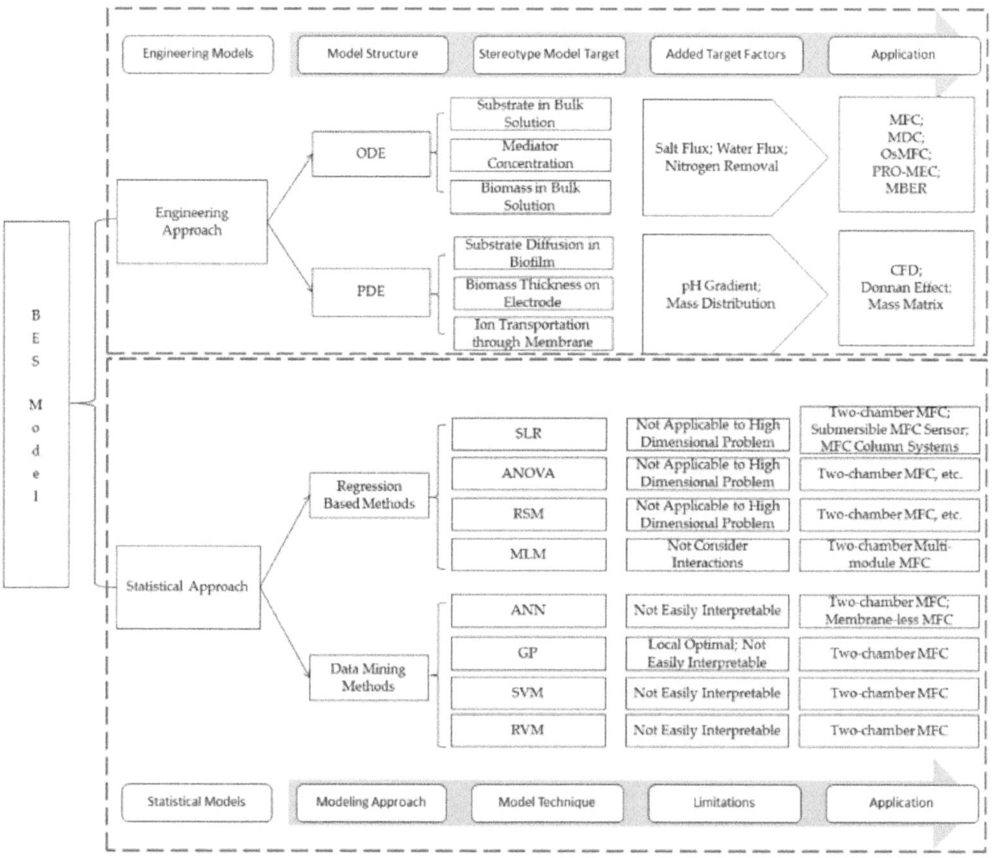

Figure 1: A bioelectrochemical systems (BES) modeling methodological tree. (ODE: ordinary differential equations; PDE: partial differential equation; MFC: microbial fuel cell; MDC: microbial desalination cell; OsMFC: osmotic MFCs; PRO-MEC: pressure-retarded osmosis/microbial electrolysis cell; MBER: membrane bioelectrochemical reactors; CFD: computational fluid dynamics; SLR: simple linear regression; ANOVA: analysis of variance; RSM: response surface methodology; MLM: Multi-task Lasso Model; ANN: artificial neural network; GP: genetic programming; SVM: support vector machine; RVM: relevant vector machine).

ENGINEERING MODELS

Overview

BES have multiple processes of physical-chemistry, microbiology and electrochemistry, which are dynamically related to each other. Such dynamic processes are governed by many key parameters, which influence the BES performance. For example, the energy recovery of a MFC is impacted by parameters about substrate concentration, substrate type and system configuration [65]. Understanding the relationships among these parameters and their dynamic processes is important to optimize the system for electricity generation and wastewater treatment [19,66,67]. Specifically, DEs are the most frequently applied techniques [52,53,54,55]. Based on the common mathematical methods used in BES, in the following, DEs are divided into ordinary differential equations (ODEs) and partial differential equations (PDEs) to discuss their motivations, advantages and limitations.

Ordinary Differential Equations

Introduction of Ordinary Differential Equations

ODE is a type of DE to describe the dynamic processes, focusing on only one variable. A classical form of ODE is shown in Equation (1) [68]:

$$F\left(x, y, y', y'', \cdots, y^{(n-1)}, y^{(n)}\right) = 0 \quad (1)$$

The objective of ODE is to find the relationships between input x and output y by describing the differential changes of output under change of input [52,68].

Ordinary Differential Equation Stereotypes in Bioelectrochemical Systems

ODE is a convenient and efficient tool to describe the relationships of various variables in BES, such as the impact of growth and decay of microorganisms on the dynamic response of the change of substrate concentration and microbial population on the anode [54,69]. ODEs allow fast computation and provide an advantageous platform for real-time process control and optimization of BES [70,71]. Some ODE stereotypes can successfully model the dynamics of substrate concentrations and various microbial populations in BES, and these ODE stereotypes are also the foundations for the future ODE development [39,54,72].

ODE stereotypes are constructed based on the principle of mass balance to illustrate the dynamic response of substrate concentration and biomass in the bulk solution in anode chamber [39]. Several parameters were considered in the ODE: the rate of substrate consumption, the rate of growth and death of biomass in the bulk solution, and the rate for the electrochemical oxidation of reduced mediator at the anode surface. These considerations have effectively tracked the source and sink for the dynamic change of substrate and biomass in the bulk solution at each moment in time, and a generic form of ODEs was constructed [39]. However, the ODE cannot predict the dynamic change of microbial biomass on the biofilm. The equations simplified the microbial community into overall biomass, which cannot specify and distinguish the contribution of each type of functional microbial species (e.g., electricigenic and acetoclastic methanogenic microorganisms) in the anode chamber [39].

More general ODE stereotypes were developed with more focus on the competition and contribution among various types of microorganisms in the anode chamber of BES, including two types of BES (*i.e.*, MFC and MEC) [54,69]. They successfully integrated multiple kinetic equations such as Monod equations to take many parameters into account, including substrate consumption rate, biomass growth rate, and even the impact of mediator on the dynamic simulation (Figure 2). The related ODE stereotype equations are listed and explained in the Table S1 in Supplementary Materials.

These ODEs could simulate the system output affected by glucose supply by constructing two-step degradation using acetate as intermediate substrate [69]. Moreover, a multi-population dynamic model involves ODEs looking into the various contributions of electricigenic and acetoclastic methanogenic microorganisms on the system performance, including substrate consumption and current generation [54,69].

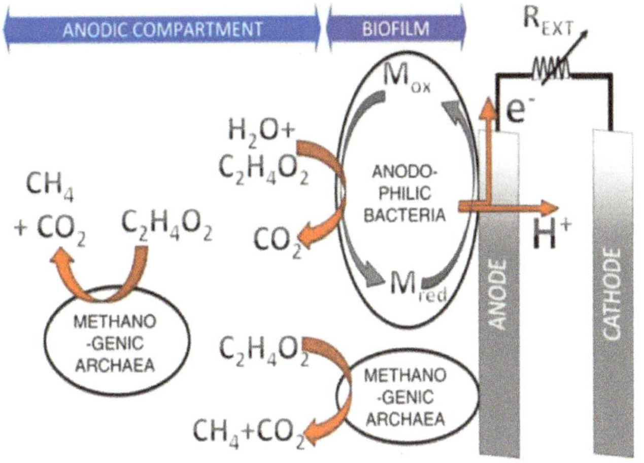

Figure 2: Conceptual MFC model showing carbon source conversion and biofilm activity. Reproduced with permission from reference [54].

Applications of Ordinary Differential Equation Stereotypes

The above ODE stereotypes are further developed for other BES, such as MDC, osmotic-microbial fuel cell (OsMFC), pressure-retarded osmosis/microbial electrolysis cell (PRO-MEC), and membrane bioelectrochemical reactors (MBERs). The ODEs applied in these BES are developed with addition of certain equations on the stereotype models mentioned before, to enrich the modeling functions for other physical and chemical processes (Table S2, Supplementary Materials). For instance, the ODEs were modified with further consideration of salt flux through the ion exchange membrane, to estimate the performance of MDC [73,74] (Figure 3). The upgraded model can predict the salt concentration in the desalination, anode and cathode chambers, and be used to study the impact of different parameters on the system performance to optimize the salt removal efficiency. The model can even predict the dynamic response of the system output by changing the influent concentration of acetate, the external resistance, and the influent salt concentration. The developed ODEs can accurately simulate the quantitative influence of different parameters on MDC performance. The best acetate flow rate, influent salt concentration, and salt solution flow rate were found according to the model output [73], and thus the developed ODEs can be applied as valuable tools for future MDC development and optimization.

Forward osmosis (FO) is reported to be competent to help MFCs to simultaneously achieve wastewater treatment, water extraction from the wastewater, and electricity generation together [75], but quantitative impact of various parameters on the performance still needs to be investigated to improve the system output. ODEs are further developed to evaluate the application value of FO in some BES applications, including OsMFCs and PRO-MEC [76,77]. Additional ODEs related to the salt flux between draw and feed solution were used to predict the performance of an OsMFC, with consideration of water and salt flux through the FO [76]. The model construction was proposed to analyze the impact of FO on the organic degradation in the anode and current generation [76]. The OsMFC model successfully validates the experimental results that higher catholyte conductivity can increase the current generation, and reveals that increased water flux through FO can decrease electrical resistance of membrane and generate higher current, which proves the important value of FO membrane in increasing current output of BES. The equations of water flux through FO were integrated into the previously developed ODE stereotype, on purpose of simulating the energy generation from the process of pressure retarding in the unit of PRO, to study the impact of given pressure on the electricity generation of the PRO unit [77]. The developed model can obtain the relations among pressure supply, electricity supply to drive MEC and the hydrogen production. The ODEs of PRO-MEC further reveal the impact of time-dependent change of proton concentration (e.g., migration and diffusion of proton to the cathode and further reduction to hydrogen gas) on the reduction potential and the current generation.

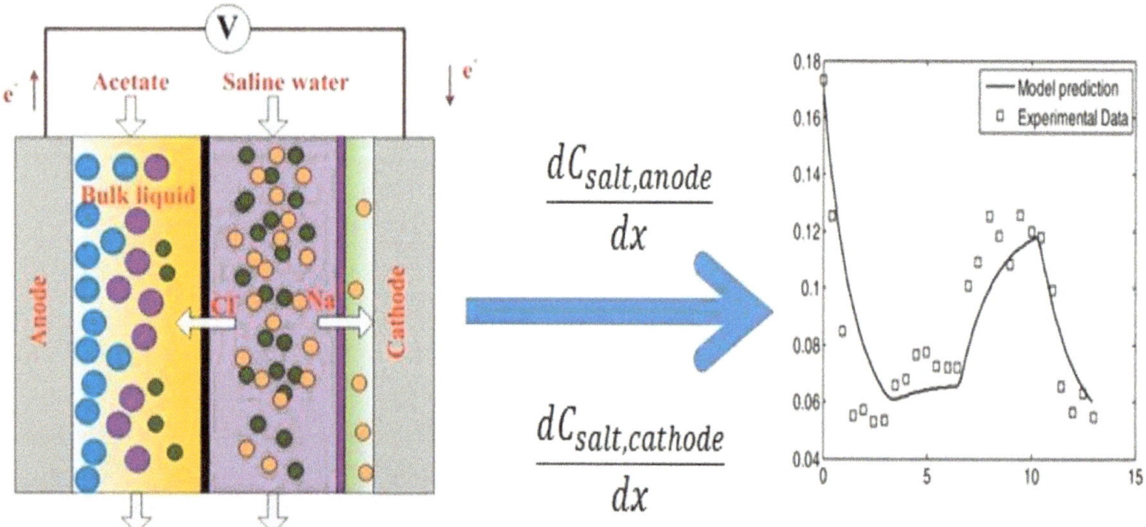

Figure 3: System configuration and model output of a MDC. Reproduced with permission from reference [73].

BES are combined with membrane bioreactor (MBRs) to form the system called MBERs for outstanding effect of wastewater treatment and nutrient removal (e.g., nitrogen) from the entire system [78]. However, quantitative analysis is still needed to improve the MBERs performance, and the ODE stereotypes were developed to simulate the system output of MBERs [79]. The model can predict the growth and loss of "cake" layer on the membrane reactor, to indicate the impact of biofouling on the system performance. The contribution of this model is to accurately model three MBERs with various configurations [80]. The versatility of the configuration application can cover many important properties in a MFC-MBR system, such as membrane fouling, nitrogen removal, cake formation and detaching based on various cross-flow rates through the membrane. Substrate diffusion in the biofilm matrix is another important factor on the biological activity attached on the electrode, and understanding of the parameter is significant to improve the activity of microorganisms on the electrode for greater electricity generation. The ODE stereotypes were modified to enrich the mathematical expressions of mass balance and substrate diffusion between the interface of biofilm and bulk solution, to simulate the biological activity inside the biofilm [44,81]. The model is able to simulate the performance of reverse MFC, especially the biofilm thickness and acetate production rate in the cathode [81]. The conception of finite difference approach was applied to make ODEs workable to simulate the mass diffusion and the impact of pH change on the substrate consumption and biofilm performance attached on the electrode (Figure 4) [44]. The updated ODEs become more powerful for analyzing the dynamic response of biofilm accompanied with the pH change on the substrate consumption and current generation.

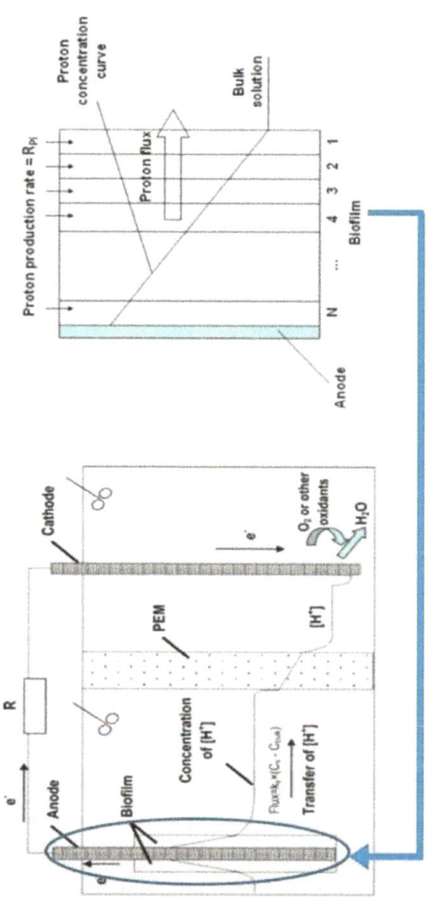

Figure 4: Proton concentration profile and concentration gradient within the biofilm. Reproduced with permission from reference [44].

Advantages and Limitations

ODEs are an easy and convenient method to reflect the macro scale dynamic response against the change of other parameters. ODEs have advantages in terms of low computational load and relatively easy to be adjusted [19,73,80].

However, ODEs also have intrinsic limitations. ODEs can only trace the time-dependent performance of one input [82]. Thus, when facing problems related to distance such as geometry of the biofilm and the distribution of ion concentration in the ion exchange membrane between anode and cathode, ODEs cannot sufficiently study the problems. Though the finite difference concept was successfully used to divide the biofilm thickness into multiple finite sections for ODE to predict the mass diffusion into or out of each section of the biofilm one by one (Figure 4), the distance-related problems cannot be handled by ODEs in general [44]. To successfully apply ODEs to perform BES modeling, multiple assumptions are needed to simplify the problems, and these assumptions are important error sources for ODEs. Most assumptions are relevant to the elimination of spatial factors, because time is already taken into account as one independent variable and spatial factor (e.g., distance) is difficult to be another independent variable. Assumptions related to the above ODE stereotypes and applications are listed and explained with the relevant key kinetic equations in Supplementary Materials. Some important examples of the assumptions are discussed as follows:

First, both anode and cathode compartments are treated as continuously stirred tank reactors with the feature of completely mixing [73,83], and biofilm is distributed uniformly attached on the electrode, and biomass retention is described by two phase growth-wash-out model [38,54,84]. This assumption may cause the delayed dynamic response of biofilm attached on the electrode against the change of environmental factors, especially the biological components such as biofilms on the electrode and bacteria in the bulk solution. For example, ODEs were successfully applied to describe the impact of catholyte conductivity (adjusted by varying NaCl concentrations) on the current generation and water flux through the membrane of OsMFC; however, when the catholyte was changed from 35 $g \cdot L^{-1}$ to 2 $g \cdot L^{-1}$ NaCl, leading to decreased conductivity and current generation, the anolyte COD changed very slightly, which was overestimated by model expectation for greater anolyte COD consumption [76]. Because microorganisms are the dominant community on the anode, the model output of COD change is expected to be strongly correlated to the change of current generation and anodophilic bacteria on anode [54,73,76,77]. However, the actual change of COD is much smaller than the modeling results when changing the experimental conditions, possibly because of delayed response of biological factors for the changing conditions in these cases [76,77].

Second, mass transportation through the membrane due to electro-migration or chemical gradient between the anode and the cathode is assumed to be sufficiently fast to ignore the thickness of ion exchange membrane [85,86]. This assumption ignores the impact of ion exchange membrane on the ion transportation, and will generate errors for the gain and loss of ions in anode and cathode chambers. For example, Donnan effects on the ion exchange membrane would lead to strong effect of ion adsorption onto the membrane, which makes electrical potential drop through the thickness of ion exchange membrane [87]. The accumulated ions on the solution/membrane interface would have repulsion effect on the same charged ions transferring through the membrane from the solution [88,89], which will have extra ion transferring impedance to cause the interruption on the ion transportation.

Other assumptions are also required to construct ODEs. For example, one can assume the ideal mixing to eliminate concentration gradient, because only one independent variable (*i.e.*, time) can be taken into account while other variables (e.g., distance) cannot be considered in ODE models. The intrinsic drawback of ODE needs other mathematical principles to solve by simultaneously considering more independent variables in the model.

Partial Differential Equations

Introduction of Partial Differential Equations

Most BES are impacted with many temporal and spatial parameters. Thus, more complex DEs are needed, which should not only consider the temporal factors, but also the spatial factors for predicting the high dimensional problems. PDEs can deal with multiple inputs simultaneously with the general form shown in Equation (2) [90]:

$$f\left(x_1, \ldots, x_n, y, \frac{\partial y}{\partial x_1}, \ldots, \frac{\partial y}{\partial x_n}, \frac{\partial^2 y}{\partial x_1 \partial x_1}, \ldots, \frac{\partial^2 y}{\partial x_1 x_n}, \ldots\right) = 0 \quad (2)$$

Partial Differential Equation Stereotypes in Bio-electrochemical Systems

Some high-dimensional problems, including biofilm thickness, microbial distribution on the electrode, and the ion transportation through the membrane, are all significant to study the working mechanisms of BES performance to enhance knowledge for performance optimization [38,63,91]. Previously, assumptions are needed for ODEs to solve the above problems. Upcoming PDEs can help analyze these assumptions that ODEs cannot deal with. Some PDE stereotype equations are also created to reveal the dynamic response of the biofilm, and the mass transportation through the ion exchange membrane in three-dimension (3D) (Table S4, Supplementary Materials) [38,39,63,91]. These PDE stereotypes lay a cornerstone for the future applications of PDEs for various objectives.

A stereotype PDE was developed by integrating Monod and Nernst equations together, to simulate the electron production rate and the local potential of the biofilm and to understand the kinetic and electrochemical property of the biofilms on substrate consumption and electrons production [38,92]. This PDE has developed a concept of conductivity of biomass biofilm to represent electrical conductivity through 3D pattern of the biofilm and to make spatial prediction for impact of biomass thickness and accumulation of inner biomass on current generation. The PDE successfully overcomes the limitation of ODE to quantitatively indicate the relation between specific detachment rate and density of biomass on the electrode.

Another PDE was assisted with Nernst-Planck equations to simulate the ion transportation through the ion exchange membrane [63]. The simulation was performed in the process of diffusion and electro-migration, and the ion transport process is shown in Figure 5. The model is able to show the ion gradient in the ion exchange membrane, which is promising to simulate ion transportation that ODEs cannot deal with. Moreover, it can describe the pH gradient in the ion exchange membrane to help explain the relation of pH gradient between two chambers and ion exchange membrane.

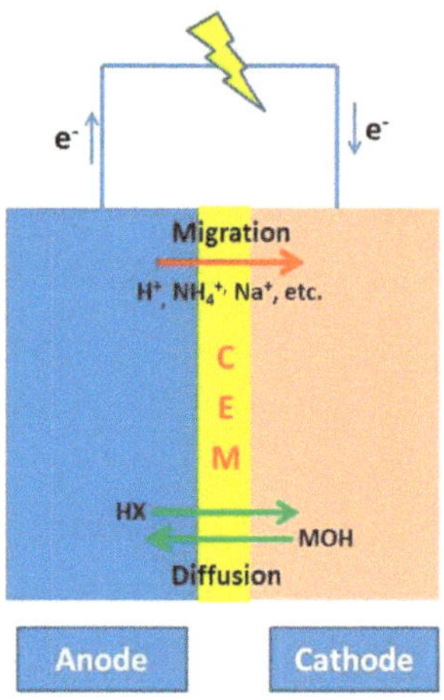

Figure 5: A schematic illustration of ion transportation across ion exchange membrane. (Note: CEM: cation exchange membrane; AEM: anion exchange membrane.)

A series of general PDEs were further developed on the basis of original ODEs. The PDEs can predict the microbial activity on the biofilm to consume the substrate for growth and electron generation, which considered substrate diffusion and the biofilm geometry together [39,72], and even the effect of pH on the model prediction [91].

Applications of Partial Differential Equation Stereotypes

The above stereotype PDEs are further developed and applied for specific problems in BES, such as impact of oxygen diffusion on substrate concentration profile in single-chamber MFCs, quantitative description of the Donnan effect, and computational fluid dynamics (CFD) to predict mass distribution inside the tubular reactor. The relevant equations are listed and explained in Supplementary Materials (Table S5).

Single-chamber MFCs are a system without ion exchange membrane to separate the anode and cathode chamber, because the design can reduce the membrane resistance for higher current generation [34,93]. However, the oxygen diffusion from the cathode to the anode will be greater in the absence of the membrane impedance, and thus the activity of the microorganisms will be impacted by the oxygen transportation [34]. Investigating the impact of transported oxygen on the system performance is important to evaluate and optimize the system structure. A previously developed stereotype PDE was modified to simulate the impact of substrate concentration, oxygen diffusion, cathode thickness (material: polytetrafluoroethylen (PTFE)), and biofilm thickness on the voltage output of the entire system [94]. It was found that at the condition of high substrate supply, the biofilm thickness had greater impact on voltage output than the PTFE layer [94].

PDE is also applied to investigate the fluid flow pattern in the BES by CFD. CFD can be applied to study the flow dynamics to improve the condition of substrate distribution to increase electricity generation [95]. Kim *et al.* applied CFD to simulate the flow dynamics in twelve MFCs with different internal structures and to analyze the impact of anode surface on the electricity generation; it was found that power generation increased exponentially with increased anode surface, due to greater bio-electrochemical reaction space [96].

The dynamic change of biofilm on the electrode is another important problem, because the biofilm growth and detaching is strongly relevant to substrate degradation and electricity generation [97,98]. A PDE is used to simulate the dynamic response of Geobacter of the biofilm and understand how the species react against the change of external operating conditions. The results reveal the importance of substrate concentration and pH gradient on the biofilm growth and electricity generation through the entire biofilm [97,98].

The simulation on the ion transportation through the ion exchange membrane can serve as a monitor to indicate the relations between the activities of specific ions and system output. Ammonium is a significant species deserving to be monitored and analyzed, because the ammonium is considered to be main carrier of protons through proton exchange membrane from anode to cathode, and the accumulation of ammonia in cathode enhances the possibility to recover the ammonia from cathode due to high pH in cathode [99,100]. A PDE model on the basis of Nernst-Planck equation is created to successfully describe the current-driving ammonia evaporation and recovery in the cathode, and the model can describe in detail the impact of different factors on the ammonia recovery including flow rate of gas in the cathode and back diffusion of ammonia from cathode to anode, which makes great contribution to improve ammonia treatment and recovery from the wastewater in BES [85].

Advantages and Limitations

The primary advantage of PDEs is the higher ability to solve the spatial problems [82]. PDEs can consider the impact of not only temporal factors, but also the spatial factors on the system performance. Thus, PDE can trace the flow dynamics and biomass growth in the angle of high dimensions in BES [39,41,101]. However, because PDEs consider the impact of more factors on the BES performance, the trade-off is the increased time required for PDE construction [82]. PDEs structure is generally more complex than ODEs, because a larger number of parameters needs identification. The complexity increases the simulation time, particularly in the multidimensional space dependent models [82].

There are still some limitations even if PDEs are used to substitute ODEs in the model structure, and the details of PDE assumptions are presented and explained in Supplementary Materials. For example, it was assumed that organic substrate is converted by reducing redox mediator, and the redox mediator is to complete the electron transferring mechanism in bulk solution and biofilm [39,91]. However, in practical application, some operational steps can make deviation between experimental data and model outputs. When voltage drops to a baseline, replacing the medium will remove some mediator from the bulk liquid, leading to abrupt concentration drop of the mediator in the solution. After medium refreshment, the change of the mediator concentration causes decreasing current in the model while experimental data indicates current increase after acetate addition [39]. The deviation exists due to certain assumptions in this particular model [91]. More advanced alternative models are required to improve the model accuracy and precision.

Model Adjustment and Validation

Sensitivity Analysis of Model Parameters

The parameters in engineering models need to be adjusted to predict the model output accurately and speed up the model construction. Sensitivity analysis can allow us to determine which parameters are the most influential factor on the model output, and select the most relevant parameters to be adjusted in less computational time [102,103]. A common method to perform the sensitivity analysis is to vary each parameter while fixing the other parameters, one factor at a time, to check the impact of the parameter on output [83,104].

Another function of sensitivity analysis is to evaluate the parameters' relative importance for the output [73,105,106]. The importance comparison among all parameters can quantitatively assist the explanation for the relationships among various factors in the system. For example, model construction consists of three parts: model fitting, validation and prediction [73,107,108]. Sensitivity analysis is one part in the process of model fitting, to determine the impact of a specific parameter on the system output [73]. In general, relatively more sensitive parameters will take priority to be adjusted and to fit the experimental data under specific situations in model fitting. In the MDC model, one parameter (k_r) has the highest sensitivity, whose physical meaning indicates how fast the internal resistance of the system responds to the change of microorganism concentration [73]. In this case, changing flow rate is applied in model fitting to obtain the experimental data and to give guidance for development of mathematical equations, assumption determination and parameter adjustment to make model structure. Because of high sensitivity, k_r significantly impacts the model accuracy and precision. High sensitivity of k_r is reasonable because the dynamic change of microorganism concentration in a MFC strongly affects the efficiency and availability of electron transferring to the anode. k_r considers and represents the overall impacts of substrate diffusion and reaction kinetics on the anode [109,110,111]. Determination of highly sensitive parameters can reduce the computational time to make satisfactory model. Once the model fitting and constructing is finished, model validation is applied to verify the model reliability, by changing operating conditions of other aspects (e.g., organic concentration) [73]. All parameters must be fixed in the model validations, except for the target parameter being studied in each test, to guarantee the model applicability [73].

Another significant parameter, mediator yield (Y_m), represents the production of intracellular mediator in the anodophilic bacteria [54]. The anodophilic bacteria in the electrode biofilm takes the electron from the substrate oxidation, reduces the intracellular mediator to finally transfer electrons to the electrode by direct contact or nanowires [54]. Higher Y_m means stronger conversion of the oxidized mediators to reduced mediators and higher electron transferring activity accompanied with higher concentration of anodophilic bacteria on the electrode [8,112,113]. Thus, determination of Y_m in BES modeling is critical to understand and quantify the impact of mediators for the current generation.

BES development would require more parameters to make an accurate model to optimize their applications. For example, submersible microbial fuel cell (SMFC) is a system to place anode in the anaerobic sediment, and the anode connects with the air-cathode to complete a closed circuit and generate electricity by degrading organics in the sediment [114]. To improve SMFC performance, it is helpful to construct an effective model to simulate the changes of DO and current generation under different conditions [115,116,117]. However, the biological aspects, such as biofouling in the environmental waters and competition among different bacteria for COD degradation and nutrient removal, need more parameters to trace their change and activity. Sensitivity analysis is a strong tool to help select the parameters. For example, biofouling formation and detachment strongly impacts electrochemical reaction on a cathode due to bacterial competition for oxygen [18,118]. To address this issue, the researchers adopted the increase rate of resistance in their model to describe the difficulty of water filtration through the membrane under biofouling [80,119]. Sensitivity analysis can also assist in parameter tuning for accelerated model construction. Other innovated applications in BES also need new parameters to analyze the system performance, such as growth rate of algal biomass for integrated photo-bioelectrochemical system (IPB), and the constant for hydrogen production rate in microbial electrolysis and desalination cells (MEDC) [120,121,122].

In summary, sensitivity analysis is an important method to identify the most influencing parameters on the model performance. It can not only reduce the working load of parameter adjustment, but also reflect the importance of certain parameters on the system output and help understand the working mechanisms and interrelations among immeasurable factors in MFC [73,123]. The current sensitivity analysis frequently applied in BES is one-factor-at-a-time method (OFAT). This method has intrinsic drawback because it cannot completely reflect the systematic impact of various parameters together on the performance. Some important and sensitive parameters might be hidden if other parameters are not set in the suitable range, which will limit the sensitive output of these actually important parameters [124]. Thus, more efficient and accurate approaches of sensitivity analysis are required to solve complicated problems [124].

Model Validation

Model validation is a significant process after finishing model construction and adjustment. A principle to validate the model is to see if experimental data can match the modeling data [54,73]. The model validation is required to enhance the model reliability and practicability. In addition, the model validation can give significant information of what physical, chemical or biological reactions or phenomena are not explained by the model, which will need further model modification. There are several standards to calculate the modeling error, such as mean square errors (MSE), root mean square error (RMSE), and coefficient of determination (R^2) [54,63,69,73].

The results of model validation of above engineering models are listed in Supplementary Materials (Table S7) (models without showing validation results are not listed), as a reference to compare the model accuracy. However, it is difficult to set a general standard to determine if the model has fit the experimental data well, because of the difference in BES configurations and operating conditions. Whether a model is acceptable will depend on both validation results and specific case conditions.

A common standard to optimize the parameter for most accurate output is to minimize the root mean square error (RMSE) by adjusting the model structure and parameters to fit the experimental data shown in Equation (3) [54]:

$$\mathrm{RMSE} = \frac{\sqrt{\frac{\sum_{i-1}^{N}\left(y_i - \widehat{y_i}\right)^2}{N}}}{\max\left(\hat{y}\right)} \tag{3}$$

Normally, RMSE below 20% is acceptable for the model validation, but acceptable level is difficult to be fixed due to variously actual cases [73,74,76,77,80,83]. For example, Yuan *et al.* [77] constructed an engineering model to successfully predict the dynamic volume profile of feed and draw solution with RMSE smaller than 2.5%; although the RMSE for a MEC model was relatively high (13%–23.6%), the predicted output was still acceptable because the error came from overestimation of organics in the model, which was related to microbial degradation of organics in anode featured with relatively higher unstableness and delayed response of biological factors than other physical attributes. There are also some cases with high RMSE or intrinsic limitations indicating that engineering models are not powerful enough to deal with specific outputs [49,63,76,80]. For example, a model was proposed to simulate the performance of a MBER with fluidized-bed system with granular activated carbon (GAC) as media [80]. A significant large RMSE of 55.7% was obtained when changing the anodic hydraulic retention time (HRT) to 5 h [80]. The huge error may come from the delayed acclimation of electrochemical active bacteria on the electrode, and it is possible that most of enhanced organics are consumed by non-electrochemical active microorganism on the GAC surface instead of electrode [80]. However, the error can be utilized to derive more impact of other factors, which is not considered in the original model structure (e.g., organic consumption by non-negligible microorganisms attached on GAC in above case).

With higher demand for model application in BES, more advanced engineering models are needed. Even engineering models themselves cannot satisfy the model demand, and another model system, statistical model, can be a complementing method to help engineering models to increase the prediction ability, which will be discussed in next section [49].

STATISTICAL MODELING

Overview

Statistical models are constructed based on the measured data from the BES system. The statistical models are useful to reveal the input–output relationships and identify the significant parameters for system quantifications, especially when the engineering models are hard to be constructed in a timely manner. The uncertainty can also be better quantified in statistical models than engineering models. On the other hand, the conclusions drawn from the statistical models needed to be validated with the engineering knowledge. A large variety of statistical models, ranging from regression based methods to data mining methods, can be used for the BES modeling. In the following, we will focus on MFCs as representative BES.

The performance of a statistical model depends heavily on the type and quality of the collected data. In general, the data can be collected from controlled experiments, observational studies, and simulations based on engineering models. Most MFC model works were performed based on controlled experiments with studying the effect of parameters on system performance [125,126,127,128,129,130,131,132,133]. More details on data collection for controlled experiments will be introduced in Section 3.2. Very limited observational studies were reported [134,135], where the data collection was performed directly during the real production of the system. Simulation data can be generated from the engineering models based on the first principles (see Section 2 for more details). In these simulation studies, the data are generated from the computer program by setting the program inputs and initial and/or boundary conditions.

Depending on the data structure, a variety of model formats can be chosen from. In general, one can select from linear models, nonlinear models and more complex data mining models [136]. A discussion on selecting the appropriate type of model is provided in Supplementary Materials. Most models used for MFC modeling are based on linear regression (LR), and these models usually have good interpretability and simple model structures [125,126,127,129,130,131,132,133] . In recent years, data mining methods were also introduced for modeling the MFC performance such as artificial neural network (ANN), genetic programming (GP), support vector machine (SVM) and relevance vector machine (RVM) [128,137,138]. These data mining methods can yield good prediction performance. However, they usually do not have an analytical form thus are difficult to be interpreted [19].

In the following, the data generation approaches are first briefly reviewed. Then, different statistical models used in the MFC studies are summarized (Table 1). The motivations, advantages and limitations of these models are discussed. Finally, model diagnostic and variable selection, which are the missing part in the literature, are also discussed.

Table 1: Summary of statistical models

Methods	Data Generation	Systems	Major Conclusions	References
SLR	OFAT	Submersible MFC	Current density increased with increase of microorganisms concentration before saturation	[130]
SLR	OFAT	MFC biosensor	Current density increased with increase of acetate concentration before saturation	[139]
SLR	OFAT	Two-chamber MFC	Voltage increased with increase of organic loading rate and flow rate at neutral pH	[140]
SLR	OFAT	Two-chamber MFC	Mainly cation species were responsible for the transport of positive charge	[141]
ANOVA, RSM	FD, CCD	Two-chamber MFC	The interaction effect was significant	[125]
ANOVA, RSM	FD	Two-chamber MFC	Two-factor and three-factor interaction effects were less significant	[127]
MLM	FD with covariates	Two-chamber MFC	Cathode moisture was significant on energy recovery but insignificant on organic removal efficiency	[142]
ANOVA, RSM	CCD	Two-chamber MFC	Quadratic effects can be significant, but interaction effects were not significant	[129]
ANOVA, RSM	CCD	Mediator-less single-chamber MFC	Interaction and quadratic effects can be significant	[126]
ANOVA, RSM	CCD	Two-chamber MFC	All the linear, square and interaction effects were significant	[132]
ANOVA, RSM	CCD	Membrane-less MFC	Interaction and quadratic effects can be significant	[133]
ANOVA	PBD	Submersible MFC	Temperature, initial pH, and conductivity were significant	[130]
ANOVA, RSM	PBD	Two-chamber MFC	Glucose, $NaHCO_3$ and KCl were significant	[131]
GP, ANN, SVM	[143]	Two-chamber MFC	GP performed best among the three models	[138]
ANN	Controlled Experiment	Membrane-less MFC	R^2 value can be as high as 0.99	[144]
RVM	UD	Two-chamber MFC	RVM performed better than SVM	[128]

[130] uses both OFAT and PBD. OFAT: one-factor-at-a-time; FD: factorial design; CCD: central composite design; PBD: Placket-Burman design; UD: uniform design.

Data Generation

MFCs have been studied via controlled experiments intensively [125,126,127,128,129,130,131,132,133,141,145]. The controlled experiments are performed in OFAT or design of experiments (DOE) fashion. In OFAT, only one factor is varied while the others are remained constant. The system performance, such as power generation and organic removal, can be measured by varying the factor under study. The effect of the varied factor is analyzed visually or by statistical models [130,139,140,141,146,147,148,149,150,151]. Note that OFAT can only study the effect of a certain parameter on the MFC performance once at a time, resulting in interpretation problems of interactions or joint effects of multiple factors on the MFC performance. Misleading results can be yielded in such circumstances [125,126,133,152]. For instance, it was found that the interaction effect of pH and buffer concentration was very significant, while a false conclusion could be made if the interaction effect was ignored [125]. In DOE, each factor will be studied with selected levels. The combinations of these levels over different factors consist of the design table. Based on the design table and measured output, the main effects and interaction effects can be studied simultaneously. Here, main effects refer to the effect of factors themselves, and interaction effects characterize the phenomenon that the effect of one factor on output changes over the level of other factor(s). Moreover, the number of experimental runs required can be significantly reduced in DOE compared with all possible combinations, or random selections of factor levels. The past MFC studies have used factorial design (FD), central composite design (CCD), box-Behnken design (BBD), Placket-Burman design (PBD), and uniform design (UD) [125,127,128,130,131,132,133,142, 153]. More general designs can be found in [152].

Figure 6 shows some of the designs for a three-factor problem, where the red points represent the settings of experiment runs and the cube represents the design space. The factors in FD usually take two levels of values, where only linear effect can be estimated (Figure 6a). In CCD and BBD, the factors will take more levels of values and the quadratic effect can be studied without adding a lot of additional experimental runs (Figure 6b,c). PBD uses a limited amount of experimental runs for screening a relatively large amount of factors (Figure 6d). In the screening experiments, the interactions between the factors are usually neglected [154]. Finally, UD uniformly assigns experimental points in the domain, and is used when the underlying model structure is unknown. See details in constructing a UD in [155]. Note that along with the experiments in DOE, other related but uncontrolled factors can be recorded. These factors are usually called covariates [152].

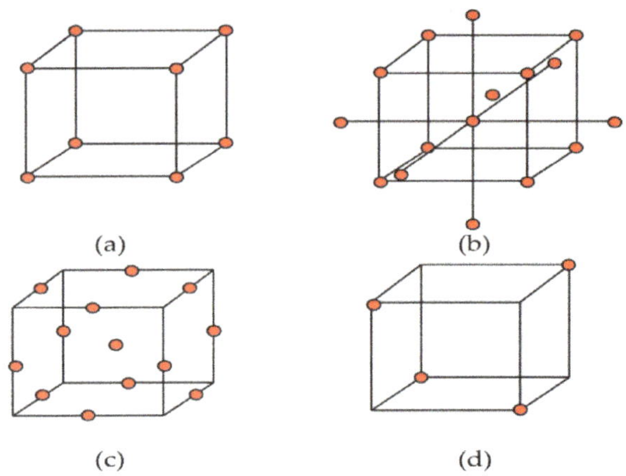

Figure 6: Design of experiments (DOE) used in the MFC studies: (a) FD; (b) CCD; (c) box-Behnken design (BBD); and (d) PBD.

The observational data, which are usually collected after the system enters the application phase, are rarely reported in the literature [134,135]. These studies did not use statistical models for the observation data collected from the field operation. The simulation studies are also widely performed in MFC literature (see Section 2 for details). In such studies, the focus was on the engineering models themselves. We would like to point out that the engineering models can generate useful simulation data, and statistical surrogate models can then be constructed based on the simulation data and real measurement data. These statistical models can be used to enhance the simulation models' prediction performance.

It is worth noting that experimental procedures vary a lot in different studies, even for the same BES configuration. The statistical models generated from one system may not be directly used for another study. Standard experimental and data collection procedures are therefore required for guaranteeing the data uniformity and quality. In BES, available data to be measured are mainly divided into two categories: electrochemistry and wastewater treatment. Cell voltage and electrode potential are significant data about electricity generation in BES. The collection and reporting procedures of these data have been previously discussed, and voltage meters, multimeter, and potentialstat are common methods to collect the data and observe the electrochemical behavior of the system under specific situation [8]. In wastewater treatment, removal efficiency of organic compounds (e.g., acetate, glucose, and COD), and other indexes (e.g., pH and total nitrogen) are significant data to be collected [76,78]. High performance liquid chromatography (HPLC) is a common method to qualitatively and quantitatively measure the organic concentration [156,157,158,159]. Chemical analysis such as COD kits (Hach Co., Loveland, CO, USA) can be applied to measure COD, ammonium, nitrate and other ion concentration in water [160,161,162]. Some calculation data such as Coulombic efficiency (CE), power density, and energy efficiency are also needed to report the efficiency of electricity generation and assess the BES performance [8,163].

Linear Regression

To model the MFC, a LR, with the form shown in Equation (4) can be used. The LR uses an additive form of inputs and an error term for the prediction of the output shown in Equation (4) [164]:

$$y_i = \beta_0 + \beta_1 x_{i,1} + \cdots + \beta_p x_{i,p} + \varepsilon_i, i = 1, \cdots, n \qquad (4)$$

where y_i is the *i*th observation of the output such as power density, $x_{i,j}$ is the *i*th observation of the *j*th input such as substrate concentration, β_0 is the intercept term, β_j is the model parameter for the *j*th input, and ε_i is the error term for the*i*th observation. The basic assumptions for any LR are that the input–output relationships can be represented in a linear form, and the error terms $\varepsilon_i, i=1,\cdots,n$ are normal, identically and independently distributed with mean zero and variance σ^2. These model assumptions need to be checked via model diagnostic after the estimation steps. However, the model diagnostic process is rarely reported in the MFC literature, which should be emphasized.

The data generated from OFAT can be studied via simple linear regression (SLR), where only one factor is used as input. For instance, the relationship between current density and microorganisms concentration was investigated [130]. It was found that the current density increased with the increase of active microorganism concentration when the active microorganism concentration was below 6.52 nmol-ATP/L (ATP: adenosine-triphosphate), and the system saturated at higher active microorganism concentrations. In another example, the association of current generation and bulk phase acetate concentration was found to be positive association at acetate concentration value less than 2.3 mM [139]. SLR was also used for studying the species that were responsible for the transport of positive charge [141], and the effect of organic loading rate, flow rate and pH on current density, voltage and volumetric power [140].

The analysis of variance (ANOVA) and response surface methods (RSM) are also based on LR analysis, which are widely used for the analysis of DOE data of MFCs. Taking a system with two factors A and B, with a and b levels respectively for example, a typical ANOVA has the form shown in Equation (5):

$$y_{i,j,k} = \mu + \alpha_i + \beta_j + (\alpha\beta)_{i,j} + \varepsilon_{i,j,k}, i = 1, \cdots, a, j = 1, \cdots, b, k = 1, \cdots, n \qquad (5)$$

where $y_{i,j,k}$ is the *k*th observation at the level *i* and *j* for A and B respectively, μ is the overall mean, α_i and β_j are main effects of A and B, $(\alpha\beta)_{i,j}$ is pair-wise interaction effect between A and B, and $\varepsilon_{i,j,k} \sim N(0,\sigma^2)$ is the error term. ANOVA can be used to identify significant effects. RSM uses a polynomial model, with the form in Equation (4), to approximate the input-output relationships. RSM can help determine the optimal settings for a response at the specified design region.

The effects of pH and catholyte buffer concentration on power output, columbic efficiency, COD change and COD removal efficiency were studied for a two-chamber MFC via ANOVA and RSM based on FD and CCD [125]. It was reported that the increase in buffer concentration would lead to an increase in power generation at a high pH level; while the opposite impact was observed at low pH level. In another word, the interaction between pH and catholyte buffer concentration was significant, which cannot be discovered by the OFAT approach [125]. The factor setting for maximal power generation was obtained and validated by physical experiments [125]. In a sediment MFC, the effects of pH, distance between electrodes and external resistance on organic removal, nitrogen removal and power generation were studied via ANOVA and RSM [127]. It was reported that organic removal, nitrogen removal and power density were optimized if controlling pH in 7.6–8.5, distance between electrode in 90–100 cm and external resistance in 0–52 ohm. The authors also concluded that two-factor and three-factor interactions were less significant, and might be negligible in the final identified mode. The ANOVA or RSM were used in several other works studying the effects of parameters on MFC performance (as summarized in Table 1) [126,129,130,131,132,133,153].

It is worth mentioning that the MFC systems are becoming more complex with successively added functions. New factors are of interest and the number of factors increases quickly for the MFC stack. For instance, new parameters such as ammonia and sulfate were detected via bipolar bioelectrodialysis, and a MFC with multiple modules was studied [116,142]. Under such circumstances, variable selection is needed. Recently, a multi-task Lasso model (MLM) was proposed for modeling the multi-module MFC system, and showed its great modeling power (Figure 7) [142].

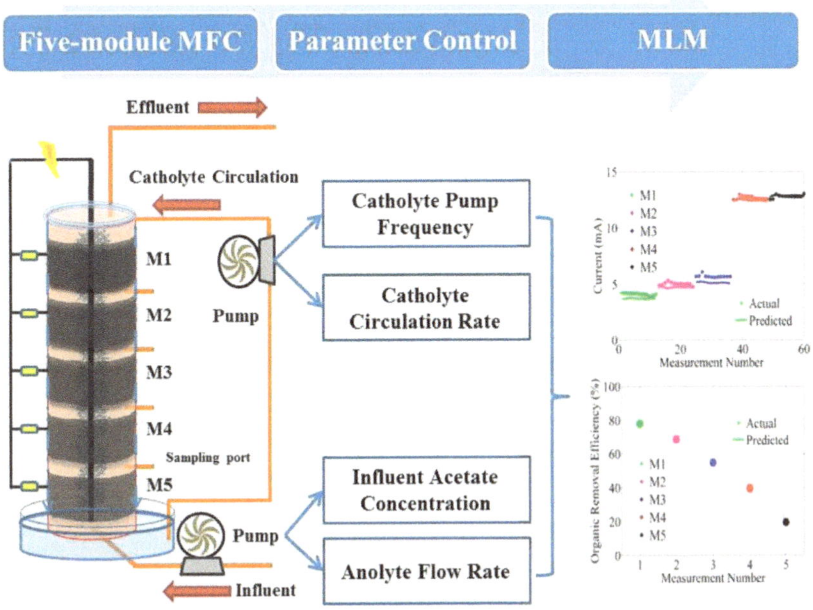

Figure 7: MLM for a two chamber multi-module MFC. Reproduced with permission from reference [142].

Data Mining Methods

Some data mining methods were applied for MFC modeling, though there were no clearly stated reasons on why these data mining methods were preferred over regression models. In general, these data mining methods can have good prediction performance, but usually do not have good engineering interpretations [19]. These data mining methods included ANN, GP, SVM and RVM.

ANN is inspired by biological neural networks and uses a number of connected "neurons" for the modeling of input-output relationships. The "neurons" can be characterized by different layers, as shown in Figure 8, where the hidden layers are composed of some function of input neurons [144]. More details can be found [165].

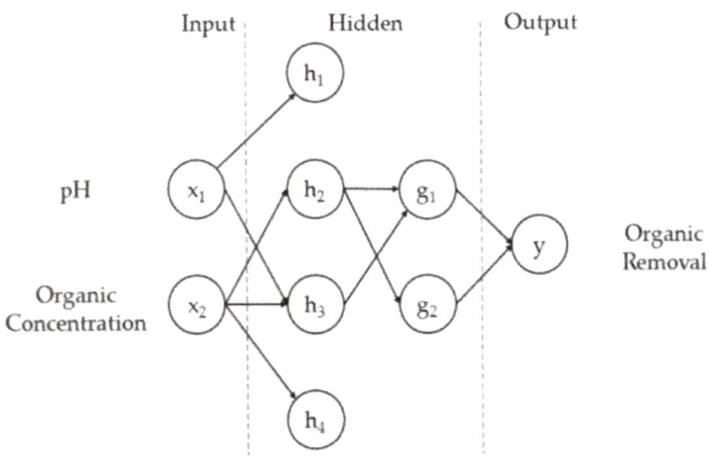

Figure 8: Illustration of ANN.

GP mimics the evolution process, and uses crossover and mutation operations for learning [138]. In GP, the solutions to a problem are represented by tree structures, and an initial structure is generated randomly. The optimal structure is learned via heuristic searching over all structures created by crossover and mutation operations until some termination criterion is met. For more details, please refer to [138].

SVM and RVM are also effective methods for predicting MFC performance. SVM maps the finite-dimensional input space to infinite-dimensional space via kernel functions, and performs the regression in the infinite-dimensional space. The learning in infinite-dimensional space is usually easier than in finite-dimensional input space [136]. RVM has very similar formulation as SVM, except that the RVM solves the problem in Bayesian framework [128]. The RVM can provide probabilistic classification, and typically use fewer kernel functions than SVM. RVM usually uses expectation maximization algorithm for learning, and may fall into local minima [166]. On the other hand, SVM uses sequential minimal optimization and is guaranteed to have a global optimum [167]. The effects of temperature and ferrous sulfate concentrations on the voltage were studied via GP, ANN and SVM [138]. The researchers adopted the data for a two-chambered MFC, and the model inputs included temperature or ferrous sulfate concentrations and time. They concluded that GP performed best among the three models based on R^2 values. However, their models contained complex transformations such as tan, tanh, sin, log, *etc.*, which are hard to be interpreted. At the same time, R^2 values can be used to describe the model fitting performance, but not the prediction performance [164]. In another example, the effects of temperature, pH and electron acceptor concentration on power density were studied via ANN for a membrane-less MFC [144]. It was reported that the R^2value could be as high as 0.99; however, the specific variable effects and the analytical model structure were not provided. A RVM was adopted for investigating the effects of ionic concentration, pH, medium nitrogen concentration and temperature on Coulombic efficiency and power density [128]. The R^2 values were larger than 0.99, and the optimized factor setting was obtained via genetic algorithm (GA), and validated by validation experiments [128].

Model Diagnostic

The model assumptions must be checked before applications. If the model assumptions are severely violated, the model inference and interpretation may have some problem [164]. However, the model diagnostic was rarely reported in MFC literature. In the following, the model diagnostic for LR will be briefly reviewed as an example.

Model diagnostic for LR is based on residuals, which are calculated as the difference between the observed value and fitted value. The common model diagnostic for LR includes: (1) whether the input-output relationship is linear; (2) whether the error terms have homogeneous variance ("identically distributed"); (3) whether the error terms are independent; (4) whether the error terms are normally distributed; and (5) whether there are outliers [164]. The model diagnostic methods include graphical methods and statistical tests. The graphic methods are more vivid and interactive, but the judgment made is subjective. On the other hand, statistical tests provide quantitative tools to evaluate the model assumptions [164]. If a certain model assumption is violated according to the model diagnostic in LR, several approaches can be used to relief the problem. For instance, the weighted least square can be used if the variance is heterogeneous. The nonlinear regression can be used if the linear model structure does not work well. Box-cox transformation can be used if the normal assumption is violated [164]. If the above remedy approaches do not work, one can use data mining methods as described in Section 3.4.

Variable Selection

Many model candidates are usually available within a given modeling framework. Given the model candidates, a certain evaluation criterion is needed to select the best candidates. These criteria include: R^2, adjusted R^2, Mallow's C_p, Akaike information criterion (AIC), Bayesian information criterion (BIC), RMSE of cross validation (CV), *etc.* [136,164]. R^2 is usually used in the MFC study to characterize how well the model fits, and p-value is used to judge the significance of variables [125,126,127,129,130,132,133]. As discussed before, R^2 may not be a good criterion to characterize the prediction performance. Moreover, the p-value is subjective and there is no consensus on how to select a threshold on variable significance [168]. More systematic variable selection methods are required.

The variable selection methods, such as subset selection, stepwise selection and model regularization, can be applied for selecting the best model. Subset selection considers all possible candidate models and select the best based on certain criterion. Such combinatorial analysis is infeasible for high dimensional problems. Stepwise selection starts from a null (empty) model or a full model (model with all the inputs), and systematically add or drop variables [169]. The stepwise selection is computationally inefficient for high dimensional problems, where regularized regression can be used. A regularized regression model is estimated by solving the objective function shown in Equation (6):

$$\sum_{i=1}^{n}\left(y_i - x_i^{T}\beta\right)^2 + \lambda\Omega(\beta), i = 1, \cdots, n \tag{6}$$

where $x_i = \left(1, x_{i,1}, \cdots, x_{i,p}\right)^{T}, \beta = \left(\beta_0, \beta_1, \cdots, \beta_p\right)^{T}$, and $\Omega(\cdot)$ is a measure of the model complexity which can take the form such as $|\beta_1|+\cdots+|\beta_p|$. The $x_{i,j}$, and β_j are defined as before. $\sum_{i=1}^{n}\left(y_i - x_i^{T}\beta\right)^2$ is a loss function to characterize the model prediction performance. λ is a tuning parameter that controls the balance between how well the model fits and how complex the model is. A family of models will be yielded by varying the tuning parameter λ. The best model candidate with the smallest MSE can be selected from a regularization path [170].

PERSPECTIVE

The BES modeling may be further improved in the aspects of variable selection, modeling with functional data, data fusion and hybrid models. With the development of more BES applications, demand for BES modeling is enhanced.

Variable Selection

Many parameters can potentially have strong correlation with the MFC performance. The number of parameters of interest could increase rapidly for a MFC system consisting of multiple modules [171]. However, due to the experimental cost, limited samples can be collected. The number of parameters of interest could be certainly larger than the number of samples. In this case, ANOVA and RSM cannot be directly applied. Variable selection method can be applied for investigating the important parameters for such systems when there are a lot of parameters but limited samples. Currently, there is very limited application of variable selection in MFC studies. In a recent work, a multi-task Lasso was adopted for identifying the important parameters for system performance characterization in a serially connected five-module MFC [142]. More works on systematic variable selection are required for understanding, monitoring and control the MFCs.

Modeling with Functional Data

The MFCs are essentially complex nonlinear dynamic systems whose performance would evolve over time [172]. The evolving history of key parameters such as temperature, pH and moisture are important for the system characterization. These measurements over time are called functional data. Generally, the dynamics of the system could be characterized by ODEs or PDEs, and validated with sampled points during operation over time. But these ODEs or PDEs are time-consuming to be constructed, and may not be accurate in the prediction. Given the measurements over time, some functional data analysis techniques could be adopted so that the important features of the system can be identified. For instance, the varying-coefficient model can be adopted for modeling a dynamic system [173].

Data Fusion

Because the MFCs, though different configurations, could contain many similar features [8], ensemble modeling strategy could integrate knowledge from an existing configuration to new systems development [174]. In this way, fewer samples will be needed for new system modeling. There will also be various differences between the field condition and lab testing condition. Therefore, the model learned from the lab measurement data may not be accurate for applications for field production. Under such a situation, ensemble modeling strategy can be used to tune the model so that it fits the need for field production [174]. The basic idea is to adopt the existing knowledge on the model format and significance of variables to the newly developed systems, and most details can be found in a previous publication [174].

Hybrid Models

Hybrid models integrate the engineering models and statistical models systematically. One example is to keep the engineering models as the mean part, but add statistical adjustments to the final model to account for the errors incurred by simplification assumptions in the engineering models. It can therefore take advantage of the strength of each type of models for system characterization, and will yield better modeling performance. Such an approach has been used for manufacturing systems and show their advantages [49,50], and may also be applied for MFC studies.

CONCLUSIONS

Mathematical modeling is a powerful tool to help understand and guide BES development for energy recovery from waste and wastewater. Both engineering models and statistical models have their merits and drawbacks. Engineering models are developed based on the biochemical mechanisms to reflect the relationships among versatile variables in MFC systems, which require knowledge in the field of physical-chemistry, biology and electrochemistry. Engineering models have advantage in quantitative analysis and explanation of the impact of certain variables that are hard to be measured. On the other hand, the statistical models have higher flexibility in the model format and are capable of dealing with the system uncertainty. The statistical models are driven by the measurement data, and easier to develop for an unknown complex system. However, the statistical models need to be interpreted by domain expert before applications.

ACKNOWLEDGMENTS

S.L. was supported by a fellowship from Water INTERface IGEP, Graduate School of Virginia Tech.

AUTHOR CONTRIBUTIONS

Shuai Luo, Hongyue Sun and Qingyun Ping wrote the paper; Zhen He and Ran Jin gave guidance and improved the paper quality. All authors contributed substantially to the work reported.

REFERENCES

1. Li, W.; Yu, H.; He, Z. Towards sustainable wastewater treatment by using microbial fuel cells-centered technologies.*Energy Environ. Sci.* 2013, *7*, 911–924.
2. Pant, D.; Singh, A.; Van Bogaert, G.; Olsen, S.I.; Nigam, P.S.; Diels, L.; Vanbroekhoven, K. Bioelectrochemical systems (BES) for sustainable energy production and product recovery from organic wastes and industrial wastewaters. *RSC Adv.* 2012, *2*, 1248–1263.
3. Wang, H.; Luo, H.; Fallgren, P.H.; Jin, S.; Ren, Z.J. Bioelectrochemical system platform for sustainable environmental remediation and energy generation. *Biotechnol. Adv.* 2015, *33*, 317–334.
4. Mohanakrishna, G.; Srikanth, S.; Pant, D. Bioelectrochemical Systems (BES) for Microbial Electroremediation: An Advanced Wastewater Treatment Technology. In *Applied Environmental Biotechnology: Present Scenario and Future Trends*; Springer: New Delhi, India, 2015; pp. 145–167. [Google Scholar]
5. McCarty, P.L.; Bae, J.; Kim, J. Domestic wastewater treatment as a net energy producer—Can this be achieved?*Environ. Sci. Technol.* 2011, *45*, 7100–7106.
6. Logan, B. Food and Light Bulbs. *Environ. Sci. Technol. Lett.* 2014, *1*, 442.
7. Wang, H.; Park, J.D.; Ren, Z.J. Practical energy harvesting for microbial fuel cells: A review. *Environ. Sci. Technol.* 2015,*49*, 3267–3277.

8. Logan, B.E.; Hamelers, B.; Rozendal, R.; Schröder, U.; Keller, J.; Freguia, S.; Aelterman, P.; Verstraete, W.; Rabaey, K. Microbial fuel cells: Methodology and Technology. *Environ. Sci. Technol.* 2006, *40*, 5181–5192.

9. Kelly, P.T.; He, Z. Nutrients removal and recovery in bioelectrochemical systems: A review. *Bioresour. Technol.* 2014,*153*, 351–360.

10. Oon, Y.-L.; Ong, S.-A.; Ho, L.-N.; Wong, Y.-S.; Dahalan, F.A.; Oon, Y.-S.; Lehl, H.K.; Thung, W.-E. Synergistic effect of up-flow constructed wetland and microbial fuel cell for simultaneous wastewater treatment and energy recovery.*Bioresour. Technol.* 2016, *203*, 190–197.

11. Santoro, C.; Soavi, F.; Serov, A.; Arbizzani, C.; Atanassov, P. Self-powered supercapacitive microbial fuel cell: The ultimate way of boosting and harvesting power. *Biosens. Bioelectron.* 2016, *78*, 229–235.

12. Abbasi, U.; Jin, W.; Pervez, A.; Bhatti, Z.A.; Tariq, M.; Shaheen, S.; Iqbal, A.; Mahmood, Q. Anaerobic microbial fuel cell treating combined industrial wastewater: Correlation of electricity generation with pollutants. *Bioresour. Technol.*2016, *200*, 1–7.

13. Qin, M.; Molitor, H.; Brazil, B.; Novak, J.T.; He, Z. Recovery of nitrogen and water from landfill leachate by a microbial electrolysis cell-forward osmosis system. *Bioresour. Technol.* 2015, *200*, 485–492.

14. Escapa, A.; Mateos, R.; Martínez, E.; Blanes, J. Microbial electrolysis cells: An emerging technology for wastewater treatment and energy recovery. From laboratory to pilot plant and beyond. *Renew. Sustain. Energy Rev.* 2016, *55*, 942–956.

15. Rosenbaum, M.; He, Z.; Angenent, L.T. Light energy to bioelectricity: Photosynthetic microbial fuel cells. *Curr. Opin. Biotechnol.* 2010, *21*, 259–264.

16. Zhang, F.; Ge, Z.; Grimaud, J.; Hurst, J.; He, Z. Long-term performance of liter-scale microbial fuel cells treating primary effluent installed in a municipal wastewater treatment facility. *Environ. Sci. Technol.* 2013, *47*, 4941–4948.

17. Zhi, W.; Ge, Z.; He, Z.; Zhang, H. Methods for understanding microbial community structures and functions in microbial fuel cells: A review. *Bioresour. Technol.* 2014, *171*, 461–468.

18. Logan, B.E. Scaling up microbial fuel cells and other bioelectrochemical systems. *Appl. Microbiol. Biotechnol.* 2010, *85*, 1665–1671.

19. Ortiz-Martínez, V.M.; Salar-García, M.J.; De Los Ríos, A.P.; Hernández-Fernández, F.J.; Egea, J.A.; Lozano, L.J. Developments in microbial fuel cell modeling. *Chem. Eng. J.* 2015, *271*, 50–60.

20. Song, H.-L.; Zhu, Y.; Li, J. Electron transfer mechanisms, characteristics and applications of biological cathode microbial fuel cells—A mini review.*Arab. J. Chem.* 2015.

21. He, Z.; Kan, J.; Mansfeld, F.; Angenent, L.T.; Nealson, K.H. Self-Sustained Phototrophic Microbial Fuel Cells Based on the Synergistic Cooperation between Photosynthetic Microorganisms and Heterotrophic Bacteria. *Environ. Sci. Technol.* 2009, *43*, 1648–1654.

22. Pendyala, B.; Chaganti, S.R.; Lalman, J.A.; Heath, D.D. Optimizing the performance of microbial fuel cells fed a combination of different synthetic organic fractions in municipal solid waste. *Waste Manag.* 2016.

23. Recio-Garrido, D.; Perrier, M.; Tartakovsky, B. Modeling, optimization and control of bioelectrochemical systems. *Chem. Eng. J.* 2016, *289*, 180–190.

24. Ki, D.; Popat, S.C.; Torres, C.I. Reduced overpotentials in microbial electrolysis cells through improved design, operation, and electrochemical characterization. *Chem. Eng. J.* 2016, *287*, 181–188.

25. Choi, S. Microscale microbial fuel cells: Advances and challenges. *Biosens. Bioelectron.* 2015, *69*, 8–25.

26. Li, X.; Abu-Reesh, I.; He, Z. Development of Bioelectrochemical Systems to Promote Sustainable Agriculture.*Agriculture* 2015, *5*, 367–388.

27. Rozendal, R.A.; Hamelers, H.V.; Rabaey, K.; Keller, J.; Buisman, C.J. Towards practical implementation of bioelectrochemical wastewater treatment. *Trends Biotechnol.* 2008, *26*, 450–459.

28. Sleutels, T.H.; Ter Heijne, A.; Buisman, C.J.; Hamelers, H.V. Bioelectrochemical systems: An outlook for practical applications. *ChemSusChem* 2012, *5*, 1012–1019.

29. He, C.; Mu, Z.; Yang, H.; Wang, Y.; Mu, Y.; Yu, H. Electron acceptors for energy generation in microbial fuel cells fed with wastewaters: A mini-review. *Chemosphere* 2015, *140*, 12–17.

30. Du, Z.; Li, H.; Gu, T. A state of the art review on microbial fuel cells: A promising technology for wastewater treatment and bioenergy. *Biotechnol. Adv.* 2007, *25*, 464–482.

31. Pant, D.; Van Bogaert, G.; Diels, L.; Vanbroekhoven, K. A review of the substrates used in microbial fuel cells (MFCs) for sustainable energy production. *Bioresour. Technol.* 2010, *101*, 1533–1543.

32. Liew, K.B.; Daud, W.R.W.; Ghasemi, M.; Leong, J.X.; Su Lim, S.; Ismail, M. Non-Pt catalyst as oxygen reduction reaction in microbial fuel cells: A review. *Int. J. Hydrog. Energy* 2014, *39*, 4870–4883.

33. Janicek, A.; Fan, Y.; Liu, H. Design of microbial fuel cells for practical application: A review and analysis of scale-up studies. *Biofuels* 2014, *5*, 79–92.

34. Liu, H.; Cheng, S.; Logan, B.E. Production of electricity from acetate or butyrate using a single-chamber microbial fuel cell. *Environ. Sci. Technol.* 2005, *39*, 658–662.

35. Kim, J.R.; Rodríguez, J.; Hawkes, F.R.; Dinsdale, R.M.; Guwy, A.J.; Premier, G.C. Increasing power recovery and organic removal efficiency using extended longitudinal tubular microbial fuel cell (MFC) reactors. *Energy Environ. Sci.*2011, *4*, 459–465.

36. Aelterman, P.; Versichele, M.; Marzorati, M.; Boon, N.; Verstraete, W. Loading rate and external resistance control the electricity generation of microbial fuel cells with different three-dimensional anodes. *Bioresour. Technol.* 2008, *99*, 8895–8902.

37. Kim, J.R.; Cheng, S.; Oh, S.-E.; Logan, B.E. Power generation using different cation, anion, and ultrafiltration membranes in microbial fuel cells. *Environ. Sci.*

Technol. 2007, *41*, 1004–1009.

38. Marcus, A.K.; Torres, C.I.; Rittmann, B.E. Conduction-based modeling of the biofilm anode of a microbial fuel cell.*Biotechnol. Bioeng.* 2007, *98*, 1171–1182.
39. Picioreanu, C.; Head, I.M.; Katuri, K.P.; Van Loosdrecht, M.C.; Scott, K. A computational model for biofilm-based microbial fuel cells. *Water Res.* 2007, *41*, 2921–2940.
40. Picioreanu, C.; Katuri, K.P.; van Loosdrecht, M.C.M.; Head, I.M.; Scott, K. Modelling microbial fuel cells with suspended cells and added electron transfer mediator. *J. Appl. Electrochem.* 2009, *40*, 151–162.
41. Picioreanu, C.; Van Loosdrecht, M.; Katuri, K.P.; Scott, K.; Head, I. Mathematical model for microbial fuel cells with anodic biofilms and anaerobic digestion. *Water Sci. Technol.* 2008, 965–971.
42. Logan, B.E.; Regan, J.M. Electricity-producing bacterial communities in microbial fuel cells. *Trends Microbiol.* 2006, *14*, 512–518.
43. Mohan, S.V.; Velvizhi, G.; Modestra, J.A.; Srikanth, S. Microbial fuel cell: Critical factors regulating bio-catalyzed electrochemical process and recent advancements. *Renew. Sustain. Energy Rev.* 2014, *40*, 779–797.
44. Zhang, L.; Deshusses, M. Application of the finite difference method to model pH and substrate concentration in a double-chamber microbial fuel cell. *Environ. Technol.* 2014, *35*, 1064–1076.
45. Wen, Q.; Wu, Y.; Cao, D.; Zhao, L.; Sun, Q. Electricity generation and modeling of microbial fuel cell from continuous beer brewery wastewater. *Bioresour. Technol.* 2009, *100*, 4171–4175.
46. Dai, J.; Li, X.; Li, B.; Wang, L. Design and modeling of an underwater energy harvesting system. In Proceedings of the International Congress on Analytical Sciences (ICAS 2011), Kyoto, Japan, 22–26 May 2011.
47. Pinto, R.; Tartakovsky, B.; Srinivasan, B. Optimizing energy productivity of microbial electrochemical cells. *J. Process Control* 2012, *22*, 1079–1086.
48. Bender, E.A. *An Introduction to Mathematical Modeling*; John Wiley & Sons: New York, NY, USA, 1978.
49. Joseph, V.R.; Melkote, S.N. Statistical adjustments to engineering models. *J. Qual. Technol.* 2009, *41*, 362–375.
50. Joseph, V.R.; Yan, H. Engineering-Driven Statistical Adjustment and Calibration. *Technometrics* 2015, *57*, 257–267.
51. Thomann, R.V. Deterministic and statistical models of chemical fate in aquatic systems. In *Applied Mathematical Ecology*; Springer: Berlin, Germany, 1989; pp. 322–351.
52. Scholz, G.; Scholz, F. First-order differential equations in chemistry. *ChemTexts* 2015, *1*, 1–12.
53. Zwillinger, D. *Handbook of Differential Equations*; Academic Press: Orlando, FL, USA, 1998.
54. Pinto, R.P.; Srinivasan, B.; Manuel, M.F.; Tartakovsky, B. A two-population bio-electrochemical model of a microbial fuel cell. *Bioresour. Technol.* 2010, *101*, 5256–5265.
55. Tebbutt, P. *Basic Mathematics for Chemists*; John Wiley & Sons: Chichester, UK, 2001.
56. Escapa, A.; Gil-Carrera, L.; García, V.; Morán, A. Performance of a continuous flow microbial electrolysis cell (MEC) fed with domestic wastewater. *Bioresour. Technol.* 2012, *117*, 55–62.
57. De Los Angeles Fernandez, M.; de Los Angeles Sanroman, M.; Marks, S.; Makinia, J.; Del Campo, A.G.; Rodrigo, M.; Fernandez, F.J. A grey box model of glucose fermentation and syntrophic oxidation in microbial fuel cells. *Bioresour. Technol.* 2015, *200*, 396–404.
58. Rimboud, M.; Desmond-Le Quemener, E.; Erable, B.; Bouchez, T.; Bergel, A. Multi-system Nernst-Michaelis-Menten model applied to bioanodes formed from sewage sludge. *Bioresour. Technol.* 2015, *195*, 162–169.
59. Torres, C.I.; Marcus, A.K.; Lee, H.-S.; Parameswaran, P.; Krajmalnik-Brown, R.; Rittmann, B.E. A kinetic perspective on extracellular electron transfer by anode-respiring bacteria. *FEMS Microbiol. Rev.* 2010, *34*, 3–17.
60. Kano, K.; Ikeda, T. Fundamentals and Practices of Mediated Bioelectrocatalysis. *Anal. Sci.* 2000, *16*, 1013–1021.
61. Holcman, D.; Yuste, R. The new nanophysiology: Regulation of ionic flow in neuronal subcompartments. *Nat. Rev. Neurosci.* 2015, *16*, 685–692.
62. Kim, J.R.; Premier, G.C.; Hawkes, F.R.; Dinsdale, R.M.; Guwy, A.J. Development of a tubular microbial fuel cell (MFC) employing a membrane electrode assembly cathode. *J. Power Sources* 2009, *187*, 393–399.
63. Harnisch, F.; Warmbier, R.; Schneider, R.; Schröder, U. Modeling the ion transfer and polarization of ion exchange membranes in bioelectrochemical systems. *Bioelectrochemistry* 2009, *75*, 136–141.
64. Walz, D.; Teissié, J.; Milazzo, G. *Bioelectrochemistry of Membranes*; Birkhäuser Verlag: Basel, Switzerland, 2012.
65. Xiao, L.; Ge, Z.; Kelly, P.; Zhang, F.; He, Z. Evaluation of normalized energy recovery (NER) in microbial fuel cells affected by reactor dimensions and substrates. *Bioresour. Technol.* 2014, *157*, 77–83.
66. Fan, L.; Zhang, J.; Shi, X. Performance Improvement of a Microbial Fuel Cell Based on Model Predictive Control. *Int. J. Electrochem. Sci.* 2015, *10*, 737–748.
67. Ki, D.; Park, J.; Lee, J.; Yoo, K. Microbial diversity and population dynamics of activated sludge microbial communities participating in electricity generation in microbial fuel cells. *Water. Sci. Technol.* 2008, *58*, 2195–2201.
68. Teschl, G. *Ordinary Differential Equations and Dynamical Systems*; American Mathematical Society: Providence, RL, USA, 2012.
69. Pinto, R.; Srinivasan, B.; Escapa, A.; Tartakovsky, B. Multi-population model of a microbial electrolysis cell. *Environ. Sci. Technol.* 2011, *45*, 5039–5046.
70. Yahya, A.M.; Hussain, M.A.; Wahab, A.; Khairi, A. Modeling, optimization, and control of microbial

electrolysis cells in a fed-batch reactor for production of renewable biohydrogen gas. *Int. J. Energy Res.* 2015, *39*, 557–572.

71. Li, J.; Rosenberger, G.; He, Z. Integrated experimental investigation and mathematical modeling of a membrane bioelectrochemical reactor with an external membrane module. *Chem. Eng. J.* 2016, *287*, 321–328.
72. Picioreanu, C.; Kreft, J.U.; Van Loosdrecht, M.C.M. Particle-Based Multidimensional Multispecies Biofilm Model. *Appl. Environ. Microbiol.* 2004, *70*, 3024–3040.
73. Ping, Q.; Zhang, C.; Chen, X.; Zhang, B.; Huang, Z.; He, Z. Mathematical model of dynamic behavior of microbial desalination cells for simultaneous wastewater treatment and water desalination. *Environ. Sci. Technol.* 2014, *48*, 13010–13019.
74. Ping, Q.; Huang, Z.; Dosoretz, C.; He, Z. Integrated experimental investigation and mathematical modeling of brackish water desalination and wastewater treatment in microbial desalination cells. *Water. Res.* 2015, *77*, 13–23.
75. Zhang, F.; Brastad, K.S.; He, Z. Integrating forward osmosis into microbial fuel cells for wastewater treatment, water extraction and bioelectricity generation. *Environ. Sci. Technol.* 2011, *45*, 6690–6696.
76. Qin, M.; Ping, Q.; Lu, Y.; Abu-Reesh, I.M.; He, Z. Understanding electricity generation in osmotic microbial fuel cells through integrated experimental investigation and mathematical modeling. *Bioresour. Technol.* 2015, *195*, 194–201.
77. Yuan, H.; Lu, Y.; Abu-Reesh, I.M.; He, Z. Bioelectrochemical production of hydrogen in an innovative pressure-retarded osmosis/microbial electrolysis cell system: experiments and modeling. *Biotechnol. Biofuels* 2015, *8*.
78. Li, J.; Ge, Z.; He, Z. Advancing membrane bioelectrochemical reactor (MBER) with hollow-fiber membranes installed in the cathode compartment. *J. Chem. Technol. Biotechnol.* 2014, *89*, 1330–1336.
79. Pimentel, G.A.; Wouwer, A.V.; Harmand, J.; Rapaport, A. Design, analysis and validation of a simple dynamic model of a submerged membrane bioreactor. *Water Res.* 2015, *70*, 97–108.
80. Li, J.; He, Z. Development of a dynamic mathematical model for membrane bioelectrochemical reactors with different configurations. *Environ. Sci. Pollut. Res.* 2015, *23*, 3897–3906.
81. Kazemi, M.; Biria, D.; Rismani-Yazdi, H. Modelling bio-electrosynthesis in a reverse microbial fuel cell to produce acetate from CO_2 and H_2O. *Phys. Chem. Chem. Phys.* 2015, *17*, 12561–12574.
82. Rabaey, K. *Bioelectrochemical Systems: From Extracellular Electron Transfer to Biotechnological Application*; International Water Association Publishing: London, UK, 2009.
83. Zeng, Y.; Choo, Y.F.; Kim, B.-H.; Wu, P. Modelling and simulation of two-chamber microbial fuel cell. *J. Power Sources*2010, *195*, 79–89.
84. Merkey, B.V.; Chopp, D.L. The performance of a microbial fuel cell depends strongly on anode geometry: A multidimensional modeling study. *Bull. Math. Biol.* 2012, *74*, 834–857.
85. Dykstra, J.; Biesheuvel, P.; Bruning, H.; Ter Heijne, A. Theory of ion transport with fast acid-base equilibrations in bioelectrochemical systems. *Phys. Rev. E* 2014, *90*.
86. Missner, A.; Kügler, P.; Saparov, S.M.; Sommer, K.; Mathai, J.C.; Zeidel, M.L.; Pohl, P. Carbon dioxide transport through membranes. *J. Biol. Chem.* 2008, *283*, 25340–25347.
87. Galama, A.; Post, J.; Stuart, M.C.; Biesheuvel, P. Validity of the Boltzmann equation to describe Donnan equilibrium at the membrane–solution interface. *J. Membr. Sci.* 2013, *442*, 131–139.
88. Wiedemann, E.; Heintz, A.; Lichtenthaler, R. Sorption isotherms of vanadium with H_3O^+ ions in cation exchange membranes. *J. Membr. Sci.* 1998, *141*, 207–213.
89. Lei, Y.; Zhang, B.; Bai, B.; Zhao, T. A transient electrochemical model incorporating the Donnan effect for all-vanadium redox flow batteries. *J. Power Sources* 2015, *299*, 202–211.
90. Leung, A. *Systems of Nonlinear Partial Differential Equations: Applications to Biology and Engineering*; Springer Science & Business Media: New York, NY, USA, 2013.
91. Picioreanu, C.; Van Loosdrecht, M.C.; Curtis, T.P.; Scott, K. Model based evaluation of the effect of pH and electrode geometry on microbial fuel cell performance. *Bioelectrochemistry* 2010, *78*, 8–24.
92. McCarty, P.; Rittmann, B. *Environmental Biotechnology: Principles and Applications*; McGraw-Hill Education: Boston, MA, USA, 2001.
93. Cheng, S.; Liu, H.; Logan, B.E. Increased performance of single-chamber microbial fuel cells using an improved cathode structure. *Electrochem. Commun.* 2006, *8*, 489–494.
94. Sirinutsomboon, B. Modeling of a membraneless single-chamber microbial fuel cell with molasses as an energy source. *Int. J. Renew. Energy Environ. Eng.* 2014, *5*, 1–9.
95. Hang, H.; Guo, Y.; Liu, J.; Bai, L.; Xia, J.; Guo, M.; Hui, M. Computational fluid dynamics modeling of an inverted frustoconical shaking bioreactor for mammalian cell suspension culture. *Biotechnol. Bioprocess Eng.* 2011, *16*, 567–575.
96. Kim, J.; Kim, H.; Kim, B.; Yu, J. Computational fluid dynamics analysis in microbial fuel cells with different anode configurations. *Water Sci. Technol.* 2014, *69*, 1447–1452.
97. Korth, B.; Rosa, L.F.; Harnisch, F.; Picioreanu, C. A framework for modeling electroactive microbial biofilms performing direct electron transfer. *Bioelectrochemistry* 2015, *106*, 194–206.
98. Wang, C.-T.; Shaw, C.K.; Hu, T.-Y. Optimization of flow in microbial fuel cells: An investigation into promoting micro-mixer efficiency with obstacle. *Tamkang J. Sci. Eng.* 2011, *14*, 25–31.
99. Cord-Ruwisch, R.; Law, Y.; Cheng, K.Y. Ammonium as a sustainable proton shuttle in bioelectrochemical

systems.*Bioresour. Technol.* 2011, *102*, 9691–9696.

100. Maurer, M.; Schwegler, P.; Larsen, T. Nutrients in urine: Energetic aspects of removal and recovery. *Nutr. Remov. Recovery* 2003, *48*, 37–46.

101. Ewing, R.E.; Wang, H. A summary of numerical methods for time-dependent advection-dominated partial differential equations. *J. Comput. Appl. Math.* 2001, *128*, 423–445.

102. Prieto, D.; Asensio, M.; Ferragut, L.; Cascón, J. Sensitivity analysis and parameter adjustment in a simplified physical wildland fire model. *Adv. Eng. Softw.* 2015, *90*, 98–106.

103. Kauschinger, B.; Kabitzsch, K.; Schroeder, S. Adjustment of Uncertain Parameters in Thermal Models of Machine Tools. In *Thermo-Energetic Design of Machine Tools*; Springer International Publishing: Cham, Switzerland, 2015; pp. 145–157.

104. Hu, J.; Wang, Y.; Cheng, A.; Zhong, Z. Sensitivity analysis in quantified interval constraint satisfaction problems. *J. Mech. Des.* 2015, *137*.

105. Rasi, M.; Rajendran, L.; Subbiah, A. Analytical expression of transient current-potential for redox enzymatic homogenous system. *Sens. Actuators B* 2015, *208*, 128–136.

106. Yang, J.; Xu, M.; Zhang, X.; Hu, Q.; Sommerfeld, M.; Chen, Y. Life-cycle analysis on biodiesel production from microalgae: Water footprint and nutrients balance. *Bioresour. Technol.* 2011, *102*, 159–165.

107. Power, M. The predictive validation of ecological and environmental models. *Ecol. Model.* 1993, *68*, 33–50.

108. Bühlmann, P.; Hothorn, T. Boosting algorithms: Regularization, prediction and model fitting. *Stat. Sci.* 2007, *22*, 477–505.

109. Manohar, A.K.; Bretschger, O.; Nealson, K.H.; Mansfeld, F. The use of electrochemical impedance spectroscopy (EIS) in the evaluation of the electrochemical properties of a microbial fuel cell. *Bioelectrochemistry* 2008, *72*, 149–154.

110. Ramasamy, R.P.; Gadhamshetty, V.; Nadeau, L.J.; Johnson, G.R. Impedance spectroscopy as a tool for non-intrusive detection of extracellular mediators in microbial fuel cells. *Biotechnol. Bioeng.* 2009, *104*, 882–891.

111. Sekar, N.; Ramasamy, R.P. Electrochemical Impedance Spectroscopy for Microbial Fuel Cell Characterization. *J. Microb. Biochem. Technol.* 2013.

112. Rabaey, K.; Boon, N.; Siciliano, S.D.; Verhaege, M.; Verstraete, W. Biofuel cells select for microbial consortia that self-mediate electron transfer. *Appl. Environ. Microbiol.* 2004, *70*, 5373–5382.

113. Adelaja, O.; Keshavarz, T.; Kyazze, G. The effect of salinity, redox mediators and temperature on anaerobic biodegradation of petroleum hydrocarbons in microbial fuel cells. *J. Hazard. Mater.* 2015, *283*, 211–217.

114. Min, B.; Angelidaki, I. Innovative microbial fuel cell for electricity production from anaerobic reactors. *J. Power Sources*2008, *180*, 641–647.

115. Zhang, Y.; Angelidaki, I. A simple and rapid method for monitoring dissolved oxygen in water with a submersible microbial fuel cell (SBMFC). *Biosens. Bioelectron.* 2012, *38*, 189–194.

116. Zhang, Y.; Angelidaki, I. Recovery of ammonia and sulfate from waste streams and bioenergy production via bipolar bioelectrodialysis. *Water Res.* 2015, *85*, 177–184.

117. Min, B.; Poulsen, F.W.; Thygesen, A.; Angelidaki, I. Electric power generation by a submersible microbial fuel cell equipped with a membrane electrode assembly. *Bioresour. Technol.* 2012, *118*, 412–417.

118. Liu, J.; Liu, L.; Gao, B.; Yang, F. Integration of bio-electrochemical cell in membrane bioreactor for membrane cathode fouling reduction through electricity generation. *J. Membr. Sci.* 2013, *430*, 196–202.

119. Liu, R.; Huang, X.; Sun, Y.F.; Qian, Y. Hydrodynamic effect on sludge accumulation over membrane surfaces in a submerged membrane bioreactor. *Process Biochem.* 2003, *39*, 157–163.

120. Xiao, L.; Young, E.B.; Berges, J.A.; He, Z. Integrated photo-bioelectrochemical system for contaminants removal and bioenergy production. *Environ. Sci. Technol.* 2012, *46*, 11459–11466.

121. Xiao, L.; Young, E.B.; Grothjan, J.J.; Lyon, S.; Zhang, H.; He, Z. Wastewater treatment and microbial communities in an integrated photo-bioelectrochemical system affected by different wastewater algal inocula. *Algal Res.* 2015, *12*, 446–454.

122. Luo, H.; Jenkins, P.E.; Ren, Z. Concurrent desalination and hydrogen generation using microbial electrolysis and desalination cells. *Environ. Sci. Technol.* 2010, *45*, 340–344.

123. Huang, Z.J.; Chu, Y.; Hahn, J. Model simplification procedure for signal transduction pathway models: An application to IL-6 signaling. *Chem. Eng. Sci.* 2010, *65*, 1964–1975.

124. Dopico, D.; Zhu, Y.; Sandu, A.; Sandu, C. Direct and Adjoint Sensitivity Analysis of Ordinary Differential Equation Multibody Formulations. *J. Comput. Nonlinear Dyn.* 2015, *10*.

125. Madani, S.; Gheshlaghi, R.; Mahdavi, M.A.; Sobhani, M.; Elkamel, A. Optimization of the performance of a double-chamber microbial fuel cell through factorial design of experiments and response surface methodology. *Fuel* 2015,*150*, 434–440.

126. Al-Shehri, A. Statistical optimization of pentachlorophenol biodegradation and electricity generation simultaneously in mediator–less air cathode microbial fuel cell. *J. Environ. Appl. Biores.* 2015, *3*, 6–15.

127. Sajana, T.K.; Ghangrekar, M.M.; Mitra, A. Effect of operating parameters on the performance of sediment microbial fuel cell treating aquaculture water. *Aquac. Eng.* 2014, *61*, 17–26.

128. Fang, F.; Zang, G.-L.; Sun, M.; Yu, H.-Q. Optimizing multi-variables of microbial fuel cell for electricity generation with an integrated modeling and experimental approach. *Appl. Enery* 2013, *110*, 98–103.

129. Hosseinpour, M.; Vossoughi, M.; Alemzadeh, I. An

efficient approach to cathode operational parameters optimization for microbial fuel cell using response surface methodology. *J. Environ. Health Sci. Eng.* 2014, *12*.

130. Zhang, Y.; Angelidaki, I. Submersible microbial fuel cell sensor for monitoring microbial activity and BOD in groundwater: Focusing on impact of anodic biofilm on sensor applicability. *Biotechnol. Bioeng.* 2011, *108*, 2339–2347.
131. Al-Shehri, A.N.; Ghanem, K.M.; Al-Garni, S.M. Statistical Optimization of Medium Components to Enhance Bioelectricity Generation in Microbial Fuel Cell. *Arab. J. Sci. Eng.* 2012, *38*, 21–27.
132. Jia, Q.; Wei, L.; Han, H.; Shen, J. Factors that influence the performance of two-chamber microbial fuel cell. *Int. J. Hydrog. Energy* 2014, *39*, 13687–13693.
133. Chen, Y.; Luo, J.; Yan, Y.; Feng, L. Enhanced production of short-chain fatty acid by co-fermentation of waste activated sludge and kitchen waste under alkaline conditions and its application to microbial fuel cells. *Appl. Enery* 2013, *102*, 1197–1204.
134. Zhang, F.; Ge, Z.; Grimaud, J.; Hurst, J.; He, Z. *In situ* investigation of tubular microbial fuel cells deployed in an aeration tank at a municipal wastewater treatment plant. *Bioresour. Technol.* 2013, *136*, 316–321.
135. Ge, Z.; Wu, L.; Zhang, F.; He, Z. Energy extraction from a large-scale microbial fuel cell system treating municipal wastewater. *J. Power Sources* 2015, *297*, 260–264.
136. Friedman, J.; Hastie, T.; Tibshirani, R. *The Elements of Statistical Learning*; Springer Series in Statistics Springer: Berlin, Germany, 2001.
137. Esfandyari, M.; Fanaei, M.A.; Gheshlaghi, R.; Mahdavi, M.A. Neural network and neuro-fuzzy modeling to investigate the power density and Columbic efficiency of microbial fuel cell. *J. Taiwan Inst. Chem. Eng.* 2015, *58*, 84–91.
138. Garg, A.; Vijayaraghavan, V.; Mahapatra, S.S.; Tai, K.; Wong, C.H. Performance evaluation of microbial fuel cell by artificial intelligence methods. *Expert Syst. Appl.* 2014, *41*, 1389–1399.
139. Tront, J.M.; Fortner, J.D.; Plotze, M.; Hughes, J.B.; Puzrin, A.M. Microbial fuel cell biosensor for *in situ* assessment of microbial activity. *Biosens. Bioelectron.* 2008, *24*, 586–590.
140. Hashemi, J.; Samimi, A. Steady state electric power generation in up-flow microbial fuel cell using the estimated time span method for bacteria growth domestic wastewater. *Biomass Bioenergy* 2012, *45*, 65–76.
141. Rozendal, R.A.; Hamelers, H.V.; Buisman, C.J. Effects of membrane cation transport on pH and microbial fuel cell performance. *Environ. Sci. Technol.* 2006, *40*, 5206–5211.
142. Sun, H.; Luo, S.; Jin, R.; He, Z. Multitask Lasso Model for Investigating Multimodule Design Factors, Operational Factors, and Covariates in Tubular Microbial Fuel Cells. *ACS Sustain. Chem. Eng.* 2015, *3*, 3231–3238.
143. Wei, L.; Yuan, Z.; Cui, M.; Han, H.; Shen, J. Study on electricity-generation characteristic of two-chambered microbial fuel cell in continuous flow mode. *Int. J. Hydrog. Energy* 2012, *37*, 1067–1073.
144. Tardast, A.; Rahimnejad, M.; Najafpour, G.; Ghoreyshi, A.; Premier, G.C.; Bakeri, G.; Oh, S.-E. Use of artificial neural network for the prediction of bioelectricity production in a membrane less microbial fuel cell. *Fuel* 2014, *117*, 697–703.
145. Oh, S.; Min, B.; Logan, B.E. Cathode performance as a factor in electricity generation in microbial fuel cells. *Environ. Sci. Technol.* 2004, *38*, 4900–4904.
146. Gil, G.-C.; Chang, I.-S.; Kim, B.H.; Kim, M.; Jang, J.-K.; Park, H.S.; Kim, H.J. Operational parameters affecting the performannce of a mediator-less microbial fuel cell. *Biosens. Bioelectron.* 2003, *18*, 327–334.
147. Zhang, F.; Jacobson, K.S.; Torres, P.; He, Z. Effects of anolyte recirculation rates and catholytes on electricity generation in a litre-scale upflow microbial fuel cell. *Energy Environ. Sci.* 2010, *3*, 1347–1352.
148. Liu, H.; Cheng, S.; Logan, B.E. Power generation in fed-batch microbial fuel cells as a function of ionic strength, temperature, and reactor configuration. *Environ. Sci. Technol.* 2005, *39*, 5488–5493.
149. Chae, K.J.; Choi, M.J.; Lee, J.W.; Kim, K.Y.; Kim, I.S. Effect of different substrates on the performance, bacterial diversity, and bacterial viability in microbial fuel cells. *Bioresour. Technol.* 2009, *100*, 3518–3525.
150. Oh, S.E.; Logan, B.E. Proton exchange membrane and electrode surface areas as factors that affect power generation in microbial fuel cells. *Appl. Microbiol. Biotechnol.* 2006, *70*, 162–169.
151. Lefebvre, O.; Tan, Z.; Kharkwal, S.; Ng, H.Y. Effect of increasing anodic NaCl concentration on microbial fuel cell performance. *Bioresour. Technol.* 2012, *112*, 336–340.
152. Wu, C.J.; Hamada, M.S. *Experiments: Planning, Analysis, and Optimization*; John Wiley & Sons: New York, NY, USA, 2011.
153. Tartakovsky, B.; Guiot, S.R. A comparison of air and hydrogen peroxide oxygenated microbial fuel cell reactors.*Biotechnol. Progress* 2006, *22*, 241–246.
154. Plackett, R.L.; Burman, J.P. The design of optimum multifactorial experiments. *Biometrika* 1946, *33*, 305–325.
155. Fang, K.-T.; Lin, D.K. Ch. 4. Uniform experimental designs and their applications in industry. *Handb. Stat.* 2003, *22*, 131–170.
156. Albert, D.B.; Martens, C.S. Determination of low-molecular-weight organic acid concentrations in seawater and pore-water samples via HPLC. *Mar. Chem.* 1997, *56*, 27–37.
157. Doyon, G.; Gaudreau, G.; St-Gelais, D.; Beaulieu, Y.; Randall, C. Simultaneous HPLC determination of organic acids, sugars and alcohols. *Can. Inst. Food Sci. Technol. J.* 1991, *24*, 87–94.
158. Guo, W.; Luo, S.; He, Z.; Feng, X. ^{13}C pathway analysis of biofilm metabolism of *Shewanella oneidensis* MR-1. *RSC Adv.*2015, *5*, 39840–39843.
159. Luo, S.; Guo, W.; Nealson, K.H.; Feng, X.; He, Z. ^{13}C pathway analysis for the role of formate in electricity

generation by *Shewanella Oneidensis* MR-1 using lactate in microbial fuel cells. *Sci. Rep.* 2016.

160. Chang, I.S.; Moon, H.; Jang, J.K.; Kim, B.H. Improvement of a microbial fuel cell performance as a BOD sensor using respiratory inhibitors. *Biosens. Bioelectron.* 2005, *20*, 1856–1859.

161. Moreno-Casillas, H.A.; Cocke, D.L.; Gomes, J.A.; Morkovsky, P.; Parga, J.; Peterson, E. Electrocoagulation mechanism for COD removal. *Sep. Purif. Technol.* 2007, *56*, 204–211.

162. Yabur, R.; Bashan, Y.; Hernández-Carmona, G. Alginate from the macroalgae Sargassum sinicola as a novel source for microbial immobilization material in wastewater treatment and plant growth promotion. *J. Appl. Phycol.* 2007, *19*, 43–53.

163. Ge, Z.; Li, J.; Xiao, L.; Tong, Y.; He, Z. Recovery of electrical energy in microbial fuel cells. *Environ. Sci. Technol. Lett.*2014, *1*, 137–141.

164. Kutner, M.H.; Nachtsheim, C.; Neter, J. *Applied Linear Statistical Models*; McGraw-Hill/Irwin: Boston, MA, USA, 2004.

165. Kalogirou, S.A. Artificial neural networks in renewable energy systems applications: A review. *Renew. Sustain. Energy Rev.* 2001, *5*, 373–401.

166. Tipping, M.E. Sparse Bayesian learning and the relevance vector machine. *J. Mach. Learn. Res.* 2001, *1*, 211–244.

167. Platt, J.C. *A Fast Algorithm for Training Support Vector Machines*; Report MSR-TR-98–14; Microsoft Research: Redmond, WA, USA, 1998.

168. Grendar, M. Is the p-value a good measure of evidence? Asymptotic consistency criteria. *Stat. Probab. Lett.* 2012, *82*, 1116–1119.

169. Neter, J.; Kutner, M.H.; Nachtsheim, C.J.; Wasserman, W. *Applied Linear Statistical Models*; Chicago/Irwin: Boston, MA, USA, 1996.

170. Friedman, J.; Hastie, T.; Tibshirani, R. Regularization paths for generalized linear models via coordinate descent. *J. Stat. Softw.* 2010, *33*, 1–22.

171. Zhuang, L.; Zheng, Y.; Zhou, S.; Yuan, Y.; Yuan, H.; Chen, Y. Scalable microbial fuel cell (MFC) stack for continuous real wastewater treatment. *Bioresour. Technol.* 2012, *106*, 82–88.

172. Woodward, L.; Tartakovsky, B.; Perrier, M.; Srinivasan, B. Maximizing power production in a stack of microbial fuel cells using multiunit optimization method. *Biotechnol. Progress* 2009, *25*, 676–682.

173. Hastie, T.; Tibshirani, R. Varying-coefficient models. *J. R. Stat. Soc. Ser. B Methodol.* 1993, *55*, 757–796.

174. Jin, R.; Deng, X. Ensemble modeling for data fusion in manufacturing process scale-up. *IIE Trans.* 2015, *47*, 203–214.

CHAPTER 12 METHANE EMISSION IN A SPECIFIC RIPARIAN-ZONE SEDIMENT DECREASED WITH BIOELECTROCHEMICAL MANIPULATION AND CORRESPONDED TO THE MICROBIAL COMMUNITY DYNAMICS

Elliot S. Friedman, Lauren E. McPhillips, Jeffrey J. Werner, Angela C. Poole, Ruth E. Ley, M. Todd Walter and Largus T. Angenent

Keywords: Microbial iron reduction, riparian zones, methanogenesis, microbial food web, bioelectrochemical systems, microbial electrochemistry

INTRODUCTION

Riparian zones, which are the areas of land adjacent to streams, are often hotspots for biogeochemical transformations (Vidon et al., 2010). These ecosystems provide valuable services by acting as buffers between land and stream to prevent nutrients and pollutants from entering aquatic environments. Within riparian zones, subsurface microbial communities play a major role in biogeochemical cycling, impacting carbon and nitrogen availability, and, by extension, plant productivity (Groffman et al., 1992; Burgin et al., 2010). Past work has examined denitrification processes in riparian zones, because these zones typically provide the last opportunity to mitigate excess nitrate in groundwater before it reaches the stream. Studies have found that the main regulators of denitrification (e.g., nitrate, anaerobic conditions, availability of other terminal electron acceptors and carbon sources) are spatially and temporally heterogeneous, which makes modeling these landscape-level processes particularly difficult (McClain et al., 2003; Groffman et al., 2009; Ranalli and Macalady, 2010; Vidon et al., 2010). Therefore, it is essential to have a full understanding of the underlying biogeochemistry to inform landscape level models, determine best management practices, and guide regulatory policies.

Bioelectrochemical systems (BESs) are engineered systems that utilize a biocatalyst, such as bacteria or enzymes, at electrodes. Microbial BESs capitalize on the ability of, for example, dissimilatory metal-reducing bacteria to respire with solid-state electrodes via extracellular electron transfer (Logan et al., 2006; Lovley, 2008; Franks and Nevin, 2010; Hamelers et al., 2010). These BESs produce power in microbial fuel cells (MFCs; He et al., 2005; Fornero et al., 2010a,b; Rosenbaum et al., 2011b); produce chemical products in microbial electrolysis cells (Rozendal et al., 2009; Villano et al., 2010;Cheng and Logan, 2011; Cusick et al., 2011; Rosenbaum et al., 2011a); remediate pollutants (Gregory and Lovley, 2005; Strycharz et al., 2010; Morris and Jin, 2012); sense environmental and chemical parameters (Chang et al., 2005; Kumlanghan et al., 2007); and produce logic gates in biocomputing devices (Li et al., 2011; TerAvest et al., 2011). In the environment, potentiostatically poised electrodes (i.e., electrodes held at a constant electrical potential using an electrical device called a potentiostat), can mimic iron(III)- and humic substance-compounds and act as the terminal electron acceptor for dissimilatory metal-reducing bacteria (Williams et al.,

Citation: Friedman ES, McPhillips LE, Werner JJ, Poole AC, Ley RE, Walter MT and Angenent LT (2016) Methane Emission in a Specific Riparian-Zone Sediment Decreased with Bioelectrochemical Manipulation and Corresponded to the Microbial Community Dynamics. Front. Microbiol. 6:1523. doi: 10.3389/fmicb.2015.01523.

2010; Zhang et al., 2010; Friedman et al., 2012,2013). The advantage of using an electrode instead of, for example, iron(III) is that the electrode can act as an inexhaustible terminal electron acceptor, which can be set precisely to the potential of interest without any other chemical interactions with the community. Thus, BESs can be a powerful tool for precise manipulation of environmental conditions for *in situ* experimentation. Here, we utilized a BES to study the interactions between microbial communities and the presence of iron(III) and humic substances as potential terminal electron acceptors under anaerobic conditions in a riparian zone.

Dissimilatory metal reduction has been shown to dominate a wide variety of anaerobic soils and sediments from the tropics to the poles (Weber et al., 2006; Dubinsky et al., 2010; Lipson et al., 2010,2013; Keller and Takagi, 2013). Thermodynamically, iron- and manganese-reduction yield less energy than denitrification but more than sulfate reduction and methanogenesis (Bethke et al., 2011; Regnier et al., 2011). Therefore, in the presence of, for example, iron(III) [or electrodes mimicking iron[III](Friedman et al., 2012, 2013)] under anaerobic conditions, methane emissions should decrease. Indeed, at least one study has clearly observed that the presence of electrodes in a MFC decreased methane production (Ishii et al., 2008), although this was conducted in the laboratory and not in natural soil environments. Thermodynamic calculations are made under ideal conditions and do not take into account the ecological and physiological factors encountered *in situ*. For example, our recent work in Arctic peat soils has shown that bioelectrochemical manipulation designed to enrich for dissimilatory metal reduction actually increased methane emissions from soils. In this case, there was likely a bottleneck in the degradation of plant organic matter, which was widened by bioelectrochemical manipulation and stimulated the production of fermentation-like products (Friedman et al., 2013).

Modeling efforts are generally concerned with landscape-level function, such as pollutant removal or greenhouse gas emissions, but the parameters used in the model are dependent on the microbial-scale processes within the ecosystem and can vary by several orders of magnitude (Schmidt et al., 2011;Palmer and Febria, 2012; Wood et al., 2012). Therefore, it is critical to elucidate the links between microbial community structure and ecosystem processes (Raes and Bork, 2008; Morales and Holben, 2011). Here, we focused on the microbial scale within a riparian zone and we combined *in situ*bioelectrochemical manipulation with quantitative measurements of community structure (assessed by 16S rRNA gene sequences), environmental parameters (i.e., soil temperature, pH), and ecosystem function (rates of CH_4 and N_2/N_2O emissions). The objective of this work is to provide a foundation for elucidating the complex relationships between microbial community structure, community function, and ecosystem function (i.e., biogeochemical cycles); ultimately, the goal is to better inform modeling efforts through a more complete understanding of these complex relationships. One of our main objectives was to find correlations between changes of the environment, the community structure, and microbial function during the experimental period for several different experimental systems at two sites, rather than to describe the microbial community (α diversity) as a snapshot measurement, which does not inform about its function (Prosser, 2015).

At two distinct sites adjacent to an agricultural field within the riparian zone of Fall Creek in Freeville, NY, we measured CH_4 and N_2/N_2O emissions from soil chambers with either potentiostatically poised electrodes (mimicking inexhaustible iron[III] or humic substances) or unpoised electrodes (control) during a sampling period of 6 weeks in the spring of 2013. We also gathered environmental data during the experiment, including: soil temperature, pH, dissolved oxygen concentration, conductivity, iron concentrations and speciation, and anion concentrations. Once per week, bacterial biofilms were harvested from both poised and unpoised electrodes for 16S rRNA gene sequencing to determine microbial community structure. Here, we demonstrated the capability of small alterations to redox conditions to impact carbon release to the atmosphere, and provide a foundation for future examinations of biogeochemical cycling using *in situ* bioelectrochemical manipulations. Our hypothesis was that placement of iron(III)/humic substance-mimicking electrodes would reduce the emissions of methane in a very specific riparian zone of Fall Creek in Freeville, NY, USA.

MATERIALS AND METHODS

Field Location and Experimental Setup

This experiment was conducted within Fall Creek, which is a third order stream in Central New York that passes through the Homer C. Thompson Vegetable Research Farm and a Cornell University facility in Freeville, NY (42°31′N, 76°20′W; McPhillips et al., 2015). Six soil chambers were installed at each of two sites that were separated by ~50 m. These sites are further denoted in the text as "upstream" or "downstream." At each site, three of the chambers contained potentiostatically poised graphite electrodes (experimental), while the other three contained unpoised graphite electrodes (control). We installed both soil chambers and electrodes by first making small incisions in the sediment with a serrated knife and then by inserting the chambers/electrodes. We performed the experiment from April 25th through June 6th, 2013.

Soil Chamber Construction

Soil chambers were constructed from 3.79 L plastic buckets (#2860, U.S. Plastic Corp., Lima, OH, USA). To create the base of the chamber, the bottom 2-cm of the bucket was removed leaving a 17-cm-long cylinder; when placed in the soil, 11-cm of the soil chamber was below the surface and 6-cm extended above the soil surface. From each chamber, an 8.5-cm × 4.5-cm section of the plastic material that would be in the subsurface area was removed and replaced with an anion exchange membrane (AMI-7001S, Membranes International, Glen Rock, NJ). To create a cap for the soil chamber, the top 4-cm was removed from another 3.79 L plastic bucket. The bottom edge of the plastic bucket was reinforced with foam insulation tape and duct tape to ensure an airtight seal between the soil chamber and the chamber cap (the cap was only used during short measurement periods). Two 0.635-cm cylindrical septa (AT6526, Fisher Scientific, USA) were affixed to the top of the chamber cap and sealed with urethane adhesive. During measurements, the chamber was vented with a 21G needle (#305129, VWR, USA) through one of the septa to prevent induced pressure differentials in the soil chamber (Hutchinson and Mosier, 1981). We took gas samples through the other septum.

Bioelectrochemical Systems

The working and counter electrodes (CE) were manufactured from medium-extruded graphite plates (GT001135, Graphite Store, Buffalo Grove, IL, USA). The CE consisted of a 6-cm × 6-cm of 0.635-cm thick block (surface area = 87.24 cm^2), while each working electrode (WE) consisted of six 6-cm × 6-cm × 0.635-cm blocks connected in parallel (total surface area = 523.44 cm^2). Having multiple electrodes in parallel for the WEs allowed for the harvesting of biofilms during different stages of the experiment. The WE and reference electrode (RE) were placed inside the chamber, while the CE was placed outside the chamber on the opposite side of the membrane to maintain an electrical connection while preventing cathodic hydrogen or methane production inside the soil chamber. Electrodes were connected to microcontroller-based potentiostats (Friedman et al., 2012) by inserting the exposed end of a 3-m length of 18-gage stranded copper wire into a 1.59-mm hole drilled in the top of each graphite block. A conductive carbon adhesive (#12664, Electron Microscopy Sciences, Hatfield, PA, USA) was used to ensure a good electrical connection (resistance < 0.5 Ω), and the junction was sealed with a urethane adhesive (#4024, Hardman, South Bend, IN, USA). The WEs in experimental chambers were poised at 0.1 V_{SHE} using a microcontroller-based potentiostat (Friedman et al., 2012) and an Ag/AgCl RE was made in house. Electrodes in unpoised chambers were left at open circuit. Current data fluctuated throughout the experimental period due to many different factors, including diurnal changes, and our data was not corrected for background chemical currents, chemical interactions, noise from electric fields, and open circuit potential for the unpoised electrodes.

Measurements

Measurements were taken every Monday and Thursday during the 6-week experimental period. We measured denitrification rates using the acetylene inhibition method, where acetylene is added to the soil to inhibit reduction of N_2O to N_2 (Freney et al., 1992; Thompson, 1996; Groffman et al., 2006), allowing quantification of N_2 and N_2O emissions through the measurement of N_2O only. Even though this method evaluates the denitrification rates (N_2 production), we indicate this in the text as N_2/N_2O emissions. We prepared and used beeswax-coated calcium carbide tablets, which react with water to form acetylene gas, as described by Thompson (1996). Six beeswax-coated calcium carbide tablets were inserted into each soil collar 45 min prior to gas sampling at a depth of 7–15 cm. We measured acetylene concentrations in the gas samples to ensure that ample acetylene (>1% v/v) was being produced to inhibit N_2O reduction (Thompson, 1996). Gas emissions were measured from the chambers by placing the cap on the chamber and collecting 12 mL of the headspace gas at four time points (0, 10, 20, and 30 min). Gas chromatography analysis of N_2O, CH_4, and C_2H_2 was performed on an Agilent 6890N gas chromatograph equipped with a HP 7694 Headspace Autosampler (Hewlett-Packard Company, Palo Alto, CA, USA). N_2O separation was performed using a Supel-Q™ PLOT capillary column (30 m × 0.32 mm; Supelco Inc., Bellafonte, PA, USA) with ultra-pure helium carrier gas (2.6 mL min^{-1}) and 95:5 Ar:CH_4 make-up gas (8.2 mL min^{-1}) and an electron capture detector set to 250°C. CH_4 and C_2H_2 separation was performed using a Carboxen 1006 PLOT capillary column (30 m × 0.32 mm; Supelco, Inc.) and a flame ionization detector set to 200°C with H_2 gas (30 mL min^{-1}), air (400 mL min^{-1}), and N_2 makeup gas (25 mL min^{-1}). The oven temperature was initially set to -22°C for 4.7 min, then increased to 30°C for 0.85 min and finally increased to 80°C for 2.5 min to allow for elution of all three gasses of interest. Calibration curves were made using serial dilutions of 1 ppm N_2O, 20 ppm CH_4, and 2.5% C_2H_2 (Airgas Inc.). Emissions were calculated as the slopes of the linear regression (concentration vs. time) curves ($R^2 > 0.9$) for each measurement period (flux).

Soil temperature, pH, dissolved oxygen, and conductivity were measured at 7-cm depth directly adjacent to each soil chamber using a portable multiparameter meter (Orion Star A329, Thermo Scientific, Pittsburgh PA, USA) and probes (ROSS Ultra Triode pH/ATC electrode, DuraProbe conductivity probe, Orion RDO probe, Thermo Scientific, Pittsburgh, PA, USA). Soil water was collected from within each soil chamber (2–7 cm below the soil surface) using porous soil moisture samplers (#220300, Rhizosphere, Wageningen, The Netherlands), vacutainers (VT6430, BD, Franklin Lakes, NJ, USA), and 21G needles (#305129, VWR, USA). These samples were analyzed for nitrate, nitrite, chloride, and sulfate concentrations using a Dionex ICS-2000 ion chromatograph with IonPac AS-18 analytical column and 25-μL sample loop. Immediately following the collection of soil water, 0.5 mL of each sample was transferred to another vacutainer containing 0.5 mL of 0.5 N HCl; these samples were analyzed for Fe^{2+} and total Fe using the ferrozine assay (Braunschweig et al., 2012).

Microbial Community Analysis

Biofilms from the WEs of both experimental and control soil chambers were collected weekly throughout the 6-week period of the experiment. Each of the six parallel WEs in every soil chamber was harvested once during the course of the experiment to obtain a time series of the microbial community. We harvested biofilms by removing the electrode from the soil and by scraping the biofilm into a 15-mL sterile centrifuge tube (#93000-026, VWR, USA) using a sterile blade. Samples were placed on wet ice immediately in the field, and then stored at -20°C until the completion of the experiment. Following biofilm harvesting, the bare electrodes were returned to the soil to keep the WE surface area constant throughout the experiment. Genomic DNA was extracted using a PowerSoil®DNA Isolation Kit (MoBio, Carlsbad, CA, USA). Bacterial extraction product was then amplified in duplicate 50-μL polymerase chain reactions (PCRs) according to Gilbert et al. (2010) using 25 cycles. In short, duplicate 50-μL PCR reactions were conducted using: 28 μL molecular grade water, 20 μL mastermix (5Prime Hot MasterMix, Catalog # 2200110, 5Prime, Fischer Scientific, USA), 0.5 μL 515f forward primer (Caporaso et al., 2012), 0.5 μL 806r barcoded reverse primer (Caporaso et al., 2012), and 1 μL template DNA. Reactions were run under the following conditions: 94°C for 3 min; then 94°C for 45 s, 50°C for 60 s, and 72°C for 90 s (repeat 25 times); 72°C for 10 min, and then hold at 4°C. Duplicate PCR products were then pooled, the presence of amplicons confirmed by gel electrophoresis, and products cleaned using the Mag-Bind® E-Z Pure Kit (Omega, Norcross, GA, USA). Cleaned product was again confirmed with gel electrophoresis and barcoded amplicons for each sample were pooled at equimolar ratios to a final concentration of 8 ng DNA μL^{-1}. The single, pooled amplicon mixture was sequenced at the Cornell University Biotechnology Resource Center using an Illumina MiSeq (2 × 250 bp, paired end). Data were quality filtered with the quantitative insights into microbial ecology (QIIME) platform (v 1.6; Caporaso et al., 2010). Forward and reverse reads were joined using fastqjoin and the data returned to the QIIME platform for operatational taxonomic unit (OTU) picking, alpha diversity, beta diversity, and further analysis. We applied a machine learning approach to identify taxa that discriminated between poised and unpoised electrode communities using the pamR package in R (Tibshirani et al., 2002), and applied a constrained correspondence analysis using the vegan package in R (Oksanen et al., 2013; R Core Team, 2013).

RESULTS AND DISCUSSION

CH_4 and N_2/N_2O Emissions

We measured CH_4 emissions bi-weekly in each of 12 chambers split between two riparian zone sites over the 6-weeks experimental period and calculated the absolute average CH_4 emission (Table1). We observed differences between the two riparian zone sites, which were 50 m apart, in regards to CH_4emissions, showing that this specific riparian zone was heterogeneous. The absolute average CH_4emissions at the upstream site were considerably higher compared to the downstream site; absolute average CH_4 emissions with the unpoised electrode (control) chambers were 1.96 ± 1.42 mg CH_4*m^{-2}*h^{-1} at the upstream site compared to 0.79 ± 0.65 mg CH_4*m^{-2}*h^{-1} at the downstream site (Table1). On average, the absolute CH_4 emissions from the poised electrode chambers at the upstream site were ~50% lower than those from the unpoised electrode chambers, but due to the large temporal error not significantly different (p = 0.55, two-tailed t-test). At, the downstream site the absolute average CH_4emissions from the poised electrode chambers were similar compared to those from the unpoised electrode chambers (p = 0.91, two-tailed t-test; Table1). However, there were large temporal trends throughout the experimental period. The CH_4 emissions for each chamber were highest at the beginning of the experiment (earlier in the spring) and decreased throughout the 6-week experimental period (data not shown).

Table 1: Average CH_4 emissions, N_2/N_2O emissions, SO_4^{2-} concentrations, and Cl^- concentrations from chambers with poised and unpoised electrodes at both the upstream and downstream sites during the experimental period

Function or environmental parameter	Upstream poised	Upstream unpoised	Downstream poised	Downstream unpoised
CH_4 emission (mg CH_4*m^{-2}*h^{-1})	0.96 ± 0.86	1.96 ± 1.42	0.93 ± 1.1	0.79 ± 0.65
N_2/N_2O emission (μg N_2O-N*m^{-2}*h^{-1})	1793 ± 1794	2138 ± 2136	2394 ± 3287	2179 ± 2873
SO_4^{2-} concentration (ppm)	15.1 ± 14.6	38.2 ± 31.2	41.4 ± 35.8	45.5 ± 32.8
Cl^- concentration (ppm)	88.5 ± 39.7	107 ± 41.2	103 ± 63.0	218 ± 142

Large error is due to temporal changes during the experimental time period; as such, in further data analysis the groups were considered as paired samples on each measurement day (i.e., poised vs. unpoised) to remove the impacts of temporal trends when determining the effects of electrode-based manipulation. Error indicates standard error. N = 54 for all measurements.

To more precisely determine the effects of potentiostatic manipulation, we normalized average CH_4emissions (mg CH_4*m^{-2}*h^{-1}) from chambers with poised electrodes against the CH_4 emissions from chambers with unpoised electrodes at each site for each measurement day, resulting in an average percentage change in CH_4 emission ratios of poised vs. unpoised electrode chambers (Figure 1A). At the upstream site, we observed a 44% reduction in normalized CH_4 emission (p = 0.017, two-tailed paired t-test) for the poised vs. unpoised electrode chambers (Figure 1A). This is in agreement with our hypothesis: placement of iron(III)/humic substance-mimicking electrodes will reduce the emission rate methane compared to the control in a riparian zone, since it is thermodynamically more attractive to sustain iron/humic substance reduction (i.e., iron(III)/humic substances as the terminal electron acceptor) than methanogenesis (i.e., CO_2 as the terminal electron acceptor). However, for the downstream site for which we observed lower soil temperatures, lower pH, more vegetation cover, and less direct sunlight, we did not observe a significantly average change in normalized CH_4 emissions between the poised vs. unpoised chambers (p = 0.61, two-tailed paired t-test; Figure 1A). Noteworthy is that the temperature difference between the upstream and downstream sites may have caused for a small error in our data (~1%). Based on our hypotheses, we anticipated lower CH_4 emissions for the poised electrode chambers. It remains, however, unclear why the normalized CH_4 emissions were indeed lower at the upstream site and not at the downstream site with poised electrodes. This could be due to environmental differences between sites (e.g., soil temperature, pH), but further work is required to determine the precise reason(s).

There were no differences in the absolute average N_2/N_2O emissions (μg N_2O-N*m^{-2}*h^{-1}) between sites or treatments (Table 1). These data were also compared using percent change based on paired sample analysis in the same manner as the normalized CH_4 emissions, and there were no significant changes between groups at either site (upstream: p = 0.051; downstream: p = 0.521, two-tailed paired t-test, Figure 1B). Regardless, the discrepancies with our CH_4 emission results from adjacent field sites show the heterogeneous character of field research, and the importance to test results in the field rather than only at the bench.

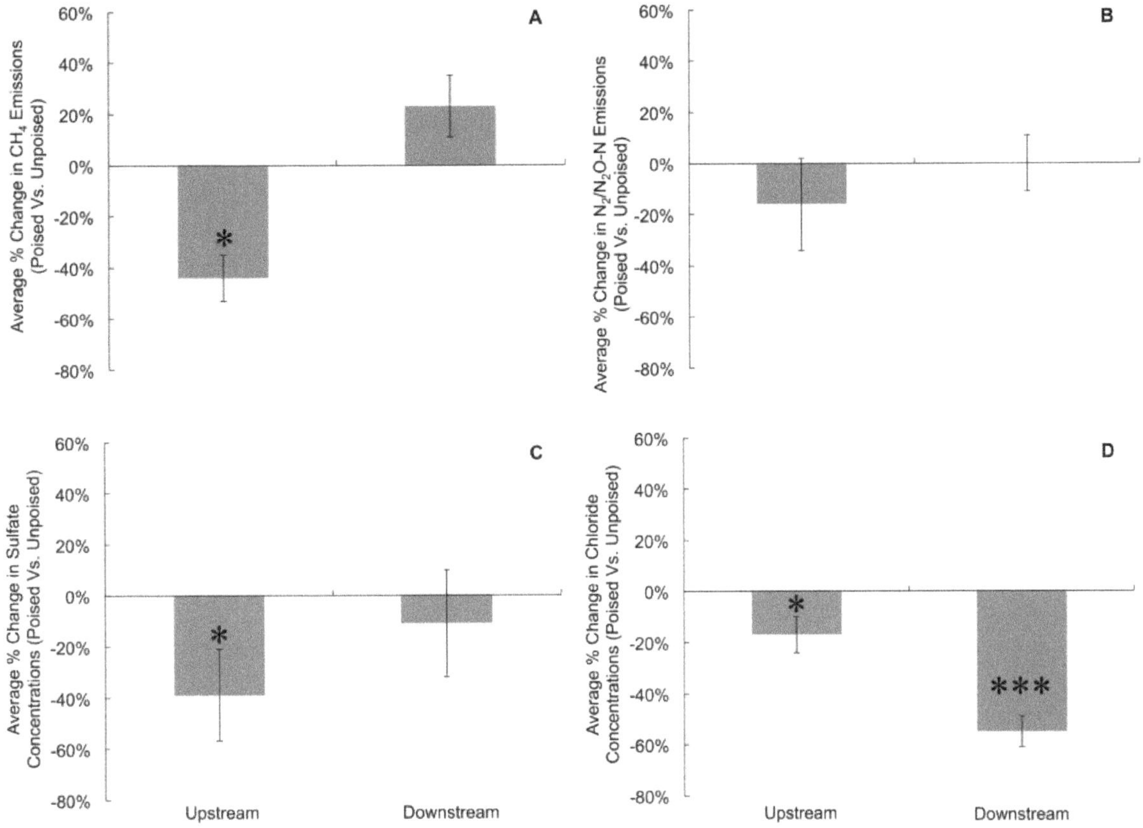

Figure 1: Average percent change in gas emissions (poised:unpoised) for CH_4(A) and N_2/N_2O (B), as well as average percent change in the concentrations of SO_4^{2-}(C) and Cl^-(D) for upstream and downstream sites during the duration of the experiment. To account for ...

Environmental Parameters

We compared several environmental parameters that we measured for upstream and downstream sites during the study period. Soil temperature generally increased during the experimental period (and thus throughout the spring season), and ranged between 6.8 and 19.1°C. On a daily basis throughout the 6-week period, the average soil temperatures at the upstream site were 1.5°C higher than that of the downstream site ($p < 0.001$, two-tailed t-test). The pH values across both sites were close to each other and near neutral, with the upstream site averaging at 7.3 ± 0.2 and the downstream site at 7.2 ± 0.2. There were no differences in soil electrical conductivity between sites or control and treatment collars; the average soil conductivity across both sites was 403 ± 165 µS/cm.

Dissolved oxygen was low across both sites due to water tables at or near the soil surface (for more than 75% of the experimental period), and both sites showed average concentrations of less than 1 mg L^{-1} dissolved oxygen. Both NO_3^- and NO_2^- were low (<1 ppm) across sites and treatments, and were below detection limit (0.1 ppm) for most samples (consistent with the relatively low N_2/N_2O emissions rates). We did not observe a difference between sites and treatments for total iron concentrations from soil pore water, which averaged 10.97 ± 9.0 µM. Approximately 95% of this iron was in the reduced form (Fe^{2+}), and there were no differences between sites or treatments in iron speciation. At the upstream site, SO_4^{2-} concentrations were lower in chambers with poised electrodes than those with unpoised electrodes, however, there was not a significant difference in absolute SO_4^{2-} concentrations at the downstream site (Table 1). These values were corrected in the same manner as the normalized CH_4 emission rates, visualizing an average reduction in normalized SO_4^{2-} concentration of 39% with poised vs. unpoised electrodes (p = 0.02, two-tailed paired t-test;Figure **1C**). It is unclear whether the difference in SO_4^{2-} concentrations at the upstream site was related to the reduction in CH_4 emissions, although sulfate reduction is more energetically favorable than methanogenesis, but less favorable than iron reduction (Regnier et al., 2011). Based on energetics, we would have anticipated a higher concentration of sulfate (less sulfate reduction) with a poised electrode. A conclusive explanation is elusive, but we can speculate: in some cases, microbial iron-reducing bacteria have been found to coincide with sulfate-reducing microbes (*Desulfovibrio* sp.;Lentini et al., 2012), and it is possible that a synergistic relationship between iron- and sulfate-reducers is responsible for the decreased sulfate concentrations in the chambers with poised electrodes at the upstream site. This co-occurrence is supported by the simultaneous normalized CH_4 emission reduction (via mimicked iron reduction with electrodes) and decreased normalized SO_4^{2-} concentration (via increased sulfate reduction) at the upstream site, while neither decrease occurred at the downstream site.

Absolute Cl^- concentrations were lower in pore water collected from chambers containing poised electrodes at both sampling sites (Table 1); these values were corrected in the same manner as normalized CH_4 emission rates. At the upstream sites, chambers with poised electrodes had normalized Cl^- concentrations that were 17% lower than those with unpoised electrodes ($p = 0.007$, two-tailed paired t-test); at the downstream site, normalized Cl^- concentrations were, on average, 55% lower than in chambers with unpoised electrodes ($p < 0.0005$, two-tailed paired t-test; Figure 1D). These lower Cl^- concentrations were a result of Cl^- transport from the inside to the outside chamber through the anion exchange membrane because of ion imbalances due to the electrochemical activity (Rozendal et al., 2006; Fornero et al., 2010b).

Microbial Community Dynamics

We obtained a total of 72 samples (69 of which were sequenced; 3 samples had too low of a DNA yield to be sequenced) from the 12 chambers during 6 weeks and achieved an average sequencing coverage of 29,782 sequences (16S rRNA gene) per sample with an average length of 251 bp. We first analyzed the beta diversity, which quantifies the differentiation between sample communities, using UniFrac distances (Lozupone and Knight, 2005). Based on the differences in CH_4 emissions and environmental parameters, we had expected microbial communities to be similar to each other if they came from the same site or treatment. However, principal coordinate analysis of both unweighted UniFrac distances and weighted UniFrac distances (weighted takes into account relative abundances of OTUs within samples) did not reveal any clustering of samples by site or treatment (Figures 2A,B). However, there was a clear clustering of samples with time, which was highlighted by UniFrac distances between the first 2 weeks (weeks 1–2) and last 2 weeks (weeks 5–6) of the experimental period (Figure 3). This result likely suggests that most of the broad differences in the microbial community phylogenetic structure were based simply on changing environmental parameters and community function with time, rather than experimental treatment.

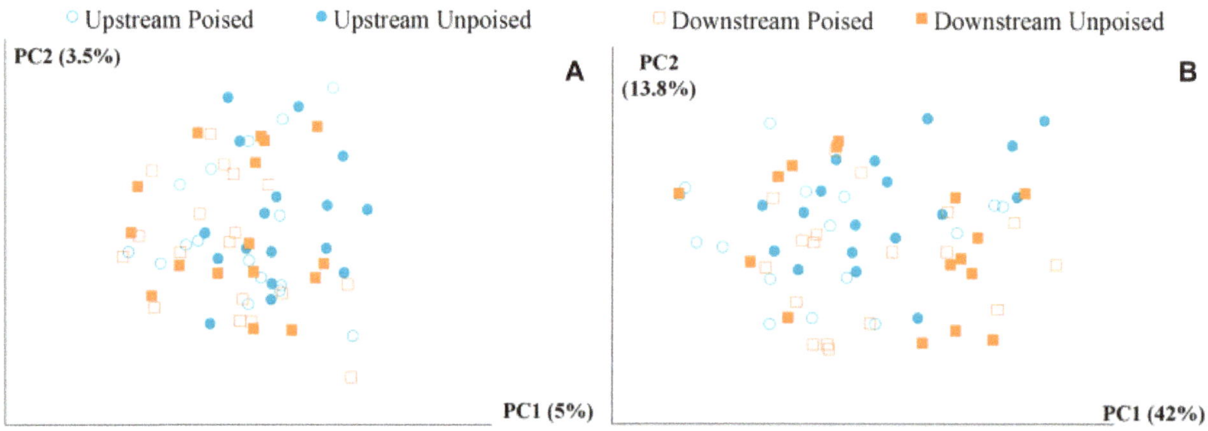

Figure 2: Principal coordinate analysis of the beta diversity of microbial communities shown using unweighted **(A) and weighted (B)**UniFrac distances. The weighted (B) method explains more variation (55.8%) in the first two principal coordinates than the unweighted ...

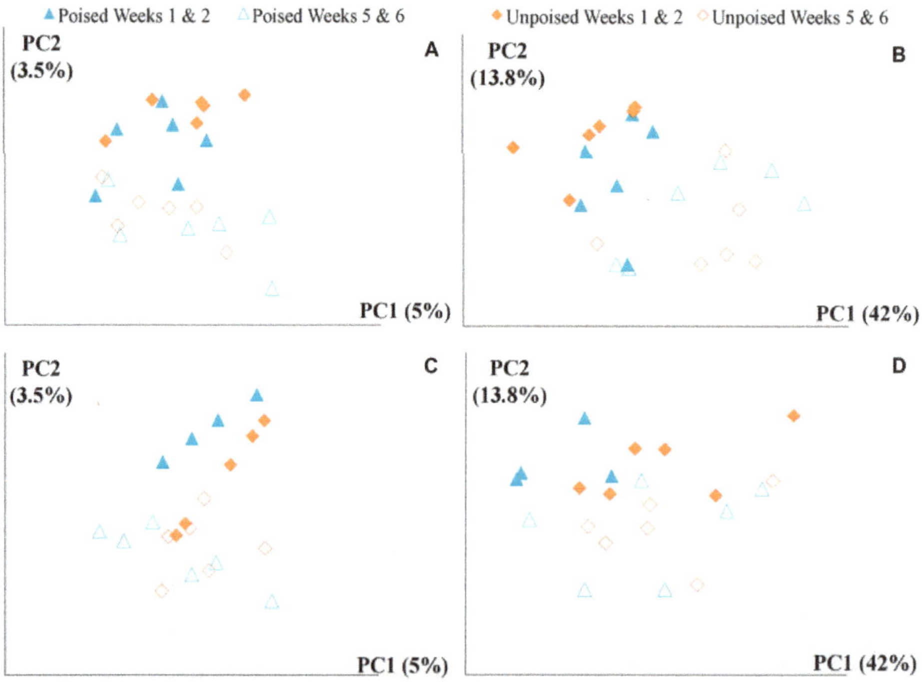

Figure 3: Principal coordinate analysis of beta diversity at upstream and the downstream sites using unweighted and weighted UniFrac distances: (A) upstream site, unweighted UniFrac; (B)upstream site, weighted UniFrac; (C) downstream site, unweighted UniFrac; ...

In each of the electrode samples, over 80% of the 16S rRNA gene sequences were classified in the phyla Proteobacteria and Bacteroidetes, with a remaining 1–6% of sequences falling in the phyla Firmicutes, Acidobacteria, Actinobacteria, Verrucomicrobia, Cyanobacteria, Chloroflexi, Planctomycetes, and Gemmatimonadetes (Figure 4). Similar to the beta diversity data analysis, we grouped the taxonomy data for weeks 1–2 and weeks 5–6 with the objective to identify phyla that changed during the experimental period of 6 weeks. At the upstream site, the community composition at the poised electrodes between weeks 1–2 and 5–6 showed a 10% increase in Proteobacteria ($p = 0.19$, two-tailed *t*-test) and a 35% decrease in Bacteroidetes ($p = 0.04$, two-tailed *t*-test) during the experimental period. Proteobacteria and Bacteroidetes populations were not different at the unpoised electrodes between weeks 1–2 and 5–6 (Figure 4A). At the downstream site, both poised and unpoised electrodes enriched for Proteobacteria between weeks 1–2 and 5–6, which was accompanied by a decrease in relative abundance of Bacteroidetes (Figure 4B). Specifically, the poised electrodes resulted in a 14% increase in Proteobacteria and a 30% decrease in Bacteroidetes ($p = 0.06$ for both phyla, two-tailed *t*-test) while the unpoised electrodes caused a 20% increase in Proteobacteria and a 50% decrease in Bacteroidetes ($p < 0.01$ for both phyla, two-tailed *t*-test) between weeks 1–2 and 5–6. The increase in Proteobacteria in both poised and unpoised treatments could be due to the presence of the graphite electrodes, which provide a conductive surface that promotes both biofilm growth and long-range electron transport through soil microenvironments (even when unpoised; Erable et al., 2011; Kato et al., 2012; Friedman et al., 2013). Thus, with this phyla richness analysis we found that three out of four comparisons of week 1–2 and week 5–6 binned data showed a similar trend of increasing Proteobacteria, but this included the poised *and* unpoised electrodes at one site.

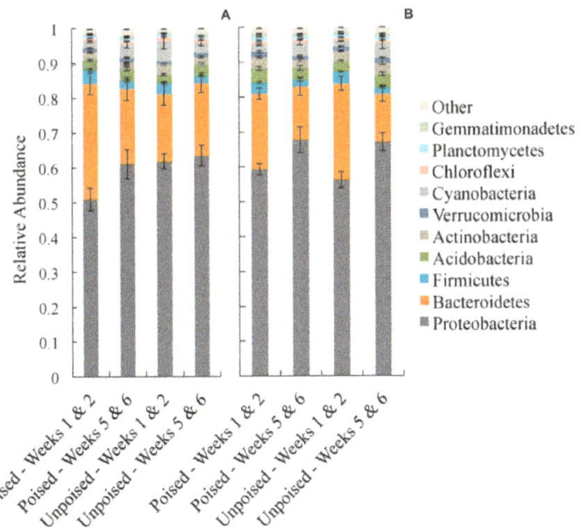

Figure 4: Average taxonomic summaries (phylum level) of the microbial communities from the upstream (A) and downstream (B) sites. Communities are grouped in the first 2 weeks (weeks 1–2) and the last 2 weeks (weeks 5–6). Error bars indicate standard ...

We used a machine learning approach to find OTUs that are predictive of samples taken from either poised or unpoised electrodes (Figure 5). We applied this to OTU abundance profiles from all samples (Figure 5A) as well as data from only the upstream site samples (Figure 5B). Upon cross-validation, the machine-learning algorithm returned a relatively high error rate (0.392 and 0.346 for both sites, and the upstream site, respectively), indicating that there were no indicator OTUs that could differentiate poised from unpoised electrode conditions. This machine learning approach confirmed our other findings from the principal coordinate analysis of UniFrac-based beta diversity (Figures 2 and 3) and the other taxonomic classification analysis (Figure 4) that the bioelectrochemical manipulation did not cause systematic, or reproducible, changes in the microbial membership.

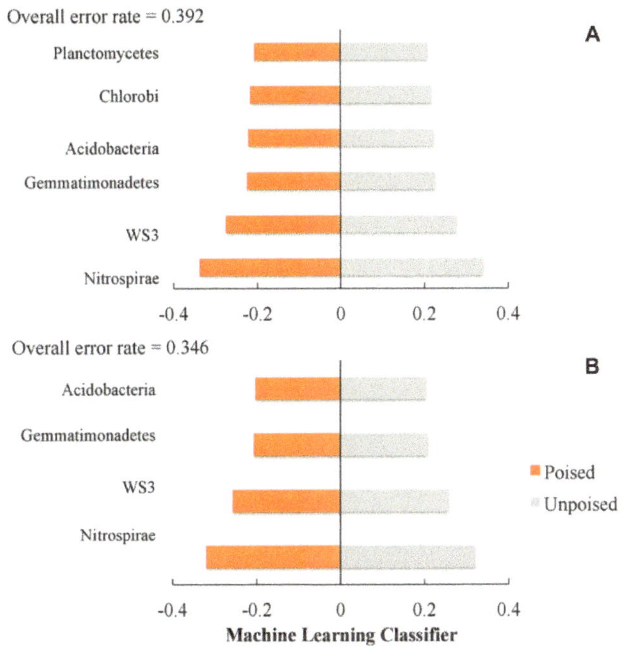

Figure 5: Machine learning analysis using phylum level taxonomy shows the phyla that are most indicative of poised or unpoised electrode communities across: (A) both sites; and (B) just the upstream site. Machine learning classifier is a nearest shrunken centroid ...

Linking Microbial Community Dynamics to Gas Emissions and Environmental Parameters

Next, we investigated whether community function (e.g., gas emission) or environmental parameters were associated with microbial community membership. We performed correspondence analysis to compare OTU abundance profiles with gas emission and environmental metadata (rarified with 500 sequences per sample; Figure 6). First, we performed principal component analysis with our OTU tables, and second we performed correspondence analysis to map environmental metadata onto the principal component results as vectors. The resulting direction of the vector (arrow) indicates that the specific variable corresponded strongly with community structure, while the resulting length of the vector indicates the magnitude of the association (i.e., longer arrows indicate a more significant weighting of the given variable in the overall ordination). When we used data from across all sites and treatments, the variables of strongest correspondence with the OTU tables were: N_2/N_2O emissions, pH, and Cl^- concentrations (Figure 6A). When limiting the analysis to the upstream site (where unpoised electrode chambers had higher CH_4 emissions), however, the CH_4 emissions were the most strongly weighted variable, with N_2/N_2O emissions, pH, Cl- concentrations, and temperature being other variables (Figure 6B). Noteworthy is that temperature and time as a vector cannot be distinguished in our work because of slowly rising soil temperatures during our experimental period.

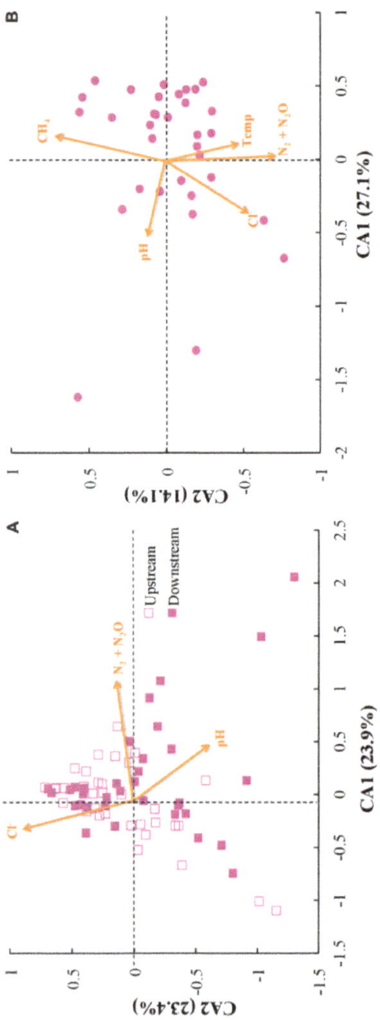

Figure 6: Correspondence analysis (CA) using rarified and filtered OTU tables, gas emissions, and environmental metadata for: (A) all samples; and (B) only samples from the upstream site. Points indicate samples while arrows indicate vectors for both gas emissions ...

Future Work Critical to Elucidate Microbial Community Dynamics in Riparian Zones

Riparian zones have the capability to buffer aquatic ecosystems from the adjacent land and are important ecosystems for watershed management, especially in agricultural areas (McClain et al., 2003; Burt et al., 2010). Many studies have attempted to directly link hydrologic parameters (e.g., water table depth, groundwater flow patterns) to both denitrification rates and greenhouse gas emissions (Vidon and Hill, 2004; Aronson et al., 2012; Audet et al., 2013; McPhillips et al., 2015). However, microbial communities are the drivers of subsurface biogeochemical transformations, and as such it is critical to quantitatively link microbial community dynamics to ecosystem level parameters and functions (Raes and Bork, 2008; Burt et al., 2010; Morales and Holben, 2011). There are often chemical or biological links (or both) between different ecosystem processes, such as temperature, groundwater flow, plant growth, and denitrification (Teh et al., 2008; Liptzin and Silver, 2009;Dubinsky et al., 2010; Vidon et al., 2010; Frei et al., 2012).

When we manipulated the soil redox environment using poised electrodes mimicking iron(III), there was a significant decrease in CH_4 emissions from the upstream site (Figure 1A), which we anticipated because the presence of a terminal electron acceptor with the redox potential of iron(III) for iron reduction is thermodynamically more favorable compared to carbon dioxide for methanogenesis. Furthermore, manipulation resulted in a decrease in SO_4^{2-} and Cl^- concentrations in soil porewater (Figures 1C,D). However, at the downstream site, which had a lower magnitude of emissions (nearly 50%) from the chambers with unpoised electrodes (i.e., controls), poised electrodes had no effect on CH_4 emissions. This is likely due to the lower overall CH_4 emission rates at this site, which would preclude inhibition via bioelectrochemical manipulation since methanogenic activity was already low.

We also observed, together with the reduction in CH_4 emission from the upstream site, an increase in the relative abundance of Proteobacteria and a decrease in Bacteroidetes (Figure 4A). However, we found no systematic link between community structure and experimental treatment (poised vs. unpoised electrodes; Figures 2 and 3). Further quantitative analysis showed that these microbial community changes corresponded with differences in the environmental parameters (pH and Cl-concentrations) and CH_4 and N_2/N_2O emissions (Figure 6).

The indirect alteration of microbial community structure and suppression of CH_4 emissions at the upstream site in conjunction with bioelectrochemical manipulation (using soil-based electrodes) demonstrates the fragile balance that governs biogeochemical cycles in these soils, and highlights the measurable impact that microbial competition has on ecosystem-scale processes. It is clear that microbial community structure and function is subject to influence from a wide array of biological, chemical, and physical factors, which in turn can have a measurable impact on the landscape level. Accurate modeling of biogeochemical processes is important for predicting responses to climate change, determining regulatory limits for anthropogenic pollutants, and designing effective best management practices (Vidon et al., 2010; Meng et al., 2011; Riley et al., 2011). As such, a deeper comprehension of subsurface microbial ecosystems, their responses to environmental conditions across spatial and temporal gradients, and their impacts on larger-scale function, is critical for improving model accuracy, and further studies are certainly warranted.

In summary, we manipulated the sediment redox environment by poising electrodes capable of being used as electron acceptors for iron(III)-reducing microbes and observed a response in ecosystem function (in this case, CH_4 emissions) at one riparian zone site. Poised electrodes provided an inexhaustible source of electron acceptor for iron reducers, and, at our upstream riparian zone site, resulted in reduced CH_4 emissions that could be a result of iron- or humic substance-reducers outcompeting methanogens for carbon sources and nutrients. However, we did not see such changes in CH_4 emissions at the downstream site. Furthermore, we showed that bioelectrochemical manipulation had minimal effects on the microbial community structure. This was confirmed by machine learning analysis, which was unable to develop an algorithm to predict sample grouping with an error rate below 30%. We did, however, find a correspondence between the community composition and the function of the microbiota (CH_4 emissions). Despite the lack of direct changes in microbial community structure at the upstream site with poised electrodes, the reduction (~45%) of CH_4 emissions together with its correspondence to the composition of microbiota suggests that the balance between competing anaerobic microbial processes can have a major impact on landscape-level processes.

Key Points:

- Potentiostatic manipulation was used to change redox dynamics in riparian soils.
- Manipulation inhibited CH_4 emissions but did not change the microbial community.
- Microbial community dynamics corresponded to environment and ecosystem function.

AUTHOR CONTRIBUTIONS

EF, LM, MW, and LA designed the study; EF and LM performed the field research; EF, LM, and AP performed the sample analysis; EF, LM, JW, AP, RL, MW, and LA analyzed the data, EF, LM, JW, AP, RL, MW, and LA wrote the manuscript.

CONFLICT OF INTEREST STATEMENT

The authors declare that the research was conducted in the absence of any commercial or financial relationships that could be construed as a potential conflict of interest.

ACKNOWLEDGMENTS

16S rRNA sequencing data are publicly available for download via MG-RAST project ID 6205 (http://metagenomics.anl.gov/linkin.cgi?project=6205). The authors would like to acknowledge Andrea Fortman, Christine Georgakakos, Helen Bergstrom, and Shane DeGaetano for assistance in conducting fieldwork. Partial funding provided by the Atkinson Center for a Sustainable Future and the Cornell University Program in Cross-Scale Biogeochemistry and Climate, which is supported by NSF-IGERT (1069193).

REFERENCES

1. Aronson E. L., Vann D. R., Helliker B. R. (2012). Methane flux response to nitrogen amendment in an upland pine forest soil and riparian zone. *J. Geophys. Res. Biogeosci.* 117 G0301210.1029/2012jg001962
2. Audet J., Johansen J. R., Andersen P. M., Baattrup-Pedersen A., Brask-Jensen K. M., Elsgaard L., et al. (2013). Methane emissions in Danish riparian wetlands: ecosystem comparison and pursuit of vegetation indexes as predictive tools. *Ecol. Indic.* 34 548–559. 10.1016/j.ecolind.2013.06.016
3. Bethke C. M., Sanford R. A., Kirk M. F., Jin Q. S., Flynn T. M. (2011). The thermodynamic ladder in geomicrobiology. *Am. J. Sci.* 311 183–210. 10.2475/03.2011.01
4. Braunschweig J., Bosch J., Heister K., Kuebeck C., Meckenstock R. U. (2012). Reevaluation of colorimetric iron determination methods commonly used in geomicrobiology. *J. Microbiol. Methods* 89 41–48. 10.1016/j.mimet.2012.01.021
5. Burgin A. J., Groffman P. M., Lewis D. N. (2010). Factors regulating denitrification in a riparian wetland. *Soil Sci. Soc. Am. J.* 74 1826–1833. 10.2136/sssaj2009.0463
6. Burt T., Pinay G., Sabater S. (2010). What do we still need to know about the ecohydrology of riparian zones? *Ecohydrology* 3 373–377. 10.1002/eco.140
7. Caporaso J. G., Kuczynski J., Stombaugh J., Bittinger K., Bushman F. D., Costello E. K., et al. (2010). QIIME allows analysis of high-throughput community sequencing data. *Nat. Methods* 7335–336. 10.1038/Nmeth.F.303
8. Caporaso J. G., Lauber C. L., Walters W. A., Berg-Lyons D., Huntley J., Fierer N., et al. (2012).Ultra-high-throughput microbial community analysis on the Illumina HiSeq and MiSeq platforms. *ISME J.* 6 1621–1624. 10.1038/Ismej.2012.8
9. Chang I. S., Moon H., Jang J. K., Kim B. H. (2005). Improvement of a microbial fuel cell performance as a BOD sensor using respiratory inhibitors. *Biosens. Bioelectron.* 20 1856–1859. 10.1016/j.bios.2004.06.003
10. Cheng S. A., Logan B. E. (2011). High hydrogen production rate of microbial electrolysis cell (MEC) with reduced electrode spacing. *Bioresour. Technol.* 102 3571–3574. 10.1016/j.biortech.2010.10.025
11. Cusick R. D., Bryan B., Parker D. S., Merrill M. D., Mehanna M., Kiely P. D., et al. (2011).Performance of a pilot-scale continuous flow microbial electrolysis cell fed winery wastewater.*Appl. Microbiol. Biotechnol.* 89 2053–2063. 10.1007/s00253-011-3130-9
12. Dubinsky E. A., Silver W. L., Firestone M. K. (2010). Tropical forest soil microbial communities couple iron and carbon biogeochemistry. *Ecology* 91 2604–2612. 10.1890/09-1365.1
13. Erable B., Etcheverry L., Bergel A. (2011). From microbial fuel cell (MFC) to microbial electrochemical snorkel (MES): maximizing chemical oxygen demand (COD) removal from wastewater. *Biofouling* 27 319–326. 10.1080/08927014.2011.564615
14. Fornero J. J., Rosenbaum M., Angenent L. T. (2010a). Electric power generation from municipal, food, and animal wastewaters using microbial fuel cells. *Electroanalysis* 22 832–843. 10.1002/elan.200980011
15. Fornero J. J., Rosenbaum M., Cotta M. A., Angenent L. T. (2010b). Carbon dioxide addition to microbial fuel cell cathodes maintains sustainable catholyte pH and improves anolyte pH, alkalinity, and conductivity. *Environ. Sci. Technol.* 44 2728–2734. 10.1021/Es9031985
16. Franks A. E., Nevin K. P. (2010). Microbial fuel cells, a current review. *Energies* 3 899–919. 10.3390/En3050899
17. Frei S., Knorr K. H., Peiffer S., Fleckenstein J. H. (2012). Surface micro-topography causes

hot spots of biogeochemical activity in wetland systems: a virtual modeling experiment. *J. Geophys. Res.* 117 G00N12 10.1029/2012jg002012

18. Freney J. R., Smith C. J., Mosier A. R. (1992). Effect of a new nitrification inhibitor (wax coated calcium carbide) on transformations and recovery of fertilizer nitrogen by irrigated wheat. *Fertil. Res.* 32 1–11. 10.1007/BF01054388

19. Friedman E. S., Miller K. E., Lipson D. A., Angenent L. T. (2013). Potentiostatically-poised electrodes mimic iron oxides and interact with soil microbial communities to alter the biogeochemistry of Arctic peat soils. *Minerals* 3 318–336. 10.3390/min3030318

20. Friedman E. S., Rosenbaum M. A., Lee A. W., Lipson D. A., Land B. R., Angenent L. T. (2012).A cost-effective and field-ready potentiostat that poises subsurface electrodes to monitor bacterial respiration. *Biosens. Bioelectron.* 32 309–313. 10.1016/j.bios.2011.12.013

21. Gilbert J. A., Meyer F., Jansson J., Gordon J., Pace N., Tiedje J., et al. (2010). The Earth Microbiome Project: meeting report of the "1st EMP meeting on sample selection and acquisition" at Argonne National Laboratory October 6(th) 2010. *Stand. Genomic Sci.* 3 249–253. 10.4056/Sigs.1443528

22. Gregory K. B., Lovley D. R. (2005). Remediation and recovery of uranium from contaminated subsurface environments with electrodes. *Environ. Sci. Technol.* 39 8943–8947. 10.1021/Es050457e

23. Groffman P. M., Altabet M. A., Bohlke J. K., Butterbach-Bahl K., David M. B., Firestone M. K., et al. (2006). Methods for measuring denitrification: diverse approaches to a difficult problem.*Ecol. Appl.* 16 2091–2122. 10.1890/1051-0761(2006)016[2091:MFMDDA]2.0.CO;2

24. Groffman P. M., Butterbach-Bahl K., Fulweiler R. W., Gold A. J., Morse J. L., Stander E. K., et al. (2009). Challenges to incorporating spatially and temporally explicit phenomena (hotspots and hot moments) in denitrification models. *Biogeochemistry* 93 49–77. 10.1007/s10533-008-9277-5

25. Groffman P. M., Gold A. J., Simmons R. C. (1992). Nitrate dynamics in riparian forests – microbial studies. *J. Environ. Qual.* 21 666–671. 10.2134/jeq1992.214666x

26. Hamelers H. V. M., Ter Heijne A., Sleutels T. H. J. A., Jeremiasse A. W., Strik D. P. B. T. B., Buisman C. J. N. (2010). New applications and performance of bioelectrochemical systems.*Appl. Microbiol. Biotechnol.* 85 1673–1685. 10.1007/s00253-009-2357-1

27. He Z., Minteer S. D., Angenent L. T. (2005). Electricity generation from artificial wastewater using an upflow microbial fuel cell. *Environ. Sci. Technol.* 39 5262–5267. 10.1021/es0502876

28. Hutchinson G. L., Mosier A. R. (1981). Improved soil cover method for field measurement of nitrous-oxide fluxes. *Soil Sci. Soc. Am. J.* 45 311–316. 10.2136/sssaj1981.03615995004500020017x

29. Ishii S. I., Hotta Y., Watanabe K. (2008). Methanogenesis versus electrogenesis: morphological and phylogenetic comparisons of microbial communities. *Biosci. Biotechnol. Biochem.* 72 286–294. 10.1271/bbb.70179

30. Kato S., Hashimoto K., Watanabe K. (2012). Microbial interspecies electron transfer via electric currents through conductive minerals. *Proc. Natl. Acad. Sci. U.S.A.* 109 10042–10046. 10.1073/pnas.1117592109

31. Keller J. K., Takagi K. K. (2013). Solid-phase organic matter reduction regulates anaerobic decomposition in bog soil. *Ecosphere* 4 art54 10.1890/ES12-00382.1

32. Kumlanghan A., Liu J., Thavarungkul P., Kanatharana P., Mattiasson B. (2007). Microbial fuel cell-based biosensor for fast analysis of biodegradable organic matter. *Biosens. Bioelectron.* 222939–2944. 10.1016/j.bios.2006.12.014

33. Lentini C. J., Wankel S. D., Hansel C. M. (2012). Enriched iron(III)-reducing bacterial communities are shaped by carbon substrate and iron oxide mineralogy. *Front. Microbiol.* 3:40410.3389/fmicb.2012.00404

34. Li Z. J., Rosenbaum M. A., Venkataraman A., Tam T. K., Katz E., Angenent L. T. (2011).Bacteria-based AND logic gate: a decision-making and self-powered biosensor. *Chem. Commun.*47 3060–3062. 10.1039/C0cc05037g

35. Lipson D. A., Jha M., Raab T. K., Oechel W. C. (2010). Reduction of iron (III) and humic substances plays a major role in anaerobic respiration in an Arctic peat soil. *J. Geophys. Res.* 115G00I06 10.1029/2009jg001147

36. Lipson D. A., Raab T. K., Goria D., Zlamal J. (2013). The contribution of Fe(III) and humic acid reduction to ecosystem respiration in drained thaw lake basins of the Arctic Coastal Plain.*Global Biogeochem. Cycles* 27 399–409. 10.1002/gbc.20038

37. Liptzin D., Silver W. L. (2009). Effects of carbon additions on iron reduction and phosphorus availability in a humid tropical forest soil. *Soil Biol. Biochem.* 41 1696–1702. 10.1016/j.soilbio.2009.05.013

38. Logan B. E., Hamelers B., Rozendal R., Schroder U., Keller J., Freguia S., et al. (2006).Microbial fuel cells: methodology and technology. *Environ. Sci. Technol.* 40 5181–5192. 10.1021/es0605016

39. Lovley D. R. (2008). The microbe electric: conversion of organic matter to electricity. *Curr. Opin. Biotechnol.* 19 564–571. 10.1016/j.copbio.2008.10.005

40. Lozupone C., Knight R. (2005). UniFrac: a new phylogenetic method for comparing microbial communities. *Appl. Environ. Microbiol.* 71 8228–8235. 10.1128/Aem.71.12.8228-8235.2005

41. McClain M. E., Boyer E. W., Dent C. L., Gergel S. E., Grimm N. B., Groffman P. M., et al. (2003). Biogeochemical hot spots and hot moments at the interface of terrestrial and aquatic ecosystems. *Ecosystems* 6 301–312. 10.1007/s10021-003-0161-9

42. McPhillips L. E., Groffman P. M., Goodale C. L., Walter M. T. (2015). Hydrologic and biogeochemical drivers of riparian denitrification in an agricultural watershed. *Water Air Soil Pollut.* 226:169 10.1007/s11270-015-2434-2

43. Meng L., Hess P. G. M., Mahowald N. M., Yavitt J. B., Riley W. J., Subin Z. M., et al. (2011).Sensitivity of wetland

methane emissions to model assumptions: application and model testing against site observations. *Biogeosci. Discuss.* 8 6095–6160. 10.5194/bgd-8-6095-2011

44. Morales S. E., Holben W. E. (2011). Linking bacterial identities and ecosystem processes: can 'omic' analyses be more than the sum of their parts? *FEMS Microbiol. Ecol.* 75 2–16. 10.1111/j.1574-6941.2010.00938.x
45. Morris J. M., Jin S. (2012). Enhanced biodegradation of hydrocarbon-contaminated sediments using microbial fuel cells. *J. Hazard. Mater.* 213 474–477. 10.1016/j.jhazmat.2012.02.029
46. Oksanen J., Blanchet F. G., Kindt R., Legendre P., Stevens H. H., Wagner H. (2013). *vegan: Community Ecology Package. R Package Version 2.0–8*. Available at: http://cran.r-project.org
47. Palmer M. A., Febria C. M. (2012). The heartbeat of ecosystems. *Science* 336 1393–1394. 10.1126/science.1223250
48. Prosser J. I. (2015). Dispersing misconceptions and identifying opportunities for the use of 'omics' in soil microbial ecology. *Nat. Rev. Microbiol.* 13 439–446. 10.1038/nrmicro3468
49. Raes J., Bork P. (2008). Molecular eco-systems biology: towards an understanding of community function. *Nat. Rev. Microbiol.* 6 693–699. 10.1038/Nrmicro1935
50. Ranalli A. J., Macalady D. L. (2010). The importance of the riparian zone and in-stream processes in nitrate attenuation in undisturbed and agricultural watersheds – a review of the scientific literature. *J. Hydrol.* 389 406–415. 10.1016/j.jhydrol.2010.05.045
51. R Core Team (2013). *R: A Language and Environment for Statistical Computing. Vienna, Austria. 2013*. Available at: http://www.R-project.org
52. Regnier P., Dale A. W., Arndt S., Larowe D. E., Mogollon J., Van Cappellen P. (2011).Quantitative analysis of anaerobic oxidation of methane (AOM) in marine sediments: a modeling perspective. *Earth Sci. Rev.* 106 105–130. 10.1016/j.earscirev.2011.01.002
53. Riley W. J., Subin Z. M., Lawrence D. M., Swenson S. C., Torn M. S., Meng L., et al. (2011).Barriers to predicting changes in global terrestrial methane fluxes: analyses using CLM4Me, a methane biogeochemistry model integrated in CESM. *Biogeosciences* 8 1925–1953. 10.5194/bg-8-1925-2011
54. Rosenbaum M., Aulenta F., Villano M., Angenent L. T. (2011a). Cathodes as electron donors for microbial metabolism: which extracellular electron transfer mechanisms are involved?*Bioresour. Technol.* 102 324–333. 10.1016/j.biortech.2010.07.008
55. Rosenbaum M. A., Bar H. Y., Beg Q. K., Segre D., Booth J., Cotta M. A., et al. (2011b).Shewanella oneidensis in a lactate-fed pure-culture and a glucose-fed co-culture with*Lactococcus lactis* with an electrode as electron acceptor. *Bioresour. Technol.* 102 2623–2628. 10.1016/j.biortech.2010.10.033
56. Rozendal R. A., Hamelers H. V., Buisman C. J. (2006). Effects of membrane cation transport on pH and microbial fuel cell performance. *Environ. Sci. Technol.* 40 5206–5211. 10.1021/es060387r
57. Rozendal R. A., Leone E., Keller J., Rabaey K. (2009). Efficient hydrogen peroxide generation from organic matter in a bioelectrochemical system. *Electrochem. Commun.* 11 1752–1755. 10.1016/j.elecom.2009.07.008
58. Schmidt M. W. I., Torn M. S., Abiven S., Dittmar T., Guggenberger G., Janssens I. A., et al. (2011). Persistence of soil organic matter as an ecosystem property. *Nature* 478 49–56. 10.1038/nature10386
59. Strycharz S. M., Gannon S. M., Boles A. R., Franks A. E., Nevin K. P., Lovley D. R. (2010).Reductive dechlorination of 2-chlorophenol by Anaeromyxobacter dehalogenans with an electrode serving as the electron donor. *Environ. Microbiol. Rep.* 2 289–294. 10.1111/j.1758-2229.2009.00118.x
60. Teh Y. A., Dubinsky E. A., Silver W. L., Carlson C. M. (2008). Suppression of methanogenesis by dissimilatory Fe(III)-reducing bacteria in tropical rain forest soils: implications for ecosystem methane flux. *Glob. Chang. Biol.* 14 413–422. 10.1111/j.1365-2486.2007.01487.x
61. TerAvest M. A., Li Z. J., Angenent L. T. (2011). Bacteria-based biocomputing with Cellular Computing Circuits to sense, decide, signal, and act. *Energy Environ. Sci.* 4 4907–4916. 10.1039/C1ee02455h
62. Thompson R. B. (1996). Using calcium carbide with the acetylene inhibition technique to measure denitrification from a sprinkler irrigated vegetable crop. *Plant Soil* 179 9–16. 10.1007/BF00011637
63. Tibshirani R., Hastie T., Narasimhan B., Chu G. (2002). Diagnosis of multiple cancer types by shrunken centroids of gene expression. *Proc. Natl. Acad. Sci. U.S.A.* 99 6567–6572. 10.1073/pnas.082099299
64. Vidon P., Allan C., Burns D., Duval T. P., Gurwick N., Inamdar S., et al. (2010). Hot spots and hot moments in riparian zones: potential for improved water quality management1. *J. Am. Water Resour. Assoc.* 46 278–298. 10.1111/j.1752-1688.2010.00420.x
65. Vidon P., Hill A. R. (2004). Denitrification and patterns of electron donors and acceptors in eight riparian zones with contrasting hydrogeology. *Biogeochemistry* 71 259–283. 10.1007/s10533-004-9684-1
66. Villano M., Aulenta F., Beccaru M., Majone M. (2010). Microbial generation of H2 or CH4 coupled to wastewater treatment in bioelectrochemical systems. *Chem. Eng. Trans.* 20 163–168.
67. Weber K. A., Achenbach L. A., Coates J. D. (2006). Microorganisms pumping iron: anaerobic microbial iron oxidation and reduction. *Nat. Rev. Microbiol.* 4 752–764. 10.1038/Nrmicro1490
68. Williams K. H., Nevin K. P., Franks A., Englert A., Long P. E., Lovley D. R. (2010). Electrode-based approach for monitoring in situ microbial activity during subsurface bioremediation.*Environ. Sci. Technol.* 44 47–54. 10.1021/Es9017464
69. Wood T. E., Cavaleri M. A., Reed S. C. (2012). Tropical forest carbon balance in a warmer world: a critical review spanning microbial– to ecosystem-scale processes. *Biol.*

Rev. 87 912–927. 10.1111/j.1469-185X.2012.00232.x

70. Zhang T., Gannon S. M., Nevin K. P., Franks A. E., Lovley D. R. (2010). Stimulating the anaerobic degradation of aromatic hydrocarbons in contaminated sediments by providing an electrode as the electron acceptor. *Environ. Microbiol.* 12 1011–1020. 10.1111/j.1462-2920.2009.02145.x

CHAPTER 13

UNDERSTANDING AMMONIUM TRANSPORT IN BIOELECTROCHEMICAL SYSTEMS TOWARDS ITS RECOVERY

Ying Liu, MohanQin, Shuai Luo, Zhen He and RuiQiao

Nitrogen removal from wastewater is of great importance to protect receiving water from eutrophication[1]. The primary form of nitrogen in wastewater is ammonia (or as a product of decomposing organic nitrogen)[2], and thus removing ammonia from municipal and industrial wastewater prior to discharge is obligatory. Ammonia can be removed from wastewater using biological or physicochemical methods such as nitrification and denitrification, ion exchange, air stripping, and anaerobic ammonium oxidation (anammox)[3,4] Because ammonia nitrogen is a key fertilizer component for agricultural applications, recovery instead of removal of ammonia from wastewater can greatly impact on both sustainable wastewater management and economics. An emerging approach for recovering ammonia is through bioelectrochemical systems (BES)[5,6].

In BES-based ammonia recovery systems, organic compounds in wastewater are oxidized by the exoelectrogens growing on an anode, producing a current through its external circuit[7]. When cation exchange membranes (CEMs) are used[8,9,10,11] the current is carried by cations such as NH_4^+ ions moving from the anolyte into the catholyte through the CEM. The NH_4^+ ions, once transported into the catholyte, are converted into NH_3 molecules because the catholyte is usually rendered basic by the production of OH^- ions therein[12,13,14]. Some of these NH_3 molecules can then be recovered through gas aeration. This type of process has been demonstrated in various BES including microbial fuel cells (MFCs) and microbial electrolysis cells (MECs). In MECs, a higher current density would greatly enhance ammonia recovery[9], and thus MECs with external power exhibit a better performance for ammonium recovery than MFCs[15]. It was reported that ammonia can be recovered from ammonium-rich wastewater, such as synthetic wastewater, urine, and swine wastewater[12,13,16,17].

While a general picture of ammonia recovery using BES is emerging, some issues remain open. In particular, the transport of NH_4^+ ions and NH_3 is not fully understood despite their essential role in ammonia recovery. Some studies indicate that NH_4^+ transport dominates the current across the CEM (i.e., for each electron passing through the external circuit, ~1 NH_4^+ ion moves from the anolyte into the catholyte through the CEM)[18], while other studies suggest that NH_4^+ ions may carry only 40% of the total current[5,12,16,19]. In addition, whether NH_4^+ ions move through the CEM via diffusion or migration is not clear. A recent study suggests that the migration of NH_4^+ ions dominates over their diffusion through a comparison of the net NH_4^+ transport under the close and open circuit conditions[9]. However, this conclusion was based on the assumption that the diffusion of NH_4^+ ions under these conditions is the same, whose validity/generality is not yet clear. In a seminal simulation study in which key transport phenomena and chemical reaction within the BES are modeled comprehensively, it was found that the diffusion of NH_3 from the catholyte to the anolyte can greatly reduce the efficiency of recovery[20]. However, the condition under which such diffusion is important has not been delineated yet. Finally, since many cations such as Na^+ ions exist in the anolyte, for a given current through the BES, they may compete with NH_4^+ ions for transport across the CEM[11,16,21]. How strong

Citation: Ying Liu, MohanQin, Shuai Luo, Zhen He and RuiQiao, Understanding Ammonium Transport in Bioelectrochemical Systems towards its Recovery, Sci. Rep. 6, 22547; doi: 10.1038/srep22547 (2016).

such competition is and how it affects the ammonia recovery is not well understood.

Another open issue is how the pH in the anolyte and catholyte is regulated. Since the pH values in the anolyte/catholyte affect how the total nitrogen content is partitioned between its two forms (NH_4^+ ions and NH_3), and the latter in turn affects the transport of NH_4^+ ions through the CEM and the recovery of NH_3 through aeration, it is essential to regulate these pH effectively. Furthermore, regulating the anolyte pH is also essential for ensuring microbial activity and hence the reliable operation of BES[8,22,23]. Maintaining a basic environment with pH >9 in the catholyte facilitates the conversion of NH_4^+ ions into NH_3, which is key for the effective recovery of ammonia by aeration. The effective control of the pH in the anolyte and the catholyte requires a thorough understanding of what governs the pH in these electrolytes[11,21,24]. While it is known that the generation of proton in the anolyte and hydroxide ions in the catholyte plays an important role in controlling the pH[11,12,20,21,24,25], how and to what extent the transport of all ions and their chemical reactions with each other affect the pH is little understood.

Resolving the above issues using experimental study alone is difficult due to the challenge in assessing details of all transport processes in the BES. Numerical modeling can address this challenge, but most prior modeling of BES focused on power production and organic removal[26,27,28,29,30]. In the only comprehensive model for BES-based ammonia recovery, the transport and chemical reactions of major species in BES was studied at a steady state[20]. The predictions of this model agree qualitatively with experimental observations and they led to critical insight into the effects of current density and membrane properties on the ammonia recovery. However, the model has not been validated by detailed comparison with experiments, and studies based on this model did not cover the issues described above, e.g., the competition of inert ions (i.e., ions that do not react with other species within the system, e.g., the Na^+ ions) with the NH_4^+ ions for transport across the CEM cannot be studied using steady state models. Here, we have studied the ammonia recovery in a representative BES – microbial electrolysis cells, by integrating experiments with simulations.

RESULTS

A bench-scale cubic shape MEC was used for experimental data collection (see Fig. 1). The wastewater fed into the anode was synthetic digestion effluent of livestock waste while deionized water was used as catholyte. The anolyte was partially replaced (150 mL) every 48 h while the catholyte was unchanged during 3 batch cycles. The variation of organic matters, NH_4^+/NH_3 concentration, pH and inert ions in both chambers were recorded for analysis and model validation.

To simulate the batch-mode operation of MECs for ammonia recovery, we extend the steady-state models developed earlier[20]. The models consider the mass conservation for all species (without loss of generality, the following species are included: Na^+, Cl^-, HAc, Ac^-, NH_4^+, NH_3, H_2CO_3, HCO_3^-, CO_3^{2-}, H^+ and OH^-), the transport of each species (diffusion and migration) across the membrane, the chemical reactions among different species, and gas-solution equilibrium. For chemical reactions, we consider mostly acid-base reactions which are essential for ammonia recovery (e.g., $NH_3 + H^+ \rightleftharpoons NH_4^+$, $H_2CO_3 \rightleftharpoons H^+ + HCO_3^-$, ... and a full list of these reactions is provided in theSupplementary Information). Using the operation parameters (e.g., the anolyte/catholyte volume V_1/V_2, the membrane surface area A, etc.) and the measured current density as input, the model can predict the time evolution of the concentration of each species in the system (hereafter, the concentration of species i in anolyte/catholyte chamber is denoted as c_i^j, with $j = 1$ and 2 for anode and cathode chamber, respectively) and the recovery of ammonia by aeration. Below we first validate the mathematical model, then investigate the above issues and how they impact the ammonia recovery in the MEC.

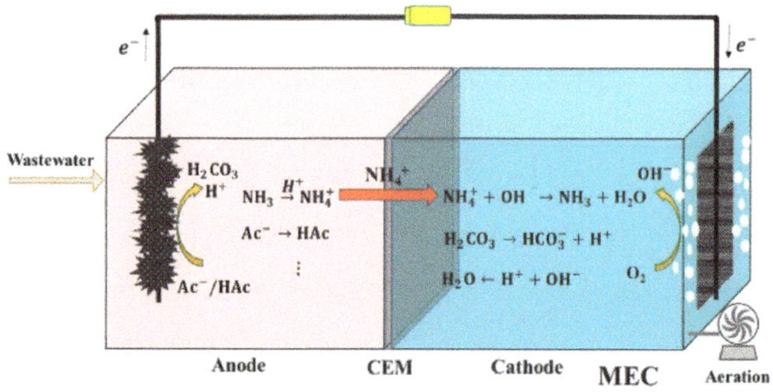

Figure 1: A sketch of the MEC-based ammonia recovery system.

Model Validation

We first validate the mathematical model for ammonia recovery using MEC by comparing its predictions against experimental data obtained under the same operation conditions. Ammonium recovery was firstly studied experimentally. During one batch cycle, the COD concentration decreased from 1089 ± 169 to 262 ± 83 mg L^{-1}, resulting in a Coulombic efficiency (CE) of 34.7 ± 5.9%. The maximum current density was 1.89 A m^{-2} (Fig. 2a). In the anolyte, the pH decreased from 7.98 ± 0.08 to 4.52 ± 0.63; in the catholyte, the pH increased dramatically initially and then stabilized at ~9.7 (Fig. 2b). The acetate concentration in the anolyte decreased from 15.6 ± 2.4 to 5.4 ± 0.9 mM, and that in the catholyte was always below the detection limit of our equipment (Fig. 2c). The latter is consistent with the fact that the diffusion loss of acetate through CMI-7000 is small[21]. Experimentally, the total input nitrogen as ammonium in each batch cycle was 9.3 ± 0.4 mmol, among which 35.7 ± 7.4% stayed in the anolyte, 0.9 ± 0.3% remained in the catholyte and 66.2 ± 2.7% was stripped out as ammonia and absorbed by the sulfuric acid. In each batch cycle (2 days), 6.1 ± 0.1 mmol of NH_3 gas was collected through aeration, resulting in a recovery rate of 10.2 ± 0.1 $g_N\,m^{-2}\,d^{-1}$ (vs. membrane surface area)[18]. Ammonia oxidation in the anode chamber might not occur because we did not detect any nitrite or nitrate in the anolyte. Numerically, the simulation predicted the total ammonia concentration in anolyte and catholyte to be 5% and 18.4% within the corresponding maximum measured values, respectively. Some Na^+ and Cl^- ions crossed into the catholyte through the CEM in each cycle. To assess their role in the net charge transport through the CEM, we computed the cumulative transport number $\bar{t}_i^+ = V_1 z_i F(c_i^1(0) - c_i^1(t)) / \int_0^t$ for these ions during each cycle. At the end of each cycle,

$\bar{t}^{+}_{Na^+} = 0.124 \pm 0.044$ and $\bar{t}^{+}_{Cl^-} = -0.021 \pm 0.004$ (Fig. 2f), indicating that overall the transport of these ions contributes to ~10% of the total current. Since the transport of all other ions except NH_4^+ ions through the CEM is very small, we thus conclude that the transport of NH_4^+ ions through the CEM accounts for ~90% of the total current in our system, which is consistent with some earlier studies[18].

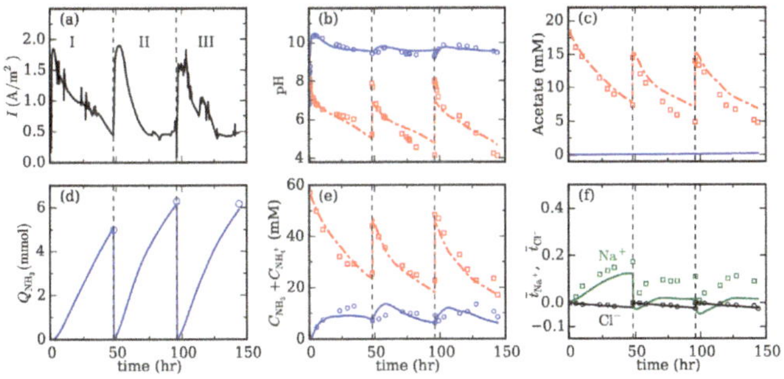

Figure 2: Comparison of experimental and model studies of ammonia recovery during three batch cycles of MEC operation. The current density measured experimentally (a) is feed into the models to predict the various observables of the anolyte and the catholyte: the pH (b), the acetate concentration (c), the amount of NH3 collected from catholyte by aeration (d), the total nitrogen content (e), and the transport of Na+ and Cl− ions out of the anolyte (f). Symbols are the experiment data and lines are model prediction. In (b–e), red lines are for anolyte and blue lines are for catholyte. In (c), the acetate concentration in catholyte is below the detection limit of our equipment and thus not shown. In (f), to highlight the relative importance of Na+/Cl− ion transport in the overall charge transport in the system, their transport is quantified using the cumulative transport number + ti (see text, = . + t 1 0 i means that the charge carried by the transport a species i is equal to the total charge passed through the membrane). A detailed summary of operating conditions and parameters (e.g., initial species concentration in anolyte/catholyte) is in provided

The above ammonia recovery operation was also simulated using the mathematical model developed. As shown in Fig. 2b–f, the model predictions agree quite well with the experimental data. In particular, the model captures the evolution of pH, acetate concentration, and total nitrogen content in the anolyte/catholyte. The recovery of NH_3 gas was also accurately captured. The model underestimates the transport of Na^+ ions across the CEM (Fig. 2f). However, given that the transport of Na^+ ions accounts for only a small fraction of the total current through the CEM and these ions do not interact with other species in the system, these small deviations are deemed acceptable.

Using the validated model, we next examine the mechanisms of MEC-based ammonia recovery using simulations and address three issues outlined earlier, i.e., the pH regulation, NH_4^+ transport and ion competition in MECs. The initial conditions in the anolyte are the same as those in the above experiment. The catholyte initially features 10 mM of Na^+ ions and accompanying carbonate group ions, which together produces a pH of ~6.8, as commonly found in experiments. Without a loss of generality, the system operates at a fixed current density of $I = 1\,A\,m^{-2}$ for 48 hours.

pH Regulation

To understand how the pH is regulated, we note that, by considering the mass conservation of proton, its concentration evolution in the anolyte follows:

$$V_1 \frac{\partial C^1_{H^+}}{\partial t} = A \cdot J'_{H^+,1} + \frac{9}{8}\frac{IA}{F} - V_1\dot{S}_{NH_3 \to NH_4^+} - V_1\dot{S}_{Ac^- \to HAc} - V_1\dot{S}_{OH^- \to H_2O} - V_1\dot{S}_{HCO_3^- \to H_2CO_3} + V_1\dot{S}_{HCO_3^- \to CO_3^{2-}} \quad (1)$$

This equation shows that the proton concentration in the anolyte is affected by several factors: the transport of proton into or out of the anolyte through the CEM (the first term on the right side), the production of proton through Faraday reactions (the second term on right), and the consumption or production of protons by various chemical reactions (the remaining terms on right), e.g., $V_1\dot{S}_{NH_3 \to NH_4^+}$ stands for the consumption of proton by the reaction $NH_3 + H^+ \to NH_4^+$. For brevity, the last two terms in Eq. (1) are hereafter lumped as $\dot{S}_{carb}$.

To understand how the pH in the anolyte is regulated, we simulated the operation of MEC under a constant current of $I = 1\,A\,m^{-2}$. Figure 3a shows that the pH in the anolyte decreases steadily during the operation. To assess the relative importance of the various factors identified above in regulating the pH, we examined each term on the right-hand side of Eq. (1) and the results are shown in Fig. 3b (all terms are normalized by the Faradaic reaction term). We observe that most of the protons generated by the Faradaic reaction were neutralized by their reactions with NH_3, the Ac^- ions, and the carbonate ions. Hence, these chemical groups essentially serve as the "buffer" to keep the pH from changing dramatically. The relative contribution of these chemical groups in regulating the pH evolves during the operation of MEC: NH_3 lost its buffer capability soon after the pH is below 7, while the carbonate ions remain effective in buffering the anolyte till ~40 hrs.

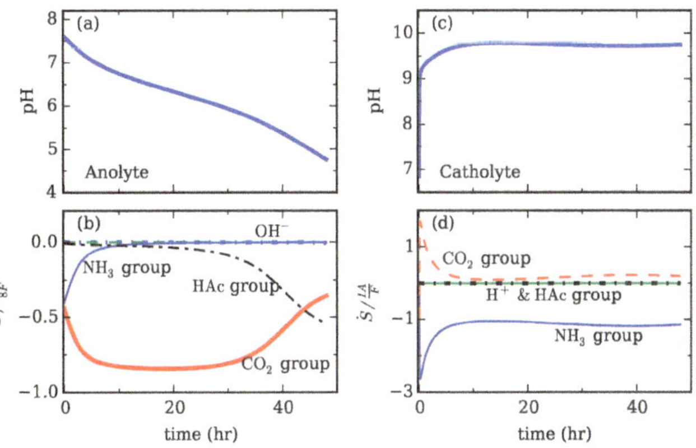

Figure 3: Regulation of pH in MEC during ammonia recovery operation.

A similar equation for the evolution of the hydroxide concentration in the catholyte was derived to understand the regulation of pH therein:

$$V_2 \frac{\partial C^2_{OH^-}}{\partial t} = AJ'_{OH^-,2} + \frac{IA}{F} - V_1\dot{S}_{HAc \to Ac^-} - V_1\dot{S}_{NH_4^+ \to NH_3} - V_1\dot{S}_{H^+ \to H_2O} - V_1\dot{S}_{H_2CO_3 \to HCO_3^-} - V_1\dot{S}_{HCO_3^- \to CO_3^{2-}} \quad (2)$$

This equation shows that the hydroxide concentration in the catholyte is affected by the transport of hydroxide into/out of the catholyte (the first term on the right side), the production of hydroxide by Faradaic reactions (the second term on right), and various chemical reactions consuming or producing hydroxide (the remaining terms on right). The last two terms are hereafter lumped as S_{carb}. Figure 3c,dshow the pH variation and the relative magnitude of each reaction term on the right-hand side of Eq. (2). We observe that, in the catholyte, the pH was predominately controlled by the Faradaic production of hydroxide and the consumption of hydroxide by their reaction with the NH_4^+ ions. Since there are few NH_4^+ ions inside the catholyte at the very beginning, the pH value rises rapidly during the earliest stage of operation. Unlike the NH_4^+ ions, the carbonate ions tend to increase the pH value because their chemical reactions inside the catholyte consume proton. Overall, the effect of carbonate ions on the pH in catholyte is limited, and this is related to their low concentration inside the catholyte. The latter is due to the limited carbonate ions inside the anolyte and their slow diffusion through the CEM. Since the pH regulation inside the catholyte depends strongly on the availability of NH_4^+ ions, and the latter in turn depends on the transport of NH_4^+ ions and NH_3 across the CEM, it is clear that the pH regulation is intimately coupled with the transport of these species.

Ammonium and Ammonia Transport

Such transport is analyzed from the perspective of how these species are removed from the anolyte. There are three ways to remove NH_3/NH_4^+ ion from the anolyte: diffusion and migration of NH_4^+ ion through the CEM and the diffusion of NH_3 through CEM. Figure 4a shows that, under the moderate electrical current considered here ($I=1\,A\,m^{-2}$), the diffusion of NH_4^+ is actually stronger than its electrical migration. This somewhat surprising result is caused largely by the different pH value inside the anolyte and catholyte. As shown in Fig. 2, the anolyte (catholyte) is strongly acidic (basic) during operation. Hence, nitrogen element exists mostly as NH_4^+ in the anolyte and as NH_3 in the catholyte. Consequently, the NH_4^+ concentration in the anolyte is always much higher than that in the catholyte (Fig. 4b), which leads to a strong diffusion flux of the NH_4^+ ions. Since the acidic (basic) pH in the anolyte (catholyte) is ultimately induced by the electrical current through the system, we see that the imposed current serves two roles: it helps drive the migration of NH_4^+ ions from the anolyte toward catholyte, and it helps set up the pH environment that favors the diffusion of NH_4^+ ion from anolyte chamber toward the catholyte chamber. The second role has not been widely recognized. However, the fact that the diffusion of NH_4^+ ions dominates over their migration suggests that this second role is at least as important as the first role. We note that, while diffusion dominates the transport of NH_4^+ under low/moderate current densities (e.g., the $I=1\,A\,m^{-2}$ considered here), electrical migration can become the dominant transport mechanism at high current densities (e.g., at $I>5\,A\,m^{-2}$)[19].

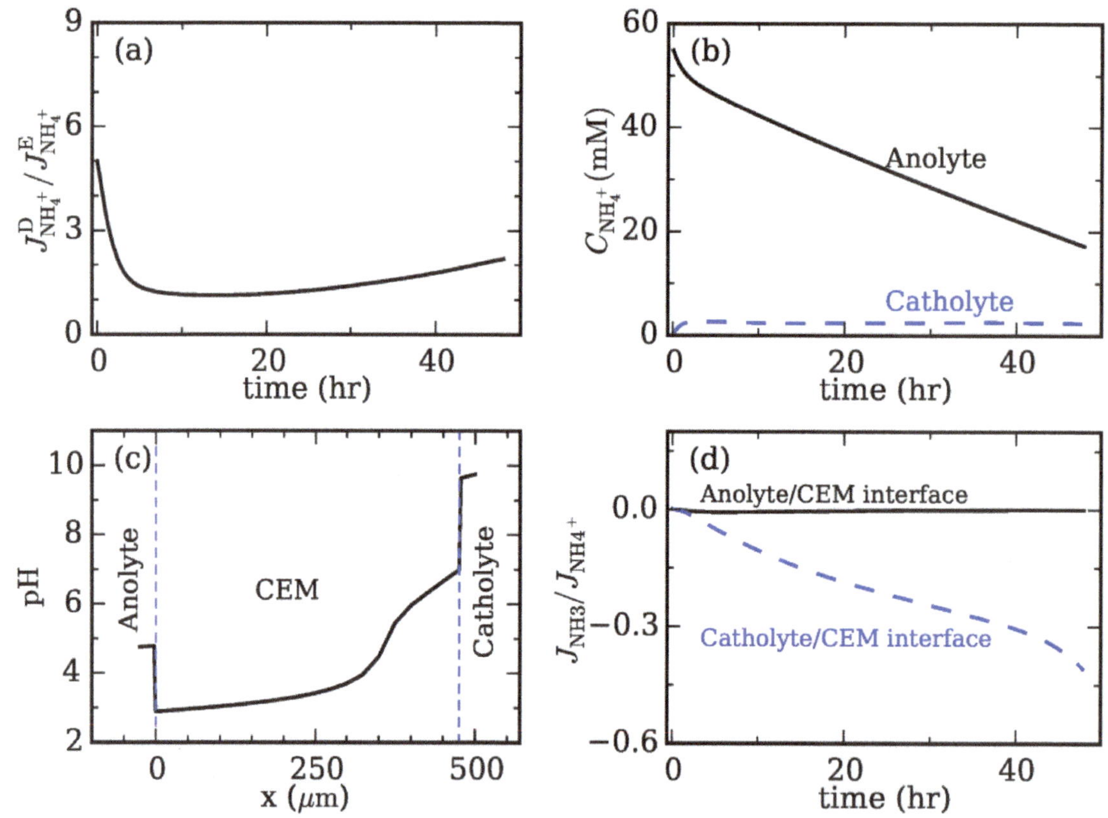

Figure 4: Mechanisms of NH_4^+/NH_3 **transport in MECs.** (a) Relative strength of NH4 + ion transport due to diffusion and migration. (b) Evolution of the NH_4^+ ion concentration in the anolyte and catholyte. (c) Distribution of pH across the CEM and at the CEM-anolyte/catholyte interfaces. (d) The diffusion flux of NH_3 toward the anolyte at the CEM-catholyte/anolyte interfaces. These diffusion fluxes are normalized by the total NH_4^+ ion flux to highlight their impact on the ammonia recovery in the catholyte. [ALL parameters in this study are the same as in Fig. 3.]

It is worth pointing out that, in absence of a net current, because of the low cation concentration in our catholyte (hence limited cation exchange between the CEM and catholyte), the removal of NH_4^+ from anolyte through transport across the CEM is minor. Indeed, both simulations and experiments suggest that ~2% of NH_4^+ could be removed from the anolyte in one batch cycle (48 hours), which is consistent with some prior studies[15].

The dominance of NH_3 over NH_4^+ ions inside the catholyte and the opposite trend in the anolyte implies that there can exist strong diffusion of NH_3 from the catholyte chamber toward the anolyte chamber. Such diffusion can potentially lower the recovery efficiency of NH_3 from the catholyte[20]. However, Fig. 4d shows that, although there is a relatively large amount of NH_3 diffuse toward the anolyte chamber at the CEM-catholyte interface, very little NH_3 crosses CEM-anolyte interface into the anolyte. This interesting phenomenon is caused by the pH environment within the CEM. As shown in Fig. 4d, within the CEM, the pH is strongly acidic at the anolyte side, which favors conversion of NH_3 to NH_4^+ ions by $NH_3 + H^+ \rightarrow NH_4^+$ reaction. Hence, the anode side of the membrane served as the "reactor": the NH_3 diffused toward the anolyte chamber was turned into NH_4^+ within the membrane and subsequently diffused back into the cathode chamber.

More generally, the relevance of NH_3 diffusion from catholyte to anolyte to its recovery can be estimated. The diffusion of NH_3 is $J_{NH_3} \sim D^m_{NH_3} C^2_{NH_3}/L$ ($D^m_{NH_3}$ ($D^m_{NH_3}$: diffusion coefficient of NH_3 in the membrane; $C^2_{NH_3}$ NH_3 concentration in the catholyte; L: membrane thickness). The evaporating flux of NH_3 in the catholyte is $J^{evap}_{NH_3} = C^2_{NH_3} Q_{inert}/A p_{tot} K_{H,NH}$[20,31]. Hence, we define a dimensionless number

$$\Pi = \frac{D^m_{NH_3} p_{tot} K_{H,NH} A}{L\, Q_{inert}} \tag{3}$$

when $\Pi \ll 1$, the effect of diffusion on ammonia recovery is negligible. For the system studied here, we have $D^m_{NH_3} = 2.1 \times 10^{-11}\ m^2/s$, $p_{tot} = 1$ bar, $K_{H,NH} = 56250$ mM/bar, L=475 μm[32], and aeration rate $Q_{inert/A} = 0.066$ mol/(m^2s). Hence, Π=0.0376, i.e., the diffusion is weak enough to be ignored, which is consistent with the result in Fig. 4c. Since Π depends only on the design and operational parameters of the MEC, it can be used conveniently during the design of MECs.

Ion Competition

For a fixed current, various cations (e.g., Na^+) in the anolyte can compete with the NH_4^+ ions for transport across the CEM, thus potentially compromising the removal of NH_4^+ ions from the anolyte. To quantitatively examine this competition, we computed the transport number of NH_4^+ and Na^+ ions at the CEM-anolyte interface, i.e., $t_i^+ = z_i F J'_{i,1}/I$ (Fig. 5a). At short time (t<9 hr), the transport number of the Na^+ ions is negative while that of the NH_4^+ ions is larger than 1.0, i.e., Na^+ ions transport from the catholyte into the anolyte, and instead of competing with the NH_4^+ ions, they *facilitate* the removal of NH_4^+ ions from the anolyte. This phenomenon weakens with increasing time. At t > ~9 hr, Na^+ ions do compete with NH_4^+ ions for transport across CEM, but the effect is still weak since the transport number of NH_4^+ ions is ~0.8 even at t=40 hr.

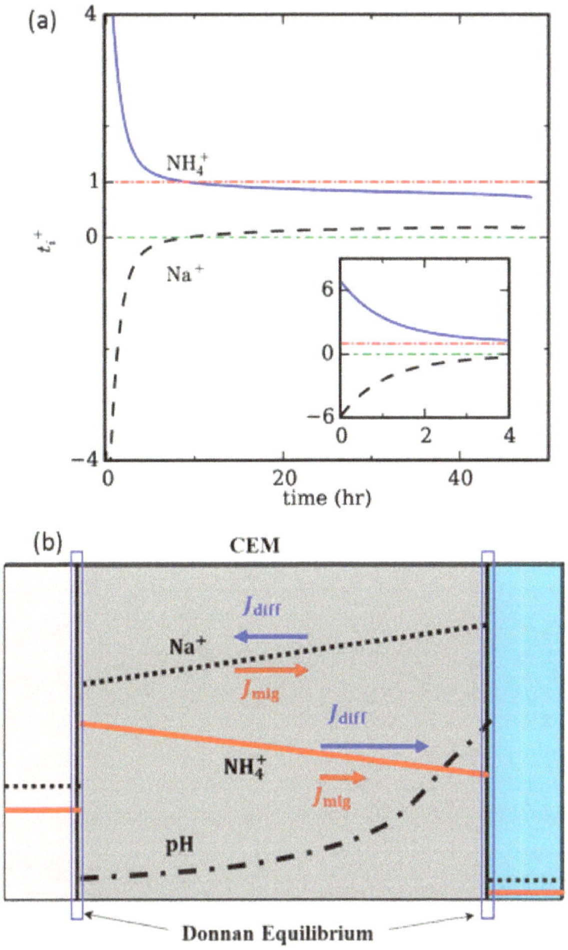

Figure 5: Ionic competition between Na^+ **and** NH_4^+.

The above surprising results are caused largely by the Donnan effects. Driven by the Donnan equilibrium, the concentrations of cations at membrane-anolyte/catholyte interfaces are elevated greatly above their concentrations in the adjacent anolyte/catholyte to balance the negative fixed charge in the CEM. For Na^+ ions, their concentration at the CEM-catholyte interface is very high because there are few other cations in the catholyte competing with them to balance the fixed negative charge in the CEM (Fig. 5b). Meanwhile, the Na^+ ions concentration at the CEM-anolyte interface is only moderately high because there are many NH_4^+ ions in the anolyte to cooperate with Na^+ ions to balance CEM's fixed charge. Consequently, the concentration of Na^+ ions within the CEM decreases from the catholyte side toward the anolyte, despite that there are more Na^+ ions in the anolyte than in catholyte. At short time, such a negative concentration gradient drives a strong diffusion flux, leading to the increase of the Na^+ ion concentration in the anolyte. Because of current conservation, more NH_4^+ ions must be removed from the anolyte than that corresponding to the current passing out of the anolyte, thus leading to a facilitated NH_4^+ ion transport, i.e., $t^+_{NH_4^+} > 1$. At longer time, the Na^+concentration at the CEM-anolyte interface increases (due to the decrease of NH_4^+ concentration in the anolyte) and the opposite happens at the CEM-catholyte interface. Hence, the diffusion of Na^+ ions is weakened. Eventually, the diffusion of Na^+ ions becomes weaker than their migration toward catholyte (Fig. 5b). However, since the migration is always partly canceled by the diffusion, Na^+ ions always compete weakly with NH_4^+ ions for transport across the CEM. As a result, the current through the system is mostly carried by the NH_4^+ ions, which is consistent with our experimental data.

The above trends in ion competition are weakly modified when the initial concentration of Na^+ ions in the anolyte changes. For higher initial concentration of Na^+ ions, the diffusion of Na^+ ions from catholyte to anolyte and its positive effect on NH_4^+ ions removal from anolyte during the early stage of operation decreases moderately, and the competition of Na^+ ions for transport across CEM sets in earlier; for lower initial concentration of Na^+ ions, the opposite occurs.

DISCUSSION

Using integrated experimental and simulation studies, we examined the NH_4^+/NH_3 transport in MECs and their coupling with the current generation, the acid-base reactions, and the transport of inert cations in the system. During MEC operation, a cascade of chemical groups regulates the pH in system to lead to an acidic (basic) environment in the anolyte (catholyte). The NH_4^+/NH_3 couple is found to play a dual role in the operation: it serves as an effective proton shuttle for the charge transport across the CEM and also as buffer agent in the anolyte and catholyte. Inert cations, even in abundance, compete rather moderately with NH_4^+ ion for transport across the CEM. The strength of the diffusion of $NH_3$from catholyte to anolyte is governed by the MEC's operating conditions and can be estimated using a dimensionless number. The findings of this study will potentially impact the ammonia recovery using BES in several aspects. First, a clear understanding of ion transport and interaction will help interpret NH_4^+ ion recovery efficiency that may be affected by the composition of wastewater. Second, delineating the role of different chemical groups in pH regulation will help guide the BES operation for optimal pH. Third, the dimensionless number of diffusion of NH_3 will help select suitable membranes. In this work, ammonium removal and recovery was examined under a high-current condition in BES, because high current generation was expected to benefit the transport of ammonium ions. Reducing current generation via lowering applied voltage and/or increasing external resistance would decrease ammonium removal/recovery.

There are still challenges to address in the future to improve the agreement between the model predictions and the experiment data. For example, there is a slight volume loss of the catholyte due to the aeration, and this effect was not considered in our model but it is likely to affect the concentration prediction in the catholyte to some extent. The bacteria activity will need to be considered in the future work. The bacteria activity in the anode could affect the balance of most of the species within the system, especially NH_4^+ and Na^+, both of which are important to ion competition.

Future studies will focus on systematic investigation of the role of current in ion transport and ammonia recovery, the interaction between electric potential and ammonium recovery, improving the numerical model by integrating current generation models into the existing model, and extension of the unsteady model to other BES systems.

MATERIALS AND METHODS

MEC Setup

A bench-scale cubic shape MEC was used in this experiment. Both the anode and the cathode chambers had the same dimension of 9 cm × 4.7 cm × 0.8 cm. The liquid volume of the anode chamber was 200 mL, while that of the cathode chamber was 180 mL. A CEM (CMI-7000, Membrane International Inc., Glen Rock, NJ, USA) with a sectional area of 42.3 cm^2 was used to separate anode and cathode chamber. The anode electrode was a carbon brush (Gordon Brush Mfg. Co., Inc., CA, USA) and the cathode electrode was a piece of 32-cm^2 carbon cloth (Zoltek Companies, Inc., MO, USA) that was coated with 5 mg cm^{-2} of Pt/C (10% wt. Pt on Carbon Vulcan, Fuel Cell Earth LLC, USA). The anode was inoculated with the anaerobic sludge from the Peppers Ferry Regional Wastewater Treatment Plant (Radford, VA, USA). To mimic the digestion effluent of livestock waste[33], the anode influent solution was prepared containing (per liter of deionized water): sodium acetate, 1.5 g; NH_4Cl, 3.0 g; $NaHCO_3$, 2.0 g; NaCl, 0.15 g; $MgSO_4$, 0.005 g; $CaCl_2$, 0.006 g; and trace elements solution[34], 1 mL. The cathode chamber was initially filled with 180 mL of deionized water. An external voltage of 0.8 V was applied to the circuit by a power supply (CSI3644A, Circuit Specialists, Inc., Mesa, AZ, USA) according to a previous study[35]. The MEC was operated in a batch mode at room temperature (~20 °C). The anolyte was partially replaced (75%) every 48 h while the catholyte was resupplied to initial volume (180 mL) when a new batch cycle started. Both the anolyte and catholyte were recirculated at a flow rate of 20 mL min^{-1}. The aeration rate was 375 mL min^{-1}. 2 mL sample were collected regularly from both chambers for measurement and 1 M H_2SO_4 was used to absorb the stripped NH_3 gas from cathode.

Measurement and Analysis

The voltage across a 1-Ω resistor (R) in the MEC circuit was recorded every 2 min by a digital multimeter (2700, Keithley Instruments Inc., Cleveland, OH, USA). The pH of liquid stream was measured by two benchtop pH meters installed in the anode and the cathode, respectively (Oakton Instruments, Vernon Hills, IL, USA and Accumet AB250, Fisher Scientific, Pittsburgh, PA, USA). The concentrations of chemical oxygen demand (COD) and ammonium nitrogen (NH_4^+-N) were measured using a DR/890 colorimeter (HACH Co., Ltd., USA) according to manufacturer's instruction. Ionic concentrations (e.g., Na^+, Cl^-) were quantified by using ion chromatography (Dionex LC20 ion chromatograph, Sunnyvale, CA, U.S.A.) equipped with an ED40 eletrochemical detector. The acetate concentration was measured in the anolyte and the catholyte samples that were filtered through 0.22 um PVDF membrane filter and by high-performance liquid chromatorgraphy (HPLC) (Shimadzu, Columbia, MD), equipped with an Aminex HPX-87H column (Bio-Rad, Hercules, CA) and refractive index detector (RID, 10A, Shimadzu). The HPLC column was kept at 65 °C, and 0.5 mM sulfuric acid solution was used as a mobile phase at flow rate of 0.6 mL min^{-1}.

Mathematical Model of BES-based Ammonia Recovery

We extend the previous steady state theory[20] to model the ammonia recovery in BES operating in the batch mode. The model considers the transport of various species across the CEM and the acid-base reactions within the entire system. Without loss of generality, the following species are included: Na^+, Cl^-, HAc, Ac^-, NH_4^+, NH_3, H_2CO_3, HCO_3^-, CO_3^{2-}, H^+ and OH^-. The system is divided into three parts: the anode chamber, the cathode chamber, and the CEM (Fig. 1). The anode and the cathode chambers are both (but separately) treated in a lumped way, and the evolution of the average concentration of a species *i* inside each chamber follows

$$V_j\frac{\partial C_i^j}{\partial t} = A \cdot J_{i,j}' - \dot{R}_{i,j}^a + \dot{S}_{i,j}^c V_j \tag{4}$$

where V_j is the volume of chamber *j* (*j*=1: anode chamber; *j*=2 cathode chamber), C_i^j is the concentration of species *i* in chamber *j*, A is the area of the CEM, and $J_{i,j}'$ is the flux of species *i* into chamber j. $\dot{R}_{i,j}^a$ is the removal of species *i* (e.g., CO_2, NH_3 etc., acetate removal due to the aeration is neglected, since it's concentration in the catholyte is always low, as shown in the study) from chamber*j* due to aeration, which is determined by assuming fast equilibrium between species dissolved in the catholyte and existing in the aeration gas. Previous study had shown that this assumption works well[20,31]. For example, the removal of NH_3 from the cathode chamber, $\dot{R}_{NH_3,2}^a$, is given by $C_{NH_3}^2 Q_{inert}/p_{tot} K_{H,NH}$, where Q_{inert} and p_{tot} are the volumetric flow rate and pressure of the inert aeration gas, respectively. $K_{H,NH}$ is the Henry's constant for NH_3 in water. $\dot{S}_{i,j}^c$ is the generation/removal rate (per volume) of the species *i* due to chemical/biological reactions, respectively. Briefly, we consider the consumption of Ac^- by the microbes, the acid-base equilibrium among different species (e.g. $NH_4^+ \leftrightarrows NH_3 + H^+, HAc \leftrightarrows Ac^- + H^+$, ...). A list of all reactions considered can be found in Table S1 in the Supplementary Information. According to the previous work[20], we assume that the acid-base reactions are fast so that the chemical equilibrium between the species involved in these reactions is always maintained.

The CEM is resolved spatially in its thickness direction (*x*=0 and *L* correspond to the CEM-anolyte and CEM-catholyte interfaces, respectively). Each species *i* at *x*=0 and *L* is always in equilibrium with that in the anode and cathode chambers, respectively (see below for more details). The distribution of each species *i* across the CEM, C_i^m, is governed by

$$\frac{\partial C_i^m}{\partial t} = \dot{S}_{i,m}^c + \nabla \cdot J_{i,m} \tag{5}$$

where $\dot{S}_{i,m}^c$ is the generation/consumption of species *i* by acid-base equilibriums and is determined in the same way as in the electrode chambers. $J_{i,m}$ is the flux of species *i* in the membrane given by the Nernst-Planck equation[20,36,37]:

$$J_{i,m} = -D_i^m\left(\frac{\partial C_i^m}{\partial x} + z_i C_i^m \frac{e}{k_B T}\frac{\partial \phi}{\partial x}\right) \tag{6}$$

where D_i^m and z_t are the diffusion coefficient in the membrane and valence of species *i*, respectively; e and $k_B T$ are the electron charge and thermal energy; φ is the electrical potential. Because the concentration of various species inside the anode/cathode chambers evolves slowly at a time scale of tens of hours but the relaxation of species concentration within the CEM is fast due to CEM's small thickness, the time dependence term in Eq. (5) is unimportant and thus dropped hereafter.

The concentration distribution of any species *i* across the anolyte/catholyte-CEM interfaces is treated as follows. For electrically neutral species, their concentrations are continuous across these interfaces. For a charged species *i*, we adopt the Donnan equilibrium condition[20,36,38], i.e.,

$$C_i^m = \exp(-z_i e\Delta\phi/k_B T) C_i^j \tag{7}$$

where Δφ is the electrical potential drop across the CEM-solution interface (usually termed the Donnan potential). Note that the Donnan potential attract cations into (repel anions from) the CEM so that the negative fixed charge inside the membrane is balanced by the free ions.

Finally, the electro-neutrality condition and the charge conservation law are enforced:

$$\omega X + \Sigma_i z_i C_i = 0 \tag{8}$$

$$I = F\sum_i z_i J_i \tag{9}$$

where w is the sign of the fixed space charge in any domain (w=0 in the anode/cathode chamber and zero charge membranes; w=-1 and +1 in cation and anion exchange membranes, respectively). X is the fixed charge density of each domain. F and I are the Faraday constant and the current density through the CEM.

Equations (4, 5, 6, 7, 8, 9), along with the equations for various chemical reactions, form a complete model of the ammonia recovery operation using BES. These nonlinear, time-dependent equations were solved using MATLAB. In our simulations, all design and operation parameters of the MEC, e.g., the volume of electrode chamber, aeration rate, and current density (Fig. 2a), are taken from that in the experiments unless otherwise mentioned. For the CEM (CMI-7000) used in this system, X was determined to be %M using the method established in prior studies[20,32,37,39]. The diffusion coefficient of various species in the CEM is usually smaller than that in bulk solutions[20,40]. The diffusion coefficients of all species are either taken from prior experimental data or by fitting the current experimental results. In the latter case, care was taken to ensure that the fitted data are within the range expected for ion diffusion in typical CEMs[9,20,21,41]. See Tables S2–S4 for a summary of all parameters used in these simulations.

ACKNOWLEDGMENTS

We thank the support from the Institute for Critical Technology and Applied Science (ICTAS) at Virginia Tech, and thank Peppers Ferry Regional Wastewater Treatment Plant for providing inocula. We thank the University Library at Virginia Tech for publication support through the open access subvention fund.

AUTHOR CONTRIBUTIONS

R.Q. and Z.H. conceived and designed the project. Y.L. performed numerical simulations; M.Q. and S.L. designed and performed the experiments. Y.L., M.Q. and S.L. analyzed the experimental and simulation results with input from Z.H., R.Q., Y.L. and M.Q. led the writing of manuscript, and everyone participated in editing the manuscript.

REFERENCES

1. Matassa S., Batstone D. J., Hülsen T., Schnoor J. & Verstraete W. Can direct conversion of used nitrogen to new feed and protein help feed the world? Environ. Sci. Technol. 49, 5247–5254 (2015).
2. Sprynskyy M., Lebedynets M., Zbytniewski R., Namieśnik J. & Buszewski B. Ammonium removal from aqueous solution by natural zeolite, Transcarpathian mordenite, kinetics, equilibrium and column tests. Sep. Purif. Technol. 46, 155–160 (2005).
3. Ahn Y.-H. Sustainable nitrogen elimination biotechnologies: a review. Process Biochem. (Amsterdam, Neth.) 41, 1709–1721 (2006).
4. Gupta V. et al. . Removal of ammonium ions from wastewater: A short review in development of efficient methods. Global Journal of Environmental Science and Management 1, 149–158 (2015).
5. Arredondo M. R. et al.. Bioelectrochemical systems for nitrogen removal and recovery from wastewater. Environ. Sci.: Water Res. Technol . 1, 22–33 (2015).
6. Kelly P. T. & He Z. Nutrients removal and recovery in bioelectrochemical systems: a review.Bioresour. Technol . 153, 351–360 (2014).
7. Wang H. & Ren Z. J. A comprehensive review of microbial electrochemical systems as a platform technology. Biotechnol. Adv. 31, 1796–1807 (2013).
8. Cheng K. Y., Kaksonen A. H. & Cord-Ruwisch R. Ammonia recycling enables sustainable operation of bioelectrochemical systems. Bioresour. Technol . 143, 25–31 (2013).
9. Haddadi S., Elbeshbishy E. & Lee H.-S. Implication of diffusion and significance of anodic pH in nitrogen-recovering microbial electrochemical cells. Bioresour. Technol . 142, 562–569 (2013).
10. Logan B. E. et al. . Microbial electrolysis cells for high yield hydrogen gas production from organic matter. Environ. Sci. Technol. 42, 8630–8640 (2008).
11. Rozendal R. A., Hamelers H. V. & Buisman C. J. Effects of membrane cation transport on pH and microbial fuel cell performance. Environ. Sci. Technol. 40, 5206–5211 (2006).
12. Kuntke P. et al. . Ammonium recovery and energy production from urine by a microbial fuel cell.Water Res . 46, 2627–2636 (2012).
13. Kim J. R., Zuo Y., Regan J. M. & Logan B. E. Analysis of ammonia loss mechanisms in microbial fuel cells treating animal wastewater. Biotechnol. Bioeng. 99, 1120–1127 (2008).
14. Haddadi S., Nabi-Bidhendi G. & Mehrdadi N. Nitrogen removal from wastewater through microbial electrolysis cells and cation exchange membrane. J. Environ. Health. Sci. Eng. 12, 48 (2014). [PMC free article]
15. Zhang F., Li J. & He Z. A new method for nutrients removal and recovery from wastewater using a bioelectrochemical system. Bioresour. Technol . 166, 630–634 (2014).
16. Kuntke P. et al. . Effects of ammonium concentration and charge exchange on ammonium recovery from high strength wastewater using a microbial fuel cell. Bioresour. Technol . 102, 4376–4382 (2011).
17. Qin M. & He Z. Self-supplied ammonium bicarbonate draw solute for achieving wastewater treatment and recovery in a microbial electrolysis cell-forward osmosis-coupled system.Environmental Science & Technology Letters 1, 437–441 (2014).
18. Cord-Ruwisch R., Law Y. & Cheng K. Y. Ammonium as a sustainable proton shuttle in bioelectrochemical systems. Bioresour. Technol . 102, 9691–9696 (2011).
19. Desloover J., Abate Woldeyohannis A., Verstraete W., Boon N. & Rabaey K. Electrochemical resource recovery from digestate to prevent ammonia toxicity during anaerobic digestion.Environ. Sci. Technol. 46, 12209–12216 (2012).
20. Dykstra J. E., Biesheuvel P. M., Bruning H. & Ter Heijne A. Theory of ion transport with fast acid-base equilibrations in bioelectrochemical systems. Phys. Rev. E 90, 013302 (2014).
21. Kim J. R., Cheng S., Oh S. E. & Logan B. E. Power generation using different cation, anion, and ultrafiltration membranes in microbial fuel cells. Environ. Sci. Technol. 41, 1004–1009 (2007).
22. He Z., Huang Y., Manohar A. K. & Mansfeld F. Effect of electrolyte pH on the rate of the anodic and cathodic reactions in an air-cathode microbial fuel cell. Bioelectrochemistry 74, 78–82 (2008).
23. Zhang L., Li C., Ding L., Xu K. & Ren H. Influences of initial pH on performance and anodic microbes of fed—batch microbial fuel cells. J. Chem. Technol. Biotechnol. 86, 1226–1232 (2011).
24. Jadhav G. S. & Ghangrekar M. M. Performance of microbial fuel cell subjected to variation in pH, temperature, external load and substrate concentration. Bioresour. Technol . 100, 717–723 (2009).
25. Kuntke P., Sleutels T., Saakes M. & Buisman C. Hydrogen production and ammonium recovery from urine by a Microbial Electrolysis Cell. Int. J. Hydrogen Energy 39, 4771–4778 (2014).
26. Zeng Y., Choo Y. F., Kim B.-H. & Wu P. Modelling and simulation of two-chamber microbial fuel cell. J. Power

Sources 195, 79–89 (2010).

27. Ping Q. et al. . Mathematical model of dynamic behavior of microbial desalination cells for simultaneous wastewater treatment and water desalination. Environ. Sci. Technol. 48, 13010–13019 (2014).

28. Peng S. et al. . Nernst-ping-pong model for evaluating the effects of the substrate concentration and anode potential on the kinetic characteristics of bioanode. Bioresour. Technol . 136, 610–616 (2013).

29. Picioreanu C., Head I. M., Katuri K. P., van Loosdrecht M. C. & Scott K. A computational model for biofilm-based microbial fuel cells. Water Res . 41, 2921–2940 (2007).

30. Qin M., Ping Q., Lu Y., Abu-Reesh I. M. & He Z. Understanding electricity generation in osmotic microbial fuel cells through integrated experimental investigation and mathematical modeling. Bioresour. Technol . 195, 194–201 (2015).

31. Powers S. E., Collins A. G., Edzwald J. K. & Dietrich J. M. Modeling an aerated bubble ammonia stripping process. J. Water Pollut. Control Fed . 59, 92–100 (1987).

32. Aeshala L., Uppaluri R. & Verma A. Effect of cationic and anionic solid polymer electrolyte on direct electrochemical reduction of gaseous CO_2 to fuel. J. CO_2 Util . 3, 49–55 (2013).

33. Park J., Jin H.-F., Lim B.-R., Park K.-Y. & Lee K. Ammonia removal from anaerobic digestion effluent of livestock waste using green alga Scenedesmus sp. Bioresour. Technol . 101, 8649–8657 (2010).

34. Angenent L. T. & Sung S. Development of anaerobic migrating blanket reactor (AMBR), a novel anaerobic treatment system. Water Res . 35, 1739–1747 (2001).

35. Yossan S., Xiao L., Prasertsan P. & He Z. Hydrogen production in microbial electrolysis cells: Choice of catholyte. Int. J. Hydrogen Energy 38, 9619–9624 (2013).

36. Higa M., Tanioka A. & Kira A. A novel measurement method of Donnan potential at an interface between a charged membrane and mixed salt solution. J. Membr. Sci . 140, 213–220 (1998).

37. Sata T. Ion exchange membranes: preparation, characterization, modification and application . (Royal Society of chemistry, 2004).

38. Galama A., Post J., Stuart M. C. & Biesheuvel P. Validity of the Boltzmann equation to describe Donnan equilibrium at the membrane–solution interface. J. Membr. Sci . 442, 131–139 (2013).

39. Sarmidi M. R. Recovery of sodium citrate and citric acid by one step electrodialysis process Ph.D thesis , Universiti Teknologi Malaysia, (2009).

40. Chen H. & Ruckenstein E. Hydrated Ions: From Individual Ions to Ion Pairs to Ion Clusters. J. Phys. Chem. B 119, 12671–12676 (2015).

41. Stenina I., Sistat P., Rebrov A., Pourcelly G. & Yaroslavtsev A. Ion mobility in Nafion-117 membranes. Desalination 170, 49–57 (2004).

CHAPTER 14

ANOXIC METABOLISM AND BIOCHEMICAL PRODUCTION IN PSEUDOMONAS PUTIDA F1 DRIVEN BY A BIOELECTROCHEMICAL SYSTEM

Bin Lai, Shiqin Yu, Paul V. Bernhardt, Korneel Rabaey, Bernardino Virdis, and Jens O. Krömer

Keywords: Anoxic metabolism, Pseudomonas putida F1, Bioelectrochemical system, Redox mediators, Extracellular electron transfer, Bio-production, Chemical feedstocks

BACKGROUND

Contrary to renewable fuels, which have steadily increased their share in the energy sector [1], bulk and speciality chemicals are still mainly derived from petroleum and natural gas. However, industrial biotechnology has significantly developed over the recent decades, and it now offers more solutions for the sustainable production of chemicals from renewable resources than ever before [2]. A range of products are currently produced in biotechnological processes [3], including enzymes, amino acids, antibiotics, alcohols, organic acids and vitamins using ever-expanding range of evolved and genetically engineered microorganisms. Such biotechnological processes, however, often face limitations based on redox balance, carbon yields or product toxicity.

A class of microbes that was recently recognised as a promising new platform for the production of chemical feedstocks (often toxic even to the microbial production strain) are the pseudomonads [4]. They have been used to produce antimicrobial aromatics such as phenol [5] and show, in comparison with other industrial organisms such as *Escherichia coli* or baker's yeast, particular advantages in solvent tolerance [6, 7]. This allows higher product concentrations of compounds with solvent properties [8–11]. In this family, *Pseudomonas putida* (*P. putida*) is regarded a model strain for studying the catabolism and synthesis of toxic aromatics. A range of aromatic compounds, such as 3-methylcatechol [12] and p-hydroxybenzoic acid [13], have been produced with *P. putida*.

Pseudomonas putida is a gram-negative, rod-shaped, flagellated, saprotrophic soil bacterium that is frequently isolated from soil contaminated with petrochemicals. It relies on oxygen as terminal electron acceptor and does not ferment [4, 14]. *Pseudomonas* catabolism is efficient in the supply of redox power [15], but with a low cellular energy demand needed for cell maintenance; in other words, it has a high net NAD(P)H generation for enzymatic reactions [16]. In *P. putida* under aerobic conditions on glucose, the most important sources for NADPH are the pentose-phosphate pathway (PPP), and Entner-Doudoroff (ED) pathway enzymes glucose-6-phosphate dehydrogenase and phosphogluconate dehydrogenase [17, 18]. Because NADPH is an important co-factor for metabolite biosynthesis [19] and dealing with toxicity [15], *P. putida* has been described as one of the promising new platform organisms adaptable for biotechnology and synthetic biology applications [4].

The strictly aerobic metabolism of *P. putida*, however, may also lead to complications when it comes to industrial applications. On the one hand, it increases the capital cost, as scaling-up of aerobic process is significantly limited by the oxygen transfer rate [20], and due to this, both the maximum and average scales of commercial aerobic fermenters are much

Citation: Bin Lai, Shiqin Yu, Paul V. Bernhardt, Korneel Rabaey, Bernardino Virdis, and Jens O. Krömer, Anoxic metabolism and biochemical production in Pseudomonas putida F1 driven by a bioelectrochemical system, doi: 10.1186/s13068-016-0452-y.

smaller in comparison with anaerobic fermenters [21]. On the other hand, aerobic production is also inseparable from substrate loss in the form of CO_2, while some anaerobic processes can achieve carbon yields close to 100 % [22]. To overcome these limitations, a range of studies of *P. putida* under oxygen-limiting conditions have been conducted aiming at developing an anaerobic mutant of *P. putida* [23]. However, the success to date has been limited, and so far only a reduced death rate of *P. putida* cells in anaerobic conditions could be achieved while limited catabolic activity was observed [18, 24].

To solve the problem of electron balancing and allow efficient anaerobic metabolism of *P. putida*, we proposed to culture the organism in the anodic compartment of a bioelectrochemical system (BES). BES were firstly proposed as technology for electricity production from biodegradable waste [25–27] and then extended to other applications including hydrogen production [28], water desalination [29], nutrient removal and recovery [30–33] as well as bio-production [30] including the conversion of CO_2 to acetate [34] and methane [35]. BESs use electrodes as electron sink or donor to drive the anoxic metabolism of microbial cells. In a recently reported study, electroactivity under different oxygen-limited conditions was observed for a recombinant *P. putida* KT2440 [36]. This bacterium was engineered to produce pyocyanin, an electron shuttle (or mediator molecule) from *Pseudomonas aeruginosa* [37].

Under the rationale that an electrode can act as electron acceptor (sink) in order to balance the intracellular energy and redox co-factors, in the present study, we investigate the mediated electron transport in a non-genetically modified *P. putida* F1 under strictly anaerobic conditions with glucose as a carbon source. We observed electrochemical activity and high yield production of 2-Keto-gluconate (2KGA), which is an industrial precursor for the production of antioxidant iso-ascorbic acid [38]. We used various mediators with broad electrochemical midpoint potentials and tested for electrochemical activity in the presence of the microbial cells. Using quantitative metabolite analysis, we show the impact of the presence of an extracellular electron acceptor on intracellular energy balance as well as the co-factor ratios of $NAD^+/NADH$ and $NADP^+/NADPH$.

METHODS

Strain and Cultivation Conditions

The strain used in this study was wild-type *P. putida* F1. Cells were cultivated in a defined mineral medium (DM9) that contained per litre: 6 g Na_2HPO_4, 3 g KH_2PO_4, 0.1 g NH_4Cl, 0.1 g $MgSO_4·7H_2O$, 15 mg $CaCl_2·2H_2O$ and 1 ml trace element solution, containing per litre: 1.5 g $FeCl_3·6H_2O$, 0.15 g H_3BO_3, 30 mg $CuSO_4·5H_2O$, 0.18 g KI, 0.12 g $MnCl_2·4H_2O$, 60 mg $Na_2MoO_4·2H_2O$, 0.12 g $ZnSO_4·7H_2O$, 0.15 g $CoCl_2·6H_2O$, 10 g EDTA (acid), and 23 mg $NiCl_2·6H_2O$. Glucose was used as the sole metabolic carbon source in all tests, and the initial medium pH was 7. Cultivation temperature was 30 °C [39]. Cell density was analysed photometrically using absorbance at 600 nm (OD_{600}) and converted to cell dry weight (CDW) by the following empirically determined conversion factor: CDW [g/L] = 0.476 × OD_{600}. Pre-cultures were prepared by picking and transferring of a single colony from a LB plate into baffled shake flasks for aerobic overnight cultivation (~16 h) in an orbital shaking incubator (2.5 cm orbit, Multitron, Infors, Bottmingen, Switzerland) at 200 rpm and 30 °C. When cell density reached OD_{600} = 0.5 (in log phase) cells were harvested by centrifugation (7000*g*, room temperature, 10 min), washed, resuspended in fresh DM9 medium, and then transferred into the main cultivations vessels. Anaerobic tests were done in 150 mL anaerobic culture bottles or BES reactors. Anaerobic conditions throughout the experiments were assured by sparging the culture medium with nitrogen. The anaerobic bottles were inoculated in an anaerobic chamber.

Bioelectrochemical System Set Up

The double-chamber BESs consisted of a double-jacketed sterilisable glass vessel with a net liquid volume of 350 mL serving as the anodic compartment. The cathodic compartment consisted of a 15 mL glass cannula inserted directly into the anode chamber. A circular cation exchange membrane (diameter 9 mm, CMI-7000, Membranes International INC., USA) was mounted at the bottom of the cannula to guarantee ionic connection between anodic and cathodic electrolytes. Carbon cloth (projected surface area of ~25 cm^2) was used as the anode electrode after pre-treatment with a modified cetyltrimethylammoniumbromide (CTAB) soaking method [40]. In brief, the carbon cloth electrodes were soaked in 2 mM CTAB solution and incubated in a shaking incubator at 40 °C, 200 rpm for 16 h. The pre-treatment was necessary to clean the carbon cloth and to improve the hydrophilicity. A titanium mesh (Kaian Metal Wire Mesh, Anping, P.R. China) was used as cathode electrode. An Ag/AgCl electrode in saturated KCl (+0.197 V vs standard hydrogen electrode) was used as reference electrode (Cat. 013457, RC-1CP, Als, Tokyo, Japan), and titanium wire (T555518, Advent Research Materials, Oxford, United Kingdom) was used as electric wire for all connections. For ease of comparison, all potentials herein are reported with respect to the standard hydrogen electrode (SHE). The working electrode potential was controlled to a set potential relatively to the reference electrode using a potentiostat (Potentiostat/Galvanostat VSP, BioLogic Science Instruments, France).The potentials were chosen based on the electrochemical characterisation of the mediators (vide infra). Riboflavin, $[Co(Sep)]Cl_3$, Fe(EDTA), thionine chloride, $Co(bpy)_3](ClO_4)_2$, $K_3[Fe(CN)_6]$ were added (1 mM) as redox mediators to the culture medium. These mediators cover a wide range of electrochemical midpoint potentials and were tested for electrochemical activity in the presence of the microbial cells. Measurements of current production over time (chronoamperometry) were used to monitor the electrocatalytic activity of the microbes. Strictly anaerobic conditions were maintained by flushing sterile nitrogen gas through the reactor headspace at a flow rate of around 30 mL/min. Dissolved oxygen concentration was confirmed to be below detection limit (<15 ppb) using an optical oxygen sensor (OXY-4 mini, PreSens, Regensburg, Germany) in preliminary tests. A condenser cooled with 4 °C H_2O was used to reduce water evaporation through the headspace due to the flushing.

Analytics and Sampling

The concentrations of glucose, gluconic acid, 2KGA and acetic acid were analysed using an Agilent 1200 high performance liquid chromatography (HPLC) system and an Agilent Hiplex H column (300 × 7.7 mm, PL1170–6830) with guard column (SecurityGuard Carbo-H, Phenomenex PN: AJO-4490). In brief, the column temperature was set to 40 °C, and analytes were eluted isocratically with 14 mM H_2SO_4 at a flow rate of 0.4 mL/min. Glucose was detected using a refractive index detector, while the carboxylates were detected with absorbance at 210 nm. Prior to injection, culture broth was collected by centrifugation at 16,000 g, 4 °C for 10 min, and the supernatant was used for HPLC analysis undiluted.

Cell extraction for Intracellular Metabolite Analysis

Samples were taken from the electrochemical reactors between 73 and 100 h when the current reached its peak value. In the controls, samples were taken at the similar time point as above when mediators were fully reduced.

For the analysis of intracellular ATP/ADP/AMP, cell pellets representing between 0.2 and 0.4 mg CDW were harvested by centrifugation at 12,000*g*, 4 °C for 2 min. Cell extractions for NAD(P) +/NAD(P)H were performed with a modified fast filtration—freeze/drying process [41, 42]. In brief, around 5 mg CDW were harvested by fast filtration (GVWP04700, 0.22-µm pore size, Millipore, Australia) and then quickly soaked into cold methanol solution (60 % v/v, −48 °C) and incubated at −48 °C for 20 min. The precise amount of CDW harvested was calculated from the optical density (as explained above) of the sample volume.

Afterwards, the extract was centrifuged at 10,000*g*, −4°C for 10 min (5810R, Eppendorf, Hamburg, Germany). Supernatants were collected and frozen in −80 °C freezer after dilution with high-purity water (R > 18 MΩ) to a final concentration of methanol <=20 % v/v). The frozen samples were lyophilised by a freeze-dryer at −20 °C and then resuspended into the reaction buffers provided by the enzymatic assay kits respectively.

For ATP/ADP/AMP analysis cell pellets were resuspended in 250 µL of ice-cold phosphate buffer (pH 7.75). Then cells were extracted by the cold trichloroaceticacid (TCA) method [17, 43]: addition of 250 µL ice-cold 5 % (w/v) TCA—4 mM EDTA and mixing (vortex, 20 s)and then incubation on ice for 20 min. Cell debris was removed by centrifugation (12,000*g*, 4 °C, 10 min) and the supernatant transferred to a new tube and kept on ice until subsequent analysis. Quantification of ATP content was conducted by a commercial bio-luminescence assay kit (LBR-S010, Biaffin, Germany) according to the manufacturer's instructions. The bioluminescence signal was quantified using a microplate reader (M200, Tecan, Switzerland) with white 96-well plate (655075, Greiner Bio-one, Germany). Prior to the assay, samples were diluted between 5 and 40 fold in 20 mM Tris-H_2SO_4 buffer (pH 7.75) containing 2 mM EDTA, for two reasons: (1) to achieve readings within the calibration curve of the assay and (2) to dilute TCA present in the extracts. TCA interferes with the optical signals and by diluting it, the assay becomes less inhibited, and the inhibitory effect can be numerically corrected. Quantification of ADP and AMP were determined indirectly by enzymatically converting them to ATP as described previously [44]. ADP was converted to ATP using pyruvate kinase (P9136, Sigma) while adenylate kinase (M5520, Sigma) was added in addition to also convert AMP. The reaction mixture was incubated at 37 °C for 15 min. Concentrations of ADP and AMP were calculated based on the difference in luminescent reading of samples with or without enzymatic conversions. Adenylate energy charge (AEC) was then calculated according to the formula:

$$AEC = (ATP + 0.5 \times ADP)/ (ATP + ADP + AMP)$$

$NADP^+$/NADPH were quantified using a commercial colorimetric assay kit (MAR038, Sigma-Aldrich, USA), and NAD^+/NADH was determined by a fluorimetric assay kit (PicoProbe™, K338-100, BioVision, USA) according to manufacturer's instructions.

Electrochemical Analysis and Calculations

The midpoint redox potentials of the soluble redox mediators were determined by cyclic voltammetry (CV) in a two–chambered electrochemical cell filled with medium containing 0.1 M KCl (counter electrode chamber) and pH 7.0 0.1 M phosphate buffer (working electrode chamber). During the measurements, the potential of the working electrode (a 2 cm^2 CTAB pre-treated carbon cloth) was swept between a low and a high limit (within a potential window between 0.9 and 1.0 V vs SHE) at a scan rate of 50 m V s^{-1}. Measurements were repeated for at least 100 cycles to guarantee reproducibility. A graphite rod was used as the counter electrode, while a reference electrode was placed into the anode chamber. During the tests, individual mediators were added at a concentration of 1 mM. The midpoint potential values (E_m) were determined as the arithmetic averages of the anodic (E_{pa}) and cathodic peak (E_{pc}) potentials as determined during the forward and backward scans of the CVs, respectively, according to equation $E_m = (E_{pa} + E_{pc})/2$.

In order to provide sufficient driving force for the oxidation of the redox mediator during normal reactor operations (chronoamperometry), the working electrode was set to a potential more positive (about 0.3 V) than the mid-point potentials as determined by CV. Control experiments with mediator but without cells did not lead to oxidation of glucose.

The coulombic efficiency (CE), i.e. the efficiency in the transfer of electric charge during the conversions was determined using the equation:

$$CE\ [\%] = Y_{electrons}/(Y_{2KGA} \times 4 + Y_{acetic\ acid} \times 4 + Y_{gluconic\ acid} \times 2) \times 100$$

where *Y* is the molar yield coefficient on glucose basis of the products (electrons, 2KGA, acetic acid, gluconic acid, etc.) multiplied by the number of electrons released during the formation of the respective product from glucose. Per mol of glucose 4 mol electrons are generated when 2KGA is formed, 2 mol electrons are generated when gluconic acid is formed, and 4 mol electrons are generated when glucose is converted to CO_2 and acetic acid. The molar yield coefficients were determined as the slope of a plot of mol product versus mol substrate converted. Reactor volume was corrected for withdrawn sample and water evaporation determined to be 0.09 mL/h at the reported headspace flushing rate.

The carbon balance was determined using the following equation:

$$CB\ [\%] = (r_{2KGA} \times 6 + r_{acetic\ acid} \times 2 + r_{gluconic\ acid} \times 6 + r_{CO2})/(r_{glucose} \times 6) \times 100$$

where r represents the specific uptake (glucose) or production rates (2KGA, acetic acid, gluconic acid and CO_2) in [mmol/(g_{CDW} h)] and multiplied by the number of carbon atoms in the respective compound.

The production of CO_2 could not be quantified because of CO_2 evolution being miniscule compared to the rate of N_2 flushing of the reactor headspace. However, the formation of acetate from pyruvate by *P. putida* will coincide with CO_2 formation, either due to the activity of the membrane-bound decarboxylating pyruvate dehydrogenase/oxidase (EC 1.2.5.1) or to the activities of other metabolic reactions such as the formation of acetyl-CoA, which can be converted to acetate either through catabolic pathways or through biosynthesis processes. The ratio of CO_2 to acetate in all these circumstances will be 1:1. Therefore the CO_2 yield was assumed to be equal to the acetate yield, major shortcomings of the carbon balance would then point to high activity of pathways producing CO_2, such as the TCA cycle or PPP. All product yield coefficients were converted to specific rates using the average planktonic biomass concentration in the systems. In fact, the contribution to the yields from cells forming biofilm was considered negligible.

RESULTS AND DISCUSSION

Redox Mediators with High Midpoint Potential Enable Anoxic Metabolism of *P. Putida* F1

Like many of its family members, *P. putida* F1 is an obligate aerobe, and no anoxic growth in glucose-based mineral medium without oxygen is observed (Fig. 1a). It was previously shown that ferricyanide could serve as electron acceptor during the oxidation of nicotinic acid by *Pseudomonas fluorescens*[45]. In fact, when the glucose-based mineral medium (DM9) was supplemented with 1 mM potassium ferricyanide as electron acceptor, *P. putida* F1 anaerobically reduced ferricyanide (oxidised form $[Fe(CN)_6]^{3-}$) to ferrocyanide (reduced form, $[Fe(CN)_6]^{4-}$) within 55–90 h, as suggested by a change in colour of the solution from yellow to green ($[Fe(CN)_6]^{3-}$) to colourless ($[Fe(CN)_6]^{4-}$). The ability to continuously utilise $[Fe(CN)_6]^{3-}$ as an electron acceptor was further tested in the BESs, where electrochemical oxidation of the $[Fe(CN)_6]^{4-}$ at the anode allowed constant regeneration of the $[Fe(CN)_6]^{3-}$ for microbial metabolism. In the absence of a mediator, *P. putida* F1 did not transfer electrons to the anode, since no catalytic current was detected when no mediator was added.

The combination of the BES and $[Fe(CN)_6]^{3-}$ as electron acceptor proved to be a feasible solution to the problem of electron transport. *P. putida* F1 was able to perform detectable anoxic metabolic conversions in the BES with an applied potential of +0.697 V vs SHE compared to the control, as indicated by a drop in medium pH (Fig. 1b). The total concentration of $[Fe(CN)_6]^{3-}$ and $[Fe(CN)_6]^{4-}$quantified by a developed optical method, was constant during the whole operating batch, confirming this chemical only served as electron mediator. The pH drop from 7.06 to 5.84 pointed towards the production of acids due to the metabolism of glucose [46]. The measured catalytic current (max current 0.066 mA cm^{-2}) confirmed electrons were released from *P. putida* F1 to the anode. This electron flux is indicative of the anoxic catabolism of glucose since this was the only electron donor in the system at sufficiently high concentration to induce a transfer of charge of over 850 °C in 218 h. The anode chamber contained 2.55 mmol glucose, and around 10 mmol electrons were released over the course of the experiment. Providing an external electron sink to the cell in the form of an anode and a mediator molecule enabled the wild-type *P. putida* F1 to stay metabolically active for over 300 h under anaerobic conditions. The concentration of planktonic cells decreased over time similar to the controls lacking the anode (Fig. 1a), but a non-homogeneous biofilm formation (unquantified) on the carbon cloth electrode could be observed during the electrochemical experiments. This not only explains, on the one hand, the drop in planktonic cells but also indicates that anaerobic growth may have been possible to a limited extent. Metabolic engineering was previously used to adapt *P. putida* to anaerobic conditions by aiming at balancing the energy and redox couples [18], but in that study, only the death rate could be reduced, while metabolic turnover remained low.

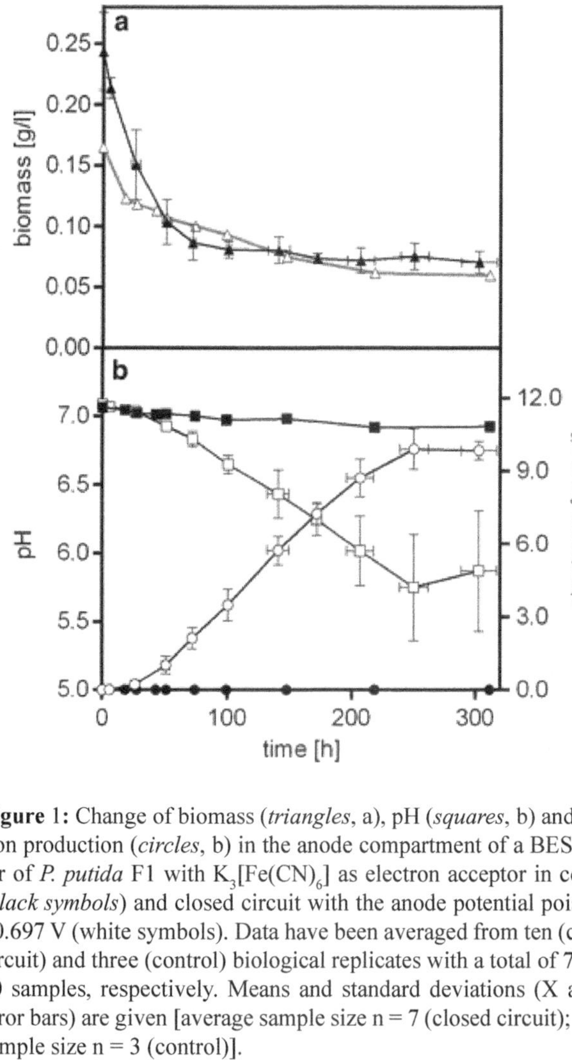

Figure 1: Change of biomass (*triangles*, a), pH (*squares*, b) and electron production (*circles*, b) in the anode compartment of a BES reactor of *P. putida* F1 with $K_3[Fe(CN)_6]$ as electron acceptor in control (*black symbols*) and closed circuit with the anode potential poised at +0.697 V (white symbols). Data have been averaged from ten (closed circuit) and three (control) biological replicates with a total of 79 and 30 samples, respectively. Means and standard deviations (X and Y error bars) are given [average sample size n = 7 (closed circuit); exact sample size n = 3 (control)].

Biochemical Production in the Presence of Different Mediators

After confirming that providing an extracellular electron sink to *P. putida* F1 in the form of electrodes and redox mediators results in production of organic acids, additional experiments were performed with the aims (1) to determine the complete product spectrum and (2) to describe the fermentation kinetics in relation to the redox potential of the mediator, through testing of compounds covering a broad range of redox potentials between approximately −0.4 and +0.4 V.

Cyclic voltammetry was used to determine the average midpoint potential values (E_m) of the mediators used in this study. Characteristic CV traces at the scan rate of 50 mV s^{-1} are reported in Fig. 2a whileE_m values are listed in Table 1. The measurements confirmed the relatively low midpoint potential of riboflavin and [Co(Sep)]Cl_3, centred at around −0.365 and −0.349 V, respectively, whereas Fe(EDTA), thionine chloride, $[Co(bpy)_3](ClO_4)_2$ and $K_3[Fe(CN)_6]$ displayed more positive E_m, centred at around 0.078, 0.208, 0.310, and 0.416 V, respectively. These values are in good agreement to those reported previously for the same compounds (e.g., see references [47–52]) thus confirming that the electrode material used in this study (carbon cloth) was suitable for the electrochemical conversion of the redox mediators tested.

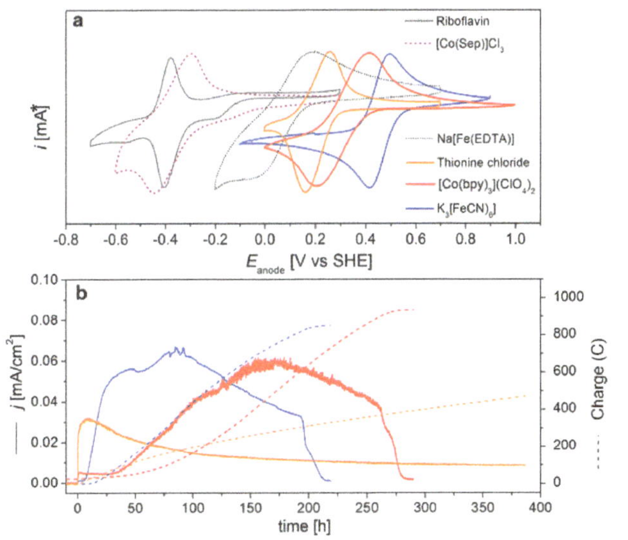

Figure 2: Electrochemical characterisation of the redox mediators used in this study by cyclic voltammetry (*CV*) (a); anodic current (*solid line*) and charge (*dash line*) production measured in the presence of mediators that show activity with *P. putida* (b). *I [mA]: each cyclic voltammogram is shown in its optimum scale to give a clear appearance for all compounds.

Table 1: Formal redox potential of tested mediator molecules and their interaction with *P. putida* F1

Redox chemical	$E_m = (E_{pa} + E_{pc})/2$ [V] vs SHE	Interact with *P. putida* F1
Riboflavin	−0.365	No
$[Co(Sep)]^{3+/2+}$	−0.349	No
$[Fe(EDTA)]^{-/2-}$	0.078	No
Thionine chloride	0.208	Yes
$[Co(bpy)_3]^{3+/2+}$	0.31	Yes
$[Fe(CN)_6]^{3-/4-}$	0.416	Yes

Neutral red [53, 54] and riboflavin [55, 56] have been previously used successfully in combination to organisms such as *E. coli* and *Shewanella oneidensis* to shuttle electrons extracellularly by coupling with the respiratory chain using NADH as the electron donor. However, when we added these mediators to the cultures of *P. putida* F1, no significant current production was observed, as was for [Co(Sep)]Cl_3 and Na[Fe(EDTA)]. On the contrary, mediators with redox potentials above 0.207 V, that is thionine chloride, $[Co(bpy)_3](ClO_4)_2$ and $K_3[Fe(CN)_6]$, demonstrated the ability to accept electrons from *P. putida* F1 cells (Fig. 2b). In fact, current output (Fig. 2b), glucose consumption and a drop in pH (Fig. 3) were observed when these mediators were present in the culturing solutions.

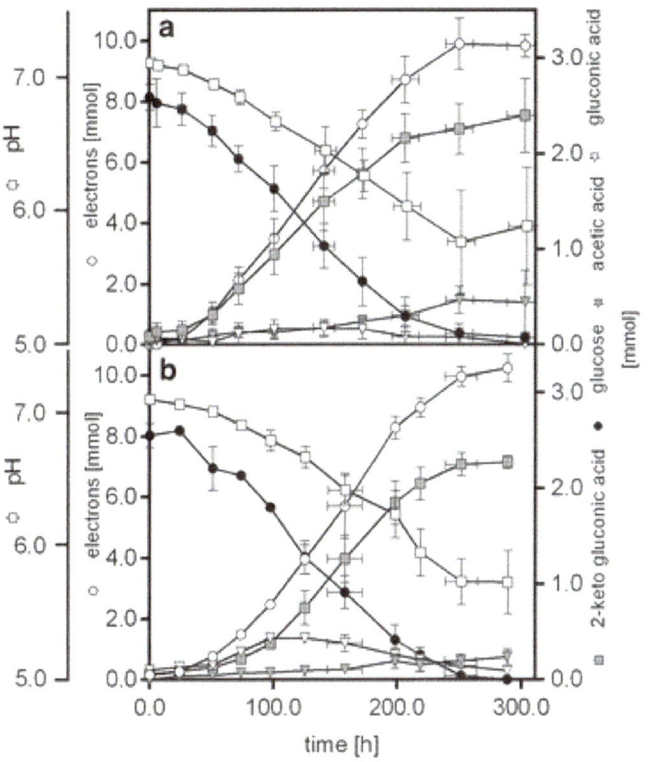

Figure 3: Total metabolite levels and pH in the anode compartment of the BES reactors with $K_3[Fe(CN)_6]$ (a) and $[Co(bpy)_3](ClO_4)_2$ (b) as mediators, respectively. The cumulative amount of electron produced during the conversions is also indicated. Data have been averaged from 10 (a) and four (b) biological replicates with a total of 79 and 36 samples, respectively. Means and standard deviations are given [average sample size in each point n = 7 (a); n = 3 (b)].

The catalytic current increased over time (Fig. 2b), despite the planktonic biomass concentration was stable after the drop during the initial 24 h. We hypothesise that the observed slow formation of a biofilm on the anode is the likely explanation for this increase along with the possibility of a change in gene expression of relevant membrane proteins involved in the mediated electron transport. Depending on available energy, both could be very slow processes.

HPLC analysis showed also the consumption of glucose by *P. putida* in the presence of each of the three mediators. However, in the case of thionine chloride, conversion rates were very low and no full substrate conversion could be reached within 400 h (Fig. 2b). Therefore, only production with $[Co(bpy)_3](ClO_4)_2$ and $K_3[Fe(CN)_6]$ was analysed in detail. For easier comparison with the produced electrons, concentrations were converted to absolute moles using the respective reactor volumes at each time point. Glucose was converted into three detectable products: 2KGA, gluconic acid and acetic acid with the former being the dominant product (Fig. 3). Gluconic acid was accumulated in the first 100 h of the cultivations and then consumed. This phenotype was much more pronounced in the case of $[Co(bpy)_3](ClO_4)_2$ where not only a higher amount of gluconic acid was observed in transition, but also a net production remained at the end of the experiment. In the case of the $K_3[Fe(CN)_6]$-mediated electron transport, all gluconic acid was consumed in the second half of the fermentation, and acetate production was also higher in this process (Fig. 3).

The carbon balances closed in both studies (Table 2) under the assumption that per mol of acetic acid, one mol of CO_2 would be released in metabolism (see M&M section). For both mediators, it was found that around 90 % of the glucose was converted to 2KGA, while twice as much acetic acid accumulated with $K_3[Fe(CN)_6]$ compared to $[Co(bpy)_3](ClO_4)_2$ (Table 2). The yields of electrons produced from glucose were comparable with both mediators, but the current profiles (Fig. 2b) indicated that the conversion rates were higher in the case of $K_3[Fe(CN)_6]$. While it is not possible to quantify the exact amounts of cells growing on the electrodes, the carbon balance indicated that growth would be minimal, which is also in agreement with our visual observations. Using the average planktonic biomass concentration over the course of the processes it was possible to calculate specific rates (Table 2). This shows that in fact the bioconversion in the presence of $K_3[Fe(CN)_6]$ was much faster and was characterised by a glucose consumption rate that was 36 % higher compared to $[Co(bpy)_3](ClO_4)_2$ and the same holds true for the rate of electron production. This rate was positively correlated to the redox potential of the mediators used; the more positive the potential, the faster the production (Table 2, Fig. 2b). Qualitatively this is expected according to Marcus' theory of electron transfer kinetics [57–59]. These results indicate that the anoxic metabolism was driven by the capability of the redox mediators to scavenge electrons from the intracellular metabolism of *P. putida* F1, since increasing conversions rates could be observed with mediators with higher potentials. Interestingly, the current density recorded immediately after the inoculation of microbes showed a different trend from the production rate of products: thionine chloride and $[Co(bpy)_3](ClO_4)_2$ could trigger electron transfer to the anode more rapidly than $K_3[Fe(CN)_6]$ in spite of having a lower redox potential than the latter (0.208 V and 0.31 V, respectively, vs 0.416 V of ferricyanide) (Fig. 2b). A similar change was also observed for coulombic efficiency, as the reactor with $[Co(bpy)_3](ClO_4)_2$ gave a higher value than when $K_3[Fe(CN)_6]$ was used (Table 2), showing that in the faster process using $K_3[Fe(CN)_6]$ more electrons are lost. Despite these losses, however, it emerges that a BES can be used to produce oxidised products under oxygen-free conditions at high yield, high purity and with minimal production of biomass, and (as it is the case here) without genetic modifications.

Table 2: Key process parameters of anaerobic glucose conversion of *P. putida* F1 in the anode compartment of a BES using $[Co(bpy)_3]^{3+/2+}$ or $[Fe(CN)_6]^{3-/4-}$ as electron acceptors with the anode potential poised at +0.697 V vs SHE

	$[Co(bpy)_3]^{3+/2+}$	$[Fe(CN)_6]^{3-/4-}$
Carbon balance (%)	99.6	97.6
Coulombic efficiency (%)	98.5	93.3
Yields [$mol_{product}/mol_{glucose}$]		
Y_{2KGA}	0.90 ± 0.03	0.90 ± 0.02
$Y_{acetic\ acid}$	0.07 ± 0.01	0.14 ± 0.01
$Y_{gluconic\ acid}$	0.31 ± 0.06	0.09 ± 0.03
	0.25 ± 0.03	0.09 ± 0.04
$Y_{electrons}$	3.94 ± 0.11	3.88 ± 0.07
Rates [mmol/(gCDW h)]		
$r_{glucose}$	−0.81 ± 0.14	−1.10 ± 0.21
$r_{acetic\ acid}$	0.06 ± 0.01	0.16 ± 0.03
r_{2KGA}	0.73 ± 0.13	0.99 ± 0.19
$r_{gluconic\ acid}$	0.25 ± 0.06	0.10 ± 0.04
	−0.20 ± 0.04	−0.10 ± 0.05
$r_{electrons}$	3.18 ± 0.55	4.28 ± 0.82

Intracellular Electron and Energy Carriers during BES-Driven Glucose Oxidation

The previous sections showed that only redox molecules whose electrochemical potentials were above 0.207 V could successfully shuttle electrons from microbes to the anode (Fig. 2a; Table 1). This potential is positive enough to oxidise a wide range of cellular redox carriers and proteins involved in the electron transport chain of *P. aeruginosa* [60], which has high similarity to the one of *P. putida*. This should enable further oxidation of carbonaceous matter, instead, our results shows that 90 % of the carbon provided accumulated as 2KGA in the BES, indicating that there is still a metabolic constraint in the cells that prevents the full utilisation of the carbon source under anaerobic conditions. In fact, no obvious growth could be observed. This is somewhat surprising, since the accumulation of acetic acid also indicated that some carbon was processed through glycolysis. This imbalance could potentially be explained by a limitation of ATP generation or imbalances of the intracellular redox couples $NAD(P)^+/NADPH$. To shed light on these possible limitations, analyses of intracellular concentrations of NADH, NAD^+, NADPH, $NADP^+$, ATP, AMP and ADP were performed for the experiments using the best performing mediator $K_3[Fe(CN)_6]$ (Fig. 3a). Concentrations of intracellular ATP were compared to cells incubated in identical medium with and without $K_3[Fe(CN)_6]$ in an anaerobic chamber. The intracellular ATP concentration in the BES was much higher than in the anaerobic conditions without an anode provided as electron acceptor (Fig. 4a). While we could not find data on anaerobic cultures of *P. putida*, the determined intracellular ATP concentrations are below 10 % of the ranges published for aerobically growing *P. putida* strains [17, 61], but well in agreement with the observed concentrations during carbon starvation [62] (note that concentrations were converted assuming 1.19 × 10^12 cfu/g_{CDW}[63]). When comparing the adenylate energy charge (AEC), which uses the ratio of ATP, ADP and AMP to estimate the relative amount of energy rich phosphate bonds, it could be observed that providing the anode and the mediator in the BES helped the cells to restore the AEC to 0.9 (Fig. 4b). The AEC should be maintained normally over 0.8 for growing microbes [44, 64] and for *P. putida*under aerobic conditions, and exponential growth values between 0.75 and 0.95 have been described [61, 65]. These findings indicate that the BES could potentially provide enough energy for growth, but carbon turnover seems to be limiting.

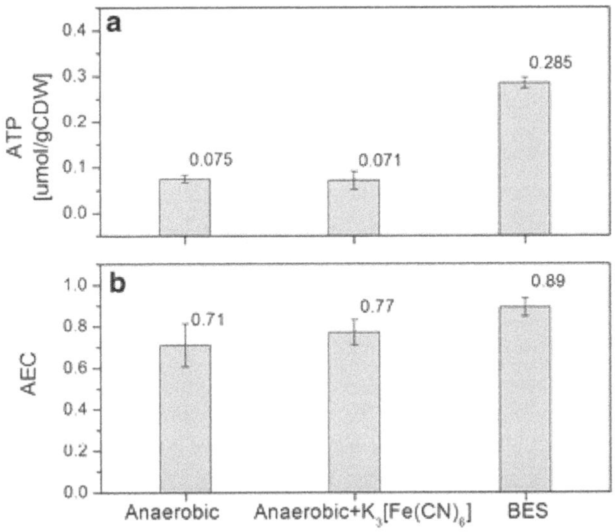

Figure 4: a Specific intracellular ATP concentration in *P. putida* F1 under anaerobic conditions in the absence or presence of $K_3Fe(CN)_6$ and electrodes. b Adenylate energy charge under the same conditions (AEC = (ATP + 0.5×ADP)/(ATP + ADP + AMP)) ...

When analysing the intracellular concentrations of redox cofactors it was observed that the ratios of NAD^+/NADH and $NADP^+$/NADPH were shifted towards the oxidised species under all conditions (Fig. 5). The ratio of NAD^+/NADH remained, however, similar for the three tested conditions. The observed ratios for NAD^+/NADH are highly similar to that observed for aerobic *P. putida* KT2440 [65], which is somewhat surprising since one would expect that in the absence of an electron acceptor (anaerobic control Fig. 5a) the NADH pool would be more reduced (It is important to note that the inverse ratio is given in the cited reference). These data show, however, that adding (electro)chemical oxidants did not alter overall NAD^+/NADH balance. This situation was quite different for the couple $NADP^+$/NADPH. In fact, the presence of oxidised mediators increased the ratio significantly (Fig. 5b), showing that the NADPH pool became more oxidised. The anaerobic control exhibited a $NADP^+$/NADPH ratio which is already more oxidised than in the case of aerobically growing *P. putida*KT2440 [65]. This could point to a limitation in the activity of the PPP for NADPH regeneration, and the imbalance of $NADP^+$/NADPH is aggravated by the presence of mediator and by the use of a BES. This imbalance may hamper the uptake and processing of carbon on the level of 2KGA and could explain the ATP concentrations that point towards carbon starvation.

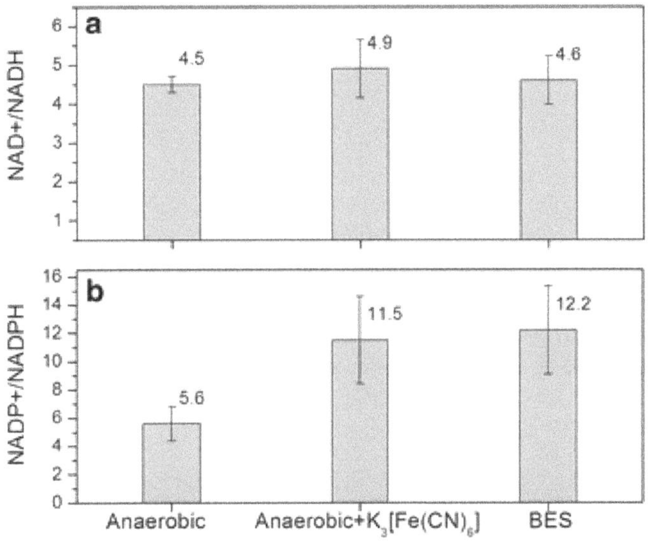

Figure 5: Determination of intracelluar pyridine nucleotide cofactors in *P. putida* under anaerobic conditions in the absence or presence of $K_3[Fe(CN)_6]$ and electrodes. Analytical samples were taken using the same procedures as those for ATP determination.

Flux Balance Analysis

Using existing knowledge of the underlying pathway stoichiometry and the measured rates from the previous sections, it is possible to conduct a flux balance analysis using a simplified model (zero growth, conversion of glucose to gluconic acid, 2KGA, acetic acid and CO_2 and electrons) (Fig. 6). The estimation of fluxes around the uptake of sugar or sugar acids, is complicated by the fact in *P. putida* transporters for the uptake of glucose, gluconic acid and 2KGA are present. We assume that the 2KGA present in the periplasm was the main C_6 molecule imported into the cytoplasm and base this assumption on the observed shift in the NADPH ratio. This unbalanced consumption of NADPH could have resulted in the high ratio of NADP +/NADPH determined in BES condition (Fig. 5b). This could also be the reason for the changed ratio in the case of anaerobic cultivation adding $K_3[Fe(CN)_6]$, because the small pool size of NADPH can experience a redox shift due to the mediator accepting electrons without a measurable difference in extracellular substrate concentrations.

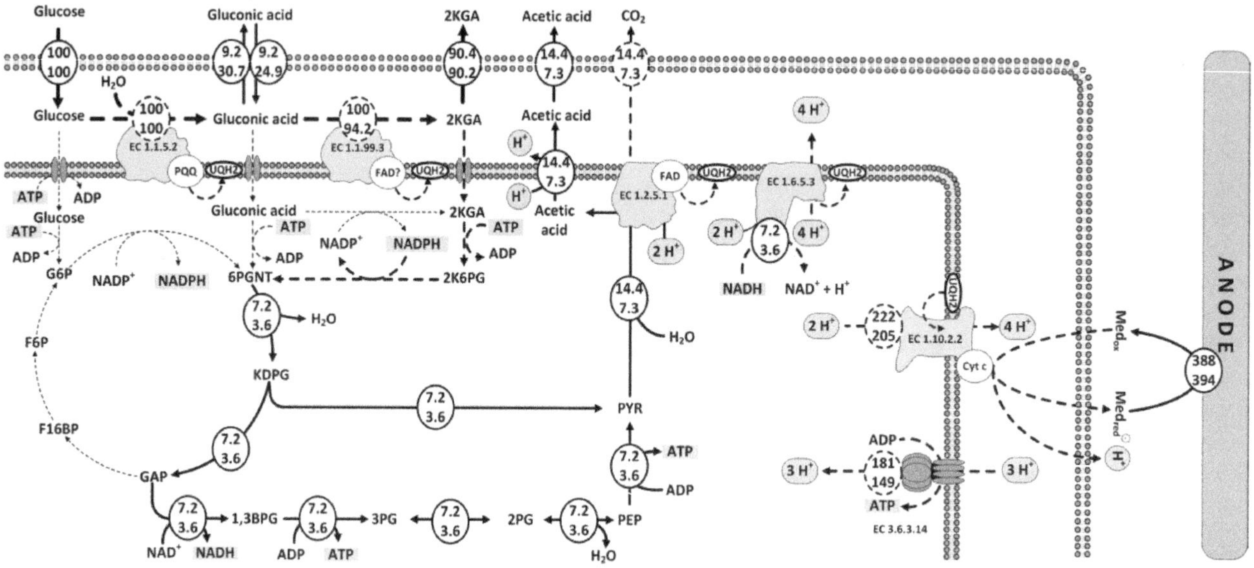

Figure 6: Estimated flux distribution in *P. putida* F1 during glucose oxidation in the anode compartment of the BES reactors with $K_3[Fe(CN)_6]$ (numbers on *top*) and $[Co(bpy)_3](ClO_4)_2$ (numbers on *bottom*) as mediators, respectively. *Solid lines* represent measured fluxes not directly deducible from mass balancing. PQQ pyrroloquinolinequinone, FAD flavin adenine dinucleotide, UQH2 reduced ubiquinones, Cyt C cytochromes C, ADP adenosine diphosphate, ATP adenosine triphosphate, $NADP^+$/NADPH nicotinamide adenine dinucleotide phosphate (oxidised / reduced), NAD^+/NADH nicotinamide adenine dinucleotide (oxidised / reduced), 2KGA 2-keto-gluconic acid, 2K6PG 2-keto-gluconic acid-6-phosphate, 6PGNT 6-phosphogluconic acid, KDPG 2-keto-3-deoxy-phosphogluconic acid, GAP glyceraldehyde-3-phosphate, F16BP fructose-1,6-bisphosphate, F6P fructose-6-phosphate, G6P glucose-6-phosphate, 1,3BPG 1,3-biphosphoglyceric acid, 3PG 3-phosphoglyceric acid, 2PG 2-phosphoglyceric acid, PEP phosphoenolpyruvic acid, PYR pyruvic acid, Med_{ox} oxidised mediator, Med_{red} reduced mediator.

The observed increase in AEC (Fig. 4) raises the question if the energy is generated through a proton gradient-driven ATP synthase and/or substrate level phosphorylation. In the case of both mediators, acetic acid could be observed as a minor by-product derived from glycolytic pathways, through which ATP could be generated on the level of phosphoglycerate kinase (Fig. 6). The reported non-growth associated maintenance (NGAM) demand for *P. putida* KT2440 has been reported to be between 0.92 $mmol_{ATP}/(g_{CDW}\ h)$ [16] and 3.96 $mmol_{ATP}/(g_{CDW}\ h)$ [66]. Since the estimation of NGAM requires growth [67] it is currently not feasible to determine it in the BES. Independent of the assumed C_6 uptake system, one mol ATP will be required for kinase reactions, while two mol ATP will be generated in glycolysis per mol C_6 substrate taken up (Fig. 6). Based on the rates for acetate production (Table 1) and due to the fact that *P. putida* does not possess an acetate kinase [18] this equates to 0.06 and 0.16 $mmol_{ATP}/(g_{CDW}\ h)$ for $[Co(bpy)_3](ClO_4)_2$ and $K_3[Fe(CN)_6]$, respectively, or 1.5–6.5 % and 4–17 % of NGAM, depending which literature value is assumed. Even if the lower value for NGAM is still an overestimation, it seems that an additional energy production must be present, to explain the ongoing metabolic activity of the cells. Since carbon balances were closed and no other products detectable, this indicates that the cells must be able to generate energy through a process coupled to the electron transport process. Under the common assumption that three protons move through the ATP synthase to generate one ATP [68] and further respecting charge balance, which means that each of the electrons donated to the anode (measured as current) would have to take a cation (proton) out of the periplasmic space for charge balance, it is possible to balance the available periplasmic protons for ATP synthase (EC 3.6.3.14). For simplicity and due to the lack of knowledge, we assume that all electron transport is mediated via the quinone pool and that the mediator interacts with the terminal oxidase (EC 1.10.2.2) containing cytochrome *c*. This is thermodynamically feasible, based on the observed redox potentials (Table 1) [60]. The fact that the NAD +/NADH ratio remains on the oxidated side (Fig. 5) also implies that the NADH generated through acetic acid synthesis (Fig. 6) could be re-oxidised through the NADH dehydrogenase complex (EC 1.6.5.3) (Fig. 6) and hence the quinone pool must be re-oxidised through the interaction with the mediator.

Using the flux analysis to estimate the rate of ATP synthesis through ATP synthase that could be derived through the available proton gradient shows that 1.2 and 1.3 $mmol_{ATP}/(g_{CDW}\ h)$ for $[Co(bpy)_3](ClO_4)_2$ and $K_3[Fe(CN)_6]$, respectively, could have been generated. This would bring the total energy available within the range of NGAM, making this scenario much more likely than energy production through acetic acid formation alone. While this is only an estimation, it supports the idea that anodic oxidation of carbohydrates in a BES could also lead to ATP generation in the cells, which in our opinion is a prerequisite for maintaining active bio-catalysts over long periods of time and crucial for the viability of BES applications for bio-production.

CONCLUSIONS

By providing an electrode and redox chemicals as extracellular electron sinks, wild-type *P. putida* F1 was able to perform anoxic metabolism, without the need for metabolic engineering and without the formation of biomass as a substrate draining by-product. The redox power from electrode and redox chemicals drove the carbon flux from glucose to 2-Keto-gluconate with a high yield of over 90 %. A survey of different redox chemicals showed that a redox potential of above 0.207 V was crucial and that reaction rates increased with increasing redox potential. Energy was generated in metabolism, but the cells remained largely unable to fully oxidise the substrates to CO_2 despite the intracellular redox co-factors being mainly oxidised. However, the study provides a proof of principle that a BES-driven bioconversion of glucose can achieve high yields, high purity and also deliver necessary energy for cell maintenance, and enables a strict aerobe to catalyse production under oxygen-free conditions for over a week. This opens the route to bi-phasic bio-processes, where the catalyst is grown under aerobic conditions and then used for anaerobic catalysis over long periods of time, without observable growth and hence drain of substrate. Combining this with metabolic engineering strategies could prove to be a powerful new way to produce bio-chemicals from renewable materials.

AUTHORS' CONTRIBUTIONS

BL performed experiments and contributed to the design, acquisition and analysis of data. SY contributed to the cell extraction and enzymatic assay development. PVB provided expertise on the organic–metal complex redox chemicals and provided the chemicals $[Co(Sep)]^{3+/2+}$, $[Fe(EDTA)]^{-/2-}$and $[Co(bpy)_3]^{3+/2+}$. KR and BV contributed to design of the study and analysis of data. JOK developed the concept of the study, contributed to data analysis and preparation of the manuscript. All the authors were involved in the drafting and editing of the manuscript, read and approved the final manuscript.

ACKNOWLEDGEMENTS

The authors thank Dr. Nicholas Coleman (University of Sydney, Australia) for providing the *P. putida*F1strain, and Dr. Manuel Plan (Metabolomics Australia, University of Queensland) for metabolite analysis. The authors acknowledge strategic research and scholarship support by the University of Queensland.

REFERENCES

1. World Energy Outlook, 2014 http://www.worldenergyoutlook.org/publications/weo-2014/.
2. Erickson B, Nelson J, Winters B. Perspective on opportunities in industrial biotechnology in renewable chemicals. Biotechnol J. 2012;7:176–185. doi: 10.1002/biot.201100069.
3. Gavrilescu M, Chisti Y. Biotechnology—a sustainable alternative for chemical industry. Biotechnol Adv. 2005;23(7–8):471–499. doi: 10.1016/j.biotechadv.2005.03.004.
4. Nikel PI, Martinez-Garcia E, de Lorenzo V. Biotechnological domestication of pseudomonads using synthetic biology. Nat Rev Micro. 2014;12(5):368–379. doi: 10.1038/nrmicro3253. [PubMed][Cross Ref]
5. Wierckx N, Ruijssenaars HJ, de Winde JH, Schmid A, Blank LM. Metabolic flux analysis of a phenol producing mutant of *Pseudomonas putida* S12: verification and complementation of hypotheses derived from transcriptomics. J Biotechnol. 2009;143(2):124–129. doi: 10.1016/j.jbiotec.2009.06.023.
6. Kuhn D, Buhler B, Schmid A. Production host selection for asymmetric styrene epoxidation:*escherichia coli* vs. solvent-tolerant *Pseudomonas*. J Ind Microbiol Biotechnol. 2012;39:1125–1133. doi: 10.1007/s10295-012-1126-9.
7. Ramos JL, Duque E, Gallegos MT, Godoy P, Ramos-Gonzalez MI, Rojas A, Teran W, Segura A. Mechanisms of solvent tolerance in gram-negative bacteria. Annu Rev Microbiol. 2002;56:743–768. doi: 10.1146/annurev.micro.56.012302.161038.
8. Nielsen D, Leonard E, Yoon S, Tseng H, Yuan C, Prather K. Engineering alternative butanol production platforms in heterologous bacteria. Metab Eng. 2009;11:262–273. doi: 10.1016/j.ymben.2009.05.003.
9. Verhoef S, Wierckx N, Westerhof RG, de Winde JH, Ruijssenaars HJ. Bioproduction ofp-hydroxystyrene from glucose by the solvent-tolerant bacterium *Pseudomonas putida* S12 in a two-phase water-decanol fermentation. Appl Environ Microbiol. 2009;75(4):931–936. doi: 10.1128/AEM.02186-08. [PMC free article]
10. Nijkamp K, Westerhof R, Ballerstedt H, de Bont J, Wery J. Optimization of the solvent-tolerant*Pseudomonas putida* S12 as host for the production of p-coumarate from glucose. Appl Microbiol Biotechnol. 2007;74:617–624. doi: 10.1007/s00253-006-0703-0.
11. Wierckx N, Ballerstedt H, de Bont J, Wery J. Engineering of solvent-tolerant *Pseudomonas putida*S12 for bioproduction of phenol from glucose. Appl Environ Microbiol. 2005;71:8221–8227. doi: 10.1128/AEM.71.12.8221-8227.2005.
12. Husken L, Beeftink R, de Bont J, Wery J. High-rate 3-methylcatechol production in *Pseudomonas putida* strains by means of a novel expression system. Appl Microbiol Biotechnol. 2001;55:571–577. doi: 10.1007/s002530000566.
13. Meijnen JP, Verhoef S, Briedjlal AA, de Winde JH, Ruijssenaars HJ. Improved p-hydroxybenzoate production by engineered *Pseudomonas putida* S12 by using a mixed-substrate feeding strategy. Appl Microbiol Biotechnol. 2011;90(3):885–893. doi: 10.1007/s00253-011-3089-6.
14. Escapa I, Garcia J, Buhler B, Blank L, Prieto M. The polyhydroxyalkanoate metabolism controls carbon and energy spillage in *Pseudomonas putida*. Environ Microbiol. 2012;14:1049–1063. doi: 10.1111/j.1462-2920.2011.02684.x.
15. Blank L, Ionidis G, Ebert B, Buhler B, Schmid A. Metabolic response of *Pseudomonas putida*during redox biocatalysis in the presence of a second octanol phase. FEBS J. 2008;275:5173–5190. doi: 10.1111/j.1742-4658.2008.06648.x.
16. Ebert BE, Kurth F, Grund M, Blank LM, Schmid A. Response of *Pseudomonas putida* KT2440 to increased NADH and ATP demand. Appl Environ Microbiol. 2011;77(18):6597–6605. doi: 10.1128/AEM.05588-11.
17. Chavarría M, Nikel PI, Pérez-Pantoja D, de Lorenzo V. The Entner–Doudoroff pathway empowers*Pseudomonas*

putida KT2440 with a high tolerance to oxidative stress. Environ Microbiol.2013;15(6):1772–1785. doi: 10.1111/1462-2920.12069.

18. Nikel PI, de Lorenzo V. Engineering an anaerobic metabolic regime in *Pseudomonas putida*KT2440 for the anoxic biodegradation of 1,3-dichloroprop-1-ene. Metab Eng. 2013;15:98–112. doi: 10.1016/j.ymben.2012.09.006.

19. Knaggs AR. The biosynthesis of shikimate metabolites. Nat Prod Rep. 2003;20(1):119–136. doi: 10.1039/b100399m.

20. Garcia-Ochoa F, Gomez E. Bioreactor scale-up and oxygen transfer rate in microbial processes: an overview. Biotechnol Adv. 2009;27(2):153–176. doi: 10.1016/j.biotechadv.2008.10.006. [PubMed][Cross Ref]

21. Hannon J, Bakker A, Lynd L, Wyman C. Comparing the scale-up of anaerobic and aerobic processes. In: Annual Meeting of the American Institute of Chemical Engineers: Salt Lake City, 2007.

22. Shukla VB, Zhou S, Yomano LP, Shanmugam KT, Preston JF, Ingram LO. Production of d(−)-lactate from sucrose and molasses. Biotechol Lett. 2004;26(9):689–693. doi: 10.1023/B:BILE.0000024088.36803.4e.

23. Costura RK, Alvarez PJJ. Expression and longevity of toluene dioxygenase in *Pseudomonas putida*F1 induced at different dissolved oxygen concentrations. Water Res. 2000;34(11):3014–3018. doi: 10.1016/S0043-1354(00)00035-X. [Cross Ref]

24. Steen A, Ütkür FÖ, Borrero-de Acuña JM, Bunk B, Roselius L, Bühler B, Jahn D, Schobert M. Construction and characterization of nitrate and nitrite respiring *Pseudomonas putida* KT2440 strains for anoxic biotechnical applications. J Biotechnol. 2013;163(2):155–165. doi: 10.1016/j.jbiotec.2012.09.015.

25. Du ZW, Li HR, Gu TY. A state of the art review on microbial fuel cells: a promising technology for wastewater treatment and bioenergy. Biotechnol Adv. 2007;25(5):464–482. doi: 10.1016/j.biotechadv.2007.05.004.

26. Lai B, Tang X, Li H, Du Z, Liu X, Zhang Q. Power production enhancement with a polyaniline modified anode in microbial fuel cells. Biosens Bioelectron. 2011;28(1):373–377. doi: 10.1016/j.bios.2011.07.050.

27. Logan BE, Rabaey K. Conversion of wastes into bioelectricity and chemicals by using microbial electrochemical technologies. Science. 2012;337(6095):686–690. doi: 10.1126/science.1217412.

28. Liu H, Grot S, Logan BE. Electrochemically assisted microbial production of hydrogen from acetate. Environ Sci Technol. 2005;39(11):4317–4320. doi: 10.1021/es050244p.

29. Cao XX, Huang X, Liang P, Xiao K, Zhou YJ, Zhang XY, Logan BE. A new method for water desalination using microbial desalination cells. Environ Sci Technol. 2009;43(18):7148–7152. doi: 10.1021/es901950j.

30. Rabaey K, Rozendal RA. Microbial electrosynthesis—revisiting the electrical route for microbial production. Nat Rev Microbiol. 2010;8(10):706–716. doi: 10.1038/nrmicro2422. [PubMed][Cross Ref]

31. Jourdin L, Freguia S, Donose BC, Chen J, Wallace GG, Keller J, Flexer V. A novel carbon nanotube modified scaffold as an efficient biocathode material for improved microbial electrosynthesis. J Mater Chem A. 2014;2(32):13093–13102. doi: 10.1039/C4TA03101F.

32. Nie H, Zhang T, Cui M, Lu H, Lovley DR, Russell TP. Improved cathode for high efficient microbial-catalyzed reduction in microbial electrosynthesis cells. Phys Chem Chem Phys.2013;15(34):14290–14294. doi: 10.1039/c3cp52697f.

33. Virdis B, Read ST, Rabaey K, Rozendal RA, Yuan Z, Keller J. Biofilm stratification during simultaneous nitrification and denitrification (SND) at a biocathode. Bioresour Technol.2011;102(1):334–341. doi: 10.1016/j.biortech.2010.06.155.

34. Nevin KP, Hensley SA, Franks AE, Summers ZM, Ou J, Woodard TL, Snoeyenbos-West OL, Lovley DR. Electrosynthesis of organic compounds from carbon dioxide catalyzed by a diversity of acetogenic microorganisms. Appl Environ Microbiol. 2011;77(9):2882–6. doi: 10.1128/AEM.02642-10.

35. Villano M, Aulenta F, Ciucci C, Ferri T, Giuliano A, Majone M. Bioelectrochemical reduction of CO2 to CH4 via direct and indirect extracellular electron transfer by a hydrogenophilic methanogenic culture. Bioresour Technol. 2010;101(9):3085–3090. doi: 10.1016/j.biortech.2009.12.077.

36. Schmitz S, Nies S, Wierckx N, Blank LM, Rosenbaum MA. Engineering mediator-based electroactivity in the obligate aerobic bacterium *Pseudomonas putida* KT2440. Front Microbiol.2015;6:284. doi: 10.3389/fmicb.2015.00284.

37. Rabaey K, Boon N, Hofte M, Verstraete W. Microbial phenazine production enhances electron transfer in biofuel cells. Environ Sci Technol. 2005;39(9):3401–3408. doi: 10.1021/es048563o.

38. Zorn H, Czermak P, Lipinski G-WvR. Biotechnology of food and feed additives. Heidelberg: Springer; 2014.

39. Alagappan G, Cowan RM. Effect of temperature and dissolved oxygen on the growth kinetics of*Pseudomonas putida* F1 growing on benzene and toluene. Chemosphere. 2004;54(8):1255–1265. doi: 10.1016/j.chemosphere.2003.09.013.

40. Guo K, Soeriyadi AH, Patil SA, Prévoteau A, Freguia S, Gooding JJ, Rabaey K. Surfactant treatment of carbon felt enhances anodic microbial electrocatalysis in bioelectrochemical systems. Electrochem Commun. 2014;39:1–4. doi: 10.1016/j.elecom.2013.12.001.

41. Bolten CJ, Kiefer P, Letisse F, Portais JC, Wittmann C. Sampling for metabolome analysis of microorganisms. Anal Chem. 2007;79(10):3843–3849.

doi: 10.1021/ac0623888.

42. Moritz B, Striegel K, De Graaf AA, Sahm H. Kinetic properties of the glucose-6-phosphate and 6-phosphogluconate dehydrogenases from *Corynebacterium glutamicum* and their application for predicting pentose phosphate pathway flux in vivo. Eur J Biochem. 2000;267(12):3442–3452. doi: 10.1046/j.1432-1327.2000.01354.x.

43. Lundin A, Thore A. Comparison of methods for extraction of bacterial adenine nucleotides determined by firefly assay. Appl Microbiol. 1975;30(5):713–721.

44. Chapman AG, Fall L, Atkinson DE. Adenylate energy charge in *Escherichia coli* during growth and starvation. J Bacteriol. 1971;108(3):1072–1086.

45. Ikeda T, Kurosaki T, Takayama K, Kano K, Miki K. Measurements of oxidoreductase-like activity of intact bacterial cells by an amperometric method using a membrane-coated electrode. Anal Chem.1996;68(1):192–198. doi: 10.1021/ac950240a.

46. Nikel PI, Chavarria M, Fuhrer T, Sauer U, de Lorenzo V. *Pseudomonas putida* KT2440 metabolizes glucose through a cycle formed by enzymes of the Entner-Doudoroff, Embden-Meyerhof-Parnas, and pentose phosphate pathways. J Biol Chem. 2015;290(43):25920–25932. doi: 10.1074/jbc.M115.687749.

47. Malinauskas A. Electrochemical study of riboflavin adsorbed on a graphite electrode. Chemija.2008;19(2):1–3.

48. Bernhardt PV, Chen KI, Sharpe PC. Transition metal complexes as mediator-titrants in protein redox potentiometry. J Biol Inorg Chem. 2006;11(7):930–936. doi: 10.1007/s00775-006-0148-z.

49. Wang ZM, Liu CX, Wang XL, Marshall MJ, Zachara JM, Rosso KM, Dupuis M, Fredrickson JK, Heald S, Shi L. Kinetics of reduction of Fe(III) complexes by outer membrane cytochromes MtrC and OmcA of *Shewanella oneidensis* MR-1. Appl Environ Microbiol. 2008;74(21):6746–6755. doi: 10.1128/AEM.01454-08.

50. Mcquillan AJ, Reid MR. Cyclic voltammetric studies of a thionine coated pyrolytic-graphite electrode. J Electroanal Chem. 1985;194(2):237–245. doi: 10.1016/0022-0728(85)85007-5.

51. O'Reilly JE. Oxidation-reduction potential of the ferro-ferricyanide system in buffer solutions.Biochim Biophys Acta. 1973;292(3):509–515. doi: 10.1016/0005-2728(73)90001-7.

52. Carter MT, Rodriguez M, Bard AJ. Voltammetric studies of the interaction of metal-chelates with DNA.2. Tris-chelated complexes of cobalt(Iii) and iron(Ii) with 1,10-phenanthroline and 2,2′-bipyridine. J Am Chem Soc. 1989;111(24):8901–8911. doi: 10.1021/ja00206a020.

53. Park DH, Zeikus JG. Electricity generation in microbial fuel cells using neutral red as an electronophore. Appl Environ Microbiol. 2000;66(4):1292–1297. doi: 10.1128/AEM.66.4.1292-1297.2000.

54. Park DH, Zeikus JG. Utilization of electrically reduced neutral red by *Actinobacillus succinogenes*: physiological function of neutral red in membrane-driven fumarate reduction and energy conservation.J Bacteriol. 1999;181(8):2403–2410.

55. Marsili E, Baron DB, Shikhare ID, Coursolle D, Gralnick JA, Bond DR. Shewanella Secretes flavins that mediate extracellular electron transfer. P Natl Acad Sci USA. 2008;105(10):3968–3973. doi: 10.1073/pnas.0710525105.

56. Yong YC, Cai Z, Yu YY, Chen P, Jiang R, Cao B, Sun JZ, Wang JY, Song H. Increase of riboflavin biosynthesis underlies enhancement of extracellular electron transfer of *Shewanella* in alkaline microbial fuel cells. Bioresour Technol. 2013;130:763–768. doi: 10.1016/j.biortech.2012.11.145.

57. Marcus RA. Electron transfer at electrodes and in solution: comparison of theory and experiment. Electrochim Acta. 1968;13(5):995–1004. doi: 10.1016/0013-4686(68)80031-3.

58. Marcus RA. On the theory of oxidation-reduction reactions involving electron transfer. I J Chem Phys. 1956;24(5):966–978. doi: 10.1063/1.1742723.

59. Bard AJ, Faulkner LR. Electrochemical methods: fundamentals and applications. New York: John Wiley; 2001.

60. Kracke F, Vassilev I, Krömer JO. Microbial electron transport and energy conservation—the foundation for optimizing bioelectrochemical systems. Front Microbiol. 2015;6:575. doi: 10.3389/fmicb.2015.00575.

61. Neumann G, Cornelissen S, van Breukelen F, Hunger S, Lippold H, Loffhagen N, Wick LY, Heipieper HJ. Energetics and surface properties of *Pseudomonas putida* DOT-T1E in a two-phase fermentation system with 1-decanol as second phase. Appl Environ Microbiol. 2006;72(6):4232–4238. doi: 10.1128/AEM.02904-05.

62. Eberl L, Givskov M, Sternberg C, Moller S, Christiansen G, Molin S. Physiological responses of*Pseudomonas putida* KT2442 to phosphate starvation. Microbiol Uk. 1996;142:155–163. doi: 10.1099/13500872-142-1-155.

63. Fakhruddin ANM, Quilty B. Measurement of the growth of a floc forming bacterium *Pseudomonas putida* CP1. Biodegradation. 2007;18(2):189–197. doi: 10.1007/s10532-006-9054-x.

64. Khlyntseva SV, Bazel' YR, Vishnikin AB, Andruch V. Methods for the determination of adenosine triphosphate and other adenine nucleotides. J Anal Chem. 2009;64(7):657–673. doi: 10.1134/S1061934809070028.

65. Martínez-García E, Nikel PI, Aparicio T, de Lorenzo V. Pseudomonas 2.0: genetic upgrading of *P. putida* KT2440 as an enhanced host for heterologous gene expression. Microb Cell Fact. 2014;13(1):1–15. doi: 10.1186/1475-2859-13-1.

66. van Duuren J, Puchalka J, Mars A, Bucker R, Eggink G, Wittmann C, dos Santos VA. Reconciling in vivo

and in silico key biological parameters of *Pseudomonas putida* KT2440 during growth on glucose under carbon-limited condition. BMC Biotechnol. 2013;13(1):93. doi: 10.1186/1472-6750-13-93.

67. Pirt SJ. The maintenance energy of bacteria in growing cultures. Proc R Soc Lond B Biol Sci. 1965;163(991):224–231. doi: 10.1098/rspb.1965.0069.

68. Berg JM, Tymoczko JL, Stryer L. Biochemistry. 5. New York: W. H. Freeman and Company; 2002.

CHAPTER 15 EFFECT OF ELECTRODE POSITION ON AZO DYE REMOVAL IN AN UP-FLOW HYBRID ANAEROBIC DIGESTION REACTOR WITH BUILT-IN BIOELECTROCHEMICAL SYSTEM

Min-HuaCui, DanCui, Hyung-Sool Lee, Bin Liang, Ai-Jie Wang & Hao-Yi Cheng

Azo dye is a serious contaminant that often found in industrial wastewaters, such as textile, paper, cosmetic and food industries[1]. Discharge of azo dye wastewaters to water bodies without appropriate treatment could cause serious environmental problems, such as aesthetic problems, deterioration of receiving water quality and restriction in light penetration[2]. Biological treatment processes have been widely used for azo dye removal[2,3], and anaerobic degradation of azo dye seems more attractive than aerobic processes due to the cost effectiveness[4,5]. However, relatively low treatment efficiency as well as slow reaction kinetics has restricted the deployment of anaerobic azo dye wastewater treatment in field.

Bioelectrochemical system (BES) possesses a great potential for transforming nitro aromatics, antibiotics, chlorophenol, halogenide and hexavalent chromium into biodegradable and less toxic products[6,7,8]. Several BES works have been tested for decolorizing azo dyes[9,10,11]. Abiotic or biotic cathodes in BESs could serve as an efficient electron donor for reduction of azo dyes. With small external power supply, azo dye decolorization efficiency and reduction rate were improved as compared to conventional anaerobic processes[9].

While BES may not replace conventional biological treatment processes, but could integrate with anaerobic treatment as a hybrid system to improve effluent quality. These hybrid systems were tested for azo dye and *p*-nitrophenol removal[12]. For instance, the removal efficiency of azo dye was enhanced by 8.2% (86.9±6.3% versus 95.1±1.5%) after installing BES in an anaerobic baffled reactor[13]. There was, however, no detailed information about the effect of BES position on azo dye removal in the hybrid anaerobic reactors, although these engineering aspects are significant for commercialization of the hybrid systems; especially, BES position between liquid and sludge zones could be critical for azo dye removal, due to microbial competition or electrode clogging issues[14,15,16]. In addition, the contribution of electrochemical reduction to the overall removal of azo dyes seems unclear.

In this study, an up-flow hybrid anaerobic digestion reactor with built-in BES was developed for azo dye wastewater treatment. With a BES reactor as the control, the roles of BES in the hybrid reactors were evaluated. In addition, the electrode positions in the hybrid process were optimized (in liquid phase or sludge phase), along with analysis of microbial community structures.

RESULTS AND DISCUSSION

Comparison of Decolorization Performance

To evaluate the effect of electrode positions on azo dye removal in the hybrid anaerobic reactors with built-in BES, three reactors equipped with different electrode positions and

Citation: Min-Hua Cui, Dan Cui, Hyung-Sool Lee, Bin Liang, Ai-Jie Wang, and Hao-Yi Cheng, Effect of electrode position on azo dye removal in an up-flow hybrid anaerobic digestion reactor with built-in bioelectrochemical system, doi: 10.1038/srep25223.

configurations were compared. The first reactor (R1 in Fig. 1A) was filled with anaerobic sludge to a half of the reactor and electrodes were immersed in liquid phase (supernatant). The second reactor (R2 in Fig. 1B) was filled with anaerobic sludge to a half of the reactor, and electrodes were placed in sludge zone. The third reactor (R3 in Fig. 1C) was equipped with two pairs of electrodes (two bundles of anodes and cathodes), which is a control for assessing the role of sludge zone in R1 and R2. To prevent the possible biotoxicity inhibition of Alizarin Yellow R (AYR) on the anode-respiring microbes, the anodes were installed at the upside of the corresponding cathodes in the three reactors.

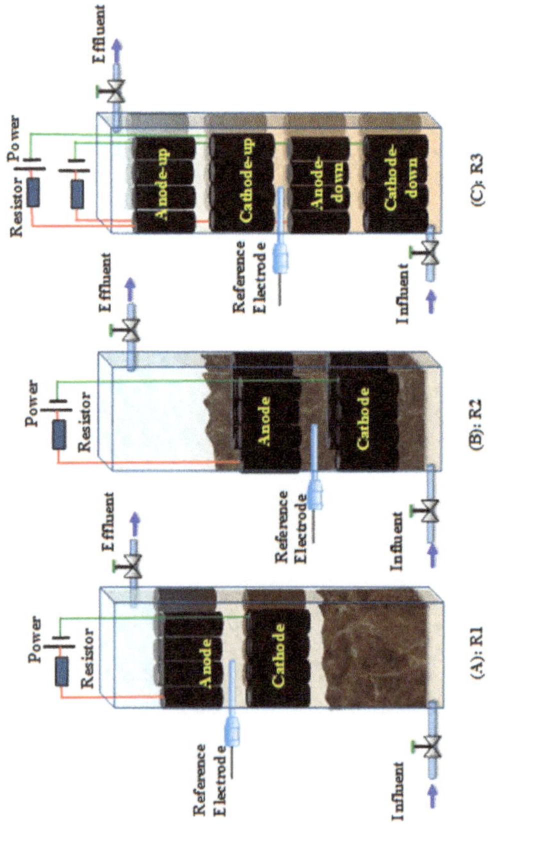

Figure 1: Schematic representation of the three anaerobic hybrid reactors: (A) R1, BES installed in liquid phase; (B) R2, BES installed in sludge phase; (C) R3, BES with two pairs of electrodes.

At open circuit, the decolorization efficiency (DE) in R1 ranged from 90.78±5.95 to 95.87±1.34% at lower AYR loading rates (ALR) (<400 g AYR/m^3·d), which fluctuated from 82.22±4.47 to 86.00±6.31% in R2 and from 77.13±4.46 to 79.46±6.52% in R3, respectively (Fig. 2A). When ALR reached to 800 g AYR/m^3·d, DE sharply dropped to 53.37±4.00% in R1. The same trend was observed in R2 and R3. At closed circuit, DE was improved for all three reactors at ALR<400 g AYR/m^3·d (Fig. 2B). Interestingly, DE in R1 showed over 90% even at higher ALR (800 g AYR/ m^3·d), while DE significantly decreased in R2 and R3 at the higher ALR. This result indicates that electrodes immersing in supernatant (liquid phase) is better than that in sludge zone. When BES was immersed in sludge zone, other microorganisms or particulate matter may seriously inhibit the metabolism of anode-respiring microorganisms. The literature reported that space competition on anodes between anode-respiring microorganisms and non anode-respiring microorganisms became important in BES fed with particulate matter including microorganisms[17]. The growth of non anode-respiring microorganism on anodes would also deteriorate mass transfer of electron donor or acceptor (AYR) in biofilms. As a result, the AYR decolorization rate increased linearly with ALR only in R1 ($y=0.9081x+11.612$, $R^2=0.9994$, y: AYR decolorization rate, x: ALR) and its maximum rate reached at 730.61±49.59 g AYR/m^3·d (Fig. 2B). High DE in open-circuit R1 indicates significant AYR reduction by fermenters or methanogens in sludge and suspension (non anode-respiring microorganisms and cathodic reduction), but consistent improvement of DE in closed-circuit R1 at high AYR loading rate clearly proves the significance of BES for AYR reduction.

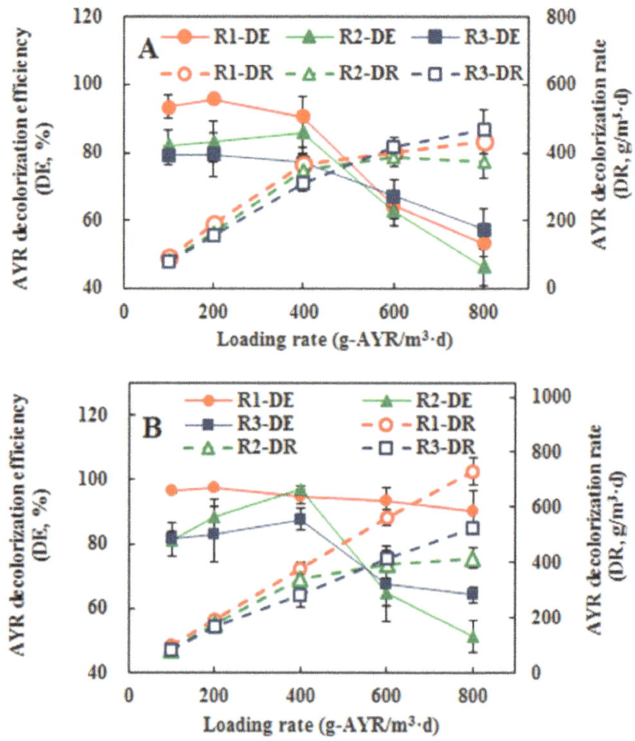

Figure 2: Comparison of AYR decolorization in different reactors: (A) AYR decolorization efficiencies and decolorization rates at open circuit; (B) AYR decolorization efficiencies and decolorization rates at closed circuit.

Products Accumulation and COD Removal

AYR degradation pathway had been proposed by previous literatures: 1 mole AYR reduction can produce 1 mol PPD and 1 mol 5-ASA theoretically[18,19]. The formation of these reduced products clearly revealed the removal of AYR was due to the reduction. The formation efficiencies (ratio of measured value and theoretical value) of PPD were over 90% among all three reactors, while the accumulated 5-ASA was less with the formation efficiencies varied from 60% to 85%. This result consists with our previous work and is likely because PPD persist in anaerobic reactor while anaerobic microorganisms, in particular of the methanogenic consortia, could mineralize 5-ASA[19,20]. As shown in Fig. 3, PPD and 5-ASA concentrations were highest at 400 g-AYR/m^3·d under open circuit and there were no distinctive differences among the three reactors. These patterns were consistent to the trend of AYR decolorization. At closed circuit, concentrations of PPD and 5-ASA were simultaneously increased in all reactors at ALR<400 g/m^3·d. When ALR increased to 800 g/m^3·d, PPD and 5-ASA accumulations in R1 kept at 69.64±6.59 and 79.00±8.70 mg/L, but sharply decreased to 37.73±6.19 and 38.66±3.45 mg/L in R2 and 46.89±3.77 and 48.23±7.99 mg/L in R3, respectively. Comparing open circuit modes with closed ones, BES improved the accumulations of two intermediates by 47.40% and 40.11% in R1, indicating more AYR reduction via bioelectrochemical reaction.

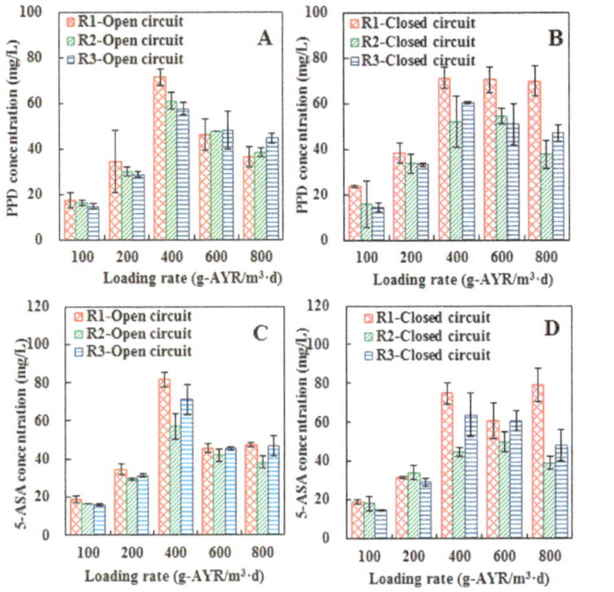

Figure 3: Concentrations of PPD and 5-ASA in different reactors: (A) PPD concentration at open circuit; (B) PPD concentration at closed circuit; (C) 5-ASA concentration at open circuit; (D) 5-ASA concentration at closed circuit.

COD removal in the three reactors tended to decrease with increasing ALR at both open and closed circuit. R1 presented superior COD removal to R2 and R3. At 100 g-AYR /m^3 d, R1 showed the COD removal efficiencies of 82.76 and 81.60% at open and closed circuit, respectively, which were nearly 15–37% higher than those in the other two reactors (Fig. 4A). At the highest ALR of 800 g/m^3 d, COD removal declined to 54.76% for R1 operated at closed circuit, but it was still 7.32% and 19.29% higher than that for R2 and R3, respectively. Poor COD removal in R3 might be due to less biomass on the electrodes. As shown in Fig. 4B, acetate acid (HAc) and propionic acid (HPr) were the main VFAs for all reactors. The other VFAs including iso- butyric acid (0.52–5.37 mg/L), n-butyric acid (0.53–5.41 mg/L), iso-valeric acid (0.67–7.77 mg/L) and n-valeric acid (0.50–3.63 mg/L) were negligible. At ALR of 800 g/m^3·d, VFAs was 106.21 mg/L for R1 at open circuit, but it significantly decreased to 62.57 mg/L at closed circuit. In comparison, VFAs for R2 and R3 (closed circuit) were relatively higher. Power supply enhanced VFAs removal by 41.09% in R1, indicating efficient VFA utilization by anode-respiring microorganisms. The lower VFAs accumulation and better COD removal in R1 also support the higher metabolic activity of anode-respiring microorganisms when electrodes are placed in the liquid phase.

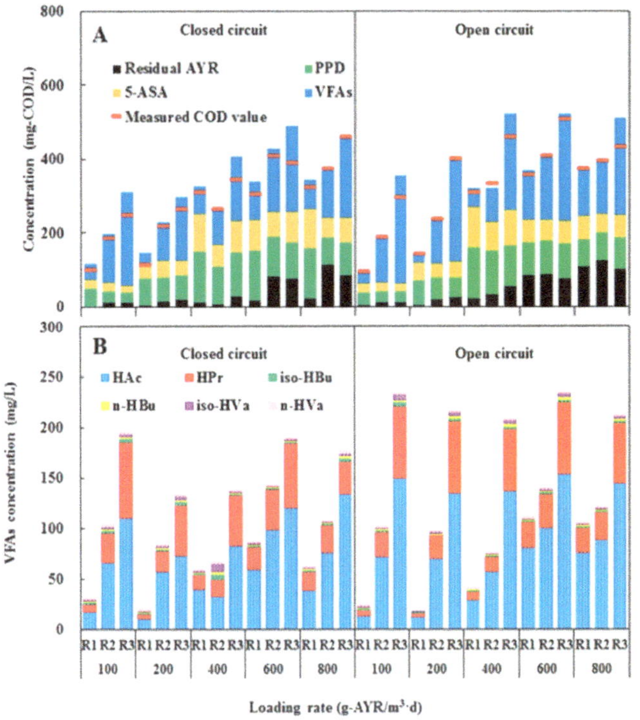

Figure 4: (A) COD balance of three reactors at open and closed circuit (AYR converts into COD with coefficient of 1.17; PPD converts into COD with coefficient of 1.93; 5-ASA converts into COD with coefficient of 1.36; VFAs converts into COD with coefficients of 1.07 for HAc, 1.51 for HPr, 1.82 for HBu, 2.04 for HVa.); (B) VFAs concentrations in effluents.

Contribution of BES to AYR Reduction

Currents and electron recovery efficiencies in the hybrid reactors (R1 and R2) were much lower than that in the sole BES (R3), as shown in Fig. 5. At ALR of 800 g/m^3 d, currents of R1 and R2 were 6.85±0.78 and 5.34±0.51 mA, respectively. Similar current was observed at an upward electrode pair in R3 (6.97±0.31 mA), but a downward electrode pair in R3 generated relatively high current (10.88±0.48 mA). Electron recovery efficiencies for R1 and R2 were 16.23±1.86%, 22.24±2.14% which were 23.23±1.02% (upward electrodes) and 36.27±1.60% (downward electrodes) for R3. It is evident that the total AYR removal was improved in the hybrid system. However, these small electron recovery efficiencies indicate that AYR reduction would mainly occur in sludge zone[13], suggesting that integration of BES with anaerobic digestion as a hybrid system actually would mitigate BES contribution to AYR removal, including R1.

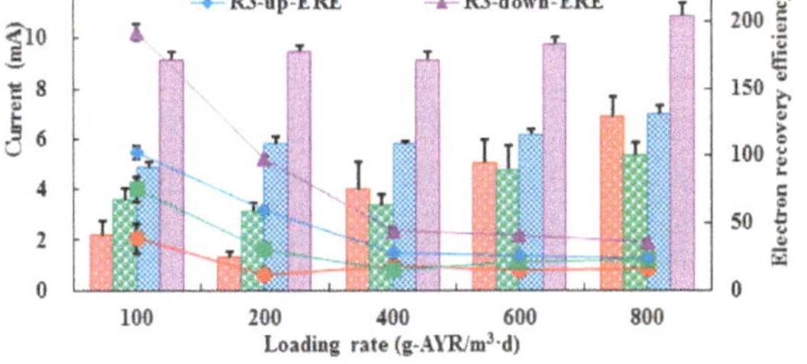

Figure 5: Currents and electron recovery efficiencies (EREs) for different pairs of electrodes.

Microbial Community

Over 20000 16S rRNA gene sequences were identified for each sample by Illumina MiSeq sequencing. As shown in Fig. 6A, operational taxonomic units (OTUs) at 3% distance were 2841, 2434 and 2771 for sludge, anode and cathode samples in R1; 1821, 2630 and 2082 for sludge, anode and cathode samples in R2. Principal component analysis (PCA) of the six samples was carried out to evaluate the difference of bacterial communities (Fig. 6B). Principal components 1 and 2 explained 54.9% and 31.6% of the total community variations, respectively. Three samples from R2 were clustered together, indicating that microorganisms in sludge significantly affected microbial communities in electrodes. In comparison, the enriched microbial communities at electrodes and sludge were different in R1, especially the anode and sludge samples.

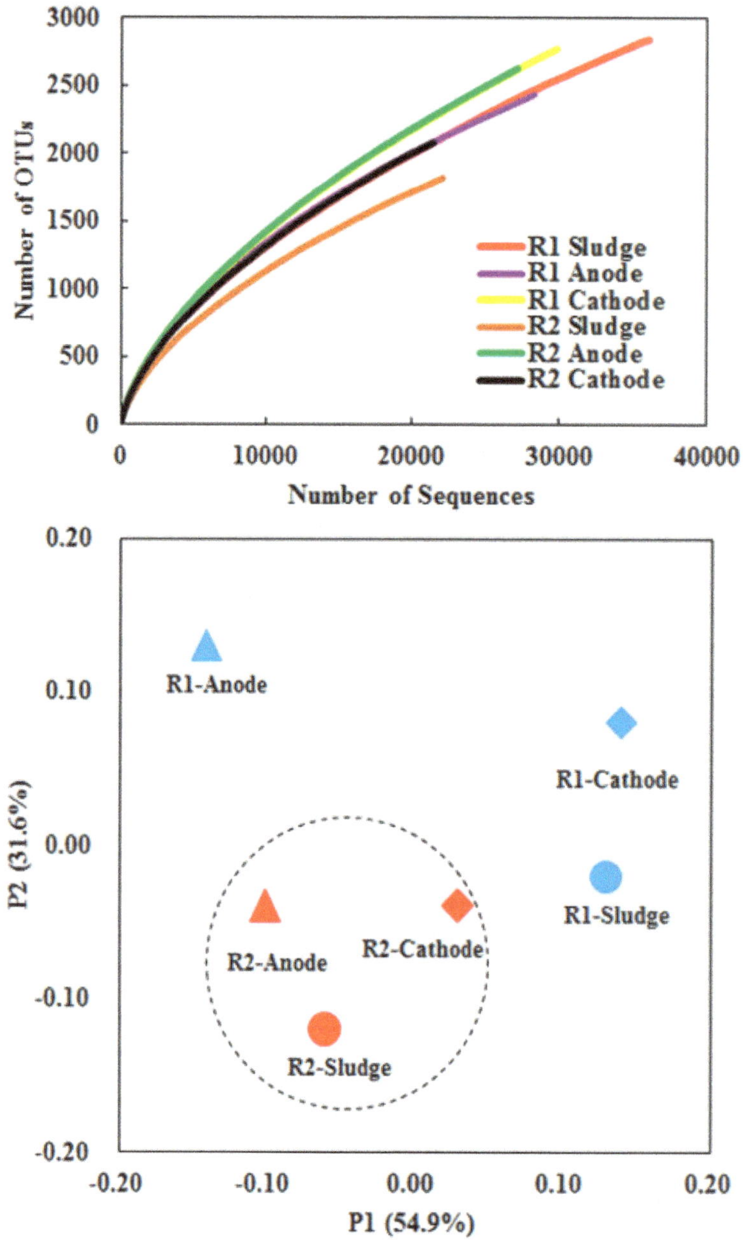

Figure 6: (A) Rarefaction curves based on the 16S rRNA gene amplicons sequencing. The OTUs were defined of 3% distances; (B) Principal component analysis (PCA) of bacterial communities from the six samples.

A total of 485 bacterial genera were identified from 6 tested samples. At least 33 genera with relative abundance >1% came from one sample (Table 1). *Enterobacter* (25.29%), *Enterococcus* (8.83%) and*Desulfovibrio* (5.93%) were abundant in R1-anode. For R2-anode, *Enterococcus* and *Streptophyta*dominated the community with relative abundance of 17.35% and 8.26%, along with *Desulfovibrio*(5.80%). Among these bacteria, *Enterobacter, Enterococcus* and *Desulfovibrio* were reported to be electrochemical active at anode and like implement the anode respiration in R1 and R2[20,21].

Table 1: Relative abundance of bacterial genera in six samples from two reactors Dominant genera (>1%) in the different samples was bolded. [a]Listed genera were >1% abundances at least one sample, genera with

	R1			R2		
Genus	Sludge	Anode	Cathode	Sludge	Anode	Cathode
Enterococcus	**15.45**	**8.83**	**6.71**	**36.08**	**17.35**	**21.07**
Enterobacter	**1.03**	**25.29**	**16.89**	**6.95**	**3.50**	**5.79**
Desulfovibrio	**6.10**	**5.93**	**6.26**	**3.96**	**5.80**	**10.72**
Levilinea	**5.26**	**2.79**	**4.81**	**2.76**	**3.08**	**5.48**
TM7_genera_incertae_sedis	**5.09**	**1.39**	**1.75**	**6.65**	**2.31**	**4.75**
Alkaliflexus	**3.40**	**1.54**	**2.16**	**3.79**	**1.88**	**4.05**
Gordonibacter	**3.31**	**2.84**	**5.36**	0.75	**4.02**	**2.38**
Bellilinea	**2.78**	**1.79**	**2.54**	**1.43**	**1.83**	**1.08**
Cloacibacillus	**2.62**	**1.96**	**2.09**	0.76	**2.24**	**2.85**
Lactococcus	**2.44**	**4.03**	**4.36**	**3.85**	**1.35**	**2.29**
Solitalea	**2.41**	0.20	**1.43**	**1.89**	0.70	**1.05**
Aminobacterium	**2.27**	0.81	**2.27**	0.39	**2.10**	0.94
Petrimonas	**2.06**	**1.16**	**1.47**	**1.17**	**1.47**	**2.19**
Saccharofermentans	**1.90**	0.44	**1.45**	0.41	**1.49**	0.86
Tissierella	**1.69**	0.12	**1.15**	**1.69**	**1.19**	**1.24**
Anaerovorax	**1.56**	0.39	0.63	0.55	0.98	0.66
Trichococcus	**1.44**	0.76	0.47	**2.92**	**2.80**	**1.04**
Smithella	**1.21**	0.91	**1.85**	0.76	**1.03**	0.30
Geobacter	**1.18**	0.49	**1.02**	0.90	**1.33**	0.43
Prolixibacter	**1.07**	0.18	0.62	0.41	0.89	0.30
Phascolarctobacterium	**1.01**	**1.09**	**1.73**	0.59	0.54	**1.41**
Proteiniphilum	0.99	0.59	0.64	**1.03**	0.68	**1.37**
Dysgonomonas	0.95	0.67	0.33	0.75	0.35	**1.16**
Comamonas	0.90	**1.08**	0.95	0.15	0.58	0.33
Bacillus	0.61	0.04	0.02	0.02	**2.84**	0.01
Klebsiella	0.52	**2.90**	**2.85**	**1.02**	**1.77**	0.78
Anaeroarcus	0.11	**1.77**	**1.10**	0.07	0.11	0.19
Achromobacter	0.10	**1.28**	0.21	0.04	0.11	1.01
Streptophyta	0.09	0.02	0.01	0.01	**8.26**	0.03
Shinella	0.03	**1.57**	0.27	0.02	0.05	0.80
Pannonibacter	0.03	**1.07**	0.38	0.01	0.03	0.06
Sphingopyxis	0.02	**1.42**	0.19	0.02	0.03	0.11
Aquamicrobium	0.02	**1.09**	0.17	0.01	0.01	0.05
Unclassified	**14.63**	**7.44**	**11.16**	**6.92**	**12.75**	**8.75**
Others[a]	**15.73**	**16.13**	**14.71**	**11.29**	**14.57**	**14.48**

Microbial community for R1-cathode was similar to R1-anode, showing *Enterobacter* (16.89%),*Desulfovibrio* (6.26%) and *Gordonibacter* (5.36%). *Enterococcus* genus (21.07%) was predominant at R2-cathode, like R2-anode, but other genera, such as *Desulfovibrio* (10.72%), *Enterobacter* (5.79%) and *Levilinea* (5.48%), were identified. The genera of *Enterobacter, Desulfovibrio* and *Enterococcus*can reduce azo dye or nitro-compounds[1,22]. Although this study did not directly ensure that the bio-cathode reduction of AYR was catalyzed by the identified bacteria, the enrichment of these bacteria on cathodes implies that BES creates a niche for the bacteria inhabiting on cathodes. The substantial increase of *Enterobacter* population on electrodes, as compared to sludge indicates that *Enterobacter sp.* plays the key roles of current generation on anodes and AYR reduction on cathodes in R1. In case of electrodes immersing into sludge (R2), *Enterococcus sp.* dominated both of the communities at the sludge and electrodes, which was likely the result of the poor segregation of communities as that was indicated in PCA analysis. The effects of segregation on azo dye decolorization as well as the current generation are still unclear and warrant further study.

METHODS

Reactor Configuration

Three identical cuboid reactors were manufactured with plexiglass, as shown in Fig. 1. Each reactor had a working volume of approximately 1.5 L with the dimensions of L 10 cm × W 5 cm × H 30 cm. A water distribution plate, equispaced with 2 mm holes, was installed at 3 cm over the bottom. Graphite fiber brushes were used for both anodes and cathodes, and each electrode consisted of 8 small graphite fiber brushes (2.5 cm in diameter and 4 cm in length, graphite fiber produced by TOHO TENAX, Co., Ltd., Japan). The distance between anode and cathode in the BES was approximately 2.5 cm. All graphite fiber brushes were pretreated according to the literature[23]. Before assembling the hybrid anaerobic reactors, all the anodes were pre- acclimated in a dual-chamber BES at a fixed anode potential of −400 mV vs. a saturated calomel electrode (SCE) (+247 mV vs. standard hydrogen electrode) controlled by a potentiostat (WMPG1000K8, Wontech International Co., Ltd., Korea). The seed sludge for inoculating the bioanode was the recycled activated sludge obtained from Taiping municipal wastewater treatment plant (Harbin, China) The titanium wire (1 mm in diameter, Baoji LiXing Titanium Group Co., Ltd., China) was coated with heat shrink tube to avoid short circuit and stretched out of the reactors as current collector. During the experiment, each pair of electrodes was operated with external resistance of 10 Ω and applied voltage of 0.5 V using a DC power supply (FDPS-180, Fudan Tianxin Scientific and Educational Instruments Co., Ltd, Shanghai, China). According to our previous work[18], 0.5 V applied voltage was sufficient for AYR decolorization. Anodes, cathodes and reference electrodes were connected to a data acquisition system (Keithley 2700, Keithley Co. Ltd., US), currents were recorded at every 10 min.

Inoculation and Operation

The anaerobic sludge for R1 and R2 were inoculated with mixture of two sources of sludge in 1:1 ratio:anaerobic sludge from a small pilot-scale anaerobic baffled reactor (ABR)[13] that had been operated for treating azo dye wastewater for more than 9 months, and anaerobic sludge from a brewery wastewater treatment facility (Harbin, China). The VSS/TSS ratio the sludge was about 0.7. R3 was not inoculated in advance (control test).

The composition of medium was glucose (500 mg/L), AYR (50–200 mg/L), phosphate buffer solution (50 mmol/L), KCl (0.13 g/L), NH_4Cl (0.31 g/L), trace element solution (10 mL/L) and vitamin solution (10 mL/L)[24]. The influent AYR loading rates (ALR) gradually increased (100, 200, 400, 600, 800 g AYR/m^3·d) by adjusting the influent AYR concentration and HRT (Table 2). Each operation condition maintained for at least two weeks. The flow rate was controlled with a peristaltic pump (BT100-1 L/YZ1515, Longer Pump Co., Ltd., China). Open and closed circuit modes were compared for the three reactors to assess the contribution of bioelectrochemical effect on AYR removal. All reactors were operated at ambient temperature (23 ± 2 °C).

Table 2: Operational conditions at different stages

Stage	1	2	3	4	5
AYR concentration (mg/L)	50	100	200	200	200
HRT (h)	12	12	12	8	6
AYR loading rate (g-AYR/m^3·d)	100	200	400	600	800
Glucose concentration (mg/L)			500		

Chemicals

AYR that contains an azo bond and a nitro group was used as the model azo dye (Commercial purity grade, Shanghai Sangon Biotech Co., Ltd., China). *p*-phenylenediamine (PPD) and 5-aminosalicylic acid (5-ASA), the main by-products of AYR reduction[18,25], were purchased from Aladdin Industrial Corporation (analytical reagent) and J&K scientific (98.5%), respectively.

Analytical Methods

Liquid samples were immediately filtered through 0.45 μm filters (Tianjin Jinteng Experiment Equipment Co., Ltd., China). AYR concentration was quantified by a UV-Vis spectrophotometer (UV-1800, Shanghai Meipuda instrument Co., Ltd., China) at a wavelength of 374 nm[18]. PPD and 5-ASA were measured by a high performance liquid chromatography (0.3% Acetate: Methanol = 9:1, HPLC, e2695, Waters Co., US) equipped with a UV-Vis detector (288 nm, model 2489, Waters Co., US) and a C18 column (5 μm; 4.6 mm × 150 mm, Symmetry, Waters Co., Ltd., US). Volatile fatty acids (VFAs) was analyzed using a gas chromatograph (GC, 6890 N, Agilent, Inc., US) equipped with a flame ionization detector (FID) and Stabilwax-DA column (30 m × 0.32 mm × 0.5 mm) at oven and injector temperatures of 60 °C and 250 °C, respectively. He carrier gas and N_2 makeup gas were used for the GC-FID. Chemical oxygen demand (COD) concentration was determined with the dichromate method.

Calculation

Azo bond cleavage and nitro-reduction decolorize AYR via both anaerobic biological and electrochemical reactions. We built a hypothesis that all electrons transferred to electrodes by anode-respiring microorganisms were used for AYR reduction. Hence, electron recovery efficiency was defined as Eq. (1) to evaluate the portions of AYR decolorization via electrochemical reduction of overall AYR decolorization.

$$\text{Electron recovery efficiency} = \frac{I}{10 \times (C_{in} - C_{eff})/M_{AYR} \times (Q \times 10^3)/(24 \times 3600) \times F} \tag{1}$$

where I is the current which was calculated from the external resistance using Ohm's law, mA; 10 is electron transfer coefficient, 1 mol AYR reduction requires 10 mol electrons (4 mol for azo bond cleavage and 6 mol for nitro-reduction), mol e^-/mol AYR; M_{AYR} is AYR's molecular weight, 287 g/mol; F is Faraday's constant, 96485 C/mol.

Biofilms Sampling and DNA Extraction

Planktonic cells in sludge zone and biofilms on electrodes (anodes and cathodes) in R1 and R2 were sampled for DNA extraction at the end of closed circuit mode operation. Planktonic cells were sampled with a sterile injector from at least three different positions and collected as one sample. Graphite fibers were cut from anodes and cathodes, respectively, and fragmented using sterile scissors. Biofilm samples taken from at least three different small brushes were combined together for DNA extraction. Each sample was put into a 1.5 mL sterile centrifuge tube with 1 mL sterile deionized water, which was then mixed by a shaker flowed by centrifuging at 13 000 rpm for 10 min with removing the supernatants. This procedure was repeated twice for maximizing DNA extraction from biomass samples. Collected pellets were stored in 2 mL sterile centrifuge tubes at −70 °C before DNA extraction. DNA extraction as well as 16S rRNA gene based Illumina MiSeq sequencing of the pellets was served by Shanghai Sangon Biotech Co., Ltd.

High-throughput 16 S rRNA Gene Illumina MiSeq Sequencing

Amplicon libraries were constructed by Illumina Miseq 2000 using bacterial universal primers 341 F (5′-CCTACACGACGCTCTTCCGATCTN-3′) and 805 R (5′- GACTGGAGTTCCTTGGCACCCGAGAATTCCA -3′). Both forward and reverse primers were added with barcode. PCR amplification, products purification and quantification, and the sequencing were carried out through the Illumina MiSeq platform in Shanghai Sangon Biotech Co., Ltd. The data analysis was as in Yang *et al.*[26].

ACKNOWLEDGMENTS

This research was supported by the National Nature Science Foundation for Distinguished Young Scholars (Grant No. 51225802), the National Nature Science Foundation of China (Grand No. 51508551), the China Postdoctoral Science Foundation (Grand No. 2015M580140), the Ministry of Environmental Protection of the People's Republic of China (Major Science and Technology Program for Water Pollution Control and Treatment) (No. 2014ZX07204-005), the "Hundred Talents Program" of the Chinese Academy of Sciences (Grant No. 29BR2013001) and the Jiangsu Key Laboratory of Chemical Pollution Control and Resources Reuse in NJUST (Grant No. 30920140122008).

AUTHOR CONTRIBUTIONS

M.H.C., D.C., H.Y.C. and A.J.W. designed the experiments; M.H.C. and D.C. conducted the experiments; H.Y.C. and A.J.W. contributed to the planning and coordination of the project; M.H.C., D.C., H.S.L. and B.L. wrote and edited the manuscript. All authors contributed to discussions about the results and the manuscript.

REFERENCES

1. Moutaouakkil A. et al. . Decolorization of azo dyes with Enterobacter agglomerans immobilized in different supports by using fluidized bed bioreactor. Curr Microbiol 48, 124–129 (2004).
2. Chen Z. B. et al. . Improving the simultaneous removal efficiency of COD and color in a combined HABMR-CFASR system based MPDW. Part 1: optimization of operational parameters for HABMR by using response surface methodology. Bioresour. Technol. 102, 8839–8847 (2011).
3. Qiu M. & Huang C. A comparative study of degradation of the azo dye C.I. Acid Blue 9 by Fenton and photo-Fenton oxidation. Desalin. Water Treat. 24, 273–277 (2012).
4. Hu D.-X. et al. . Performance of a novel HABR–CFASR system for the biological treatment of mixed printing and dyeing wastewater (MPDW). Desalin. Water Treat. 52, 5553–5562 (2013).
5. Şen S. & Demirer G. N. Anaerobic treatment of real textile wastewater with a fluidized bed reactor. Water Res. 37, 1868–1878 (2003).
6. Liang B. et al. . Accelerated reduction of chlorinated nitroaromatic antibiotic chloramphenicol by biocathode. Environ. Sci. Technol. 47, 5353–5361 (2013).
7. Feng H. et al. . Enhanced removal of p-fluoronitrobenzene using bioelectrochemical system.Water Res. 60C, 54–63 (2014).
8. Kong D. et al. . Cathodic degradation of antibiotics: Characterization and pathway analysis.Water Res (2015).
9. Mu Y., Rabaey K., Rozendal R. A., Yuan Z. G. & Keller J. Decolorization of Azo Dyes in Bioelectrochemical Systems. Environ. Sci. Technol. 43, 5137–5143 (2009).
10. Wang Y. Z. et al. . Accelerated azo dye removal by biocathode formation in single-chamber biocatalyzed electrolysis systems. Bioresour. Technol. 146, 740–743 (2013).
11. Wang Y. Z. et al. . Electrode as sole electrons donor for enhancing decolorization of azo dye by an isolated Pseudomonas sp. WYZ-2. Bioresour. Technol. 152, 530–533 (2014).
12. Shen J. et al. . Coupling of a bioelectrochemical system for p-nitrophenol removal in an upflow anaerobic sludge blanket reactor. Water Res. 67C, 11–18 (2014).
13. Cui D. et al. . Enhanced decolorization of azo dye in a small pilot-scale anaerobic baffled reactor coupled with biocatalyzed electrolysis system (ABR-BES): A design suitable for scaling-up.Bioresour. Technol. 163C, 254–261 (2014).
14. Li W.-W., Yu H.-Q. & He Z. Towards sustainable wastewater treatment by using microbial fuel cells-centered technologies. Energy Environ. Sci. 7, 911–924 (2013).
15. Zhou M., Chi M., Luo J., He H. & Jin T. An overview of electrode materials in microbial fuel cells. J. Power Sources 196, 4427–4435 (2011).
16. Wei J., Liang P. & Huang X. Recent progress in electrodes for microbial fuel cells. Bioresour. Technol. 102, 9335–9344 (2011).
17. Dhar B. R. & Lee H.-S. Evaluation of limiting factors for current density in microbial electrochemical cells (MXCs) treating domestic wastewater. Biotechnol. Rep. 4, 80–85 (2014).
18. Cui D. et al. . Azo dye removal in a membrane-free up-flow biocatalyzed electrolysis reactor coupled with an aerobic bio-contact oxidation reactor. J. Hazard. Mater. 239–240, 257–264 (2012).
19. Razo Flores E., Luijten M., Donlon B. A., Lettinga G. & Field J. A. Complete biodegradation of the azo dye azodisalicylate under anaerobic conditions. Environ. Sci. Technol. 31, 2098–2103 (1997).
20. Mohan Y., Kumar S. M. M. & Das D. Electricity generation using microbial fuel cells. Int. J. Hydrogen Energy 33, 423–426 (2008).
21. Borole A. P., Mielenz J. R., Vishnivetskaya T. A. & Hamilton C. Y. Controlling accumulation of fermentation

inhibitors in biorefinery recycle water using microbial fuel cells. Biotechnol. Biofuels 2, (2009). [PMC free article]

22. Yoo E. S., Libra J. & Wiesmann U. Reduction of azo dyes by desulfovibrio desulfuricans. Water Sci. Technol. 41, 15–22 (2000).

23. Feng Y., Yang Q., Wang X. & Logan B. E. Treatment of carbon fiber brush anodes for improving power generation in air–cathode microbial fuel cells. J. Power Sources 195, 1841–1844 (2010).

24. Wolin E. A., Wolin M. J. & Wolfe R. S. Formation of methane by bacterial extracts. J. Biol. Chem 238, 2882–2886 (1963).

25. Cui D. et al. . Efficient azo dye removal in bioelectrochemical system and post-aerobic bioreactor: Optimization and characterization. Chem. Eng. J. 243, 355–363 (2014).

26. Yang C. et al. . Freezing/thawing pretreatment coupled with biological process of thermophilic Geobacillus sp. G1: Acceleration on waste activated sludge hydrolysis and acidification.Bioresour. Technol. 175C, 509–516 (2014).

CHAPTER 16

A PORTABLE LOW-POWER ACQUISITION SYSTEM WITH A UREASE BIOELECTROCHEMICAL SENSOR FOR POTENTIOMETRIC DETECTION OF UREA CONCENTRATIONS

Wei-Jhe Ma, Ching-Hsing Luo, Jiun-Ling Lin, Sin-Houng Chou, Ping-Hung Chen, Mei-Jywan Syu, Shin-Hung Kuo, and Shin-Chi Lai

Keywords: Portable, battery-driven, bioelectrochemical signal acquisition, urea, potentiometric, immobilized urease, MEMS

INTRODUCTION

The urea concentration in human blood, called blood urea nitrogen (BUN), is significant for the assessment of normal renal function. BUN, a metabolic waste product of protein, is secreted by the kidneys and excreted from the body via urine. The BUN of a healthy adult ranges between 6 and 21 mg/dL (1–3.5 mM). Renal failure and related complications, such as acute or chronic urinary tract obstruction, hepatic failure and nephritic syndrome, could causes in urea concentration [1,2,3,4]. Currently, more than two thousand people per million worldwide rely on hemodialysis. Normally, hemodialysis is required when the urea concentration in serum is above 180 mg/dL. BUN is temporarily raised in some diseases or human conditions such as liver dysfunction, upper gastro-intestinal bleeding, hydropenia, or hepatitis. Thus, it is essential to develop a device that can measure urea concentration rapidly and precisely.

The urease catalytic reaction mechanism for urea is as follows:

$$NH_2CONH_2 + 3H_2O + H^+ \xrightarrow{urease} HCO_3^- + OH^- + 2NH_4^+ \quad (1)$$

According to Equation (1), the determination of urea concentration depends on the formation of ammonia or the change of pH [5]. In this manner, an electrochemical urea sensor can accurately measure the urea concentration.

A recent surge of research into bioelectrochemical sensors has concentrated on sensor fabrication via micro-electro-mechanical system (MEMS) technology. Many MEMS-based bioelectrochemical sensors for different applications have been described, such as *E. coli* bacteria detection [6], nitric oxide sensing [7], glucose detection [8], and DNA hybridization [9]. Each of these papers targeted the urea concentration via fabrication of a bioelectrochemical sensor by the MEMS technique.

In the past few decades, electrochemical sensors have been widely applied in various research fields owing to their high selectivity and sensitivity, rapid analysis, simplicity, easy fabrication and low cost [10,11]. These electrochemical sensors serve as transducers that convert a chemical concentration into an analog signal for measurement by an electronic device. Additionally, by physical entrapment and covalent bonding [12,13,14,15,16,17,18], urease has been immobilized onto a working electrode to achieve electropolymerization [19,20]. So far, all research efforts have been aimed at immobilizing urease within different kinds of polymeric matrices and with minimum leakage and loss of activity. To measure the urea concentration, various studies designed electrochemical urea sensors based on conductometry, potentiometry and

Citation: Wei-Jhe Ma, Ching-Hsing Luo, Jiun-Ling Lin, Sin-Houng Chou, Ping-Hung Chen, Mei-Jywan Syu, Shin-Hung Kuo and Shin-Chi Lai, A Portable Low-Power Acquisition System with a Urease Bio-electrochemical Sensor for Potentiometric Detection of Urea Concentrations, doi: 10.3390/s16040474.

amperometry with flow injection analysis (FIA). In recent years, much attention has been given to various urea biosensor designs, such as a solid-state urea biosensor [21] in the potentiometric mode, a gold interdigitated microelectrode urea biosensor based on polyhydroxybutyrate substrate in the impedimetric mode [22], and a carbon-black electrode urea sensor based on the amperometric mode [23,24,25]. Li *et al.* [26] developed a conducting polymer hydrogel amperometric biosensor for measuring human metabolites, such as uric acid, cholesterol and triglycerides, in a wide linear range. They reviewed the design and fabrication of such an electrode—Including different-dimensional nanostructures and biosensor applications [27,28]. More generally, these studies highlighted the importance of urea biosensors for bioelectrochemical sensing applications.

Recently, increasing attention has been devoted to research into personal health-care due to the globally aging population and chronic diseases. To meet the demands of personal health-care, the miniaturization of medical monitoring instruments into portable devices has become an important issue for health-care applications. Some commercial products based on the concept of point-of-care (home-care) service as well as long-term monitoring have already begun to appear on the market, such as portable and wearable blood glucose meters and continuous electrocardiogram (ECG) monitors [29,30,31,32,33,34]. In line with this trend, the main aim of the present study was to develop a rapid biomedical monitoring device.

Over the past few decades, much research has focused on portable biomedical devices that offer simple and reliable biomedical detection. In this regard, bioelectrochemical acquisition systems can acquire signals quickly and accurately. Some studies have used discrete components and evaluation board platforms to realize bioelectrochemical acquisition applications [35,36,37,38,39], while other works have concentrated on application-specific integrated circuit (ASIC) chips in single-, dual-, or multi-type electrochemical biosensor-integrated device studies [40,41,42,43]. In the present study, a bioelectrochemical acquisition system prototype device was realized as a biosensor for the kidney function. More specifically, the present paper proposes a bioelectrochemical monitoring prototype for the analysis of renal function by targeting urea concentration both in rapid detection and long-term monitoring. In this study, to increase the autonomy of users and improve the quality of detection, we focused on speed, convenience, performance and low-power consumption. Figure 1 schematically depicts our proposed bioelectrochemical acquisition prototype device for rapid analysis of renal function. As shown, it comprises a MEMS-based urea sensor, a front-end circuit, a signal-processing unit, an RS-232 driver, a ZigBee module, and custom-developed software for the monitoring. The proposed urea bioelectrochemical sensor based on a silicon substrate is fabricated by a MEMS technique and responds in the potentiometric mode. The proposed front-end circuit is linked to a 12-bit ADC, which converts the analog signal into a digital sequence. Furthermore, the microcontroller unit processes and transmits the data to a computer or any receiving device by two modes, namely a ZigBee RF module in wireless mode and a UART in cable mode. Finally, the custom-developed software features a user-friendly interface for data display and storage.

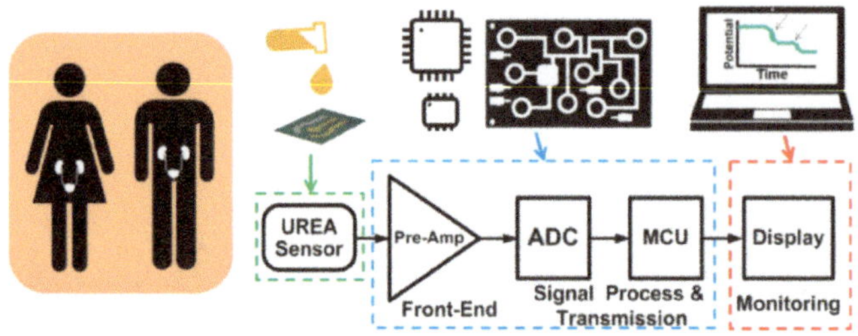

Figure 1: The proposed bioelectrochemical acquisition system prototype.

To sense urea, the bioelectrochemical module is composed of three electrodes: A Au working electrode, a Au counter electrode, and a Ag/AgCl reference electrode. The gold and silver-based electrodes are manufactured by a MEMS process. The silver-based reference electrode is oxidized by an after-process treatment to form a Ag/AgCl electrode, after which the immobilization of urease is performed by physical entrapment using electropolymerization.

The experimental results revealed that the system was able to achieve excellent performance in the calibration of urea concentration against potentiometric change. Accordingly, it would be suitable for application in rapid analysis and monitoring of renal function, for instance, in future development of a low-power wireless biotelemetry system-on-a-chip (SoC) device.

MATERIALS AND METHODS

Fabrication of the MEMS-Based Bioelectrochemical Sensor

The proposed design of the electrochemical sensor for urea detection involved a three-electrode system fabricated by a MEMS technique. The electrochemical module consisted of three electrodes on a silicon substrate as the urea sensing area, comprising a Au working electrode, a Au counter electrode and a Ag/AgCl reference electrode. The sensor's working electrode is the one that plays the principal role in urea sensing. Next, the counter electrode of the sensor was constructed from another gold electrode, while the reference electrode was a pseudo (quasi) Ag/AgCl electrode [44,45]. The gold on the electrodes was deposited via e-beam evaporation, with the metallization utilized for interconnecting the metal traces and testing pads.

The fabrication process involved three steps: Standard photolithography, thin-film deposition and lift-off. A 4" phosphorus-doped silicon wafer (100) with a thickness of 500 μm was used as the silicon substrate (Ultimate Materials Technology Co., Pingtung, Taiwan). The substrate was first cleaned using the standard RCA cleaning steps. Afterwards, a low-stress 1.0 μm-thick Si_3N_4 layer was deposited on the silicon substrate as the isolation layer on each side via a low-pressure chemical vapor deposition system (LPCVD) (NFC-NCTU, Hsinchu, Taiwan). The operation conditions to deposit the low-stress Si_3N_4 thin film were: SiH_2Cl_2 (85 sccm), NH_3 (175 sccm), 50 mtorr pressure, and a temperature of 850 °C.

The first step in fabricating the three-electrode layer from the MEMS process involved spin-coating the AZ1500 photoresist with a thickness of 1.5 ctrode layer from the

MEMS proces$_3$N$_4$), and then setting the pattern with a Mask Aligner (MA 150CC, SUSS Micro-Tec, Garching, Germany). Then, a Ti (15 nm) layer was deposited to enhance the adhesion of Au to the silicon surface, after which e-beam evaporation was used to deposit the Au layer (120 nm). The pattern on the Au layer was then created by a standard lift-off process. In this manner, both the Au electrodes namely the working and counter electrodes were fabricated. Also the fabrication of the Ag electrode (120 nm) followed the above-mentioned process. The metal-deposition conditions are summarized in Table 1, with each layer deposited in two steps. The metal-deposition rate of the first step was 10 times slower than that of the second step. However, the two-step metal-deposition procedure was used because it gave superior adhesion and film quality. After finishing the micro-fabrication of the three electrodes, the Ag electrode was oxidized in a $FeCl_3$ (100 mM) solution to produce the Ag/AgCl reference electrode. Finally, the fabrication process was completed by washing the three electrodes in distilled water. A flow chart of the entire process is shown in Figure 2.

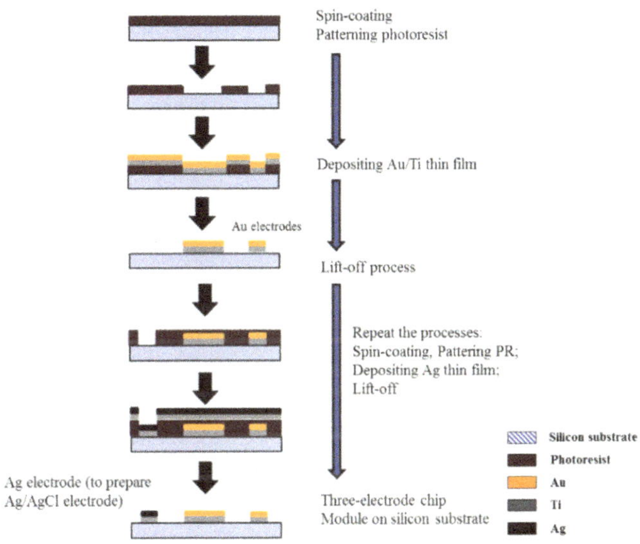

Figure 2: Three-electrode chip molding process flow.

Table 1: Deposition conditions of metal layers (Ti, Au and Ag) on the substrate

	Ti	Au	Ag
1st Deposition rate	0.01 nm/s	0.01 nm/s	0.01 nm/s
1st Thickness	5 nm	5 nm	5 nm
2nd Deposition rate	0.1 nm/s	0.1 nm/s	0.1 nm/s
2nd Thickness	10 nm	115 nm	115 nm
Vacuum		3×10^{-6} torr	

Electrochemical Detection of Urea Concentration

The potentiometric mode was used to detect the urea concentration. Three electrodes, namely the working, counter and reference electrodes, were needed to perform the electropolymerization. The conductive materials not only enhanced the conductivity of the electrodes, but also trapped the urease enzyme which detected the urea. Cyclic voltammetry was used to trigger the electropolymerization for the formation of the conductive polymeric matrix on the Au working electrode. By physical entrapment, urease (E.C. 3.5.1.5, from Jack Beans, Type III, Sigma-Aldrich Inc., St. Louis, MO, USA, crude enzyme without further purification or pretreatment before use) in a concentration of 5.0 mg/mL was immobilized within the conductive polymeric matrix. Furthermore, urease was mixed with aniline (ANI) monomer (Sigma-Aldrich Chemie GmbH, Buchs, Switzerland) and *o*-phenylenediamine (*o*-PD; Sigma-Aldrich Chemie GmbH), after which cyclic voltammetric polymerization was initialized for the synthesis of poly(ANI-co-*o*-PD), the polymer, poly(aniline-co-*o*-phenylenediamine) as well as for the immobilization of urease. Polyaniline is a well-known electroconducting polymer in a linear chain. However, it is difficult to firmly entrap urease only by polyaniline. Therefore, *o*-phenylenediamine was used to add a branching structure into the polymer matrix for enhancing the immobilization of urease. The scan range was from −0.2 to 1.0 V, with a scan rate of 50 mV/s over 10 cycles (*i.e.*, 20 segments). To detect the urea concentration, urease catalyzed the reaction of urea into ammonia, as shown in formula (1), so that a corresponding potential-change signal was obtained. Instead of the three-electrode system, only the working and reference electrodes (a two-electrode process) were needed to detect the urea concentration. The open-circuit potential (OCP) between the working and reference electrodes was measured and correlation of the urea concentration was established based on the Nernst equation.

Low-Power Battery-Driven Bioelectrochemical Acquisition System

The block diagram of the battery-driven portable potentiometric device designed in this work is shown in Figure 3. The open-circuit potential (OCP) obtained from the sensing electrode (*i.e.*, the polymer immobilized urea on the MEMS-base substrate) was amplified and filtered by the front-end circuit, which included an instrumentation amplifier and a low-pass filter. The low-pass filter was composed of an operation amplifier and passive components. Next, the analog signals were converted and processed by a commercial MCU, which included a 12-bit ADC. Afterwards, the digital signals were transmitted to a laptop computer by RS-232 serial communication protocol (MAX3232E-Q1, Texas Instruments Inc., Dallas, TX, USA) in cable mode, or an IEEE 802.15.4/ZigBee-compliant transceiver in wireless mode. Finally, the data were further plotted and stored by the custom-developed software.

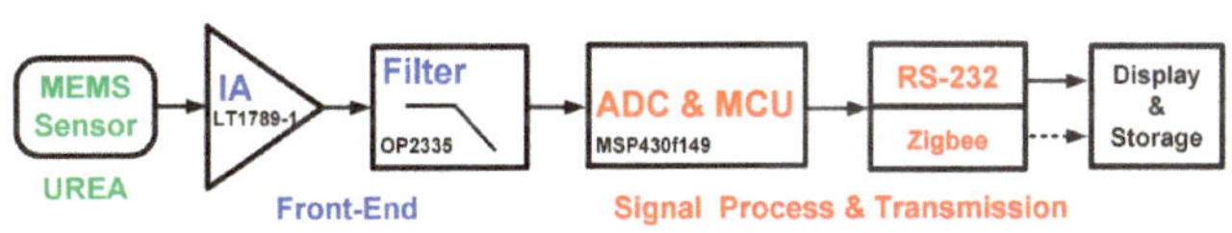

Figure 3: Block diagram of the proposed bio-electrochemical acquisition system.

Bioelectrochemical Readout Circuit

Herein, the urea detection was mainly aimed at potential responses between the electrochemical cell working electrode (WE) and the reference electrode (RE). The potential response of the proposed urea sensor is called the open-circuit potential, which involves no current flow through the WE and RE during the experiments; accordingly, the input impedance of the terminals connected to the WE and RE must be as high as possible. In addition, the electrochemical cell WE and RE of the urea sensor were linked to the proposed bioelectrochemical readout circuit for the bioelectrochemical signal acquisition.

Figure 4 schematically depicts the proposed readout circuit, which includes two parts: The front-end amplifier and low-pass filter. The front-end amplifier of the readout circuit must possess the following characteristics: a high common mode rejection ratio (CMRR), high input impedance, low-power consumption, and low noise for processing the bioelectrochemical signals. Therefore, a high-performance instrumentation amplifier (LT1789-1, Linear Technology, Milpitas, CA, USA) was adopted as the front-end readout circuit of the system. The utilized amplifier had a CMRR of 100 dB and a power supply rejection ratio (PSRR) of 100 dB. The WE and RE were connected to the differential inputs of the instrumentation amplifier, with R_G used to tune the gain for the readout potential change of the urea concentration. Accordingly, the gain of LT1789-1 can be represented as:

$$Gain = \left[1 + (\frac{200k\Omega}{R_G})\right] \quad (2)$$

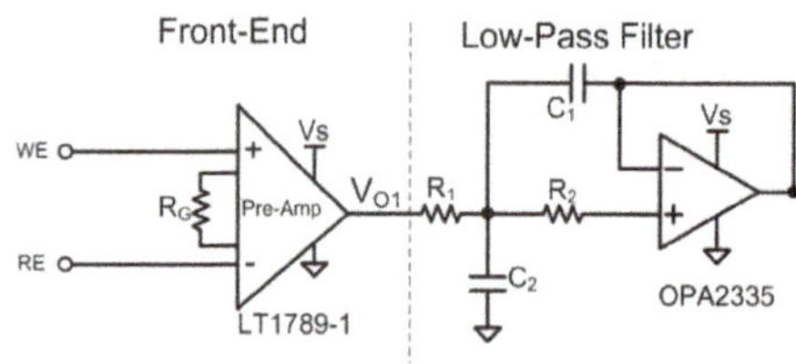

Figure 4: The proposed bio-electrochemical readout circuit for OCP measurement.

Following the instrumentation amplifier circuit, a low-pass filter with a low cut-off frequency was selected to suppress the out-of-band noise and to provide better a signal-to-noise ratio (SNR). In addition, the low-pass filter was designed with an operational amplifier (OPA2335, Texas Instruments Inc., Dallas, TX, USA) to reduce the high frequency noise.

The gain of this instrumentation amplifier has the advantage that it can be tuned for various bioelectrochemical experiments by trimming the resistor value and the bandwidth of the filters can be designed to suit the system. The output of the readout circuit was connected to the input of the MSP430f149 ADC (Texas Instruments Inc., Dallas, TX, USA) for data processing and the UART unit for data transmission.

The LT1789-1 features low-power consumption, a high CMRR, high PSRR, high accuracy, rail-to-rail input and output range, and a wide supply voltage (2.2–36 V). Considering the purpose of the portable low-power system, the instrumentation amplifier (LT1789-1) was designed for low-voltage supply operation with a quiescent current of 95 μA.

Microcontrol Unit Usage

The proposed peripheral devices included an ADC and an MCU. Therefore, an ultra-low-power high-performance 16-bit RISC microcontroller (MSP430f149, Texas Instruments Inc., Dallas, TX, USA) with a 12-bit built-in ADC was selected for the system [46]. This integrates 2 KB of SRAM, and 48 KB + 256 bytes of flash memory for code programing. As mentioned above, the microcontroller featured a powerful 16-bit RISC CPU, 16-bit registers and constant generators that allow maximum code efficiency. Since the resolution of the built-in ADC is 12 bits, it is suitable for biomedical instruments and biophysical measurements.

The bioelectrochemical analog signals extracted by the analog front-end circuit are converted into digital data by the built-in ADC. Next, the MCU processes the digital signals and sends them to the computer for data display and storage via the UART or radio frequency module. The personal computer then acquires the signals through the peripheral devices and processes them using the custom-develop software.

The architecture of the commercial low-power MSP430f149 MCU consists of two built-in 16-bit timers, a 12-bit ADC, two UARTs, 48 I/O pins and small size and weight. In addition, it has one active mode and five software selectable power-saving modes; thus, it is optimized to achieve an extended battery life for portable measurement applications. In power-saving mode, the digitally controlled oscillator (DSC) can wake up from a low-power mode to active mode in less than 6 μs, including performing the functions of power-management control, enabling instrumentation amplifier, ADC and processing the signals. Furthermore, under the active mode, the active current is 280 μA at 1 MHz, while the idle current is 1.6 μA in standby mode.

Peripheral Device for Signal Monitoring and Storage

The UART interface is widely used for serial communication and its transport protocol is simpler than that of the universal serial bus (USB). Although UART has relatively slow data transmission, it is sufficient for this work due to the long response time of the sensors.

The peripheral devices of the proposed system transmit the acquired bioelectrochemical signal to the computer for data display and storage. In front of the peripheral device, the MCU processes and sends the signals via a UART or wireless transmission solution. Furthermore, a personal computer or laptop computer acquires the signals through the peripheral devices and displays them by using system designed software. The present study employed two modes of data transmission for the proposed data acquisition, as described below.

In cable mode, the selected MSP430f149 MCU has two UART interfaces; thus, the proposed portable device can be connected to a computer via a USB, a UART port and/or a RS-232 serial protocol. A multi-channel RS-232 line driver/receiver (MAX3232E-Q1, Texas Instruments Inc. Texas Instruments Inc., Dallas, TX, USA) was used to meet the requirements of TIA/EIA-232-F (RS-232) and to provide the electrical interface between the asynchronous communication controller and the serial-port connector.

In wireless mode, the chosen wireless module was an IEEE 802.15.4/ZigBee-compliant transceiver. It contains a powerful 8-bit 8051 MCU and an RF single-chip (CC2530, Texas Instruments Inc., Dallas, TX, USA) 2.4 GHz IEEE 802.15.4-compliant RF transceiver. The RF transceiver module is employed as a communication transceiver in bidirectional telemetric devices. Accordingly, a wireless sensor network can be built from the transceiver modules for various applications. The transmission range of the transceiver is 100 m (LOS). Additionally, the receiver unit receives UART formatted data transmitted via a USB port on a UART/USB interface IC (FT232RL, FTDI Chip Inc., Glasgow, Scotland).

The proposed custom-developed software was written in Visual Basic and Lab-View to provide a user-friendly interface to display the data in real-time and to automatically upload the data to a database via the Internet.

RESULTS AND DISCUSSIONS

Urea Disposable Sensing Microchip

A three-electrode system functioning as a urea sensor, as well as a system for detecting urea concentration, was fabricated by a MEMS technique. A schematic diagram of the fabricated three microelectrode urea sensor prototype is shown in Figure 5a. As can be seen, the diameter of the sensing area is 2.4 mm, while those of the counter and reference electrodes are both 1.0 mm. The total area of the urea sensor is 70 mm^2, with an actual sensing area of 4.78 mm^2 on the working electrode. Additionally, polydimethylsiloxane (PDMS) was used to create a reservoir for the urea samples. Detection of the urea concentration was carried out once the urease enzyme was immobilized on the Au working electrode via the formation of a film of copolymer poly(ANI-co-*o*-PD), *i.e.*, poly(aniline-*co*-*o*-phenylenediamine) (*i.e.*, A photograph of the fabricated three-electrode electropolymerization urea sensing chip is presented in Figure 5b, with the proposed sensor integrated with the IC heterogeneously. Therefore, the proposed sensor was attached to a printed circuit board (PCB) and the reservoir channel holding the working electrode was made by PDMS, as shown in Figure 5c.

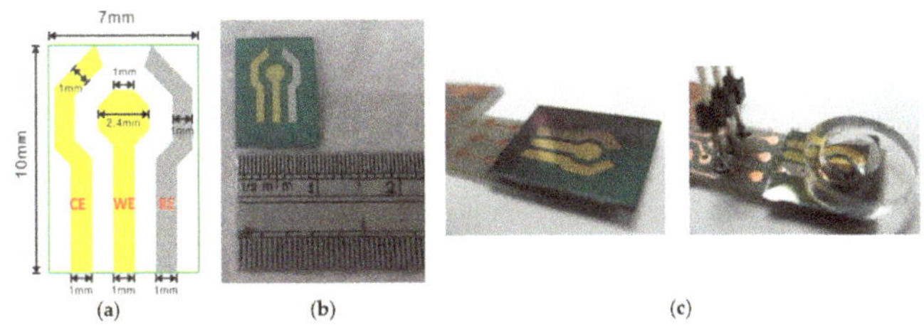

Figure 5: (a) Schematic diagram of the proposed urea sensor chip; (b) Photo of the urea sensor; (c) Photos of the disposable urea sensor integrated with the PCB (Left) and PDMS (Right).

SEM Images of the Poly Film Fabricated on the Au Electrode

The chemical structure of poly(ANI-co-*o*-PD) is shown in Figure 6. Polyaniline can only provide electro-conductivity; however, it has a linear chain structure. Therefore, we used *o*-phenylenediamine to graft with aniline so that the copolymer thus fabricated could more firmly entrap urease. Figure 7a–c present the SEM images of the Au surface, the morphology of polyaniline and the morphology of the poly(ANI-co-*o*-PD) adhered onto Au, respectively. Obviously, the presence of *o*-phenylenediamine together with aniline, after electropolymerization, formed a kind of network structure, which could be seen on the surface of the copolymer film thus created, which then enhanced the immobilization effect of urease onto the electrode.

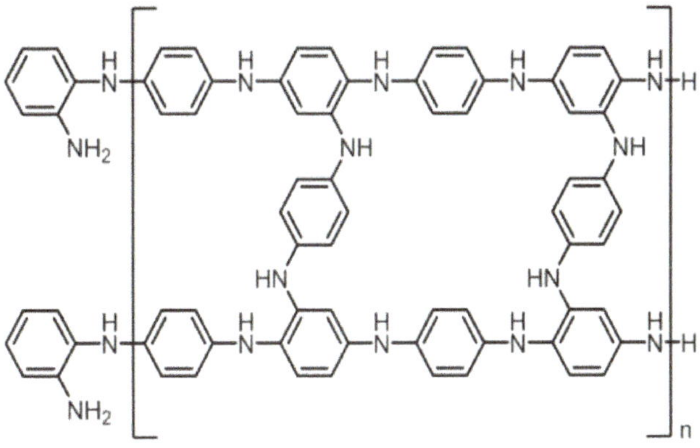

Figure 6: The chemical structure of poly(aniline-co-*o*-phenylenediamine).

Figure 7: SEM images of the sensing surface of (a) Au; (b) polyaniline electro-fabricated onto Au; and (c) poly (aniline-co-*o*-phenylenediamine) on Au.

The Proposed Bioelectrochemical Signal Acquisition System

Portable biomedical instruments for the purpose of home-care application are becoming increasingly important. The proposed acquisition system presented in this paper was implemented and integrated with a MEMS urea sensor, a low-power bioelectrochemical readout circuit, an ultra-low-power MCU, a low-power RS-232 driver/receiver, and custom-developed software, with the entire system shown in Figure 8. The size of the proposed system device is 6.0 × 4.3 cm^2 and the height is approximately 2 cm.

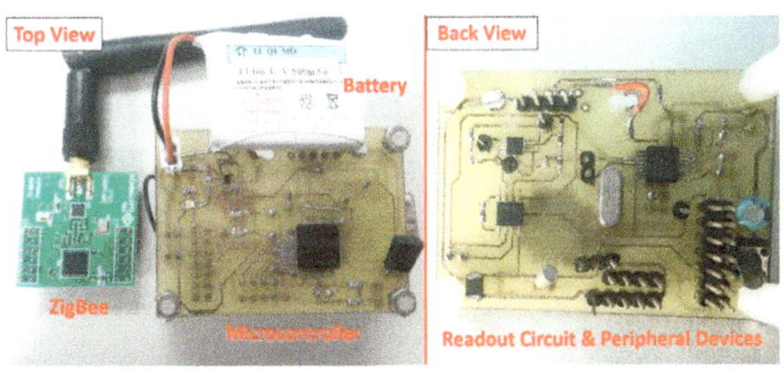

Figure 8: Realization of the proposed system device.

The proposed device and sensor for urea concentration detection are described in the following. First, the working and reference electrodes of the urea sensor were connected to the dual inputs of the readout circuit, and the open-circuit potential between them was measured; then, the potential signal from the sensor was amplified and linked to the ADC for signal conversion. The gain was set to 40 dB for this active component. The chosen MCU was programmed to initialize the built-in ADC, clock system and UART. The reference voltage provided for the built-in ADC and sampling timer was 3.3 V. In addition, the sampling timer (SHT_0) and the sampling time ($t_{sample} = 4 \cdot t_{ADC12CLK} \cdot 8$) were chosen for signal processing. The MCU clock was produced by an external 4 MHz crystal oscillator. The baud rate of the UART was 9600 bps and in cable mode, the peripheral device included an MCU and UART/RS232 communication chip (MAX3232E-Q1). The MCU obtains biomedical analog signals by using the proposed processing chip and transmits the signals via UART to the PC. Furthermore, in wireless mode, the device contains a MCU and wireless ZigBee module (CC2530); accordingly, the MCU can transmit biomedical signals wirelessly.

The factors of long-term monitoring, low-power consumption, low cost, and compact size must be taken into consideration. In cable mode, the total power consumption of the proposed design is 12.42 mW with a supply voltage of 3.3 V. Consequently, the power dissipation of the proposed system device is considered a low-power design and therefore suitable for battery-driven operation and long-term monitoring. The MCU can transmit the biomedical signals wirelessly with a power dissipation of 32.62 mW. A wireless solution is more appropriate for portable monitoring applications, despite consuming more power. Moreover, this system is very stable and can be continuously operated for at least 4 days with a 500 mA·h battery in cable mode; by contrast, the device lifetime in wireless mode is about 2 days. Table 2 summarizes the power consumption of the proposed readout system device in dual mode.

Table 2: Power consumption of the proposed readout system device in dual mode

Cable Mode	Front-End	MCU	MAX3232	Regulator
P.C. (mW)	0.32	7.88	3.31	0.92
% of Total P.C.	2.57%	63.40%	26.63%	7.40%
Total Power	12.43 mW			
Device lifetime	4–5 days (A 3.7V Li-ion 500 mA·h battery)			
Wireless mode	Front-end	MCU	ZigBee	Regulator
P.C. (mW)	0.32	7.88	23.5	0.92
% of Total P.C.	0.98	24.15%	72.04%	2.82%
Total Power	32.62 mW			
Device lifetime	1–2 days (A 3.7V Li-ion 500 mA·h battery)			

The proposed battery-driven device is powered by a 3.3 V regulated external power supply. Therefore, the power supply of this device consisted of a +3.3 V voltage regulator commercial IC, TPS76933 and bulk decoupling capacitors. The TPS769 family is a low-dropout-voltage and low-noise regulator; accordingly, it is suitable for portable applications. The TPS769 also features a logic-enabled sleep mode to shut down the regulator, thereby reducing quiescent current to 1 μA, which makes the IC appropriate for use in battery-powered devices. A rechargeable 3.7 V Li-ion 500 mA·h battery pack provided the power to the proposed battery-driven device.

Detection of Urea Concentration by the Proposed System Device and Sensor

In this paper, electrochemical analysis experiments were conducted with the proposed system device and urea electrochemical sensor. The urea concentration measurement was performed by the sequential injection of urea up to a working volume of 50 μL. The urea concentration of the proposed system device was calibrated by using a commercial potentiostat (CHI-611C, CH Instruments Inc., Tennison Hill Drive Austin, TX, USA) as the standard. Figure 9 shows the time profile of the corresponding voltage changes in response to the sequential injections of urea. Figure 10 shows the potential variations against the urea concentrations obtained. The minimum urea concentration required that could be detected was experimentally found to be 3.16×10^{-4} M (1.89 mg/dL) with the pH set at 7.4. Furthermore, a sensitivity of 31.12 mV/log [M] could be achieved in the urea concentration range of 3.16×10^{-4} M to 3.16×10^{-2} M (1.89–189.93 mg/dL) with a precision of 0.995 ($R^2 = 0.995$). With a sample volume of only 10 μL, accurate measurement of the urea concentration could be obtained in less than 100 s.

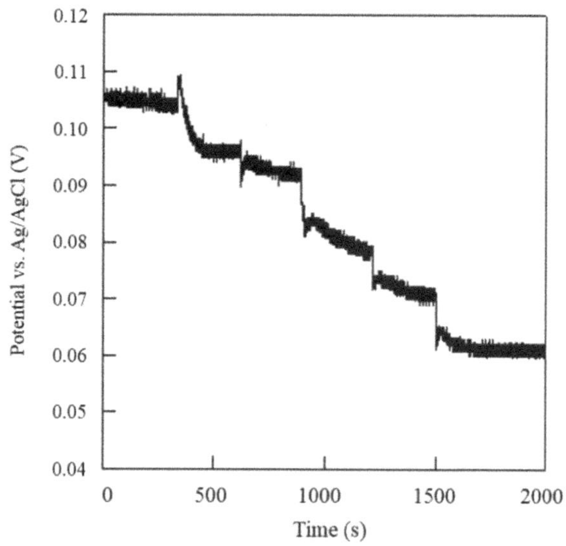

Figure 9: Potential time profile in response to the injection of urea solution.

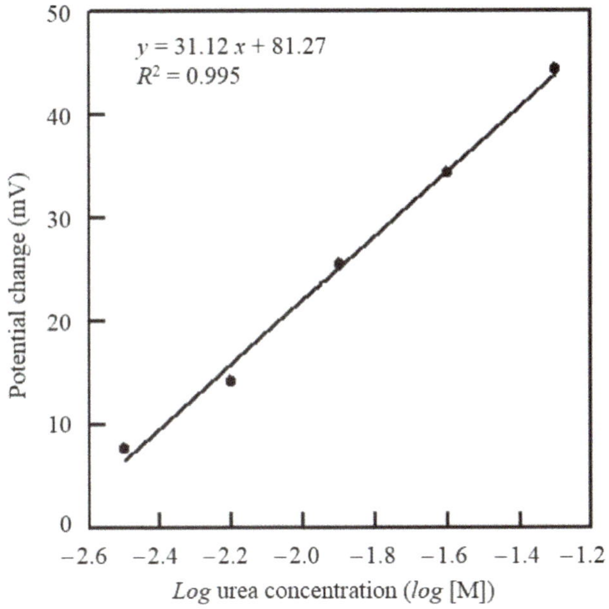

Figure 10: Characteristic curve of urea concentration in test samples.

According to the Nernst equation, the sensing results can be reasonably explained. Moreover, the results confirm the successful measurement of urea concentrations by this portable device. We further compared the sensing results in this work with our previous work [47]. The major differences are listed as follows. First, the proposed battery-driven portable device in the present study replaces the function of a potentiostat, while in our previous work the sensing system was operated by a potentiostat. Second, the MEMS in our previous work was designed with a microchannel for the pumping of microfluid to the sensing area, while in the present work the MEMS has no pumping action. A comparison of the sensing performance for the present work and our previous work is listed in Table 3. With an injection volume of 10 μL to a working volume of 50 μL whole serum, the linear calibration had a sensitivity of 1.59 ± 0.47 mV/log [M] for serum urea concentration, which was similar to [47]. The experimental results of the integrated device and sensor are summarized in Table 4, while Table 5compares the differences between the system proposed in the present work and the design described by Wang *et al.* [40].

Table 3: Comparison between the proposed sensor and previous work

Specification	[47]		This Work	
Background Solvent	water	serum	water	serum
Sensitivity (mV/log [M])	28.68 ± 0.01	2.71 ± 0.56	31.12	1.59 ± 0.47
Sampling Time (s)	400	400	100	100
Response Time (s)	~800	~1200	~100	~500
Sample Volume (mL)	0.20	0.20	0.01	0.01
Working Volume (mL)	5	5	0.05	0.05

Table 4: Specifications results of the proposed system and sensor

Key Item	Specification	
System Power Supply	3.3 V	
Power Consumption (Cable mode & Wireless mode)	Cable 12.42 mW	ZigBee 32.62 mW
Linearity (R^2)	0.995	
Sensitivity (mV/log [M])	31.12	
UART Baud rate	9600	
Device Size	6.0 × 4.3 cm^2	
Sensor Size	0.7 × 1.0 cm^2	

Table 5: Comparison between our system and previous work

Specification	[40]	This Work
Circuit Composition	CMOS/FPGA	Commercial Chip
Power Supply	1.8V/3.3 V	3.3 V
Linearity	0.998	0.995
Power consumption	157.25 mW	32.62 mW
Data transmission	433 MHz/ISM	2.4 GHz/ZigBee

CONCLUSIONS

To achieve the requirements of low-power consumption, high portability, and low cost, we proposed a low-power battery-driven portable device in this paper. The performance of this device was successfully confirmed via its integration with a disposable urea sensor. The proposed system device includes the following modules: A low-supply-voltage IA, 2nd-order low-pass filter, low-power microcontroller, a RS-232 driver/receiver, an IEEE 802.15.4/ZigBee-compliant transceiver and custom-developed software. The potential response results of the urea sensor show that the sensitivity is 31.12 mV/log [M], which was achieved in the urea concentration range of 3.16×10^{-4} M to 3.16×10^{-2} M (1.89–189.93 mg/dL). This low-power portable device can not only quantify the urea concentration, but also stably measure urea signals for at least four days running on a 3.7-V Li-ion battery. Consequently, this low-power portable device holds potential and its use in biosensors for home health care applications is feasible.

ACKNOWLEDGMENTS

The authors gratefully acknowledge the financial support on this work by Ministry of Science and Technology (MOST) of Taiwan, under grant number MOST 104-2218-E-006-022. The effort contributed by Yu-Sung Chang (both are Master graduates from Chemical Engineering, National Cheng Kung University) in the initial stage of this work were also acknowledged. We also appreciate National Nano Device Laboratories (NDL) and Center for Micro/Nano Science and Technology (CMNST) for providing all apparatus and resources of the MEMS process.

AUTHOR CONTRIBUTIONS

W.J. Ma carried out the fabrication of the circuit device and participation of the cross-field experiment. S.H. Kuo carried out the fabrication of MEMS material experiments. J.L. Lin carried out the fabrication of sensing material experiments and SEM images. S.H. Chou carried out extra experiments to verify and provided to the reviewer. P.H. Chen carried out the urea concentration calibration experiments. M.J. Syu designed and assigned the experiments, discussed the data. S.C. Lai assigned the digital part of the system experiment. C.H. Luo assigned the system and discuss MEMS material experiment. W.J. Ma collected the experiment data and wrote the paper. S.C. Lai, M.J. Syu and C.H. Luo supervised the research, critically reviewed and revised the paper.

REFERENCES

1. Aronson D., Mittleman M.A., Burger A.J. Elevated blood urea nitrogen level as a predictor of mortality in patients admitted for decompensated heart failure. Am. J. Med. 2004;116:466–473. doi: 10.1016/j.

amjmed.2003.11.014.

2. Ronco C., Bellomo R., Homel P., Brendolan A., Dan M., Piccinni P., Greca G.L. Effect of different doses in continuous veno-venous hemofiltration on outcomes of acute renal failure: A prospective randomized trial. Lancet. 2000;356:26–30. doi: 10.1016/S0140-6736(00)02430-2.

3. Tiwari A. A novel nanocomposite matrix based on silylated chitosan and multiwall carbon nanotubes for the immobilization of urease. J. Inorg. Organomet. Polym. Mater. 2009;19:361–366. doi: 10.1007/s10904-009-9280-x.

4. Bozgeyik I., Senel M., Cevik E., Abasıyanık M.F. A novel thin film amperometric urea biosensor based on urease-immobilized on poly (*N*-glycidylpyrrole-*co*-pyrrole) Curr. Appl. Phys. 2011;11:1083–1088. doi: 10.1016/j.cap.2011.01.041.

5. Ho W.O., Krause S., McNeil C.J., Pritchard J.A., Armstrong R.D., Athey D., Rawson K. Electrochemical Sensor for Measurement of Urea and Creatinine in Serum Based on ac Impedance Measurement of Enzyme-Catalyzed Polymer Transformation. Anal. Chem. 1999;71:1940–1946. doi: 10.1021/ac981367d.

6. Gau J.J., Lan E.H., Dunn B., Ho C.M., Woo J.C.S. A MEMS based amperometric detector for *E. coli* bacteria using self-assembled monolayers. Biosens. Bioelectron. 2001;16:745–755. doi: 10.1016/S0956-5663(01)00216-0.

7. Chang C.W., Maduraiveeran G., Xu J.C., Hunter G.W., Dutta P.K. Design, fabrication, and testing of MEMS-based miniaturized potentiometric nitric oxide sensors. Sens. Actuators B Chem.2014;204:183–189. doi: 10.1016/j.snb.2014.06.108.

8. Kudo H., Sawada T., Kazawa E., Yoshida H., Iwasaki Y., Mitsubayashi K. A Flexible and wearable glucose sensor based on functional polymers with Soft-MEMS techniques. Biosens. Bioelectron.2006;22:558–562. doi: 10.1016/j.bios.2006.05.006.

9. Cha J., Han J.I., Choi Y., Yoon D.S., Oh K.W. DNA hybridization electrochemical sensor using conducting polymer. Biosens. Bioelectron. 2003;18:1241–1247. doi: 10.1016/S0956-5663(03)00088-5.

10. Eggenstein C., Borchardt M., Diekmann C., Gründig B., Dumschat C., Cammann K., Knoll M., Spener F. A disposable biosensor for urea determination in blood based on an ammonium-sensitive transducer. Biosens. Bioelectron. 1999;14:33–41. doi: 10.1016/S0956-5663(98)00103-1.

11. Yin L.T., Lin Y.T., Leu Y.C., Hu C.Y. Enzyme immobilization on nitrocellulose film for pH-EGFET type biosensors. Sens. Actuators B Chem. 2010;148:207–213. doi: 10.1016/j.snb.2010.04.042.

12. Senillou A., Jaffrezic-Renault N., Martelet C., Cosnier S. A miniaturized urea sensor based on the integration of both ammonium based urea enzyme field effect transistor and a reference field effect transistor in a single chip. Talanta. 1999;50:219–226. doi: 10.1016/S0039-9140(99)00122-8.

13. Koncki R., Radomska A., Głąb S. Potentiometric determination of dialysate urea nitrogen. Talanta.2000;52:13–17. doi: 10.1016/S0039-9140(99)00346-X.

14. Jin J.H., Pake S.H., Lee C.W., Min N.K., Hong S.I. Fabrication of amperometric urea sensor based on nano-porous silicon technology. J. Korean Phys. Soc. 2003;42:S735–S738.

15. Singh M., Verma N., Garg A.K., Redhu N. Urea biosensors. Sens. Actuators B Chem.2008;134:345–351. doi: 10.1016/j.snb.2008.04.025.

16. Syu M.J., Chang Y.S. Ionic effect investigation of a potentiometric sensor for urea and surface morphology observation of entrapped urease/polypyrrole matrix. Biosens. Bioelectron. 2009;24:2671–2677. doi: 10.1016/j.bios.2009.01.036.

17. Herlem B.G., Lakard S., Antoniou A., Fahys B. Urea potentiometric biosensor based on modified electrodes with urease immobilized on polyethylenimine films. Biosens. Bioelectron. 2004;19:1641–1647.

18. Lakard B., Magnin D., Deschaume O., Vanlancker G., Glinel K., Demoustier-Champagne S., Nysten B., Jonas A.M., Bertrand P., Yunus S. Urea potentiometric enzymatic biosensor based on charged biopolymers and electrodeposited polyaniline. Biosens. Bioelectron. 2011;26:4139–4145. doi: 10.1016/j.bios.2011.04.009.

19. Das G., Yoon H.H. Amperometric urea biosensors based on sulfonated graphene/polyaniline nanocomposite. Int. J. Nanomed. 2015;10:55–66. [PMC free article]

20. Jia W.Z., Su L., Lei Yu. Pt nanoflower/polyaniline composite nanofibers based urea biosensor.Biosens. Bioelectron. 2011;30:158–164. doi: 10.1016/j.bios.2011.09.006.

21. Pan C.W., Chou J.C., Sun T.P., Hsiung S.K. Solid-state urea biosensor based on the differential method. IEEE Sens. J. 2006;6:269–275.

22. Slaugther G. A gold interdigitated microelectrodes fabricated on polyhydroxybutyrate substrate for the determination of urea using impedimetric measurements. IEEE Sens. J. 2012;12:821–826. doi: 10.1109/JSEN.2011.2159786.

23. Laurinavicius V., Razumiene J., Gureviciene V. Bioelectrochemical conversion of urea on carbon black electrode and application. IEEE Sens. J. 2013;13:2208–2213. doi: 10.1109/JSEN.2013.2250711.

24. Chou N.H., Chou J.C., Sun T.P., Hsiung S.K. Measurement and comparison of potentiometric selectivity coefficients of urea biosensors based on ammonium ion-selective electrodes. IEEE Sens. J.2005;5:1362–1368. doi: 10.1109/JSEN.2005.857881.

25. Mahadeva S.K., Kim J. Porous tin-oxide-coated regenerated cellulose as disposable and low-cost alternative transducer for urea detection. IEEE Sens. J. 2013;13:2223–2228. doi: 10.1109/JSEN.2013.2250497.

26. Li L., Wang Y.Q., Pan L.J., Shi Y., Cheng W., Shi Y., Yu G.H. A nanostructured conductive hydrogels-based biosensor platform for human metabolite detection. Nano

Lett. 2015;15:1146–1151.

27. Li L., Shi Y., Pan L.J., Shi Y., Yu G.H. Rational design and applications of conducting polymer hydrogels as electrochemical biosensors. J. Mater. Chem. B. 2015;3:2920–2930. doi: 10.1039/C5TB00090D.

28. Zhao Y., Liu B., Pan L.J., Yu G.H. 3D nanostructured conductive polymer hydrogels for high-performance electrochemical devices. Energy Environ. Sci. 2013;6:2856–2870. doi: 10.1039/c3ee40997j.

29. Vering T., Adam S., Drewer H., Dumschat C., Steinkuhl R., Schulze A., Siegel E.G., Knoll M. Wearable microdialysis system for continuous *in vivo* monitoring of glucose. Analyst. 1998;123:1605–1609. doi: 10.1039/a800017d.

30. Tang Z.P., Du X.G., Louie R., Kost G.J. Effect of drugs on glucose measurements with handheld glucose meters and a portable glucose analyzer. Am. J. Clin. Pathol. 2000;113:75–86. doi: 10.1309/QAW1-X5XW-BVRQ-5LKQ.

31. Ahmadi M.M., Jullien G.A. A wireless-implantable microsystem for continuous blood glucose monitoring. IEEE Trans. Biomed. Circuits Syst. 2009;3:169–180. doi: 10.1109/TBCAS.2009.2016844.

32. Lucani D., Cataldo G., Cruz J., Villegas G., Wong S. A portable ECG monitoring device with bluetooth and holter capabilities for telemedicine applications; Proceedings of the 28th Annual International Conference on Engineering in Medicine and Biology Society; New York, NY, USA. 29–30 August 2006; pp. 5244–5247.

33. Wen C., Yeh M.F., Chang K.C., Lee R.G. Real-time ECG tele-monitoring system design with mobile phone platform. Measurement. 2008;41:463–470. doi: 10.1016/j.measurement.2006.12.006.

34. Jaw F.S., Tseng Y.L., Jang J.K. Modular design of a long-term portable recorder for physiological signals. Measurement. 2010;43:1363–1368. doi: 10.1016/j.measurement.2010.07.015.

35. Chou N.H., Chou J.C., Sun T.P., Hsiung S.K. Study on the disposable urea biosensors based on PVC-COOH membrane ammonium ion-selective electrodes. IEEE Sens. J. 2006;6:262–268. doi: 10.1109/JSEN.2006.870166.

36. Sekhar P.K., Brosha E.L. Trace detection of 2, 4, 6-trinitrotoluene using electrochemical gas sensor. IEEE Sens. J. 2015;15:1624–1629. doi: 10.1109/JSEN.2014.2364519.

37. Li H., Mu X., Yang Y., Mason A.J. Low power multimode electrochemical gas sensor array system for wearable health and safety monitoring. IEEE Sens. J. 2014;14:3391–3399. doi: 10.1109/JSEN.2014.2332278.

38. Nandakumar V., Bishop D., Alonas E., LaBelle J., Joshi L., Alford T.L. A Low-Cost Electrochemical Biosensor for Rapid Bacterial Detection. IEEE Sens. J. 2011;11:210–216. doi: 10.1109/JSEN.2010.2055847.

39. Kim C.S., Lee C.H., Fiering J.O., Ufer S., Scarantino C.W., Nagle H.T. Manipulation of microenvironment with a built-in electrochemical actuator in proximity of a dissolved oxygen microsensor. IEEE Sens. J. 2004;4:568–575. doi: 10.1109/JSEN.2004.832857.

40. Wang W.S., Huang H.Y., Chen S.C., Ho K.C., Lin C.Y., Chou T.C., Wu C.H., Wang W.F., Wu C.F., Luo C.H. Real-time telemetry system for amperometric and potentiometric electrochemical sensor.Sensors. 2011;11:8593–8610. doi: 10.3390/s110908593. [PMC free article]

41. Chung W.Y., Yeh M.H., Chen J.C., Hsiung S.K. Design of a low-voltage instrumentation amplifier for enzyme-extended-gate field effect transistor based urea sensor application; Proceedings of the First IEEE International Workshop on Electronic Design, Test and Applications; Christchurch, New Zealand. 29–31 January 2002; pp. 177–180.

42. Wang J.Q., Chou J.C., Sun T.P., Hsiung S.K., Hsiung G.B. pH-based potentio-metrical flow injection biosensor for urea. Sens. Actuators B Chem. 2003;91:5–10. doi: 10.1016/S0925-4005(03)00160-6.

43. Ma W.J., Huang H.Y., Luo C.H. A low power analog front-end (AFE) circuit dedicated for driving bio-electrochemical sensors and peripheral device; Proceedings of the IEEE Biomedical Circuits and Systems (BioCAS) Conference; Hsinchu, Taiwan. 28–30 November 2012; pp. 120–123.

44. Lambrechts M., Sansen W. Biosensors: Microelectrochemical. Devices. 1st ed. Institute of Physics Publishing; Bristol, UK: 1992.

45. Bard A.J., Faulkner L.R. Electrochemical Methods: Fundamentals and Applications. 2nd ed. Wiley; Hoboken, NJ, USA: 2001.

46. Texas Instrument MASP430F149 Texas Instrument Mixed Signal Microcontroller, 2011. [(accessed on 18 November 2015)]. Available online: http://www.ti.com/lit/ds/symlink/msp430f149.

47. Huang C.J., Lin J.L., Chen P.H., Syu M.J., Lee G.B. A multi-functional electrochemical sensing system using micro-fluid technology for detection of urea and creatinine. Electrophoresis.2011;32:931–938. doi: 10.1002/elps.201000679.

CHAPTER 17

REAL-TIME DYNAMIC ADSORPTION PROCESSES OF CYTOCHROME C ON AN ELECTRODE OBSERVED THROUGH ELECTROCHEMICAL HIGH-SPEED ATOMIC FORCE MICROSCOPY

Kouta Takeda, Takayuki Uchihashi, Hiroki Watanabe, Takuya Ishida, Kiyohiko Igarashi, Nobuhumi Nakamura, Hiroyuki Ohno

INTRODUCTION

Understanding dynamic protein behavior on the electrode interface could enhance the efficiency of developing bioelectronics devices, such as biosensors and biofuel cells. To understand the correlation between molecular dynamic behavior and the electrochemical response of proteins adsorbed on thc electrode, numerous studies have used spectro-electrochemical techniques that combine electrochemical measurements and various spectroscopy techniques, such as surface-enhanced resonance Raman (SERRS) [1–3] and surface-enhanced infrared absorption (SEIRA) [4–6]. Electrochemical quartz crystal microbalance (EQCM) and surface plasmon resonance (SPR) measurements also have been used to study dynamic molecular adsorption processes [7–9]. These techniques provide excellent opportunities to study redox protein mechanisms and provide overall information concerning the interface between proteins and an electrode; however, the local structure is not obtained, due to a lack of spatial resolution. Moreover, sometimes spectroscopic approaches suffer from the limitation where only protein chromophores are observed. Thus far, ex- or in-situ AFM imaging with electrochemical measurements have been employed to characterize protein adsorption properties on electrodes [10–12]. Furthermore, integrated AFM and scanning electrochemical microscopy (SECM) were developed to obtain in-situ information on electrochemical process at a liquid-solid interface [13, 14]. However, due to a limitation of temporal resolution of conventional AFM, it was difficult to directly monitor rapid protein adsorption processes and the relevant electrochemical processes. In contrast, high-speed atomic force microscopy (HS-AFM) was recently established, which allowed us to directly visualize several dynamic protein events, such as structural changes, molecular interactions, diffusion processes, and molecule adsorption onto a solid surface [15–20].

In this paper, we report the development of an electrochemical AFM based on HS-AFM, which was constructed in the laboratory and can analyze electrochemical measurements with the advantage of fast imaging. This approach could allow direct visualization of dynamic biological molecule behavior at the electrode with simultaneous electrochemical measurements. We applied electrochemical HS-AFM to observe the cytochrome c (cyt c) electrochemical reaction and adsorption processes on a self-assembled monolayers (SAMs)-modified Au electrode surface. Cyt c plays a crucial role as an electron transfer protein in eukaryotic and many prokaryotic respiratory chains. Cyt c is the most extensively studied protein in bioelectrochemistry [21] because cyt c is often regarded as a model metalloprotein for studying electron transfer within and between proteins. It is well known that cyt celectrostatically immobilizes on SAMs

Citation: Takeda K, Uchihashi T, Watanabe H, Ishida T, Igarashi K, Nakamura N, et al. (2015) Real-Time Dynamic Adsorption Processes of Cytochrome c on an Electrode Observed through Electrochemical High-Speed Atomic Force Microscopy. PLoS ONE 10(2): e0116685. doi:10.1371/journal.pone.0116685.

of carboxylic acid-terminated alkanethiols via lysine residues surrounding the cyt c heme crevice and exhibits a reversible voltammetric response [22–27]. Numerous studies on direct electrochemistry of cyt c adsorbed on a SAM-modified [25] electrode have employed various surface analysis techniques [5, 9, 11, 28–33]. Until now, however, cyt c molecules adsorbed on electrodes have not been directly observed through real-time imaging using microscopy techniques. This study demonstrates, to our knowledge, for the first time real-time visualization of cyt c adsorption processes on an 11-mercaptoundecanoic acid (MUA)-modified Au electrode with simultaneous electrochemical measurements.

MATERIALS AND METHODS

Preparation of the SAM-Modified Gold Electrode

A gold (111) substrate was used as the working electrode. The gold was deposited on a freshly cleaved mica substrate through vapor deposition and immersed in a 1 μM ethanol solution composed of MUA (Sigma) for 16 h at room temperature after flame annealing. The resulting gold (111)-mica substrates were thoroughly rinsed with ethanol and a 10 mM sodium phosphate buffer solution (pH 7.0) to remove the physically adsorbed thiol molecules. The SAM-modified gold (111)-mica substrate was affixed to the HS-AFM glass sample stage using an electricity-conducting adhesive.

The Electrochemical HS-AFM Measurements

We used a laboratory-constructed high-speed AFM apparatus [15, 16]. The cantilevers (Olympus) used were 6–7 μm long, 2 μm wide, and 90 nm thick. The spring constant was 0.1–0.2 N/m, and the resonant frequency and quality factor in an aqueous solution were 0.7–1 MHz and ~2, respectively. For AFM imaging, the free oscillation amplitude was ~2 nm, and the set-point amplitude was approximately 90% of the free oscillation amplitude. An amorphous carbon tip was grown on the original tip through electron-beam deposition. For simultaneous electrochemical measurements, a Pt wire counter electrode and Ag wire reference electrode were incorporated into the HS-AFM cantilever holder. The electrochemical experiments were conducted using an ALS electrochemical analyzer (Model 1200b). The potentials cited in this paper refer to the Ag wire electrode with the potential −150 mV vs. Ag/AgCl (3 M NaCl) in 10 mM sodium phosphate buffer solution (pH 7.0). We performed the experiment to determine the potential of the Ag wire vs. Ag/AgCl by using 2,6-dichlorophenolindophenol. This experiment was conducted under equal conditions (10 mM sodium phosphate buffer solution at pH 7.0). The current densities were calculated with respect to the geometric electrode area 1.76 × 10^{-6} m^2 (1.5 mm diameter gold (111)-mica substrates). The measurements were collected at room temperature. Before the measurements, oxygen was purged from the buffer solution through bubbling with highly purified nitrogen for 30 min. The electrochemical HS-AFM (EC-HS-AFM) measurements began when a 0.5 μL aliquot of 1 mM cyt c from horse heart (Sigma, USA) was added to a 10 mM sodium phosphate buffer solution (pH 7.0) at the total volume 70.5 μL. The final cyt cconcentration was 7.1 μM.

RESULTS AND DISCUSSION

Cyclic voltammetric measurements using our newly constructed EC-HS-AFM were performed to confirm the redox behavior of the 0.1 mM 2,6-dichlorophenolindophenol (DCPIP) in 50 mM pH 7.0 HEPES buffer with a gold electrode. Because reversible redox peaks from DCPIP were observed in a diffusion system, EC-HS-AFM was established as the potential control for the sample stage.

Fig. 1A shows an AFM image of the MUA-modified gold electrode and a cross-section profile with an arbitrarily configured line in the image. The SAM-modified gold surface was flat with a well-packed monolayer and confirmed small pits. This image was consistent with typical AFM images of 11-mercaptoundecanol or MUA SAMs on a Au (111) substrate [34]. The EC-HS-AFM measurements were collected on the MUA-modified gold electrode as the sample stage after cyt c was injected into the analytical solution. Fig. 1 presents AFM images of cyt c molecules adsorbing on the MUA-modified gold surface collected at 2 frame s^{-1} . The cyt c molecules began to adsorb on the MUA-modified gold surface approximately 340 s after injection. The data generated from the time cyt c was added through the beginning of adsorption might be due to cyt c molecules approaching the nearby surface through diffusion. Randomly adsorbing cyt c molecules on the electrode were gradually observed from 340 s to 430 s. Cyt c molecules also frequently disappear immediately after adsorption during this period, which suggests that cyt c molecules rapidly move onto the surface when a small number of molecules adsorb. Next, rapid and remarkable cyt c molecule adsorption was observed from 430 s to 490 s followed by a full adsorption on the electrode surface after 500 s. After 500 s, no noticeable morphologically changes were observed in the AFM images likely because the electrode was full covered by molecules, and the cyt cmolecules did not form multilayers.

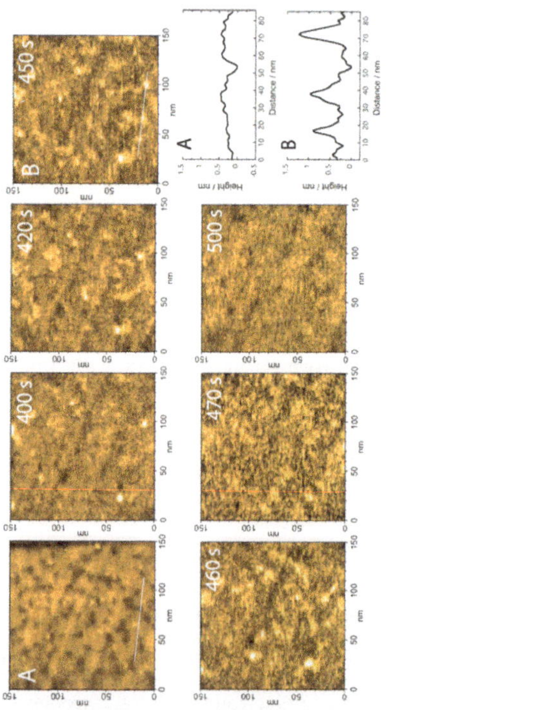

Figure 1: AFM images show (A) the MUA-modified gold surface and (B) cyt cadsorbed on the MUA SAM at 450 sec.

Continuous AFM images of adsorbing cyt c molecules with real-time labels. Frame rate, 2 frames/s; image area, 150 × 150 nm^2. The cross sections of the images in (A) and (B), along the short white line, are shown in the lower right.

http://dx.doi.org/10.1371/journal.pone.0116685.g001

Cyclic voltammograms (CVs) were sequentially collected while cyt c adsorption was monitored using EC-HS-AFM and are shown from 337.5 s to 517.5 s in Fig. 2. CVs for saturated cyt cadsorption on the gold electrode with an MUA SAM showed the cathodic and anodic peak potentials −122 and −102 mV (vs. Ag wire), respectively, at 522 s. The cyt c midpoint potential was approximately −112 mV (vs. Ag wire). As depicted in Fig. 2, current redox peaks for cyt cwere also observed and increased from 340 s to 480 s simultaneously with the cyt cadsorption processes in the AFM images. The peak currents were steady after 500 s. An EC-HS-AFM experiment of the SAM unmodified gold electrode (bare gold surface) was conducted for comparisons of the MUA-modified gold electrode.

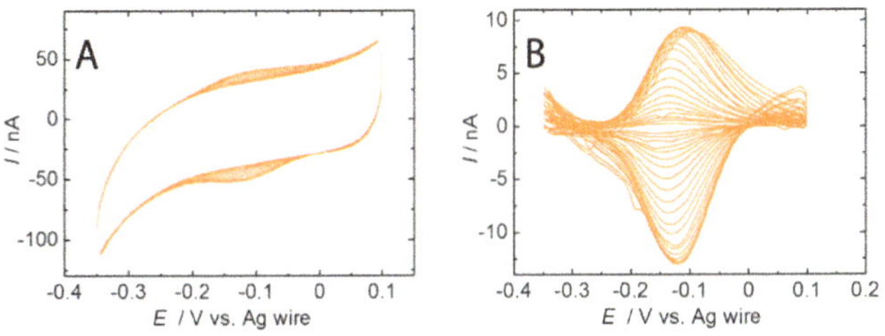

Figure 2: CVs of cyt c **adsorbed on the electrode.**

(A) CVs of cyt c molecules adsorbed on the MUA electrode in a 10 mM phosphate buffer solution (pH 7.0) from 333 to 522 s (from −0.35 V to 0.1 V, each segment is 4.5 s). (B) Background-subtracted CVs from the voltammogram from 313 to 324 s and 333 to 522 s (from inside to outside, each segment is 4.5 s). The voltammograms were collected at a scan rate of 100 mVs^{-1}. Ag wire was used as a reference electrode.

http://dx.doi.org/10.1371/journal.pone.0116685.g002

The root mean square (RMS) roughness R_{RMS}, which is given by the standard deviation of height, for the commonly used AFM images were used as roughness profile ordinates and plotted against time in Fig. 3A. Fig. 3B shows a schematic explanation of the cyt c adsorption processes in Fig. 3A. The R_{RMS} variations indicate that cyt c adsorbed on the electrode. The R_{RMS} values exponentially increased beginning at approximately 430 s and peaked at 470 s. At approximately the R_{RMS} peak time, cyt c adsorption was not saturated, but cyt c covered almost half of the sorbable sites on the electrode surface. After 470 s, theR_{RMS} values immediately decreased because more cyt c molecules adsorbed more on the electrode surface, which likely filled a gap between the immobilized cyt c molecules. We simulated the evolution time for the R_{RMS} values from random adsorption on a 50 × 50 site based on a change in the number of adsorbing molecules in proportion to the number of the active adsorption sites. The simulation model results show that R_{RMS} values increase logarithmically with increased surface coverage and reach a maximum at almost half of the coverage. In Fig. 3A, it is noteworthy that the R_{RMS} values an exponentially-increased with time, which indicates that the cyt c adsorption process has a cooperative effect. This result suggests that the cooperative effect includes potential interactions, such as electrostatic or hydrophobic interactions between the immobilized cyt c molecules. Reports show that protein adsorption is a cooperative process [35, 36]. The slight change in R_{RMS}values from approximately 500 s to 550 s indicates saturated cyt c adsorption. We conclude that this result supports protein monolayer formation on the MUA-modified gold electrode as suggested in the AFM. The number of the electrochemically active cyt c molecules adsorbed onto the electrode (Γ) was estimated by integrating the observed CV curve reduction peak using the following equation: $\Gamma = Q/nFA$. Q is the charge involved in the reaction, n is the number of electrons involved in the redox process (for cyt c-Fe(II) oxidation, n = 1), F is the Faraday constant, and A is the electrode surface area. As depicted in Fig. 3A, the time dependence of the Γ values was also plotted using a sigmoid and the R_{RMS} values; these results suggest cooperative cyt c adsorption. The value Γ at the steady state after 500 s was calculated as 11 pmol cm^{-2} (the largest value in several times), which is consistent with the experimental and theoretical values for full monolayer coverage at 13 pmol cm^{-2} [24, 29]. This value also agrees with that of the integrated the peak currents from CV, 9.1 pmol cm^{-2} [27]. In addition, Nakano et al. showed that the QCM analysis and AFM imaging supported the cyt cmonolayer formation on the MUA surface [11]. The finding of the cyt c monolayer formation agrees with our results of CV and HS-AFM experiments. The difference between our results and their report on the value of electroactive surface coverage could be explained as the way of cyt c immobilization. The cyt c was covalently immobilized on MUA surface in their report, while it was immobilized by the electrostatic interaction in our case. The QCM analysis in their report showed that the value of apparent cyt c surface concentration was 28 ± 12 pmol cm^{-2}, which was larger than those calculated from CVs (7.2 ± 4.8 pmol cm^{-2}) in their report and our data (11 pmol cm^{-2}). In the case of the QCM analysis, Furusawa et al. demonstrated that apparent larger frequency changes by the protein immobilization onto the gold surface in the aqueous solutions (ΔF_{water}) were observed compared with those in the dry air phases (ΔF_{air}) owing to the interaction with the surrounding water molecules [37]. Myoglobin (Mw: 16.9 kDa), for example, exhibited apparent mass ratio 2.1 [$(-\Delta F_{water})/(-\Delta F_{air})$]. It could happen in the case of cyt c.

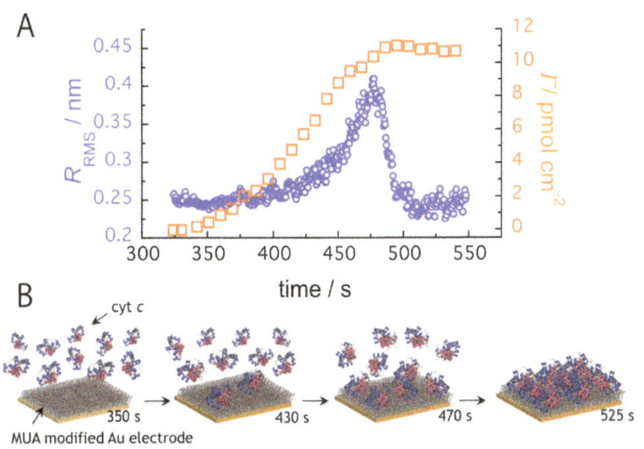

Figure 3: Time-course analysis of R_{RMS} **and** Γ **values.**

(A) Time evolution of R_{RMS} values from the HS-AFM images (open circle) and the level of electrochemically active cyt c (Γ) from the cyclic voltammograms (open square). (B)

Schematic of the adsorbing cyt c molecules on the MUA-modified electrode at each time point.

http://dx.doi.org/10.1371/journal.pone.0116685.g003

The Γ values compared with the R_{RMS} in Fig. 3A show that both parameters simultaneously increased until 470 s. At approximately 500 s, the Γ values reached the maximum and theR_{RMS} was returned to the initial value, respectively. On the other hand, although the Γ values began to increase at approximately ~400 s, the R_{RMS} increased after ~400 s. This observation is presumably due to unimmobilized cyt c molecules on the electrode that were not imaged using HS-AFM because the movement was too rapid, but these molecules which are located adjacent the electrode could exhibit the electrochemical response.

In addition, the peak potential positively shifted with the increasing Γ values. A change in cyt c redox potential is induced by several factors, the heme axial ligands [38], the heme environment polarity [39], and the heme solvent accessibility [40] as well as due to the electrostatic interactions between heme propionates and amino acid residues [41, 42]. Recent research has shown that heme propionate hydration shifts the reduction potential of approximately 50 mV toward more positive values [42]. When there is more hydrophobicity and a localized structural change due to well-packed cyt c adsorption on the electrode with MUA SAM, a positive cyt c redox potential shift should be observed that originates with the environment surrounding the heme. The localized structural change of the environment surrounding the heme could modulate several factors that affect redox potential. Further studies are necessary to discern the heme conditions using spectroscopic techniques.

For cyt c adsorption on the saturated electrode, 20 μL of the electrolyte solution of EC-HS-AFM was extracted, and 20 μL of a 4 M NaCl solution was injected into the analytical solution (NaCl at the final concentration 1.13 M) to generate an analytical solution with a high salt concentration. Cyt c molecules desorbed from the MUA-modified gold electrode surface immediately after adding the NaCl solution as shown Fig. 4A . Most of the cyt cmolecules desorbed, and the surface with exposed MUA SAMs was observed after 120 s. Similarly, the R_{RMS} decreased at approximately 120 s but was steady after 120 s as shownFig. 4B. Certain ordinates of the points were unusual as shown in Fig. 4B; they are due to spike-like noise in AFM images with an abrupt increase in ionic strength. Fig. 4C shows a schematic representation of the desorbing cyt c molecules on the MUA SAM-modified electrode. The results indicate that cyt c molecules desorb due to weaker electrostatic adsorption on electrode-coated MUA SAMs at higher ionic strengths.

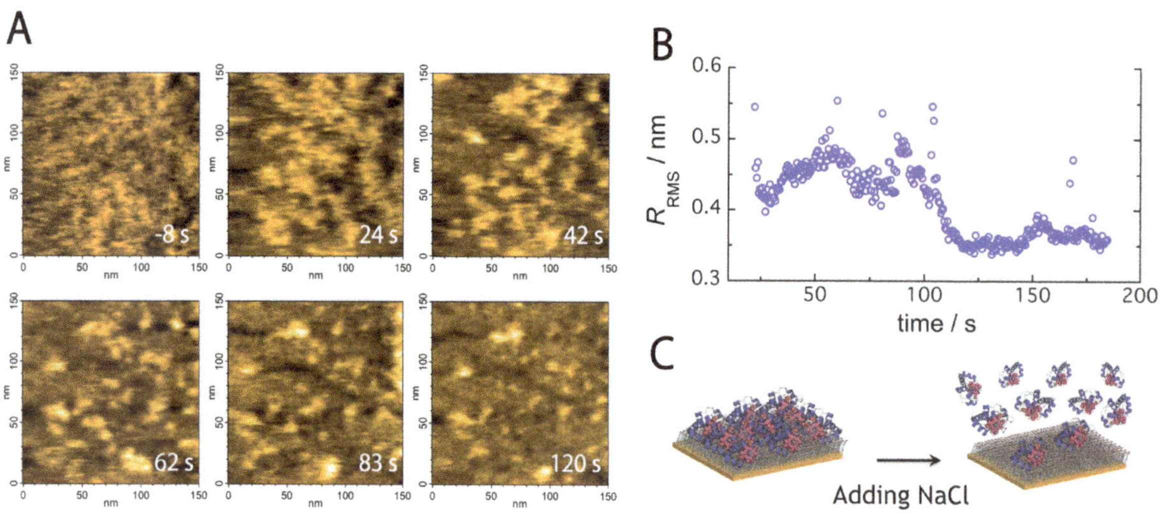

Figure 4: Real-time cyt c **desorption processes from the MUA-modified gold electrode.**

(A) Continuous AFM images of desorbing cyt c molecules. Frame rate, 2 frames/s; image area, 150 × 150 nm^2. (B) Time evolution of R_{RMS} values at the higher ionic strengths. (C) Schematic of desorbing cyt c molecules from the MUA-modified electrode.

http://dx.doi.org/10.1371/journal.pone.0116685.g004

CONCLUSIONS

In summary, we altered the HS-AFM sample stage to control the electrical potentials and constructed EC-HS-AFM equipment that simultaneously allows direct visualization of dynamic behavior with nanoscale resolution and electrochemical measurements. Using this EC-HS-AFM approach, this study demonstrates, to our knowledge, for the first time direct observation of adsorption processes and the cyt c electron transfer reaction on a SAM-modified gold electrode in real-time. Cyt c molecules were fully adsorbed at approximately 160 s with increasing redox currents from cyt c. The details of cooperative adsorption have not been clarified; however, the findings herein could lead to new, in-depth discoveries on the cyt cadsorption processes. EC-HS-AFM also enables us to directly visualize the structure and dynamics of molecules in response to an applied potential. This use of EC-HS-AFM significantly facilitates studies aimed at understanding the dynamic behavior of biomolecules on an electrode interface and contributes to the further development of bioelectronics.

ACKNOWLEDGMENTS

The authors are grateful to Ms Patricia McGahan of the Tokyo University of Agriculture and Technology for English proofreading during preparation of this paper.

AUTHOR CONTRIBUTIONS

Conceived and designed the experiments: HO TU KI NN. Performed the experiments: KT TU TI. Analyzed the data: KT TU KI NN. Contributed reagents/materials/analysis tools: KT TU HW KI NN. Wrote the paper: KT TU NN.

REFERENCES

1. Cotton TM, Schultz SG, Van Duyne RP (1980) Surface-enhanced resonance raman-scattering from cytochrome c and myoglobin adsorbed on a silver electrode. J Am Chem Soc 102: 7960–7962. doi: 10. 1021/ja00547a036
2. Murgida DH, Hildebrandt P (2008) Disentangling interfacial redox processes of proteins by SERR spectroscopy. Chem Soc Rev 37: 937–945. doi: 10. 1039/b705976k. pmid:18443679
3. Murata K, Suzuki M, Nakamura N, Ohno H (2009) Direct evidence of electron flow via the heme c group for the direct electron transfer reaction of fructose dehydrogenase using a silver nanoparticle-modified electrode. Electrochem Commun 11: 1623–1626. doi: 10. 1016/j. elecom. 2009. 06. 012
4. Ataka K, Giess F, Knoll W, Naumann R, Haber-Pohlmeier S, et al. (2004) Oriented attachment and membrane reconstitution of His-tagged cytochrome c oxidase to a gold electrode: in situ monitoring by surface-enhanced infrared absorption spectroscopy. J Am Chem Soc 126: 16199–16206. doi: 10. 1021/ja045951h. pmid:15584756
5. Wisitruangsakul N, Zebger I, Ly KH, Murgida DH, Ekgasit S, et al. (2008) Redox-linked protein dynamics of cytochrome c probed by time-resolved surface enhanced infrared absorption spectroscopy. Phys Chem Chem Phys 10: 5276–5286. doi: 10. 1039/b806528d. pmid:18728870
6. Ataka K, Heberle J (2007) Biochemical applications of surface-enhanced infrared absorption spectroscopy. Anal Bioanal Chem 388: 47–54. doi: 10. 1007/s00216-006-1071-4. pmid:17242890
7. Goda T, Miyahara Y (2012) Interpretation of protein adsorption through its intrinsic electric charges: a comparative study using a field-effect transistor, surface plasmon resonance, and quartz crystal microbalance. Langmuir 28: 14730–14738. doi: 10. 1021/la302977s. pmid:22992058
8. Fleming BD, Praporski S, Bond AM, Martin LL (2008) Electrochemical quartz crystal microbalance study of azurin adsorption onto an alkanethiol self-assembled monolayer on gold. Langmuir 24: 323–327. doi: 10. 1021/la702511w. pmid:18041855
9. Paulo Tde F, de Sousa TP, de Abreu DS, Felicio NH, Bernhardt PV, et al. (2013) Electrochemistry, surface plasmon resonance, and quartz crystal microbalance: an associative study on cytochrome c adsorption on pyridine tail-group monolayers on gold. J Phys Chem B 117: 8673–8680. doi: 10. 1021/jp400694k. pmid:23819458
10. Boussaad S, Tao NJ (1999) Electron transfer and adsorption of myoglobin on self-assembled surfactant films: An electrochemical tapping-mode AFM study. J Am Chem Soc 121: 4510–4515. doi: 10. 1021/ja990117r
11. Nakano K, Yoshitake T, Yamashita Y, Bowden EF (2007) Cytochrome c self-assembly on alkanethiol monolayer electrodes as characterized by AFM, IR, QCM, and direct electrochemistry. Langmuir 23: 6270–6275. doi: 10. 1021/la063697w. pmid:17461603
12. Zhang L, Jiang X, Wang E, Dong S (2005) Attachment of gold nanoparticles to glassy carbon electrode and its application for the direct electrochemistry and electrocatalytic behavior of hemoglobin. Biosens Bioelectron 21: 337–345. doi: 10. 1016/j. bios. 2004. 10. 021. pmid:16023961
13. Kueng A, Kranz C, Lugstein A, Bertagnolli E, Mizaikoff B (2003) Integrated AFM-SECM in tapping mode: simultaneous topographical and electrochemical imaging of enzyme activity. Angew Chem Int Ed Engl 42: 3238–3240. doi: 10. 1002/anie. 200351111. pmid:12876731
14. Kranz C, Kueng A, Lugstein A, Bertagnolli E, Mizaikoff B (2004) Mapping of enzyme activity by detection of enzymatic products during AFM imaging with integrated SECM-AFM probes. Ultramicroscopy 100: 127–134. doi: 10. 1016/j. ultramic. 2003. 10. 004. pmid:15231302
15. Ando T, Kodera N, Takai E, Maruyama D, Saito K, et al. (2001) A high-speed atomic force microscope for studying biological macromolecules. Proc Natl Acad Sci USA 98: 12468–12472. doi: 10. 1073/pnas. 211400898. pmid:11592975
16. Uchihashi T, Kodera N, Ando T (2012) Guide to video recording of structure dynamics and dynamic processes of proteins by high-speed atomic force microscopy. Nat protoc 7: 1193–1206. doi: 10. 1038/nprot. 2012. 047. pmid:22635111
17. Shibata M, Yamashita H, Uchihashi T, Kandori H, Ando T (2010) High-speed atomic force microscopy shows dynamic molecular processes in photoactivated bacteriorhodopsin. Nat Nanotechnol 5: 208–212. doi: 10. 1038/nnano. 2010. 7. pmid:20154686
18. Uchihashi T, Iino R, Ando T, Noji H (2011) High-speed atomic force microscopy reveals rotary catalysis of rotorless F_1-ATPase. Science 333: 755–758. doi: 10. 1126/science. 1205510. pmid:21817054
19. Igarashi K, Uchihashi T, Koivula A, Wada M, Kimura S, et al. (2011) Traffic jams reduce hydrolytic efficiency of cellulase on cellulose surface. Science 333: 1279–1282. doi: 10. 1126/science. 1208386. pmid:21885779
20. Ando T, Uchihashi T, Kodera N (2013) High-speed AFM and applications to biomolecular systems. Annu Rev Biophys 42: 393–414. doi: 10. 1146/annurev-biophys-083012-130324. pmid:23541159
21. Fedurco M (2000) Redox reactions of heme-containing metalloproteins: dynamic effects of self-assembled monolayers on thermodynamics and kinetics of cytochromec electron-transfer reactions. Coordin Chem Rev 209: 263–331. doi: 10. 1002/chin. 200104296
22. Tarlov MJ, Bowden EF (1991) Electron-transfer reaction of cytochrome c adsorbed on carboxylic acid terminated alkanethiol monolayer electrodes. J Am Chem Soc 113: 1847–1849. doi: 10. 1021/ja00005a068

23. Song S, Clark RA, Bowden EF, Tarlov MJ (1993) Characterization of cytochromec/alkanethiolate structures prepared by self-assembly on gold. J Phys Chem 97: 6564–6572. doi: 10. 1021/j100126a037

24. Avila A, Gregory BW, Niki K, Cotton TM (2000) An electrochemical approach to investigate gated electron transfer using a physiological model system: cytochrome cimmobilized on carboxylic acid-terminated alkanethiol self-assembled monolayers on gold electrodes. J Phys Chem B 104: 2759–2766. doi: 10. 1021/jp992591p

25. Wei J, Liu H, Khoshtariya DE, Yamamoto H, Dick A, et al. (2002) Electron-transfer dynamics of cytochrome c: a change in the reaction mechanism with distance. Angew Chem Int Ed Engl 41: 4700–4703. doi: 10. 1002/anie. 200290021. pmid:12481331

26. Niki K, Hardy WR, Hill MG, Li H, Sprinkle JR, et al. (2003) Coupling to lysine-13 promotes electron tunneling through carboxylate-terminated alkanethiol self-assembled monolayers to cytochrome c. J Phys Chem B 107: 9947–9949. doi: 10. 1021/jp035392l

27. Jin B, Wang GX, Millo D, Hildebrandt P, Xia XH (2012) Electric-field control of the pH-dependent redox process of cytochrome c immobilized on a gold electrode. J Phys Chem C 116: 13038–13044. doi: 10. 1021/jp303740e

28. Fujita K, Suzuki C, Ohno H (2003) Non-contact spectral analysis of cytochrome c on carbon electrodes with optical waveguide spectroscopy. Electrochem Commun 5: 47–50. doi: 10. 1016/s1388-2481(02)00531-3

29. Feng ZQ, Imabayashi S, Kakiuchi T, Niki K (1997) Long-range electron-transfer reaction rates to cytochrome c across long- and short-chain alkanethiol self-assembled monolayers: electroreflectance studies. J Chem Soc Faraday Trans 93: 1367–1370. doi: 10. 1039/a605567b

30. Murgida DH, Hildebrandt P (2001) Active-site structure and dynamics of cytochrome cimmobilized on self-assembled monolayers—A time-resolved surface enhanced resonance raman spectroscopic study. Angew Chem Int Ed Engl 40: 728–731. pmid:11241605 doi: 10. 1002/1521-3773(20010216)40:4<728::aid-anie7280>3. 3. co;2-g

31. Hobara D, Imabayashi S, Kakiuchi T (2002) Preferential adsorption of horse heart cytochrome c on nanometer-scale domains of a phase-separated binary self-assembled monolayer of 3-mercaptopropionic acid and 1-hexadecanethiol on Au(111). Nano Lett 2: 1021–1025. doi: 10. 1021/nl0256863

32. Ataka K, Heberle J (2004) Functional vibrational spectroscopy of a cytochrome cmonolayer: SEIDAS probes the interaction with different surface-modified electrodes. J Am Chem Soc 126: 9445–9457. doi: 10. 1021/ja048346n. pmid:15281838

33. Murgida DH, Hildebrandt P (2004) Electron-transfer processes of cytochrome c at interfaces. New insights by surface-enhanced resonance raman spectroscopy. Acc Chem Res 37: 854–861. doi: 10. 1021/ar0400443. pmid:15612675

34. Hiasa T, Kimura K, Onishi H (2012) Hydration of hydrophilic thiolate monolayers visualized by atomic force microscopy. Phys Chem Chem Phys 14: 8419–8424. doi: 10. 1039/c2cp40252a. pmid:22509496

35. Luo Q, Andrade JD (1998) Cooperative adsorption of proteins onto hydroxyapatite. J Colloid Interf Sci 200: 104–113. doi: 10. 1006/jcis. 1997. 5364

36. Deng J, Sun M, Zhu J, Gao C (2013) Molecular interactions of different size AuNP-COOH nanoparticles with human fibrinogen. Nanoscale 5: 8130–8137. doi: 10. 1039/c3nr02327c. pmid:23884371

37. Furusawa H, Ozeki T, Morita M, Okahata Y (2009) Added mass effect on immobilizations of proteins on a 27 MHz quartz crystal microbalance in aqueous solution. Anal Chem 81: 2268–2273. doi: 10. 1021/ac802412t. pmid:19228044

38. Mao J, Hauser K, Gunner MR (2003) How cytochromes with different folds control heme redox potentials. Biochemistry 42: 9829–9840. doi: 10. 1021/bi027288k. pmid:12924932

39. Kassner RJ (1973) A theoretical model for the effects of local nonpolar heme environments on the redox potentials in cytochromes. J Am Chem Soc 95: 2674–2677. doi: 10. 1021/ja00789a044. pmid:4348492

40. Stellwagen E (1978) Haem exposure as the determinate of oxidation-reduction potential of haem proteins. Nature 275: 73–74. doi: 10. 1038/275073a0. pmid:683346

41. Moore GR (1983) Control of redox properties of cytochrome c by special electrostatic interactions. FEBS letters 161: 171–175. doi: 10. 1016/0014-5793(83)81001-1. pmid:6311622

42. Bortolotti CA, Amadei A, Aschi M, Borsari M, Corni S, et al. (2012) The reversible opening of water channels in cytochrome c modulates the heme iron reduction potential. J Am Chem Soc 134: 13670–13678. doi: 10. 1021/ja3030356. pmid:22873369

CHAPTER 18

TOWARDS A NOVEL BIOELECTROCATALYTIC PLATFORM BASED ON "WIRING" OF PYRROLOQUINOLINE QUINONE-DEPENDENT GLUCOSE DEHYDROGENASE WITH AN ELECTROSPUN CONDUCTIVE POLYMERIC FIBER ARCHITECTURE

Johannes Gladisch, David Sarauli, Daniel Schäfer, Birgit Dietzel, Burkhard Schulz & Fred Lisdat

INTRODUCTION

The organic nature of conductive electroactive polymers such as polyaniline, polypyrrole or polythiophene provides a good basis for the construction of electrodes that enable efficient communication with biosystems[1,2,3]. As a consequence of this fact they became significantly important as materials for the development of biosensors, since they can serve as immobilization matrices or represent transducing layers, the conductivity of which can be modified by a binding event[4,5,6]. Moreover, they can represent labels for binding partners or act as molecular wiring agents connecting biocatalysts with electrodes[4,5,6].

Particularly polyanilines have been devoted attention since they show some advantages over other conducting polymers. They are of low-cost and easy to synthesize, being environmentally and thermally stable[7,8,9]. Furthermore, they can form dense, ordered or differently structured films, which possess a defined charge and are capable of binding or entrapment of biomolecules[10,11,12,13,14]. Customizing polyanilines' properties by introducing of functional groups, e.g. upon sulfonation, one could not only increase their conductivity and solubility, but also improve their redox activity over a wide pH range, compared to unsubstituted polyanilines[7]. While sulfonated polyanilines are soluble in majority of solvents, unsubstituted ones often require additional substances for a proper processability[7]. In addition, the negative charge of the sulfonic acid group may be advantageous for the interaction with proteins or enzymes[13]. Several studies suggest that sulfonated polyanilines provide a suitable environment for the immobilization of redox proteins such as cytochrome c[15,16,17] and hemoglobin[18] as well as various redox enzymes, as e.g. sulfite oxidase[19], peroxidase[20,21], urease[22], glucose oxidase[23], or tyrosinase[24]. For some enzymes they also allow a wiring of the enzyme with the electrode[25,26,27].

As electrochemical biosensor properties are dependent on the overall sensing surface, an enlarged electrode area is desirable. The enlargement can be achieved, for instance, by roughening the electrode or by creating three-dimensional structures. For this purpose different techniques are applicable, including chemical vapor deposition, electro-polymerization or immobilization of nanoparticles. A rather new technique with this respect is electrospinning, by means of which

Citation: Gladisch, J. et al. Towards a novel bioelectrocatalytic platform based on "wiring" of pyrroloquinoline quinone-dependent glucose dehydrogenase with an electrospun conductive polymeric fiber architecture. Sci. Rep. 6, 19858; doi: 10.1038/srep19858 (2016).

vastly enlarged surfaces can be achieved by the fabrication of architectures consisting of polymer fibers with diameters ranging from a few hundred nanometers to micrometers[28,29]. Many applications for electrospun fiber mats from medical diagnostics, and tissue engineering to materials for super capacitors or air filters have already been shown[30,31,32,33]. Also examples for biosensor applications of electrospun fiber mats have been reported[34,35,36,37]. However until now the polymeric fiber networks are mainly used as an interface to accommodate a large amount of molecules. Therefore, it would be desirable to develop the application of the technique further in order to combine biomolecule fixation with the electrochemical connection to the electrode.

Common polymers applied in electrospinning are polyacrylonitrile (PAN), polyethylene oxide (PEO) or polyvinyl alcohol. Modification of the fiber properties can not only be achieved by spinning mixtures of different polymers[30,38,39], but also by addition of small particle-type of materials such as e.g carbon nanotubes to the polymer solution to be spun. The latter approach is also relevant when high conductivity is desired[38,40,41]. In this study we demonstrate the fabrication of conducting 3D polymer fiber networks by electrospinning of 3 different chemically synthesized sulfonated polyanilines together with non-conducting PAN. These electrodes have been initially characterized by means of microscopy and electrochemistry of a small-size, inorganic redox couple. Subsequently, the enzyme pyrroloquinoline quinone-dependent glucose dehydrogenase (PQQ-GDH) has been covalently bound to the polymer electrodes and the shuttle-free bioelectrocatalytic behavior of the system has been investigated.

RESULTS AND DISCUSSION

Sulfonated polyanilines have been chosen for the electrospinning process because of their conductive properties and since the polymers directly interact with the enzyme PQQ-GDH allowing direct electron transfer between the substrate reduced enzyme and the polymer[42]. Three different sulfonated polyanilines (PABSA [poly(3-aminobenzoic acid-co-3-aminobenzenesulfonic acid)], PAPASA [poly(3-aminobenzeneacetic acid-co-3-aminobenzenesulfonic acid)], PANABMSA [poly(3-aminobenzoic acid-co-2-methoxyaniline-5-sulfonic acid-co-aniline)], Fig. 1) have been used. Apart from sulfonic acid groups, the polymers exhibit carboxylic acid groups, which allow a covalent fixation of the enzyme PQQ- GDH. Differences among the polymers occur due to the presence of different substitutions on the aniline ring (namely methoxy-, ethylcarboxy- and carboxy- groups) or the presence of unsubstituted aniline during copolymerisation.

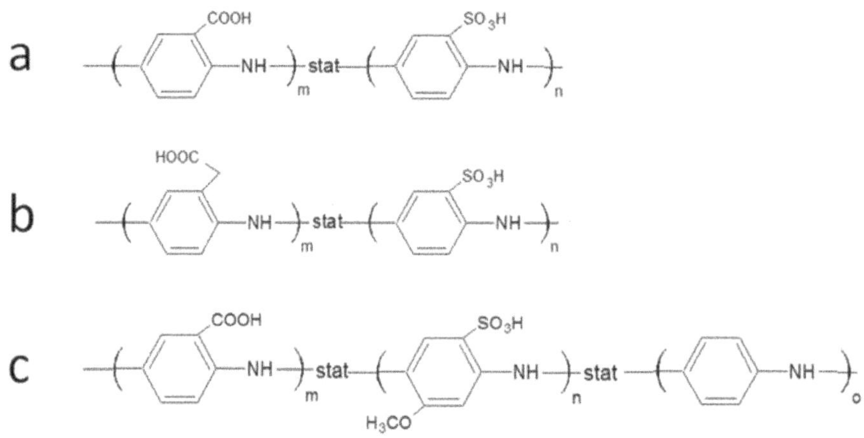

Figure 1: Chemical structure of different sulfonated polyanilines used: (a) PABSA, (b) PAPASA, (c) PANABMSA.

Characterization of the Copolymers

The synthesized polymers have been characterized by means of gel permeation chromatography (GPC), thermogravimetral analysis (TGA), and IR spectroscopy. The results are presented in Table 1, and the IR spectra are shown in Fig. 2. There, the bands in the range of 3000–3500 cm^{-1} correspond to stretching vibration N-H bonds for all polymers (not shown). The most prominent absorption band of the carbonyl group of carboxylic acids is found between 1704 cm^{-1} and 1689 cm^{-1}. Further bands at 1558–1585 cm^{-1} and 1488–1506 cm^{-1} correspond to the ring stretching vibration of the quinoid and benzenoid rings. All polymers show a more intense benzenoid band compared to the quinoid band. The C-N stretching mode gives rise to a band near 1448–1451 cm^{-1}. The band at 1293–1300 cm^{-1} is related to the C-N stretching vibration of aromatic amines associated with para-linked aniline units[43]. The bands observed at 1143–1158 cm^{-1} are primarily due to the aromatic C-H in-plane deformation vibration. The intense band near 1031 cm^{-1} is attributed as the symmetric stretching vibration of S=O. The sharp bands at 612–615 cm^{-1} and 687–693 cm^{-1} are associated with the C-S and S=O stretching vibrations. Bands of the C-H out-of-plane bending vibrations at 869–878 cm^{-1}, 825–830 cm^{-1} and 796–798 cm^{-1}corresponding to 1,2,4 and 1,4-substituted benzene rings indicating that the monomers in the polymers are connected in head-to-tail fashion[44]. The presence of tetra1,2,4,5 substituted benzene rings has been identified by the appearance of characteristic bands at 851 cm^{-1}

Table 1: Composition, sulfonation grade, yields, molecular weights and the polydispersity index (PDI) values of the copolymers used in this study

Polymers	C [%]	H [%]	N [%]	S [%]	S/N	Yield [%]	M_n [g/mol]	M_w [g/mol]	PDI
PAPASA	49,88	3,86	7,22	3,66	0,22	30	7500	11000	2,06
PANABMSA	51,19	3,78	9,12	2,58	0,12	42	n. d.	n. d.	n. d.
PABSA	48,79	3,88	7,72	4,24	0,24	19	8400	11200	1,30

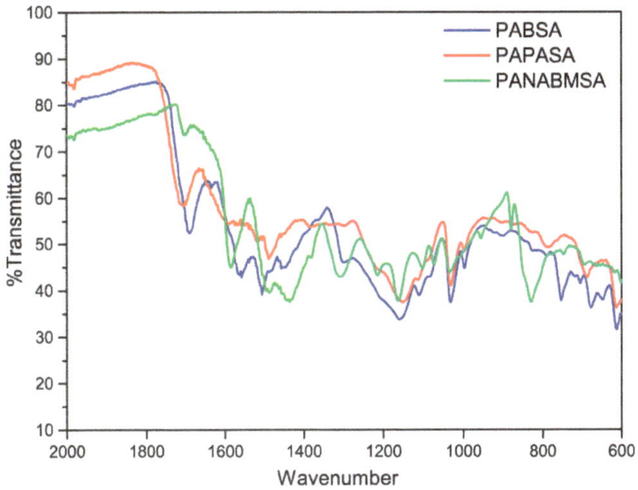

Figure 2: IR-spectra of PABSA, PAPASA and PANAMBSA polymers.

Fabrication and Characterization of the 3D Electrodes

Attempting to electrospin the polyanilines, we have found that the used polyanilines cannot be processed in electrospinning alone, since electrospraying occurs, probably due to relatively low viscosity and molecular weight of these polymers. Thus, mixtures of sulfonated polyanilines and polyacrylonitrile (PAN) have been used for the electrode fabrication. In order to get a 3D network allowing both electrochemical and optical investigations, the electrospun fiber networks have beenprepared on planar indium tin oxide (ITO) electrodes. After preparation they have been initially characterized by light microscopy in order to check the general quality of polymer networks and to evaluate their thickness (see Supporting Information, Figure S1). Subsequently, the polymer electrodes are characterized by means of scanning electron microscopy.

Figure 3 presents light microscopy and SEM images of PAN/PAPASA, PAN/PANAMBSA and PAN/PABSA composites fabricated from a mixed polymer solution in DMSO. Prior to the spinning process all polymer-containing solutions have been ultrasonicated. Once dispersed, they remain stable in DMSO for at least 3 weeks; neither sedimentation nor aggregation of polymers have been observed. A flow rate of 2 µl/min is chosen for the electrospinning procedure at almost no dripping. Depending on the chosen polyaniline, the polymer fiber mats show a brownish (Figure S1: a1 - PAN/PAPASA and c1 - PAN/PABSA) or green-bluish (Figure S1: b1 - PAN/PANABMSA) color, which is in agreement with the different redox states (and thus color) of the polyanilines after synthesis. The polymer fiber networks have found to be uniform in their appearance. This is valid for the mats from different polymeric blends and also among multiple electrodes modified with the same polymer mix. Only in some few cases fiber agglomeration could be observed - particularly for fibers made of PAN and PANABMSA, most probably because of the higher viscosity of PANAMBSA compared to other polymers.

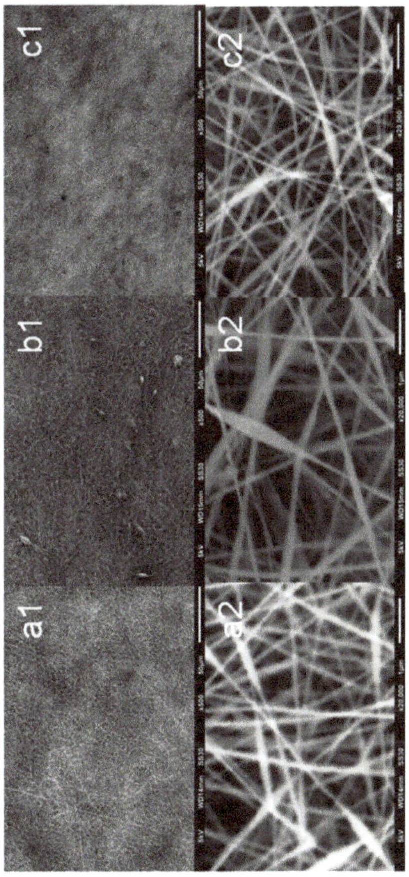

Figure 3: SEM images of the fiber morphology in the electrospun mats of (a) ITO-PAN/PAPASA, (b) ITO-PAN/PANABMSA, (c) ITO-PAN/PABSA, whereas 1 and 2 correspond to different magnifications used (with lines represent 50 µm and 1 µm, respectively).

Full size image

Figure 4 shows a cut edge of a PAN/PABSA polymer fleece prepared in a 15 min electrospinning process indicating a proper mat thickness in the range of few micrometers. The averaged diameters of the fibers are shown in Table 2. In case of PAN/PANABMSA the mean diameters as well as the standard deviations are slightly larger. Moreover, even when the fibers of all polymer mixtures show a slightly beaded structure, the averaged bead diameters were significantly larger in case of PAN/PANABMSA. As the bead formation is influenced by many parameters, especially concentration and voltage, the relatively high voltage of 25 kV might be the reason for bead formation45·46. As for the interaction with water, all mats showed a sponge-like behavior. For instance, when contact angle measurements are conducted, the polymer fiber mats give small angles mostly about 30 °.

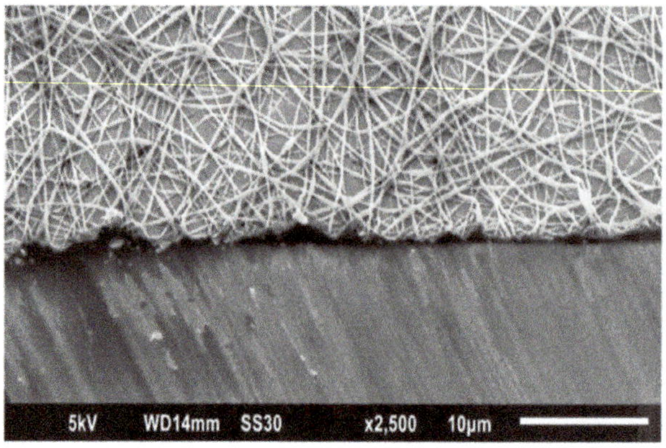

Figure 4: SEM image of an ITO-PAN/PABSA mat after 15 minutes electrospinning.

Table 2: Averaged diameter and bead diameter of different fiber mats as a result of the characterization by electron microscopy (n=3)

	ITO-PAN/PAPASA	ITO-PAN/PANABMSA	ITO-PAN/PABSA
diameter [µm]	0,13 ± 0,02	0,15 ± 0,04	0,12 ± 0,02
bead diameter [µm]	0,31 ± 0,08	0,51 ± 0,15	0,30 ± 0,05

Electrochemical Characterization of the Fiber Mats

In order to evaluate whether the electrospun network can operate as electrode interface cyclic voltammetric investigations in the presence of the $[Fe(CN)_6]^{4-/3-}$ redox couple have been performed. In comparison to the bare ITO electrode, which served as the basic material, a clearly enhanced conversion of the redox ions is found for all polymer fiber electrodes. This indicates on the one hand sufficient conductivity and on the other hand a surface which is suited to electrocatalytic conversions. Comparing different polymer electrodes (Fig. 5 and Table 3), PAN/PABSA gives the highest current values. PAN/PAPASA and PAN/PABSA show smaller peak separations than PAN/PANABMSA (all electrospun for 15 min). Moreover, these peak separations are all lower than for a bare ITO electrode indicating a faster kinetics. In case of the electrodes with different spinning durations (and thus different thicknesses - studied for PAN/PABSA), both 5 and 30 min spun electrodes show a higher peak separation compared to 15 min. Concerning the E_f value, no difference between all the different preparations have been determined. Furthermore, heterogeneous electron transfer rate constants (k_s) are calculated according to the Nicholson and Shain approach (Table 3)47. The lowest k_s value has been determined in case of the bare ITO electrodes, demonstrating that the 3D polymer mats are beneficial in terms of electron transfer rate. Comparing the values for different polymers electrospun for 15 minutes, PAN/PABSA shows the lowest value and smallest error, whilst both k_s values for PAN/PAPASA and PAN/PANABMSA are slightly higher but with higher errors. Taking the peak current maxima and the kinetics in account one can conclude that all three polymer networks provide a suitable interface for electrochemical conversions.

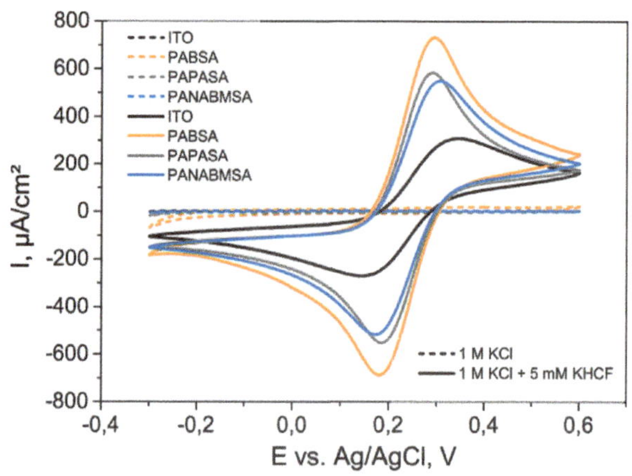

Figure 5: Cyclic voltammograms of differently modified electrodes obtained at a scan rate of 100 mV/s. Measurements were conducted with 1 M KCl and 1 M KCl + 5 mM $K_4[Fe(CN)_6]$/ $K_3[Fe(CN)_6]$ respectively. The polymer fibers were deposited on ITO surfaces by means of electrospinning for 15 minutes.

Table 3: Results of the electrochemical investigations (peak separations, formal potentials and heterogeneous electron transfer rates) of the fiber mats on ITO surfaces performed at 100 mV/s in 1 M KCl + 5 mM $K_4[Fe(CN)_6]$/ $K_3[Fe(CN)_6]$ prepared by electrospinning for 15 min

electrodes	ΔE_p [V]	E_f [V]	$k_s \times 10^3$, [cm/s]
PAN/PAPASA	0,104 ± 0,007	0,239 ± 0,002	9,5 ± 3,6
PAN/PANABMSA	0,139 ± 0,043	0,244 ± 0,004	8,9 ± 3,6
PAN/PABSA	0,113 ± 0,006	0,239 ± 0,002	6,4 ± 1,1
bare ITO	0,186 ± 0,052	0,244 ± 0,002	5,6 ± 2,5

Bioelectrocatalysis of PQQ-GDH at the Polymer Fiber Network

To evaluate the effect of the electrospun networks as efficient conductive supports for enzyme immobilization, the resulting electrodes have been modified by the enzyme PQQ-GDH. The enzyme has been chosen since different applications of PQQ-GDH with surface structures have been demonstrated as e.g. in sensing, substance release and bioenergetics[48,49,50]. Here, the enzyme has been covalently bound to the carboxylic acid groups on the aniline rings (Fig. 1). The attractive feature of the sulfonated polyaniline is the conductivity at neutral pH. Moreover, we have already reported on an efficient reaction of PQQ-GDH with these kind of polymers in solution but also on the communication with electrodes modified by similar sulfonated polyaniline copolymers[26]. The approach of covalent fixation of the enzyme in this study has been mainly chosen to ensure a good stability of the enzyme electrodes.

Figure 6 summarizes cyclic voltammograms of electrospun PAN/PABSA, PAN/PAPASA and PAN/PANAMBSA electrodes in the absence and in the presence of the substrate glucose. In the absence of glucose no redox conversion of polymers PABSA and PAPASA can be observed, whereas a clear redox couple with a formal potential about E=+0.14 V vs. Ag/AgCl can be measured for PANAMBSA, being in good agreement with already reported values for sulfonated polyanilines[26,51,52]. The different voltammetric behavior might be attributed to different initial redox states of copolymers after synthesis (PABSA and PAPASA are in fully oxidized pernigraniline state, whereas PANAMBSA is in a half-oxidized emeraldine salt state) and has also been found in previous studies[8,26,53,54]. Nevertheless, in all three cases significant bioelectrocatalytic currents appear upon addition of glucose. When a bare ITO electrode is incubated in a PQQ-GDH solution and subsequently studied in glucose containing buffer no bioelectrocatalysis could be found, demonstrating the role of the electrospun polymer matrix in the enzymatic reaction.

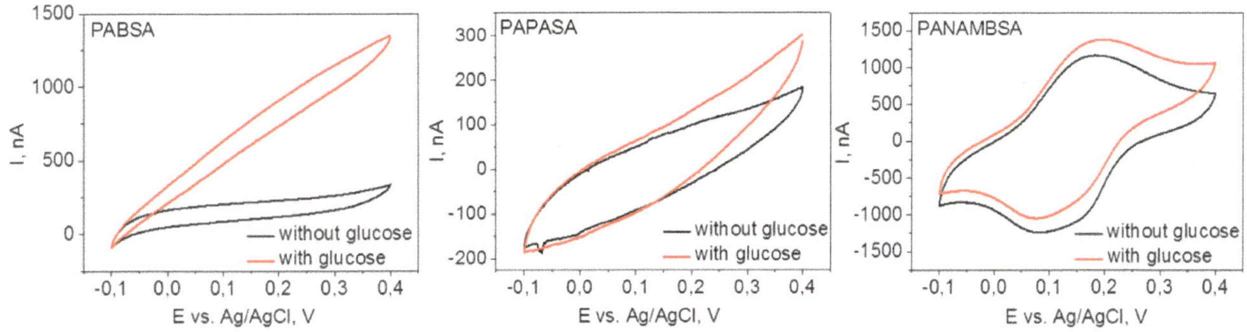

Figure 6: CVs of ITO-PAN/PABSA, ITO-PAN/PAPASA and ITO-PAN/PANAMBSA with covalently fixed PQQ-GDH without (black lines) and after addition of 5 mM of glucose (red lines). Experimental conditions: [glucose] = 5 mM in 20 mM MES buffer at pH 6, t = 25 °C, scan rate 5 mV/s.

Table 4: Start potentials of bioelectrocatalysis and current intensities after glucose addition (5 mM) at 0,35 V vs Ag/AgCl of ITO-PAN/PABSA, ITO-PAN/PAPASA and ITO-PAN/PANAMBSA electrodes prepared by 15 minutes electrospinning

electrodes	starting potential, *E* vs. Ag/AgCl, V	current intensity, nA
ITO-PAN/PABSA	−0.1	980 ± 50
ITO-PAN/PAPASA	−0.05	98 ± 30
ITO-PAN/PANAMBSA	+0.1	360 ± 40

However, the bioelectrocatalytic signals differ both by their starting potentials and the current intensities (Table 4). If a bioelectrocatalytic current after addition of the substrate is observed at potentials corresponding to the redox conversion of the polymer, mediation can be concluded. In case of ITO-PAN/PANAMBSA electrodes this mechanism might be suggested, since the bioelectrocatalytic current appears at E=+0.1 V vs. Ag/AgCl indicating that the oxidation process of the polymer may trigger the bioelectrocatalysis. However, for ITO-PAN/PABSA and ITO-PAN/PAPASA bioelectrocatalysis is observed at lower potentials (Table 4), leading to the conclusion that a direct electron transfer between the suitably oriented enzyme and polymer-modified electrode surface is valid. This is in agreement with already reported starting potentials of PQQ-GDH catalysis at electrodes modified with polymers[13,26,55,56] or other surfaces.[48,49,58,59]

From the data in Table 4 it is obvious, that depending on used polymers, significantly different intensities of catalytic currents can be observed. ITO-PAN/PAPASA electrodes exhibit lowest bioelectrocatalytic activity. In contrast the intensity of the. glucose-induced catalytic current in the case of PAN/PABSA is found to be 10fold higher than that of ITO-PAN/PAPASA, making ITO-PAN/PABSA electrodes to the most promising candidate for efficient enzyme connection to electrodes and furthermore for potential biosensorial utilization.

In order to optimize the performance of PQQ-GDH-modified 3D polymer mats the electrospinning time has been varied from 15 via 30 to 60 minutes (prior to enzyme fixation). The bioelectrochemical results for these electrodes are summarized in Table 5. It can be clearly concluded that the increase in electrospinning time leads to a slight decrease in catalytic activity of the enzyme electrodes, demonstrating that thinner polymer mats are more suitable for the preparation of electroactive electrodes. Although thicker polymer layers can provide more interaction sites for the immobilization of the enzyme the transport of glucose may limit the overall process. The slight decrease in catalytic current may also be attributed to the large distances for electron transport within the rather thick polymeric mats. Here one has to take into account that a large portion of the fibers is composed of the non-conducting PAN polymer. Thus, the efficient bioelectrocatalysis obtained with these fibers from polymeric blends can be considered a real success.

Table 5: Electrospinning time and its influence on catalytic current intensities of ITO-PAN/PABSA, ITO-PAN/PAPASA and ITO-PAN/PANAMBSA electrodes in the presence of 5 mM glucose determined at +0,35 V vs Ag/AgCl (n=3)

electrodes	electrospinning time, min	current intensity nA
ITO-PAN/PABSA	30	807 ± 30
	60	630 ± 40
ITO-PAN/PAPASA	30	74 ± 20
	60	51 ± 10
ITO-PAN/PANAMBSA	30	293 ± 50
	60	226 ± 30

Figure 7 demonstrates a calibration curve obtained for ITO-PAN/PABSA electrodes (n=3) by evaluating the oxidation current as a function of the glucose concentration at E=+0.35 V vs Ag/AgCl. For glucose concentrations from 0.0025 to 1 mM a linear response is found. From 2 mM to 8 mM a saturation behavior is observed. An apparent K_m value of 0.45 mM glucose can be calculated based on the Michaelis-Menten kinetics. The concentration dependent behavior shows that the system allows a well-defined detection of glucose, but it also demonstrates that that at higher glucose concentration the enzyme electrode cannot follow anymore and the electron withdrawal from the reduced enzyme becomes rate limiting.

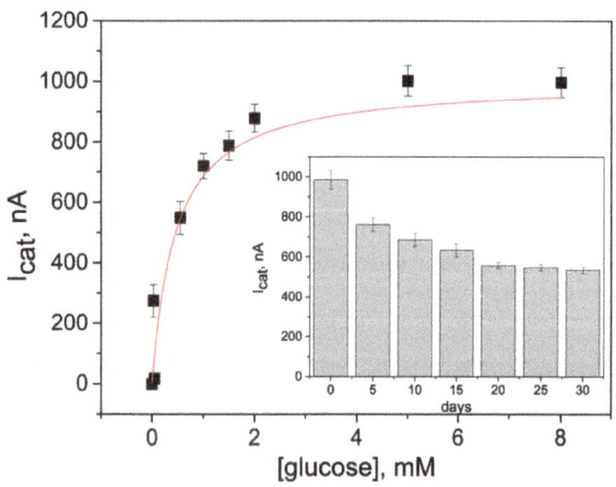

Figure 7: Catalytic current response of the ITO-PAN/PABSA/PQQ-GDH electrode (15 minutes electrospinning time) as a function of the glucose concentration. Bioelectrocatalytic signals were registered from the respective cyclic voltammograms at E=+0.35 V vs Ag/AgCl. Inset: Stability of the electrode response towards 5 mM glucose. Experimental conditions: 20 mM MES buffer pH 6, CV at 5 mV s^{-1}, n=3.

Considering the blood glucose level in the reference range between 3.6– 5.6 mM one can conclude that the present biosensing system has a potential to be implemented to monitoring the glucose level in human blood. It has also to be explicitly noted here that glucose detection can be performed at low potentials because bioelectrocatalysis already starts at −0.1 V vs Ag/AgCl. This is advantageous in complex media, in which other redox-active substances are present. However, limited glucose selectivity of PQQ-GDH is still an issue to be overcome with this respect60.

Since long-term stability is an important parameter for the evaluation of the performance of any sensing system, we have traced the stability of our ITO-PAN/PABSA-PQQ-GDH electrodes by testing their activity in glucose solution, after the electrodes were kept at 4 °C in 20 mM MES, pH 6 when not in use. Inset in Fig. 7 shows that the catalytic current density response is maintained over 65% of its initial value after 30 days. The signal decline might be caused by loss in PQQ-GDH activity rather than by leakage of the enzyme from the electrode. Although such kind of bioelectrocatalytic platform is mostly desirable for a single-use application, the improved long-term stability makes this type of architecture promising to be considered for the construction of glucose detection systems.

CONCLUSION

In this report the successful fabrication of conducting 3D polymer fiber networks by electrospinning of 3 different, chemically synthesized sulfonated polyanilines together with non-conducting polyacrylonitrile has been demonstrated. Microscopic characterizations of PAN/PABSA, PAN/PAPASA and PAN/PANAMBSA mats have shown that the fibers are uniformly distributed within the network on the ITO surface. The electrochemical characterization of these electrodes in the presence of hexacyanoferrate (II/III) reveals that the polymeric network allows electrochemical conversions to occur. Compared to a bare ITO electrode higher Faradayic currents and faster electron transfer are found.

Moreover, the polymer fibers provide a surface for fixation of active enzyme as exemplified with PQQ-GDH. Significant bioelectrocatalytic currents occur upon addition of glucose to 3D polymer mats with the covalently-bound enzyme. Depending on the choice of the sulfonated polyanilline mediatorless bioelectrocatalysis can be obtained, starting at quite low potentials; i.e., no large overpotential is needed to drive the reaction. A well-defined concentration dependent response is observed for PAN/PABSA. In addition, the mild and well reproducible fabrication process and the active role of the polymer film in withdrawing electrons from the reduced PQQ-GDH lead to a system with high stability. Thus, such electrospun networks developed by making use of a sulfonated polyaniline and its interaction with PQQ-GDH can be considered as a promising candidate for biosensorial purposes.

METHODS

Materials

If not otherwise stated, all reagents and solvents have been used without further pretreatment. Dimethyl sulfoxide

(DMSO), $CaCl_2$, 2-(N-morpholino)ethanesulfonic acid (MES), polyacrylonitrile and potassium ferricyanide were provided by Sigma Aldrich (Sigma, Aldrich, Sigma Aldrich, Fluka). Acetone, ethanol and potassium ferrocyanide were bought from Roth (CARL ROTH GmbH und CO. KG, Germany). 3-Aminobenzoic acid (3-AB) (Sigma-Aldrich Chemicals Co., Germany) and 3-aminobenzenesulfonic acid (3-ABS) (Sigma-Aldrich Chemicals Co. ,Fluka, Germany) were recrystallized from water (3-AB: m.p. 176–177 °C; 3-ABS: m.p. 288 °C). 3-Aminobenzeneacetic acid (3-ABA), 3-Amino-4-methoxybenzenesulfonic acid (MAS) (Sigma-Aldrich Chemicals Co., Germany), ammonium persulfate (APS), HCl, DMF, and sodium hydroxide (Sigma-Aldrich Chemicals Co., Fluka, Germany) were used as purchased. Ultra clear water was produced with a SG water ultra pure water unit.

COPOLYMER SYNTHESIS

PAPASA

was prepared by chemical oxidative polymerization according the following procedure: 1 g (0,006 mol) 3-aminobenzeneacetic acid and 1,14 g (0,006 mol) 3-aminobenzenesulfonic acid were dissolved in 27 ml 0,5 M NaOH. 2,97 g (0,013 mol) ammonium persulfate dissolved in 9 ml water added slowly under continuous stirring for more than 30 min at room temperature. The reaction mixture was stirred for further 24 hrs at room-temperature to complete the reaction. The precipitated product was filtered, washed with aqueous 1.2 M HCl and water and dried under vacuum at 80 °C for 24 h.

PANABMSA

was synthesized as follows: A 1:1 mixture of 80 ml 0,5 M NaOH containing 8,1 g (0,04 mol) 3-amino-4-methoxybenzene sulfonic acid, 10 ml 0,5 M NaOH containing 0,55 g (0,004 mol) 3-aminobenzoic acid and 3,72 g (0,04 mol) Aniline was stirred at room temperature. 12,3 g (0,054 mol) ammonium persulfate dissolved in 30 ml water added slowly for more than 30 min. The reaction mixture was stirred for further 24 hrs at room-temperature to complete the reaction. The precipitated product was filtered, washed with aqueous 1.2 M HCl and water and dried under vacuum at 80 °C for 24 h.

PABSA

was synthesized as previously reported26.

Polymer Characterization

The molecular weights were determined by gel permeation chromatography (GPC) using a Waters apparatus provided with refraction and UV detectors and PL Mixed Gel C Column. Measurements were carried out in DMF using polystyrene standards. IR spectra were recorded with Bio-Rad FTX 3000MX spectrometer. ^{1}H-NMR measurements were carried out with a Bruker Instrument. Thermogravimetral analysis (TGA) was performed by Perkin Elmer TG 7 apparatus.

Preparation of the polymer solution

For the preparation of the polymer solution 50 mg/ml polyacrylonitrile and 10 mg/ml of a polyaniline derivative were dissolved in DMSO and ultrasonicated for 11 hours.

Electrospinning

Prior to the preparation of the polymer fibers, the rectangular indium tin oxide (ITO) coated glass slides with surface resistivity 15–25 Ω sq^{-1}(obtained from Sigma-Aldrich, Taufkirchen, Germany) were cleaned processing DMSO-water-acetone-water-ethanol solvent steps (15 min ultrasonication for each step). The cleaned electrodes were then placed into the cabin of the electrospinning apparatus EC-CLI (IME Technologies, The Netherlands), and the polymer solution was electrospun onto their surfaces from a distance of 35 cm at a voltage of 25 kV with 2 µl/min spinning rate. The temperature was controlled to be 25 °C and the humidity of 40%. For the spinning process a 27G1" needle (B Braun Melsungen, Germany) was used. The electrospinning procedure was conducted for different times from 5 to 60 min.

Characterization of polymer fiber mats

First the coated electrodes were investigated by means of light microscopy (Olympus DSX500, Olympus Deutschland GmbH, Germany) to confirm the overall structure and uniformity. Additionally the polymer fiber mat coated electrodes where investigated with an electron microscope (JEOL JSM-6510 SEM, JEOL Germany GmbH, Germany) at a voltage of 5 kV. Images where taken at different areas of the electrodes and the diameters of the fibers were measured. Contact angle measurements were conducted with a Dataphysics OCA15EC machine (DataPhysics GmbH, Germany).

Preparation of the enzyme solution

Soluble glucose dehydrogenase (apoGDH) was gained as a kind gift from Roche. The enzyme was dissolved in 5 mM MES buffer in the presence of 1 mM $CaCl_2$, and the pH was adjusted to 5. apoGDH was reconstituted with PQQ in a ratio of 1:1 according to Olsthoorn and Duine57. For this purpose, sGDH and PQQ were incubated together for 1 h at room temperature in the dark. Aliquots were stored at −20 °C. Prior to each measurement, the specific activity of the reconstituted enzyme was determined to be 2200±30 U mg^{-1} by using 2,6-dichlorophenolindophenol as an electron acceptor.

Enzyme binding

The ITO electrodes on which the electrospun polymer network has been fabricated, have been used to covalently bind PQQ-GDH. In order to fix the enzyme covalently, firstly the polymer-modified electrodes were placed in an EDC/NHS-solution (100 mM/25 mM) in 5 mM MES + 1 mM $CaCl_2$, pH 6.5 for 15 minutes and after that washed 3 times with the same buffer before enzyme incubation. For the incubation purpose polymer electrodes have been placed in enzyme solution for 1 h, afterwards washed with buffer prior measurements.

Electrochemistry

The electrodes were characterized electrochemically by means of cyclic voltammetry using a CH Instruments 1230b in a custom made 1 ml measurement cell. A polymer-modified ITO electrodes with the effective electrode area of 0.32 cm^2 (Figure

S2, Supporting Information) were used as working electrodes, together with an Ag/AgCl/1 M KCl reference (Biometra, Germany) and a Pt-wire counter electrodes. Electrochemical characterizations were performed in the presence of $K_4[Fe(CN)_6]$/ $K_3[Fe(CN)_6]$ (5 mM) in 1 M KCl.

Bioelectrochemical measurements were performed at room temperature in the same cell using a 20 mM MES buffer pH 6 as electrolyte. The catalytic response was studied with addition of different glucose concentrations to the solution. Cyclic voltammetric experiments were carried out with a μAutolab Type II device (Metrohm, The Netherlands). The scan rate was set to 5 mV s^{-1}. The potential range was chosen as between −0.4 and +0.4 V vs. Ag/AgCl. Data analysis was performed using GPES software (General Purpose for Electrochemical System, Eco Chemie, The Netherlands).

ACKNOWLEDGEMENTS

The work was partially supported by the BMBF Germany (project 03IS2201I).

REFERENCES

1. Balint, R., Cassidy, N. J. & Cartmell, S. H. Conductive polymers: Towards a smart biomaterial for tissue engineering. *Acta Biomater.* 10, 2341–2353 (2014).
2. Guiseppi-Elie, A. Electroconductive hydrogels: Synthesis, characterization and biomedical applications. *Biomaterials* 31, 2701–2716 (2010).
3. Wallace, G. G. & Kane-Maguire, L. A. P. Manipulating and monitoring biomolecular interactions with conducting electroactive polymers. *Adv. Mater.* 14, 953–960 (2002).
4. Ates, M. A review study of (bio)sensor systems based on conducting polymers. *Mater. Sci. Eng. C-Mater. Biol. Appl.* 33, 1853–1859 (2013).
5. Higgins, M. J., Molino, P. J., Yue, Z. L. & Wallace, G. G. Organic Conducting Polymer-Protein Interactions. *Chem. Mater.* 24, 828–839 (2012).
6. Sassolas, A., Blum, L. J. & Leca-Bouvier, B. D. Immobilization strategies to develop enzymatic biosensors. *Biotechnol. Adv.* 30, 489–511 (2012).
7. Jaymand, M. Recent progress in chemical modification of polyaniline. *Prog. Polym. Sci.* 38, 1287–1306 (2013).
8. Kang, E. T., Neoh, K. G. & Tan, K. L. Polyaniline: A polymer with many interesting intrinsic redox states. *Prog. Polym. Sci.* 23, 277–324 (1998).
9. Stejskal, J., Sapurina, I. & Trchova, M. Polyaniline nanostructures and the role of aniline oligomers in their formation. *Prog. Polym. Sci.* 35, 1420–1481 (2010).
10. Dhand, C., Das, M., Datta, M. & Malhotra, B. D. Recent advances in polyaniline based biosensors. *Biosens. Bioelectron.* 26, 2811–2821 (2011).
11. Gizzie, E. A., LeBlanc, G., Jennings, G. K. & Cliffel, D. E.Electrochemical preparation of photosystem I-polyaniline composite films for biohybrid solar energy conversion. *ACS Appl. Mater. Iinterfaces* 7, 9328–9335 (2015).
12. Pan, L. J. *et al.* Hierarchical nanostructured conducting polymer hydrogel with high electrochemical activity. *Proc. Natl. Acad. Sci. USA.* 109, 9287–9292 (2012).
13. Sarauli, D., Xu, C. G., Dietzel, B., Schulz, B. & Lisdat, F. A multilayered sulfonated polyaniline network with entrapped pyrroloquinoline quinone-dependent glucose dehydrogenase: tunable direct bioelectrocatalysis. *J. Mater. Chem. B* 2, 3196–3203 (2014).
14. Zhai, D. Y. *et al.* Highly Sensitive Glucose Sensor Based on Pt Nanoparticle/Polyaniline Hydrogel Heterostructures. *ACS Nano* 7, 3540–3546 (2013).
15. Sarauli, D. *et al.* Insights into the formation and operation of polyaniline sulfonate/cytochrome c multilayer electrodes: contributions of polyelectrolytes' properties. *Phys. Chem. Chem. Phys.* 12, 14271–14277 (2010).
16. Wegerich, F., Turano, P., Allegrozzi, M., Mohwald, H. & Lisdat, F.Electroactive Multilayer Assemblies of Bilirubin Oxidase and Human Cytochrome C Mutants: Insight in Formation and Kinetic Behavior. *Langmuir* 27, 4202–4211 (2011).
17. Zhang, L., Jiang, X., Niu, L. & Dong, S. J. Syntheses of fully sulfonated polyaniline nano-networks and its application to the direct electrochemistry of cytochrome c. *Biosens. Bioelectron.* 21, 1107–1115 (2006).
18. Zhou, B. *et al.* Facile interfacial electron transfer of hemoglobin mediated by conjugated polymers. *Int. J. Mol. Sci.* 6, 303–310 (2005).
19. Sarauli, D. *et al.* Thin films of substituted polyanilines: interactions with biomolecular systems. *Soft Matter* 8, 3848–3855 (2012).
20. Tatsuma, T., Ogawa, T., Sato, R. & Oyama, R. Peroxidase-incorporated sulfonated polyaniline-polycation complexes for electrochemical sensing of H_2O_2. *J. Electroanal. Chem.* 501, 180–185 (2001).
21. Yu, X., Sotzing, G. A., Papadimitrakopoulos, F. & Rusling, J. F.Wiring of enzymes to electrodes by ultrathin conductive polyion underlayers: Enhanced catalytic response to hydrogen peroxide.*Anal. Chem.* 75, 4565–4571 (2003).
22. Sioniewska, A. & Palys, B. Supramolecular polyaniline hydrogel as a support for urease. *Electrochim. Acta* 126, 90–97 (2014).
23. Lee, K. P., Komathi, S., Nam, N. J. & Gopalan, A. I. Sulfonated polyaniline network grafted multi-wall carbon nanotubes for enzyme immobilization, direct electrochemistry and biosensing of glucose. *Microchem J.* 95, 74–79 (2010).
24. Mangombo, Z. A., Baker, P., Iwuoha, E. & Key, D. Tyrosinase biosensor based on a boron-doped diamond electrode modified with a polyaniline-poly(vinyl sulfonate) composite film.*Microchim. Acta* 170, 267–273 (2010).
25. Sarauli, D. *et al.* Interaction of Fructose Dehydrogenase with a Sulfonated Polyaniline: Application for Enhanced Bioelectrocatalysis. *ACS Catal.* 5, 2081–2087 (2015).
26. Sarauli, D., Xu, C. G., Dietzel, B., Schulz, B. & Lisdat, F. Differently substituted sulfonated polyanilines:

The role of polymer compositions in electron transfer with pyrroloquinoline quinone-dependent glucose dehydrogenase. *Acta Biomater.* 9, 8290–8298 (2013).

27. Xu, S. & Minteer, S. D. Pyrroloquinoline Quinone-Dependent Enzymatic Bioanode: Incorporation of the Substituted Polyaniline Conducting Polymer as a Mediator. *ACS Catal.* 4, 2241–2248 (2014).
28. Bhardwaj, N. & Kundu, S. C. Electrospinning: A fascinating fiber fabrication technique. *Biotechnol. Adv.* 28, 325–347 (2010).
29. Long, Y. Z., Yu, M., Sun, B., Gu, C. Z. & Fan, Z. Y. Recent advances in large-scale assembly of semiconducting inorganic nanowires and nanofibers for electronics, sensors and photovoltaics. *Chem. Soc. Rev.* 41, 4560–4580 (2012).
30. Chaudhari, S. *et al.* Electrospun polyaniline nanofibers web electrodes for supercapacitors. *J. Appl. Polym. Sci.* 129, 1660–1668 (2013).
31. Hughes, L. A., Gaston, J., McAlindon, K., Woodhouse, K. A. &Thibeault, S. L. Electrospun fiber constructs for vocal fold tissue engineering: Effects of alignment and elastomeric polypeptide coating. *Acta Biomater.* 13, 111–120 (2015).
32. Kerr-Phillips, T. E. *et al.* Electrospun rubber fibre mats with electrochemically controllable pore sizes. *J. Mater. Chem. B* 3, 4249–4258 (2015).
33. Lu, W. J., Sun, J. S. & Jiang,X. Y. Recent advances in electrospinning technology and biomedical applications of electrospun fibers. *J. Mater. Chem. B* 2, 2369–2380 (2014).
34. Bourourou, M. *et al.* Chemically reduced electrospun polyacrilonitrile-carbon nanotube nanofibers hydrogels as electrode material for bioelectrochemical applications. *Carbon* 87, 233–238 (2015).
35. Jose, M. V., Marx, S., Murata, H., Koepsel, R. R. & Russell, A. J.Direct electron transfer in a mediator-free glucose oxidase-based carbon nanotube-coated biosensor. *Carbon* 50, 4010–4020 (2012).
36. Mondal, K., Ali, M. A., Agrawal, V. V., Malhotra, B. D. & Sharma, A.Highly Sensitive Biofunctionalized Mesoporous Electrospun TiO2 Nanofiber Based Interface for Biosensing. *ACS App. Mater. Interfaces* 6, 2516–2527 (2014).
37. Su, X. F. *et al.* A New Amperometric Glucose Biosensor Based on One-Step Electrospun Poly(Vinyl Alcohol)/Chitosan Nanofibers. *J. Biomed. Nanotechnol.* 9, 1776–1783 (2013).
38. Raeesi, F., Nouri, M. & Haghi, A. K. Electrospinning of polyaniline-polyacrylonitrile blend nanofibers. *e-Polymers*, 1, 1350–1362 (2009).
39. Lin, Y. W. & Wu, T. M. Fabrication of water-soluble polyaniline/poly(ethylene oxide)/carbon nanotube electrospun fibers. *J. Appl. Polym. Sci.* 126, E123–E129 (2012).
40. Attout, A., Yunus, S. & Bertrand, P. Electrospinning and alignment of polyaniline-based nanowires and nanotubes. *Polym. Eng. Sci.*48, 1661–1666 (2008).
41. Llorens, E. *et al.* Nanomembranes and Nanofibers from Biodegradable Conducting Polymers. *Polymers* 5, 1115–1157 (2013).
42. Schubart, I. W., Gobel, G. & Lisdat, F. A pyrroloquinolinequinone-dependent glucose dehydrogenase (PQQ-GDH)-electrode with direct electron transfer based on polyaniline modified carbon nanotubes for biofuel cell application. *Electrochim. Acta* 82, 224–232 (2012).
43. Trchova, M. & Stejskal, J. Polyaniline: The infrared spectroscopy of conducting polymer nanotubes (IUPAC Technical Report). *Pure Appl. Chem.* 83, 1803–1817 (2011).
44. Roy, B. C., Gupta, M. D., Bhowmik, L. & Ray, J. K. Studies on water soluble conducting polymer–Aniline initiated polymerization of m-aminobenzene sulfonic acid. *Synth. Met.* 100, 233–236 (1999).
45. Deitzel, J. M., Kleinmeyer, J., Harris, D. & Tan, N. C. B. The effect of processing variables on the morphology of electrospun nanofibers and textiles. *Polymer* 42, 261–272 (2001).
46. Luo, C. J., Nangrejo, M. & Edirisinghe, M. A novel method of selecting solvents for polymer electrospinning. *Polymer* 51, 1654–1662 (2010).
47. Nicholson, R. S. & Shain, I. Theory of stationary electrode polarography. Single scan and cyclic methods applied to reversible, irreversible and kinetic systems. *Anal. Chem.* 36, 706–723 (1964).
48. Gamella, M., Guz, N., Mailloux, S., Pingarron, J. M. & Katz, E.Activation of a Biocatalytic Electrode by Removing Glucose Oxidase from the Surface-Application to Signal Triggered Drug Release. *ACS Appl. Mater. Interfaces* 6, 13349–13354 (2014).
49. Halamkova, L. *et al.* Implanted Biofuel Cell Operating in a Living Snail. *J. Am. Chem. Soc.* 134, 5040–5043 (2012).
50. Katz, E. *et al.* Substance Release Triggered by Biomolecular Signals in Bioelectronic Systems. *J. Phys. Chem. Lett.* 6, 1340–1347 (2015).
51. Dominis, A. J., Spinks, G. M., Kane-Maguire, L. A. P. & Wallace, G. G. A de-doping/re-doping study of organic soluble polyaniline.*Synth. Met.* 129, 165–172 (2002).
52. Tawde, S., Mukesh, D. & Yakhmi, J. V. Redox behavior of polyaniline as influenced by aromatic sulphonate anions: cyclic voltammetry and molecular modeling. *Synth. Met.* 125, 401–413 (2001).
53. Lee, K. *et al.* Metallic transport in polyaniline. *Nature* 441, 65–68 (2006).
54. Strounina, E. V., Shepherd, R., Kane-Maguire, L. A. P. & Wallace, G. G. Conformational changes in sulfonated polyaniline caused by metal salts and OH. *Synth. Met.* 135, 289–290 (2003).
55. Sarauli, D. *et al.* 3D-Electrode Architectures for Enhanced Direct Bioelectrocatalysis of Pyrroloquinoline Quinone-Dependent Glucose Dehydrogenase. *ACS Appl. Materi. Interfaces* 6, 17887–17893 (2014).
56. Gobel, G., Schubart, I. W., Scherbahn, V. & Lisdat, F. Direct electron transfer of PQQ-glucose dehydrogenase

at modified carbon nanotubes electrodes. *Electrochem. Commun.* 13, 1240–1243 (2011).

57. Olsthoorn, A. J. J. & Duine, J. A. Production, characterization, and reconstitution of recombinant quinoprotein glucose dehydrogenase (soluble type; EC 1.1.99.17) apoenzyme of Acinetobacter calcoaceticus. *Arch. Biochem. Biophys.* 336, 42–48 (1996).

58. Flexer, V., Durand, F., Tsujimura, S. & Mano, N. Efficient Direct Electron Transfer of PQQ-glucose Dehydrogenase on Carbon Cryogel Electrodes at Neutral pH. *Anal. Chem.* 83, 5721–5727 (2011).

59. Flexer, V. & Mano, N. Wired Pyrroloquinoline Quinone Soluble Glucose Dehydrogenase Enzyme Electrodes Operating at Unprecedented Low Redox Potential. *Anal. Chem.* 86, 2465–2473 (2014).

60. Frias, J. P., Lim, C. G., Ellison, J. M. & Montandon, C. M. Review of Adverse Events Associated With False Glucose Readings Measured by GDH-PQQ-Based Glucose Test Strips in the Presence of Interfering Sugars. *Diabetes Care* 33, 728–729 (2010).

INDEX

L

M

N

O

P

Q

R

S

T

U

V

W

X